Managing Festivals for Destination Marketing and Branding

Sharad Kumar Kulshreshtha
Department of Tourism and Hotel Management, North-Eastern Hill University, Shillong, India

IGI Global
PUBLISHER of TIMELY KNOWLEDGE

A volume in the Advances in Marketing, Customer Relationship Management, and E-Services (AMCRMES) Book Series

Published in the United States of America by
IGI Global
Business Science Reference (an imprint of IGI Global)
701 E. Chocolate Avenue
Hershey PA, USA 17033
Tel: 717-533-8845
Fax: 717-533-8661
E-mail: cust@igi-global.com
Web site: http://www.igi-global.com

Library of Congress Cataloging-in-Publication Data

Names: Kulshreshtha, Sharad Kumar, 1978- editor.
Title: Managing festivals for destination marketing and branding / editor, Sharad Kumar Kulshreshtha.
Description: Hershey, PA : Business Science Reference, 2023. | Includes bibliographical references and index. | Summary: "Managing Festivals for Destination Marketing and Branding addresses the most current and promising parts of tourism-centric festivals, which are held in numerous tourist areas throughout the world. It links tourism festivals around the world as a catalyst for destination marketing and identity. Covering topics such as destination brand equity, social media networks, and motivations and expectations of tourists, this premier reference work is a dynamic resource for business executives and leaders, brand managers, event managers, festival managers, government officials, students and educators of higher education, librarians, researchers, and academicians"-- Provided by publisher.
Identifiers: LCCN 2022043753 (print) | LCCN 2022043754 (ebook) | ISBN 9781668463567 (hardcover) | ISBN 9781668463574 (paperback) | ISBN 9781668463581 (ebook)
Subjects: LCSH: Culture and tourism. | Festivals--Management. | Place marketing.
Classification: LCC G156.5.H47 M258 2023 (print) | LCC G156.5.H47 (ebook) | DDC 910.68/8--dc23/eng20221229
LC record available at https://lccn.loc.gov/2022043753
LC ebook record available at https://lccn.loc.gov/2022043754

This book is published in the IGI Global book series Advances in Marketing, Customer Relationship Management, and E-Services (AMCRMES) (ISSN: 2327-5502; eISSN: 2327-5529)

British Cataloguing in Publication Data
A Cataloguing in Publication record for this book is available from the British Library.

Advances in Marketing, Customer Relationship Management, and E-Services (AMCRMES) Book Series

Eldon Y. Li
National Chengchi University, Taiwan & California Polytechnic State University, USA

ISSN:2327-5502
EISSN:2327-5529

Mission

Business processes, services, and communications are important factors in the management of good customer relationship, which is the foundation of any well organized business. Technology continues to play a vital role in the organization and automation of business processes for marketing, sales, and customer service. These features aid in the attraction of new clients and maintaining existing relationships.

The Advances in Marketing, Customer Relationship Management, and E-Services (AMCRMES) Book Series addresses success factors for customer relationship management, marketing, and electronic services and its performance outcomes. This collection of reference source covers aspects of consumer behavior and marketing business strategies aiming towards researchers, scholars, and practitioners in the fields of marketing management.

Coverage

- Text Mining and Marketing
- Electronic Services
- Relationship Marketing
- Social Networking and Marketing
- CRM in financial services
- Telemarketing
- Ethical Considerations in E-Marketing
- Customer Relationship Management
- Data mining and marketing
- Mobile Services

IGI Global is currently accepting manuscripts for publication within this series. To submit a proposal for a volume in this series, please contact our Acquisition Editors at Acquisitions@igi-global.com or visit: http://www.igi-global.com/publish/.

The Advances in Marketing, Customer Relationship Management, and E-Services (AMCRMES) Book Series (ISSN 2327-5502) is published by IGI Global, 701 E. Chocolate Avenue, Hershey, PA 17033-1240, USA, www.igi-global.com. This series is composed of titles available for purchase individually; each title is edited to be contextually exclusive from any other title within the series. For pricing and ordering information please visit http://www.igi-global.com/book-series/advances-marketing-customer-relationship-management/37150. Postmaster: Send all address changes to above address.

Titles in this Series

For a list of additional titles in this series, please visit: www.igi-global.com/book-series

Global Agricultural and Food Marketing in a Global Context Advancing Policy, Management, and Innovation
Aluwani Maiwashe-Tagwi (University of South Africa, South Africa) Ailweli Solomon Mawela (University of South Africa, South Africa) and Phineas Khazamula Chauke (University of South Africa, South Africa)
Business Science Reference • ©2023 • 236pp • H/C (ISBN: 9781668447802) • US $250.00

Global Perspectives on the Strategic Role of Marketing Information Systems
Jose Melchor Medina (Tamaulipas Autonomous University, Mexico) Miguel Sahagun (High Point University, USA) Jorge Alfaro (Universidad Catolica del Norte, Chile) and Fernando Ortiz-Rodriguez (Tamaulipas Autonomous University, Mexico)
Business Science Reference • ©2023 • 320pp • H/C (ISBN: 9781668465912) • US $250.00

Handbook of Research on Consumer Behavioral Analytics in Metaverse and the Adoption of a Virtual World
Pantea Keikhosrokiani (School of Computer Sciences, Universiti Sains Malaysia, Malaysia)
Engineering Science Reference • ©2023 • 400pp • H/C (ISBN: 9781668470299) • US $325.00

Enhancing Customer Engagement Through Location-Based Marketing
Amandeep Singh (Chitkara Business School, Chitkara University, India) Amit Mittal (Chitkara Business School, Chitkara University, India) and Murat Unanoglu (İstanbul Aydin University, Turkey)
Business Science Reference • ©2023 • 300pp • H/C (ISBN: 9781668481776) • US $250.00

Big Data Marketing Strategies for Superior Customer Experience
Jose Ramon Saura (Rey Juan Carlos University, Spain)
Business Science Reference • ©2023 • 305pp • H/C (ISBN: 9781668464540) • US $240.00

Handbook of Research on the Interplay Between Service Quality and Customer Delight
Sarmistha Sarma (Institute of Innovation in Technology and Management, India) and Neha Gupta (IBCS, SOA University (Deemed), India)
Business Science Reference • ©2023 • 438pp • H/C (ISBN: 9781668458532) • US $295.00

Cases on Social Justice in China and Perspectives on Chinese Brands
Youssef Elhaoussine (Beijing Normal University-Hong Kong Baptist University United International College, China) and Lulu Wang (Sinotrend Consulting, China)
Business Science Reference • ©2023 • 276pp • H/C (ISBN: 9781668449554) • US $215.00

IGI Global
PUBLISHER of TIMELY KNOWLEDGE

701 East Chocolate Avenue, Hershey, PA 17033, USA
Tel: 717-533-8845 x100 • Fax: 717-533-8661
E-Mail: cust@igi-global.com • www.igi-global.com

This book is to dedicated my loving mother
Late Smt. Sushma Kulshreshtha

Editorial Advisory Board

Table of Contents

Foreword *by Wayne W. Smith* xvii

Foreword *by Craig Webster* xviii

Preface xx

Acknowledgment xxviii

Chapter 1
The Role of Festivals in Destination Branding 1
Emrah Keskin, Faculty of Tourism, Nevşehir Hacı Bektaş Veli University, Turkey
Nevres Sezen, Nevşehir Hacı Bektaş Veli University, Turkey
Özge Kaya, Nevşehir Hacı Bektaş Veli University, Turkey

Chapter 2
Antecedents of Participation in Music Festivals: The Role of Congruence Between the Sponsor and the Festival 26
Sara Santos, Research Centre in Digital Services (CISeD), Polytechnic of Viseu, Portugal
Pedro Espírito Santo, Research in Management and Economics (CARME), School of Technology and Management, Portugal & Polytechnic of Leiria, Portugal
Luísa Augusto, Research Centre in Digital Services (CISeD), Polytechnic of Viseu, Portugal

Chapter 3
Why Do Festivals Fail? A Case Study of Sonepur Cattle Fair Festival 43
Sushant Kumar, Indian Institute of Management Lucknow, India

Chapter 4
The Effect of Perceived Authenticity on Satisfaction, Behavioural Intentions, and Expenditure 57
Ester Napolitano, University of Sassari, Italy
Antonio Salvatore Loriga, University of Sassari, Italy
Giacomo Del Chiappa, University of Sassari, Italy

Chapter 5
Managing Festivals With Smart Technologies and Their Contribution Towards Destination Promotion 76
Sahil Sharma, Himachal Pradesh University, India
Jeet Dogra, Indian Institute of Tourism and Travel Management (IITTM), Noida, India

Chapter 6
The Role of Culinary Festivals in Building Destination Brand Equity 95
Davut Kodas, Mardin Artuklu University, Turkey

Chapter 7
Festivals as a Resource of Heritage Tourism: A Bibliometric Analysis Using the Vosviewer 110
Ruchika Kulshrestha, Amity University, Gurgaon, India
Surabhi Srivastava, Om Kothari Institute of Research and Management, Kota, India

Chapter 8
Gastronomic Festival Tours in Turkey: Motivations and Expectations of Tourists 127
Suat Akyürek, Gümüşhane University, Turkey

Chapter 9
Local Fairs: Festivals and Sustainable Tourism Development in Rajasthan 145
Anukrati Sharma, University of Kota, India
Shruti Arora, University of Kota, India
Debendra Nath Dash, Mahatma Gandhi National Council of Rural Education, India

Chapter 10
Karaga Jatre of Karnataka: A Tryst With History, Anthropology, Belief, Culture, and Tourism 156
Shaheed Khan, Dharthi NGO, India
Freeda Maria Swarna M., Dharthi NGO, India

Chapter 11
Social Media Networks as an Effective Way to Promote Events 172
Betül Kodaş, Mardin Artuklu University, Turkey
Ela Nazlı Köz, İstanbul Gedik Üniversitesi, Turkey

Chapter 12
Conceptual Framework to Position Devika River (Gupt Ganga) as a Religious Festival Tourist Destination 186
Biswabhusan Pradhan, Central University of Jammu, India
Ayush, Central University of Jammu, India
Krishma Rana, Central University of Jammu, India
Amit Gangotia, Central University of Jammu, India

Chapter 13
Destination Marketing and Branding of Nashik: A Case Study of Sula Wine Festival....................... 196
Gayatri Gawde, Amity University, Mumbai, India
Sachin Kumar Behera, Amity University, Mumbai, India
Sharad Kumar Kulshreshtha, Department of Tourism and Hotel Management, North-Eastern Hill University, Shillong, India

Chapter 14
Destination Branding Through Tourism Festivals of Khasi Hills and Garo Hills Region Meghalaya in India: Case of Cherry Blossom Festivals and Me.gong Festival 210
Sharad Kumar Kulshreshtha, Department of Tourism and Hotel Management, North-Eastern Hill University, Shillong, India
Iadonlang Tynsong, Martin Luther Christian University, India
Ashok Kumar, Institute of Hotel Management, Catering Technology, and Applied Nutrition, India

Chapter 15
Value Creation of Destinations by Destination Management Organisations Through Strategic Planning and Promotion of Fairs and Festivals of NE India.. 227
Angshumi Goswami, North-Eastern Hill University, India

Chapter 16
Long-Term Forecasting of Tourism Festivals Post COVID-19 244
Rachel Chanel Clarke, Trigo Networks, USA

Chapter 17
Investigation of Local People's Perceptions of Festival Tourism With Cultural Interaction Dimensions 263
Sibel Sü Eröz, Kırklareli University, Turkey
Kaplan Uğurlu, Kırklareli University, Turkey
Aydemir Ay, Trakya University, Turkey

Chapter 18
Destination Branding and Mega Events: A Case of FIFA World Cup 2022 and Sultanate of Oman. 285
Mohit Kukreti, Independent Researcher, Oman
Amitabh Mishra, University of Technology and Applied Sciences, Oman
Aarti Dangwal, Chitkara University, India
Vishal Jain, University of Technology and Applied Sciences, Oman
Veena Tewari, University of Technology and Applied Sciences, Oman

Compilation of References 303

About the Contributors 356

Index 364

Detailed Table of Contents

Foreword *by Wayne W. Smith* xvii

Foreword *by Craig Webster* xviii

Preface xx

Acknowledgment xxviii

Chapter 1
The Role of Festivals in Destination Branding 1
Emrah Keskin, Faculty of Tourism, Nevşehir Hacı Bektaş Veli University, Turkey
Nevres Sezen, Nevşehir Hacı Bektaş Veli University, Turkey
Özge Kaya, Nevşehir Hacı Bektaş Veli University, Turkey

Destinations are areas visited by tourists, containing natural/cultural attractions and openings to experience. The brand is an important reason for tourists to choose the destination. Destination branding contributes to the recognition of the destination and attracting more people. Events are organizations with a specific purpose, time, place, and theme predetermined. Festival events are events that have a specific theme in destinations and experiences which are made; this increases attractiveness, and contributes to destination marketing. Festivals are very important in increasing the recognition of destinations and attracting visitors. In this study, the contributions of festivals to the branding of destinations are written. As a result of the research, it was concluded that culture/art, gastronomy, and sports festivals are effective in destination branding. For this reason, festivals are events that contribute to the promotion of destinations, provide positive feedback about the destination, and encourage people to experience the destination.

Chapter 2
Antecedents of Participation in Music Festivals: The Role of Congruence Between the Sponsor and the Festival 26
Sara Santos, Research Centre in Digital Services (CISeD), Polytechnic of Viseu, Portugal
Pedro Espírito Santo, Research in Management and Economics (CARME), School of Technology and Management, Portugal & Polytechnic of Leiria, Portugal
Luísa Augusto, Research Centre in Digital Services (CISeD), Polytechnic of Viseu, Portugal

In the marketing literature, the desire or need to participate in a music festival is triggered by various reasons. Some sponsorship forms have associated brands with music festivals. One of the counterparts of the sponsorship is the possibility of associating the name of the brand with the designation of the name of the festival—that is, the brand acquiring the naming right of the festival. However, this theme has aroused the academy's interest, and studies that show the importance of the festivals in connection with the sponsoring brand. Thus, this study (N = 306) was carried out through a research model, tested through the PLS-SEM methodology, which showed that cultural exploration and socialization have effects on trust and brand connection and that the congruence between the sponsoring brand and the festival's characteristics play a mediating role in this relationship. Therefore, recommendations are suggested that contribute to the theoretical-practical deepening of new forms of persuasive communication, such as sponsorship, applied to the creative and cultural industries.

Chapter 3
Why Do Festivals Fail? A Case Study of Sonepur Cattle Fair Festival 43
Sushant Kumar, Indian Institute of Management Lucknow, India

This chapter explores the incidences and causes of failure of festivals and other threatening issues that might damage an event or the organization responsible for organizing the event. Literature often offers examples of success stories of festivals, but failures are more difficult to identify, evaluate, and document. A greater understanding of potential antecedents of failures will surely equip managers to avoid consequences and plan for contingencies. This chapter utilizes anecdotal evidence from various festivals to set the stage. This is followed by adoption of case study research design to explore the cattle fair festival organized at Sonepur, Bihar, India. Qualitative data have been collected from various sources along with field notes to analyze the reason of failure of festival. Findings are discussed in several themes that offer guidelines and framework to managers which helps in understanding and utilizing the findings in context of any festivals.

Chapter 4
The Effect of Perceived Authenticity on Satisfaction, Behavioural Intentions, and Expenditure 57
Ester Napolitano, University of Sassari, Italy
Antonio Salvatore Loriga, University of Sassari, Italy
Giacomo Del Chiappa, University of Sassari, Italy

The chapter investigates the influence of socio-demographics, travel-related variables, and perceived authenticity on attendee satisfaction, behavioural intentions, and expenditure level at a small-scale cultural event. A structured questionnaire was developed, and data was collected on-site during the "Autunno in Barbagia" event (Sardinia, Italy). Factor analysis was applied to items measuring authenticity, satisfaction, and behavioural intentions, and the resulting factor scores were entered in the OLS models. Findings show that objective and existential authenticity significantly and positively influence all the dependent variables. In addition, the three dependent variables were significantly influenced by certain travel-related variables and socio-demographic characteristics of attendees. Findings help deepen the scientific debate devoted to analysing the influence of socio-demographics, travel-related variables, and perceived authenticity on attendee behaviour at a small-scale event. Managerial implications are discussed and suggestions for further research are outlined.

Chapter 5
Managing Festivals With Smart Technologies and Their Contribution Towards Destination Promotion 76
Sahil Sharma, Himachal Pradesh University, India
Jeet Dogra, Indian Institute of Tourism and Travel Management (IITTM), Noida, India

This paper presents a scholarly synthesis of contemporary smart event technologies under a clear three phase event classification. The constructed conceptualisation is designed as an ideal synthesis of tech basics, utilisation benefits, and real examples. In the pre-event phase, uses of 3D event diagramming software are discussed. The actual-event phase provides a comprehensive discourse on - beacon technology, geo fencing, radio frequency identification (RFID), near field communication (NFC), wrist band, mobile app, chat bots, virtual reality (VR), and augmented reality (AR) - using some real examples from iconic international events/festivals such as Coachella, Tomorrowland, Glastonbury, etc. Lastly, in the post-event phase basics and uses of smart event data is explained.

Chapter 6
The Role of Culinary Festivals in Building Destination Brand Equity 95
Davut Kodas, Mardin Artuklu University, Turkey

The current chapter theoretically aims to reveal the relationship between culinary festivals and destination branding, focusing on studies investigating aspects of culinary festivals such as festival experience, quality, and motivation in the relationship between the culinary festivals and destination brand equity. In this chapter, the role and importance of culinary festivals in developing destination brand equity are theoretically discussed through reviewing hospitality and tourism literature. Theoretical results of the present chapter indicate that some topics of culinary festival studies such as food experience, food quality, and food festival motivation from a tourist perspective have played an essential role in how increasing core constructs of destination brand equity which included destination brand image, destination brand awareness, perceived quality, and brand loyalty.

Chapter 7
Festivals as a Resource of Heritage Tourism: A Bibliometric Analysis Using the Vosviewer 110
Ruchika Kulshrestha, Amity University, Gurgaon, India
Surabhi Srivastava, Om Kothari Institute of Research and Management, Kota, India

Rich and colorful festivals are a witness to our diversified arts and heritage around the world. Festivals are increasingly finding a place in every segment with integrity, where tangible and intangible heritage treasures are combined. To reveal the potential for tourism through promoting festivals as a resource of heritage, this research uses bibliometric analysis to carry out a systematic analysis of the literature. From the Web of Science data over the last 10 years (2012–2022), 199 English-language publications were gathered for analysis using VOS viewer. The most cited articles, contributions to heritage festivals research by various authors, nations, organizations, journals, and so on for innovations in the tourism sector, provided useful insights. This research entails a thorough examination of recent advancements in festival tourism, as well as a discussion of the researchers' future plans.

Chapter 8
Gastronomic Festival Tours in Turkey: Motivations and Expectations of Tourists............................ 127
Suat Akyürek, Gümüşhane University, Turkey

The demand for gastronomy tourism has been increasing in recent years, and gastronomy tours and festivals are marketed as an important gastronomic product in meeting this demand. However, research on the motivation and expectations of tourists participating in gastronomy festivals are still not sufficient in the literature. In this study, it is aimed to determine the motivation and expectations of tourists who participated in the tours organized under the name of gastronomic festival tours. Face to face interviews were carried out with 23 tourists who participated in three different gastronomic festival tours arranged in Turkey. In the light of the findings, the motivations of tourists participating in gastronomic festival tours were evaluated within the scope of push and pull factors. Expectations of tourists from gastronomic festival tours were classified into five categories as general services, relations with other people, environmental conditions and internal expectations, which were especially for products/activities that are the focal point of the tours.

Chapter 9
Local Fairs: Festivals and Sustainable Tourism Development in Rajasthan 145
Anukrati Sharma, University of Kota, India
Shruti Arora, University of Kota, India
Debendra Nath Dash, Mahatma Gandhi National Council of Rural Education, India

Festivals and fairs abound throughout India. More festivals are observed in this nation than everywhere else in the globe. Social and cultural festivals and fairs have become a tool for promoting regional economies, extending the tourist season, and developing the tourism industry. The effects of these fairs and festivals matter for the local economy as well as the tourism industry. The culture, traditions, and festivals provide an explanation of the "way of life of a place" and aid in our understanding of a certain locale. To discover more about the customs and traditions of the host town, many tourists go to attend a variety of festivals. The chapter has therefore concentrated on different regional fairs and festivals of Rajasthan that have not yet been explored with the intention of enhancing the local economy, and such chances occasionally go unused. The recommendation would therefore be to invest in festival development and community involvement while providing tourists with distinctive experiences.

Chapter 10
Karaga Jatre of Karnataka: A Tryst With History, Anthropology, Belief, Culture, and Tourism 156
Shaheed Khan, Dharthi NGO, India
Freeda Maria Swarna M., Dharthi NGO, India

Karaga Jatre/procession is observed in Thigalarapete in Bangalore, India with religious fervour by a community known to be warriors and gardeners who have part of history for more than four hundred years. The scale is huge, with close to half a million people attending the eleven-day festival in the heart of the city of Cosmopolitan Bangalore, now known as Bengaluru. It is unique because in the canvas of India, the facets of communal harmony come to the core through this festival and so many others. However, with the changes in time and modernity appearing with new generations coming to the fore in greater numbers, will we as a people be able to hold on to the cultural ethos of a time that has gotten us to freeze in history and heritage? The researchers ask the primordial question regarding the future of this festival, and if it will be in the same shape and form. How should the federal and provincial governments intervene to ensure the festival, the Jatre moves ahead, binding us as humanity.

Chapter 11
Social Media Networks as an Effective Way to Promote Events 172
Betül Kodaş, Mardin Artuklu University, Turkey
Ela Nazlı Köz, İstanbul Gedik Üniversitesi, Turkey

Today, we have witnessed many events and festivals organized by different countries throughout the years. Some of these are organized locally, while the others are international. Social media networks are one of the essential marketing strategies which included advertising and promotion for event organizers. In this chapter, social media platforms, which can be used in promoting events and festivals, will be discussed conceptually by making a comprehensive literature review based on relevant literature. After reviewing the literature, this chapter will report how social media networks can be used as a promotional strategy for the event industry. So, this chapter aims to analyze in depth the relationship between social media platforms and promotions of events, and assessed how social media applications can be used to promote tourism events and festivals. At the end of the chapter, some practical implications will be presented to events marketers about social media tools used effectively to build effective promotional strategies for events and festivals.

Chapter 12
Conceptual Framework to Position Devika River (Gupt Ganga) as a Religious Festival Tourist Destination 186
Biswabhusan Pradhan, Central University of Jammu, India
Ayush, Central University of Jammu, India
Krishma Rana, Central University of Jammu, India
Amit Gangotia, Central University of Jammu, India

In Hinduism, rivers are considered holy and sacred as they have various historical, religious, and social importance. The aim of this research is to explore the historical, religious, and socio-economic importance of the Devika River (Gupt Ganga) and the surrounding area and its potential for being developed as a religious festival tourism destination. This study employed qualitative method to explore the potential of research sites by using both primary data and secondary data. The results of this study revealed that the conservation of rivers is very important to save this culture and history. Also, the importance of festivals for the regional development through tourism was highlighted. Furthermore, it has been found that Devika River and the surrounding area has enormous potential to be developed as a religious festival tourist destination, but lack of planning and support from the government and minimal participation from other tourism stakeholders has always deprived this destination from exploring its true potential.

Chapter 13
Destination Marketing and Branding of Nashik: A Case Study of Sula Wine Festival 196
Gayatri Gawde, Amity University, Mumbai, India
Sachin Kumar Behera, Amity University, Mumbai, India
Sharad Kumar Kulshreshtha, Department of Tourism and Hotel Management, North-Eastern Hill University, Shillong, India

This chapter presents a case study, highlighting the importance of wine tourism with its connection with the Sula Wine Festival that happens in Nashik, Maharashtra. This case study presents an exploratory study of the Sula Wine Festival, with a focus on educating readers on its prominent features, USP, and attractions for the people interested in wine festivals. The study further explains some of the distinct advantages of Nashik such as its soil, location, and wide range of attractions nearby. The key objective of the case study is to make the readers aware of the power of wine festivals in tourism and encourage the blending of tourism and wine consumption for marketing and branding Nashik.

Chapter 14
Destination Branding Through Tourism Festivals of Khasi Hills and Garo Hills Region Meghalaya in India: Case of Cherry Blossom Festivals and Me.gong Festival 210
Sharad Kumar Kulshreshtha, Department of Tourism and Hotel Management, North-Eastern Hill University, Shillong, India
Iadonlang Tynsong, Martin Luther Christian University, India
Ashok Kumar, Institute of Hotel Management, Catering Technology, and Applied Nutrition, India

Meghalaya is the most beautiful and scenic state among the northeastern regions. In Sanskrit, 'Meghalaya' means 'Abode of Clouds.' This state has blessed with natural and cultural attractions. Meghalaya tourism has conducted two more yearly tourism festival associated with their nature and culture to attract more tourists Khasi and Garo hills region, a sincere initiative for destination branding through events. This chapter is an attempt to showcase the potential these two tourism festivals, i.e., Cherry Blossom Festivals and Me'gong Festivals, which are associated with festival destination marketing of these two regions of Meghalaya. The purpose of this chapter is to investigate the emergence of such tourism festivals with mix of traditions, culture, heritage, indigenous characterizes of local destination, in order to transform it into a festival and attract a large number of tourists to build-up iconic destination branding.

Chapter 15
Value Creation of Destinations by Destination Management Organisations Through Strategic Planning and Promotion of Fairs and Festivals of NE India.. 227
Angshumi Goswami, North-Eastern Hill University, India

This chapter aims to add value to available existing literature in term of highlighting the catalytic roles and responsibilities of destination management organisations (DMOs). Moreover, it aims to study the involvement of DMOs in strategic planning and promotion of local fairs and festivals. Destination management organisations (DMOs) act as torchbearers of every global and local tourism destination and are indispensable when promoting a destination through events and festivals. In this context, the sustainable development of the destination is imperative and a crucial factor for prospective visitors and guests. This chapter shall comprehensively discuss DMOs and their underlying roles and responsibilities. Moreover, it shall highlight the operational activities undertaken by city-level DMOs in relation to the strategic planning and promotion of events being hosted. This study shall be purely based on reviews done on existing literature supported by cases of major fairs and festivals specifically with reference to North Eastern India.

Chapter 16
Long-Term Forecasting of Tourism Festivals Post COVID-19 244
Rachel Chanel Clarke, Trigo Networks, USA

Forecasting festivals and destination tourism does not involve a secret recipe held by industry gatekeepers but is the applying tools, techniques, and systems. The significance that the COVID-19 pandemic contributes to the greater landscape of destination festivals is multifaceted as time progresses. Sales determinants are affected by a change in the labor force, logistic delays, health and safety concerns, and many factors still to be monitored and readapted through forecasting. Changes in social interactions, such as norms of personal space between family, friends, and strangers, have shifted, causing festival planners to be aware of directional signals that affect festival gatherings (McClure et al., 2020). Long-term forecasting tools assess the cultural shifts of consumer segmentation, revealing the adverse effects that COVID-19 has on demographic trends. Consequently, this chapter provides context while identifying forecasting methodologies and a framework for festivals and destination tourism post-COVID-19.

Chapter 17
Investigation of Local People's Perceptions of Festival Tourism With Cultural Interaction Dimensions 263
Sibel Sü Eröz, Kırklareli University, Turkey
Kaplan Uğurlu, Kırklareli University, Turkey
Aydemir Ay, Trakya University, Turkey

Festivals are events that enable the transmission of past ties to the future without severing them. Festivals serve as a conduit for cultural exchange between the residents of a region and the local and international visitors who attend the festivals. The primary objective of this study is to investigate the perceptions of Edirne, Turkey locals regarding festival tourism with cultural exchange and cultural balance interaction dimensions, as well as to establish whether the dimensions of cultural interaction differ based on demographic variables. A total of 390 individuals were surveyed to obtain data for the study, and SPSS 22.0 was used to analyze the acquired data. This study reveals that festivals significantly contribute to regional tourism and the local population, as well as to the development of intercultural communication, and that the sub-dimensions of cultural exchange and cultural balance differ depending on the type of events.

Chapter 18
Destination Branding and Mega Events: A Case of FIFA World Cup 2022 and Sultanate of Oman. 285
Mohit Kukreti, Independent Researcher, Oman
Amitabh Mishra, University of Technology and Applied Sciences, Oman
Aarti Dangwal, Chitkara University, India
Vishal Jain, University of Technology and Applied Sciences, Oman
Veena Tewari, University of Technology and Applied Sciences, Oman

Tourist attractions are often centered on mega sports events. Since the mega events have the potential to generate new revenue and enhance the appeal of the destination similarly, the FIFA World Cup 2022 in Qatar presented a unique opportunity for Oman to leverage the mega sporting event to promote its tourism industry globally. Oman used this mega event to brand its tourism industry by collaborating with FIFA, hosting fan zones, effectively using social and print media, highlighting accessibility and accommodation, and promoting local tourism attractions. The mega event resulted in increased international arrivals, revenue generation, and an enhanced image of Oman in the Middle East. The chapter is an attempt to explore various initiatives taken by the Sultanate of Oman to attract international visitors. Further details will review the methods adopted by public and private sectors in Oman in leveraging the recent mega event of FIFA World Cup 2022 to become a preferred global destination.

Compilation of References 303

About the Contributors 356

Index 364

Foreword

Festivals have been part of humanity for centuries. Their original intention was to be used as a tool to get communities to come together to celebrate, commemorate or provide support for neighbours, family, and friends. Over the years, festivals have grown in size and stature and, while retaining much of the original intent, have added the additional element of sharing community with visitors. The organic nature of festivals allows others from different areas of the world to share and better understand the community they are visiting. In that vein, festivals have become an increasingly important tourism industry component. They often shape perceptions of the place and space in potential visitors' minds. This is why this text is so valuable. It shapes the discussion of how to properly manage the marketing and branding of festivals so that they retrain the authentic nature and adequately reflect the community in which they reside so that visitors get an accurate indication of what to expect.

The text begins with an excellent chapter on the role of festival in destination branding. This chapter substantially discusses how cultural/art, gastronomy and sports festivals influence destination branding. Further insight into these specific types of festivals are found in chapters on music festivals (Chapter 2), gastronomic & wine festivals (Chapters 8 & 14), religious festivals (Chapter 13), and local fairs (Chapters 15 & 16). The chapter on "Festivals as a Resource of Heritage Tourism: A Bibliometric Analysis Using the Vosviewer" (Chapter 7) provides a solid bibliometric analysis of the academic literature from 2012-2022.

Another text element provides valuable insight into why festivals fail and how perceived authenticity affects event satisfaction, behavioural intentions and expenditures (Chapters 3, 4, & 17). The chapter on why festivals fail (Chapter 3) is beneficial and often not discussed within the literature. The chapter on "Long-Term Forecasting of Tourism Festivals Post COVID-19" (Chapter 17) provides insights into how decreased labour force, logistic delays and health and safety concerns are being projected—these critical chapters to understanding festival planning and developed in modern times.

Another area that is delivered in the text is examining the role of social media (Chapter 12) and SMART technologies (Chapter 5) in destination marketing and branding in relation to festivals. The chapter on smart technologies (Chapter 5) provides an excellent discussion on the role of Beacon Technology, Geo-Fencing, Radio Frequency Identification (RFID), Near Field Communication (NFC), Wrist Bands, Mobile Apps, Chat Bots, Virtual Reality (VR), and Augmented Reality (AR) and their potential uses for festival delivery, market intelligence and message creation and delivery. The discussion on the role of social media (Chapter 12) analyzes the relationship between social media platforms and promotions of events and assesses how social media applications can be used to promote tourism events and festivals. These two chapters provide essential insights into how technology is shaping the future of festival delivery and marketing.

Wayne W. Smith
Ted Rogers School of Hospitality and Tourism Management, Toronto Metropolitan University, Canada

Foreword

As the world emerges from the global pandemic that had slowed down a great deal of the global economy, tourism and its associated industries are returning to normalcy and we can expect a vibrant revival of almost all components of the tourism and hospitality industry. While we all look in a cautious way to the return of many of the industries that give life pleasure and joy, we see that tourism and hospitality have bounced back quickly, even if there are some new challenges that seem somewhat novel, in the post-pandemic economic ecosystem. One of those industries and activities that was hurt the hardest during the global pandemic was the event industry, of which festivals are one major component. This book comes at a great time in economic history and is a useful research resource for many people during this historical period in which festivals and events are becoming increasingly understood as an important part of the tourism and hospitality ecosystem. This book offers lessons/insight into the use of festivals and events as a platform to encourage destination development, destination marketing, and eventually economic development that is beneficial to the host community.

A person needs not be an expert in the study of tourism nor hospitality to comprehend that festivals and events are a central player in the tourism product of many destinations, as Getz and Page (2016) point out. However, the academic scholarship that links festivals and events with the development and the marketing of destinations has lagged behind a great deal, despite the significant growth of the festival and event industries in recent decades, as Mair and Weber (2019) correctly identify. Dr. Sharad Kumar Kulshreshtha, as the editor, should be congratulated with excellent timing for this project, since it fills in the long-term need in the scholarship for a useful reference that can contribute to an academic subfield that is underdeveloped relative to the massive commercial strides that have occurred in recent years and the short-term need to have scholarship that can assist in inspiring destination marketers and branding for the recovery from the global pandemic.

The book has many chapters that readers will find useful for various reasons. Some of the chapters are theoretical, some are more empirical, but they can all be used for inspiration, as well as a reference for practical purposes. For example, Chapter 1 stands out as a chapter not only setting the tone for the book but also explaining why festivals and events are critical to understand in terms of branding and marketing. A reader of that particular chapter would not have to be very imaginative to understand that the basic concept of the chapter can go a long way to inspiring the integration of festivals and events into the branding of a destination to encourage economic development. The other chapters use various different methods and methodologies to inform readers about festivals and events and how they can be understood (and better leveraged) for the development of destinations.

The world of the hospitality and tourism scholar needed this book to come into being and be available to scholars and Dr. Sharad Kumar Kulshreshtha and IGI Global made it happen. The final outcome of the hard work of the editor is a book with relevant chapters that will assist in giving both scholars and practitioners insight into how the booming industry of festivals and events can be harnessed in an effective way for destination marketers to encourage economic development. The use of various cases taken from different contexts, countries, and types of events (wine festivals, religious festivals, and other types of events/festivals) should be of use to scholars, practitioners, and students. I envision that this book will have a substantial impact upon the field, since the topic is obviously important, the scholarship comes from varied approaches, and it comes at a time when the knowledge and ideas from the book are most in need, not just for the short-term needs of recovery from a global pandemic but also for the long-term needs of the field, as research on the topic has lagged behind the massive growth of festivals and events globally. It is also impressive what an international collaboration it is with scholars from so many different countries, giving the book some more value, given the diversity of insights.

My congratulations to Dr. Sharad Kumar Kulshreshtha and IGI Global for bringing this book into existence. It is both relevant, needed, and insightful. However, I must also congratulate the reader, as this book the reader has chosen to read should give the reader a great deal of insight into an important component of tourism/hospitality and will inform and inspire in many different ways.

Craig Webster
Ball State University, USA

REFERENCES

Getz, D., & Page, S. J. (2016). Progress and prospects for event tourism research. *Tourism Management*, *52*, 593–631. doi:10.1016/j.tourman.2015.03.007

Mair, J., & Weber, K. (2019). Event and festival research: A review and research directions. *International Journal of Event and Festival Management*, *10*(3), 209–216. doi:10.1108/IJEFM-10-2019-080

Preface

OVERVIEW

Festivals are currently among the fastest developing cultural events in the world as well as are treated as an important element in promotion of cities and regions, a tourist asset and product, attracting tourists and encouraging them to spend their money (Cudny, 2014a). Festivals symbolized by a kaleidoscope of experiences that finds meaning through the lives of the participants (Goldblatt, 2002). Festivals always symbolize its own unique value, and identity most associated with numbers of celebration of events, happenings, memories, with inclusion of local, regional, and national rituals, ceremonies, ethics and have full of rejoice with energy and enthusiasm among the people of respective destinations. Every festival associated to anyone who lives in the respective societies, regions, countries with their respective culture, religion agriculture, education, business, and sports. Festivals all across the world represent the joy, recreation, and traditions of their different societies and cultures. There is surfeit of reasons to commemorate and organise such festivals in respective worldwide. The most primitive festivals in the history of mankind were correlated to plenteous Agri-produces by communities and celebrate share joy, happiness with singing and dancing togetherness. There is an ancient, mediaeval, and modern history to the celebration of such events or festivals, which have been held in various locations around the world from time to time. Every festival has its own distinct features festivescapes, character, charms, appeal, attraction, and experiences that all are very closely associated to culture, customs, issues, core values, creativity, cuisines, connectedness, and so on. All of these factors combine to create a one-of-a-kind selling offer for specific destinations. Some festivals are associated with specific locations. The current term "managing festivals for destination marketing and branding" will be the most essential area of expertise in this regard. Festivals always create value addition for destinations, attracts visitors or tourists and promoting tourism in a destination by improving the lives of local communities and create a positive image and branding of the destination.

IMPACT OF FESTIVALS

The festival has many genuine factors: civilizational (including sociological, psychological, and ecological components), economic, including developments in the tourist industry, political evolution, and finally – advances in technology (Cudny, 2014b). Festival tourism is expressed through its multifaceted impacts on different elements of tourist space as well as has some negative impacts (dysfunctions) (Cudny, 2013). This negative impact most of time outcomes of the various issues, economic impact i.e., inflated price, opportunity cost, financial loss; social impact i.e., negative community image, bad behavior, social dislocation, loss of amenities, environmental impact i.e., pollution, noise destruction, traffic congestion, etc. (Hall, 1989). Destinations must carefully take all essential considerations as well as their discharge their responsibility to local community, and hosting taken care sustainability, community involvement and destination-marketing aspects of festivals and events (Mohanty et al., 2022). The community issues, sustainability and environmental issues have some keys concerns of festivals worldwide (Gursoy et al., 2020).

FESTIVALS AND DESTINATION MARKETING: A GLOBAL SCENARIO

The Government, tourism boards, tourism departments, and tourism destination management organizations (DMOs) are all aware of the importance of increasing footfall in their respective destinations by organising innovative and thematic tourism festivals and attempting to make destinations more lively while also providing opportunities for local residents to earn a living. Every time visitors visit a destination, these developing tourism-centric festivals provide a lasting feeling of something unusual. Such festival attractions entice local and international visitors to come see and experience these festive locales. Putting on a festival like this can serve to popularize and strengthen the tourist economy, as well as promote employment, entrepreneurship, and tourism destination branding for the location. The faster pace of information technology driven tourism sustainable tourism development, social media, accessibility and so on to cover fairs, festivals and events (Rahman et al., 2022). Local festival has emerged as to celebrating local culture, culinary and local product to develop sustainable destination (Botti et al., 2018).

OBJECTIVE OF THE BOOK

The aim of this book is to address the most current and promising parts of tourism-centric festivals, which are held in numerous tourist areas throughout the world. In terms of destination marketing and branding, this is the most dynamic method.

The focus of this book's publishing will be on showcase and project festivals as a magnetic force for destination marketing and branding, as well as the global development of the visiting economy. The title of this book, *Managing Festivals for Destination Marketing and Branding*, will link tourism festivals around the world as a catalyst for destination marketing and identity. It focuses on how tourism destination planners and industry policymakers, tourism departments, tourism NGOs, and destination marketing organisations (DMOs) market their intangible heritages and festivals as destination promotion and brand building strategies for local, regional, and national destinations.

TARGET AUDIENCE

This book will cover topics such as event business management and operations, event management, festival management, festival services i.e., travel, tourism, hospitality management, volunteer management, project management, destination marketing, festival entrepreneurship, festive visitor economy, ethnic and cultural festivals, sport event management, environment and sustainability issues for festivals, and global event studies, etc. This book is aimed at scholars, researchers, academicians, festival event policymakers, and tourist industry practitioners who are interested in tourism festivals, events, and festival management.

ORGANISATION OF THIS BOOK

This book is the collection of 18 diverse chapters from a various domain, i.e., tourism hospitality and events as an area of expert scholarly contributed from various countries, with various thematic aspects high lightening intangible events features of various different destinations worldwide with major emphasis on destination marketing and branding. This book offers an interesting fact about the local and global festivals in empirical, qualitative, and case study mode with in-depth and systematic manner highlighting current and emerging issues of festivals and destination management study.

Chapter 1: The Role of Festivals in Destination Branding

This chapter describes about the key role of festivals in destination branding and creating unique experiences to visit destination. Festivals are very important in increasing the recognition of destinations and attracting visitors. In this study, the significant contribution of festivals to the branding of destination as well as marketing Chapter concluded that managing culture/art, gastronomy and sports festivals are some effective marketing and promotion strategies in destination branding. Therefore, these festivals contribute to the promotion of destinations, as well as provide positive feedback about the destination and encourage visitors to experience such unique destinations.

Chapter 2: Antecedents of Participation in Music Festivals – The Role of Congruence Between the Sponsor and the Festival

This chapter explains the various reasons responsible to trigger any festival specially music festivals led by sponsorship forms have associated brands with music festivals. Consequently, this study (N = 306) was carried out through a research model, tested through the PLS-SEM methodology, which showed that cultural exploration and socialization have effects on trust and brand connection and that the congruence between the sponsoring brand and the festival's characteristics play a mediating role in this relationship which develop a new form of persuasive communication, such as sponsorship, applied to the creative and cultural industries.

Chapter 3: Why Do Festivals Fail? A Case Study of Sonepur Cattle Fair Festival

This chapter explores the incidence and causes of failure of festival and other threatening issues that might damage an event or the organization responsible for organizing the event. This chapter also utilizes anecdotal evidence from various festivals to set the stage. This is followed by adoption of case study research design to explore the cattle fair festival organized at Sonepur, Bihar, India. Qualitative data have been collected from various sources along with field notes to analyze the reason of failure of festival. Findings are discussed in several themes that offer guidelines and framework to managers which helps in understanding and utilizing the findings in context of any festivals.

Chapter 4: The Effect of Perceived Authenticity on Satisfaction, Behavioural Intentions, and Expenditure

This chapter investigates the influence of socio-demographics, travel-related variable and perceived authenticity on satisfaction, behavioural intentions and expenditure of attendees at a small-scale cultural event.Factor analysis was applied to items relating to authenticity, satisfaction and behavioural intentions and the factor scores were entered in the OLS models. Findings show that both objective and existential authenticity significantly and positively influence all the dependent variables. Findings contribute to deepen the scientific debate devoted to analysing the influence of socio-demographics, travel-related variables and perceived authenticity on participants' behaviour at a small-scale event. Managerial implications and limitations are discussed and suggestions for further research are given.

Chapter 5: Managing Festivals With Smart Technologies and Its Contribution Towards Destination Promotion

This chapter presents a scholarly synthesis of contemporary smart event technologies under a clear three phase event classification. The constructed conceptualisation is designed as an ideal synthesis of tech basics, utilization benefits, and real examples. In the pre-event phase uses of 3D Event Diagramming Software is discussed. The actual-event phase provides a comprehensive discourse on—Beacon Technology, Geo Fencing, Radio Frequency Identification (RFID), Near Field Communication (NFC), Wrist Band, Mobile App, Chat Bots, Virtual Reality (VR), and Augmented Reality (AR)—using some real examples from iconic international events/festivals such as Coachella, Tomorrowland, Glastonbury, etc. Lastly, in the post-event phase basics and uses of smart event data is explained.

Chapter 6: The Role of Culinary Festivals in Building Destination Brand Equity

This chapter theoretically aims to reveal the relationship between culinary festivals and destination branding, focusing on studies investigating aspects of culinary festivals such as festival experience, quality, and motivation in the relationship between the culinary festivals and destination brand equity apart of that the role and importance of culinary festivals in developing destination brand equity are theoretically discussed through reviewing hospitality and tourism literature. Theoretical results of the present chapter indicates that some topics of culinary festival studies such as food experience, food quality, and food festival motivation from a tourist perspective have played an essential role in how increasing core constructs of destination brand equity which included destination brand image, destination brand awareness, perceived quality, and brand loyalty.

Chapter 7: Festivals as a Resource of Heritage Tourism – A Bibliometric Analysis Using the Vosviewer

This chapter reveals the potential for tourism through promoting festivals as a resource of heritage, this research uses bibliometric analysis to carry out a systematic analysis of the literature. From the Web of Science data over the last 10 years (2012–2022), 199 English-language publications were gathered for analysis using VOS viewer. The most cited articles, contributions to heritage festivals research by various authors, nations, organizations, journals, and so on for innovations in the tourism sector, provided useful insights. This research entails a thorough examination of recent advancements in festival tourism, as well as a discussion of the researchers' future plans.

Chapter 8: Gastronomic Festival Tours in Turkey Motivations and Expectations of Tourists

This chapter emphasis about the demand for gastronomy tourism has been increasing in recent years and gastronomy tours and festivals are marketed as an important gastronomic product in meeting this demand. In this study, it is aimed to determine the motivation and expectations of tourists who participated in the tours organized under the name of gastronomic festival tours. In the light of the findings, the motivations of tourists participating in gastronomic festival tours were evaluated within the scope of push and pull factors. Expectations of tourists from gastronomic festival tours were classified into five categories as general services, relations with other people, environmental conditions and internal expectations; which were especially for products/activities that are the focal point of the tours.

Chapter 9: Local Fairs – Festivals and Sustainable Tourism Development in Rajasthan

This chapter focuses on social and cultural festivals and fairs which have to be considered as a tool for promoting regional economies, by supporting the tourism industry. The centre of study is to find the effects of these fairs and festivals matter for the local economy as well as the tourism industry. The chapter has therefore concentrated on different regional fairs and festivals of Rajasthan that have not yet been explored with the intention of enhancing the local economy, and such chances occasionally go unused. The recommendation would therefore be to invest in festival development and community involvement while providing tourists with distinctive experiences.

Chapter 10: Karaga Jatre of Karnataka – A Tryst With History, Anthropology, Belief, Culture, and Tourism

This chapter based on Karaga Jatre/procession which is observed in Thigalarapete in Bangalore, India with religious fervour by a community known to be Warriors and Gardeners who have part of history for more than four hundred years. Here the researchers raise the primordial question, future of this festival, will it be in the same shape and form. How should the Federal and Provincial Governments intervene to ensure the festival, the Jatre moves ahead, binding us as humanity.

Chapter 11: Social Media Networks as an Effective Way for the Promotion of Events

This chapter highlights social media platforms which can be used in promoting events and festivals will be discussed conceptually by making a comprehensive literature review based on relevant literature. After reviewing the literature, this paper will report how social media networks can be used as a promotional strategy for the event industry. So, this chapter aims to analyze in depth the relationship between social media platforms and promotions of events and assessed how social media applications can be used to promote tourism events and festivals. At the end of the chapter, some practical implications will be presented to events marketers about social media tools used effectively to build effective promotional strategies for events and festivals.

Chapter 12: Conceptual Framework to Position Devika River (Gupt Ganga) as a Religious Festival Tourist Destination

This chapter employed qualitative method to explore the potential of research sites by using both primary data and secondary data. The results of this study revealed that the conservation of river is very important to save our culture and history. Also, the importance of festivals for the regional development through tourism was highlighted. Furthermore, it has been found that Devika River and the surrounding area has enormous potential to be developed as a religious tourist destination, but lack of planning and support from the government and minimal participation from other tourism stakeholders has always deprived this destination from exploring its true potential.

Chapter 13: Destination Marketing and Branding of Nashik – A Case Study of Sula Wine Festival

This chapter presents a case study, highlighting the importance of wine tourism with its connection with the Sula Wine Festival that happens in Nashik, Maharashtra. This case study presents an exploratory study of the Sula Wine Festival, with a focus on educating readers on its prominent features, USP, and attractions for the people interested in Wine Festivals. The study further explains some of the distinct advantages of Nashik such as its soil, location, and wide range of attractions nearby. The key objective of the case study is to make the readers aware of the power of wine festivals in tourism and encourage the blending of tourism and Wine Consumption for marketing and branding Nashik.

Chapter 14: Destination Branding Through Tourism Festivals of Khasi Hills and Garo Hills Region Meghalaya in India – Case of Cherry Blossom Festivals and Me.gong Festival

This chapter is an attempt to showcase the potential these two tourism festivals i.e., Cherry Blossom Festivals and Me.gong Festival which associated with destination marketing of these two regions of Meghalaya. The case study method has applied to study the potential destination marketing and branding.]The purpose of this chapter is to investigate the emergence of such tourism festivals with blends of traditions, culture, heritage, indigenous characterizes of local destination, which in able to transformed into a festival and attract a large number of tourists to build-up iconic destination branding.

Chapter 15: Value Creation of Destinations by Destination Management Organisations Through Strategic Planning and Promotion of Fairs and Festivals of NE India

This chapter comprehensively discusses about DMOs and their underlying roles and responsibilities. This chapter aims to add value to available existing literatures in term of highlighting the catalytic roles and responsibilities of Destination Management Organisations (DMOs). Moreover, it shall highlight the operational activities undertaken by city-level DMOs in relation to the strategic planning and promotion of events being hosted. This study shall be purely be based on reviews done on existing literature supported by cases of major fairs and festivals specifically with reference to North Eastern India.

Chapter 16: Long-Term Forecasting of Tourism Festivals Post COVID-19

This chapter provides context that incorporates forecasting methodologies for festivals and destination tourism post-COVID-19. The impact that the COVID-19 pandemic contributes to the greater landscape of destination festivals is multifaceted as time progresses. Sales forecasting are affected by a decreased labor force, logistic delays, health and safety concerns, and many factors still to be monitored and re-adapted through forecasting. Long-term forecasting tools assess the cultural shifts of the consumer segmentation, revealing the adverse effects that COVID-19 has on generational and demographic trends.

Chapter 17: Investigation of Local People's Perceptions of Festival Tourism With Cultural Interaction Dimensions

This chapter reveals that festivals significantly contribute to regional tourism and the local population, as well as to the development of intercultural communication, and that the sub-dimensions of cultural exchange and cultural balance differ depending on the type of events. The primary objective of this study is to investigate the perceptions of Edirne, Turkey locals regarding festival tourism with cultural exchange and cultural balance interaction dimensions, as well as to establish whether the dimensions of cultural interaction differ based on demographic variables.

Chapter18: Destination Branding and Mega Events – A Case of FIFA World Cup 2022 and Sultanate of Oman

The proposed chapter is an attempt to explore various initiatives taken by the Sultanate of Oman to attract international visitors. Further details will review the methods adopted by public and private sectors in Oman in leveraging the recent mega event of FIFA World Cup 2022 to become a preferred global destination. Oman used this mega event to brand its tourism industry by collaborating with FIFA; hosting Fan Zones; effectively using Social and Print Media; highlighting accessibility and accommodation; and promoting local tourism attractions. The mega event resulted in increased international arrivals, revenue generation, and an enhanced image of Oman in the Middle East.

CONCLUSION

This book provides a comprehensive overview of the most important concepts of festival management destination branding. It identifies that festivals act as catalyst for destination development and destination marketing to attract more numbers of visitor as well as create unique destination identity associated with nature, commerce, culture and community.

Finally, the book will be a valuable resource for event, festivals, tourism and hospitality and destination marketing domains focus the needs of research scholar, academicians, and industry practitioners, event management and festivals planners, it will also be useful for tourism and culture departments, local organizations those are planning to organize festivals for such rural and urban destination branding.

Sharad Kumar Kulshreshtha
Department of Tourism and Hotel Management, North-Eastern Hill University, Shillong, India

REFERENCES

Botti, A., Monda, A., & Vesci, M. (2018). Organizing Festivals, Events and Activities for Destination Marketing. In Tourism Planning and Destination Marketing. Emerald Publishing Limited. doi:10.1108/978-1-78756-291-220181010

Cudny, W. (2013). Festival Tourism – The Concept, Key Functions and Dysfunctions in the Context of Tourism Geography Studies. *Geographical Journal, 65*(2), 105-118.

Cudny, W. (2014). The Phenomenon of Festivals: Their Origins, Evolution, and Classifications, Anthropos 2014. www.jstor.org/stable/pdf/43861801.pdf

Goldblatt, J. (2002) *Special Events: Twenty First Century Global Event Management.* John Wiley & Sons, Inc.

Gursoy, D., Nunkoo, R., & Yolal, M. (Eds.). (2020). *Festival and Event Tourism Impacts* (1st ed.). Routledge. doi:10.4324/9780429274398

Hall, C. M. (1989). Hallmark events and the planning process. In G. J. Syme, B. J. Shaw, D. M. Fenton, & W. S. Mueller (Eds.), *The Planning and Evaluation of Hallmark Events.* Avebury.

Mohanty, P. (2022). *Managing Events and Festivals Post-COVID-19: Maintaining the Status Quo or Resetting for the Sustainability Agenda.* CABI International. doi:10.1079/9781789248685.0002

Rahman, M. K., Rana, M. S., Akter, S., & Hassan, A. (2022). Technology Innovation as a Destination Branding Tool for Festivals and Events Tourism. In A. Hassan (Ed.), *Technology Application in Tourism Fairs, Festivals and Events in Asia.* Springer. doi:10.1007/978-981-16-8070-0_8

Acknowledgment

First my sincere gratitude to divine almighty to his blessings to successfully able to complete this book. Afterward, I would like to extraordinary debt I owe to my respected teacher (Guru) mentor Professor Umendra Narayan Shukla who always inspire me and support to continue this journey and my respected father Shri Jiwan Prakash Kulshreshtha for their endless affection, support and encouragement I able to this book project.

My most sincere thanks to Professor Wayne W. Smith and Dr. Craig Webster who have given their valuable words as 'Foreword' to this book. I'm also immensely grateful to all editorial advisory board members Professor Dimitrios Buhalis, Professor Maximiliano Korstanje-, Professor Catheryn Khoo, Professor U.N. Shukla, Professor. Aydoğan Aydoğdu, Professor Ian Yeoman, Dr. Sofia Borda de Água de Almeida, Professor Sudhir Rana, and Mr. Sanjay Nadkarni.

My sincere thanks to all my valuable contributors as author and co-authors who have shown their interest, enthusiasm with continuous hard work from proposal to final chapter submission to book project.

I owe an enormous sincere gratitude to all reviewers who gave their constructive in-depth and insight review comments on one or more chapters of this book.

I would like to extend my special thanks to all faculties of Department of Tourism and Hotel Management, North-Eastern Hill University colleagues Prof. Saurabh Kumar Dixit, Dr. Punit Gautam and Dr. Benjamin Franklin Lyngdoh, Head of Department for share their valuable suggestions and support as well as thank to our research scholars Ashok, Iadon, Sachin and Angshumi as well as my office staff Kong Bigi and Bah Badap.

Apart of that my other university colleagues from Bio-medical Department, NEHU, Dr Nitin Sahai and Dr. Sudip Paul for their continuous support this work.

I would like to offer my special thanks to Lindsay Wertman, President of IGI Global who has considered this book project on this theme, as well as I am particularly grateful for the assistance given by Elizabeth Barrantes, and Melissa Wagner, development team, marketing production team, and marketing team of IGI Global Publication, U.S.A.

Most importantly, I wish to thank my loving and supportive wife, Sakshi, and my two loving children, Atharv (Son), Aashvi (Daughter), who provide unending inspiration.

Chapter 1
The Role of Festivals in Destination Branding

Emrah Keskin
Faculty of Tourism, Nevşehir Hacı Bektaş Veli University, Turkey

Nevres Sezen
Nevşehir Hacı Bektaş Veli University, Turkey

Özge Kaya
Nevşehir Hacı Bektaş Veli University, Turkey

ABSTRACT

Destinations are areas visited by tourists, containing natural/cultural attractions and openings to experience. The brand is an important reason for tourists to choose the destination. Destination branding contributes to the recognition of the destination and attracting more people. Events are organizations with a specific purpose, time, place, and theme predetermined. Festival events are events that have a specific theme in destinations and experiences which are made; this increases attractiveness, and contributes to destination marketing. Festivals are very important in increasing the recognition of destinations and attracting visitors. In this study, the contributions of festivals to the branding of destinations are written. As a result of the research, it was concluded that culture/art, gastronomy, and sports festivals are effective in destination branding. For this reason, festivals are events that contribute to the promotion of destinations, provide positive feedback about the destination, and encourage people to experience the destination.

INTRODUCTION

Destinations are defined as areas with various touristic products and destination products can be named with various slogans, logos or names. Naming destination products creates brand perception. For this reason, many destinations strive to create brand value through their products (Çiçek & Pala, 2017). Today, many destination managers aim to become a brand. In order to achieve this goal, ideas are put

DOI: 10.4018/978-1-6684-6356-7.ch001

forward on how the destination will be positioned in the minds of the visitors and which designs will be realized. In line with the determined target, destinations first consider the travel needs of the visitors in order to create a brand. Therefore, in reaching the goal of creating brand value, various designs were realized by taking into account the expectations of the visitors from the trip. Fulfilling expectations from travel creates a positive image for the destination. The positive image perception of the visitors towards the destination and the continuity of this perception create positive results for the branding goals of the destination (Özaltaş Serçek & Serçek, 2015; Akturan & Oğuztimur, 2016). In the branding of destinations, first of all, a unique image and identity of the destination is created. With the researches and studies carried out by destination managers, it is examined which product will be branded in line with the identity of the destination and which branding strategies the destination can apply (Keller, 1998).

While branding destinations, which product will contribute to branding, what the expectations of local people and tourists are, the relationship of people with brand theory and financial capabilities are taken into consideration. In line with these issues, successful branding is achieved by advertising with an effective marketing plan in tourism (Morgan, Anette & Roger, 2004; Pike, 2005; Bagaeen, 2007). Destinations differentiate from their competitors as a result of branding. In order to see the positive effects of branding, it is necessary to create a quality brand (Ritchie & Ritchie, 1998; Pike & Bianchi, 2016; Fathabadi, Nejad, & Alizadeh, 2017). Through branding, tourists visiting destinations can be diversified, tourism products can be developed and problems in destination management can be reduced (Fyall, Wang & Garrod, 2012).

Events are one of the touristic organizations that are included in marketing strategies in tourism as a part of destination attractiveness. Events are organizations that are influenced by the culture of societies and affect other societies, have various themes, play a role in attracting tourists and branding goals of the region (Jackson, 2008; Amara, 2017). Festival events, which offer authentic and different experiences, affect the emotions of the visitors and enable them to create a positive image towards the destination. The positive image perception that occurs through festival activities can brand the destination (Amara, 2017). Festival events are a good competitive element for destinations, a source of finance, an event that showcases regional identity, and one of the factors contributing to regional development. An effective plan considering these factors makes the festival a brand. Since destinations with festivals that are branded in tourism are branded on behalf of the festival, festival events have an important power in the branding of destinations (Getz, 2008; Lopez & Leenders, 2018).

BACKGROUND

Branding is an important factor in ensuring the sustainability of the business, achieving financial goals, creating positive behavioral intentions and guaranteeing the future of the business. Effective promotion of the brand allows businesses or destinations to be perceived as an attractive place in the human mind (Clark, Doraszelski & Draganska, 2009; Dogan at al. 2019; Yılmaz and Çizel, 2000). Building trust is a factor in the sustainability of a business's brand. Through a sense of trust and quality experiences, people can stay loyal to businesses or brands. This is very important for the sustainability of brands. Especially when considered in terms of destinations, a destination that has provided confidence in people's minds can also come to a positive place in terms of branding (Olimpia, 2007; Hankinson, 2007; Vukic, Kuzmanovic & Vukic, 2016). Activities are one of the elements that provide attractiveness to people and brand the destination by making them attractive (Horng & Tsai, 2012; Wong, Wan & Qi, 2015). Events are one of the elements that show the power of the destination towards branding, introduce the destination to other people, and serve to reveal the cultural identity of the destination (Getz, 1997; Lee and Lee, 2009). Activities have a positive effect on people. These effects play a role in a good perception of destinations by positively affecting people's behavioral intentions. When destinations are well remembered, events have a role in destination branding as it affects the branding of destinations (O'Sullivan and Jackson, 2002; Knott, Fyall and Jones, 2015; Knott, Fyall and Jones, 2017). Festival events are one of the events that provide branding of destinations. Many studies conducted in previous years have proven that festivals have a role in destination branding (Leenders, 2010; Dimitrova & Yoveya, 2014; Türten, 2021).

FOCUS OF THE STUDY

In this study, it is aimed to investigate the effect of festival activities on destination branding and to make a theoretical contribution to the studies carried out in previous years. Festivals are events that make destinations attractive, change the behavior of visitors in a positive way and ensure that the destination has an identity, make the destination valuable and become a symbol of some destinations. The festival events are aimed at the organization of visitors to have fun or think around a theme, so when the goals are achieved, the destination where the festival is held becomes important along with the festival. One of the most important elements for the recognition and promotion of destinaion is that it has memorable activities. When the goals are achieved in the festivals, it contributes to making the destination a valuable place by positively affecting the visitors. That's why festivals are valuable events for destinations. Therefore, this study is important because of its theoretical contribution to demonstrating the contribution of festivals to destination branding.

Destination Concept

Places where tourists travel, host and perform touristic activities have been called destinations (Türkay, 2014). Destination areas can be places such as villages, cities, regions and islands where local people and tourists come together, but they are places that offer easy access to visitors, have both natural and cultural richness and create attractiveness (Tinsley & Lynch, 2001; Medlik, 2003; Uğur, Gökkaya & Acar, 2018). Accessibility, package tours, facilities, activities and auxiliary services are the main elements that make up the destination (Buhalis, 2000). Since these elements of destinations are evaluated socially and culturally by tourists, they are important in terms of destination attractiveness (Güripek, 2013).

Destinations have some marketing plans and competitive strategies. In the realization of these plans and strategies, factors such as the cultural heritage of the destination, the quality of the service provided and the quality of the personnel are taken into consideration, and it is aimed to increase the demand for the destination in this direction (Aydemir, Saylan & Aydoğmuş, 2014). Destinations make many innovations, activities or changes to create a different image from their competitors. These changes and developments are also the factors that make destinations brands. For this reason, the branding of destinations is beneficial both in gaining competitive advantage and in the marketing of touristic products (Atay & Dülgaroğlu, 2017).

Factors Affecting Destination Selection

Social, economic and cultural characteristics are the main factors that affect tourists' destination selection (Beerli & Martin, 2004). Natural resources and environmental factors of destinations play an important role in the formation of touristic demand. The features of these resources and the events organized are the features that make the destination attractive. The diversity and richness in the attractive features of the destination are effective in the choice of the destination (Güngördü, 2007; Özdemir, 2014). One of the criteria that is effective in the choice of destination is the image owned and each destination has its own unique image. After a touristic visit, a conceptual or emotional image of the destination may occur. This image perception can directly affect the destination preferences of tourists (Leiesen, 2000; Baloğlu & Mangaloğlu, 2001; Wang, 2003). Pricing policies applied by destinations are also important in their preference. In pricing, the location of the destination area and its products are taken into account. Marketing plans are made for a specific target audience in accordance with the characteristics of the destination. Thus, it is ensured that the destination choices of individuals are affected through marketing plans designed to attract the attention of individuals in the target audience (Ersun & Arslan, 2011).

The cultural attractions of the touristic destinations or the entertainment and leisure activities they develop later contribute to the quality time of the visitors. Water sports, fishing, hiking and many other activities are important in the touristic experience. Experiencing through activities makes the destination attractive for tourists and plays an important role in destination preference (Emekli, 2008; Breiby & Slatten, 2018).

The Importance of Branding

A brand is a name, term, sign, symbol, shape or combination that identifies the goods or services of a seller or group of sellers and tries to distinguish them from their competitors (Wood, 2000). Factors such as symbol, name, company, feature, expectation and perception play a role in the formation of the brand (Tokol, 1994). Symbols used in the brand provide the interaction between the customer and the product. Through the symbol, customers remember the brand, the product and the features that come to mind (Kotler, 2000).

Customer expectations are taken into account in the formation of the brand because expectations play a role in the attitude towards the brand. (Deanna & Robert, 1998; Tanses, 1999). Innovations must be followed for a successful brand (İlgüner, 2006). Strong brands arouse the desire to buy, create a sense of trust, provide recognition, increase the quality of the product or service (Ak, 2011). Some of the experiences made at the destination are remembered by the visitors. These experiences, which will remain as memorable experiences in the minds of visitors, are the elements that make the destination a strong brand. (Unal, 2001).

Branding is a planned promotion method (Connor, 2006). The main purpose is to be recognized and for this, advertising and communication features are developed. Ensuring sustainability in the brand shows the signs of branding (Deniz, 2013; Robertson, 1998). Changes in touristic destination preferences in recent years have led destination managers to branding. In order to become active in tourism, some cities have evaluated their historical, cultural and natural resources as touristic products, used them in touristic marketing and succeeded in branding by differentiation. The unique features of the branded cities are recognized by national and int ernational tourists. For this reason, branding is important for destinations to be recognized and sustainable (Yalım, 2018; Oğan & Yasak, 2020).

Brand Management and Image

Brand management is determining the brand and presenting the market in line with the targets. The brand manager is the person who differentiates the brand and establishes the relationship between the product and the customer. (Ambavale & Surtı, 2015). While determining the strategy in brand management, the innovations in the market and the creative designs of the business are taken into consideration. (Alkin, Bulu & Kaya, 2007).

As a result of individuals' own experiences, the experiences of others, and the knowledge they have acquired, a perception of image for a brand occurs (Yükselen & Güler, 2009). The effect of the destination on visitors is expressed as the destination brand image.Today, many destinations offer unique experiences and build brand image. A destination that offers unique experiences creates a positive brand image.(Özaltaş Serçek & Serçek, 2015; Özilhan Özbey & Başer, 2015).

A positive perception of the brand image can be created in the minds of consumers. For this, the products in the destination must have attractive features. Individuals who purchase the destination service should be aware of these features (Fırat & Kömürcüoğlu, 2015; Avcılar & Kara, 2015). While the ideas about the destination brand image are formed in the minds of the tourists, the elements that are attractive to the visitors are effective. These attractive factors are important in the destination image or in the marketing process of the destination (Manhas, Manrai & Manrai, 2016).

Since tourism is a global industry, competition between destinations is quite high. Vitality in tourism is important in marketing the destination effectively. For this, branding studies are carried out. Destination unique features are important for branding as they influence the emotions and behavior of tourists. Therefore, differentiation and branding are achieved through unique features.(Ruiz-Real et al. 2020; González-Mansilla et al. 2019; Giannopoulos et al. 2021; Shi et al. 2022).

Local festivals and other cultural events are important in branding. Because event experiences give tourists information about the brand. Therefore, a strong image benefits the destination's competition. It also benefits strong and positive image, sustainability and tourist demand. (Camarero, Garrido & Vicente, 2010; Kim, Moon & Choe, 2016; Kim, Choe & Petrick, 2018; Cardoso et al. 2019).

The display of events is an attractive element for tourists to visit, as it demonstrates the local culture and authenticity of the region. These events, organized to attract the attention of tourists, play a role in the branding of destinations. Therefore, such activities enable tourists to perceive the brand image (Song & Cheung, 2010; Ferns & Walls, 2012; Domínguez-Quintero et al. 2020; Zhang et al. 2021).

Factors Affecting Brand Awareness and Brand Preference

People who are regular buyers of a product are brand aware people. These people can easily distinguish the product they will buy from other brands. (Aaker, 1991; Keller, 2003). To ensure the sustainability of brand awareness, businesses work on advertising and marketing communications. In this way, it can keep it in the mind of the consumer for a long time (Clark, Doraszelski & Draganska, 2009; Homburg, Klarmann, & Schmitt, 2010). In addition, the brand enables the business to achieve its financial goals. It positively affects consumers' repurchase intentions and contributes to the sustainability of the services offered (Doğan et al. 2019). If the benefit in terms of the destination is mentioned, the brand ensures that the destinations are recognized and perceived as a quality place and increases the attractiveness of the region (Yılmaz & Çizel, 2000).

The ability to create a brand provides economic advantages to the business, gains customer satisfaction and loyalty (Lee, 2015). Customer expectations are important in demanding the created brand. Meeting expectations creates positive emotions in consumers and the business gains an advantage (Janiszewska & Insch, 2012). Through brand knowledge and brand experience, consumers make evaluations in their minds. As a result of the evaluation, some brands are deleted in the mind of the consumer, and some brands take a permanent place. Brands that remain in the minds of consumers affect brand preferences (Roberts & Lattin, 1991; Mathur, Moschis & Lee, 2003).

Social, psychological and personal characteristics of individuals are formed depending on their experiences. As a result of this effect, individuals carry out various ideas for products and services. The opinions of individuals affect the decision to buy the brand and their behavior. Therefore, individual characteristics has an effect on brand preference (Keskin & Yıldız, 2010; Sethi & Chawla, 2014; Durmaz, 2017). Behavioral patterns can be developed according to their cultural characteristics. This also plays a role in their purchasing behavior. Therefore, the effect of individuals' cultural characteristics on brand preference is also seen (Ghosh et al. 2016).

Destination Branding and Events

Branding in tourism destinations is very important in recent years. Influencing the emotions of visitors and gaining their trust is important for international branding (Crockett & Wood, 1999; Vukic, Kuzmanovic & Vukic, 2016). After the destination experience, unforgettable experiences and emotional bond remain in the minds of visitors and these experiences play a role in branding. (Akdoğan 2017; Kocaman, 2012). The contributions of the producers of local products, namely the local people, affect branding. Since branded destinations create trust in tourists, branding features are evaluated in the marketing process of destinations (Olimpia, 2007; Hankinson, 2007).

Brand perception is formed after visitors become aware of the assets in the destination. After the brand perception is formed, the visitor understands that the destination is special and important (Mandariš, 2016; Terziš & Đališ, 2019; Brzakoviš et al. 2019; Periš & Mandariš, 2020). Events are one of the destination branding elements that provide a positive perception of the destination. These events provide recognition and experience of local features (Brown et al. 2004; Horng & Tsai, 2012; Deng & Li, 2014; Wong, Wan & Qi, 2015). Destinations that do not have local and natural features gain visitors by organizing events. Because events strengthen the image of destinations, gain visitors and provide an advantage in marketing. Therefore, events are important for the destination. (Getz, 1997).

Festival events have a positive effect on destination branding (Chalip & Costa, 2005; Kim & Morrsion, 2005; Lee & Lee, 2009). Since the events affect the perception towards the destination positively, the destination managers used the events as a branding tool and benefited from the effects of the events on the positive perception of the destination (Wells, 2014; Lee et al. 2017; Kruger & Viljoen, 2021).

Some destinations preferred branding through events and engaged in differentiation studies (Getz, 2008; Capriello & Rotherham, 2011; Lorenzini, Calzati & Giudici, 2011). Activities are used to position the destinations well in the minds of the visitors. Destination managers benefit from the activities in positioning the destination, making it attractive, differentiating, influencing tourists and creating repeat intention. (O'Sullivan & Jackson, 2002). Emotional changes may occur in visitors' perceptions of the destination after the event. In case of emotional impact, the destination perception of visitors also changes (Ekinci & Hosany, 2006; Mossberg & Getz, 2006; Hernández-Mogollón, Duarte & Folgado-Fernández, 2018). As the events, which are the symbols of the destination, provide a positive perception in the global sense, the destinations benefit from the effect of the events on branding. Establishing a global connection between the visitor and the destination makes the destination advantageous in the tourism market and creates an opportunity for self-development (Knott, Fyall & Jones, 2015; Knott, Fyall & Jones, 2017).

The Effects of Festival Events on Destination Branding

Activity experiences are activities in which individuals purchase services and interact socially (Horner & Swarbrook, 2005; Wood & Masterman, 2008; Bowdin et al. 2012). Festival events contribute to the promotion of the destination, increase its attractiveness and encourage visitors to experience more (Janeczko, Mules & Ritchie, 2002; Chalip & Green, 2003). For this reason, festivals are among the important touristic activities.

Festival events add value to the economic recovery and attractiveness of destinations. Especially in gastronomy-themed festivals, the local products of the destination, which are obtained naturally and prepared with local methods, are exhibited and marketed. Many destinations have their own local products, and these products, which are exhibited through the organization of various events in different places, help the destinations to be recognized and maintain their local characteristics (Smith & Costello, 2009; Spiller, 2012). Various activities are periodically added to the tourism activities of destinations through festivals. In this way, visitor demand and earnings can be increased, the length of stay can be extended, and the image can be strengthened (Hall, 1992; McCartney & Osti, 2007). Aktürk and Akbaba (2021) stated in their study that festivals are an important step in creating a destination brand.

The purpose of festival tourism planning is to organize organizations that will allow people outside the destination area to visit the festival. Festival events are held at a certain time. Some festivals make destinations perceived attractive only during the periods they are organized and they are also effective in the increase in revenues during these periods. (Visser, 2005; Çoban & Süer, 2018). Festival events are held within the scope of certain themes. These themes can be gastronomy, music, dance, sports or cinema. Since festivals are an event that serves the development and recognition efforts of the destination, they have been an important element of branding in tourism (Çoban & Süer, 2018).

Gastronomy Festivals

Local food is an essential part of touristic experiences. Local food experiences are motivating and tourist-attracting experiences. For this reason, one of the reasons for organizing gastronomic events is to meet the local dining experience expectations of gastronomy tourists (Hjalager & Corigliano, 2000; Santich, 2004). Local food and beverages are introduced, experienced and observed through gastronomic events. Such activities are usually held in small-scale local areas because they are for local products and contribute to the development of regional tourism (Bjeljac, 2010; Stojanoviš et al. 2013). Authentic gastronomic products represent the destination and play a role in tourists recommending the destination. Thus, gastronomic events have a positive effect on developing gastronomic tourism, promoting the destination and branding the destination (Mandariš et al. 2017; Čavić & Mandarić, 2021).

Displaying and advertising local foods and allowing tourists to experience them increases the attraction of the destination and the curiosity of the participants. Food events are extremely important activities for tourists whose main purpose is to focus on the dining experience. For this reason, destinations organize planned and successful food events in order to satisfy the tourists whose main purpose is to experience food, and to attract the attention of other tourist groups (Getz, 2000; Eventbrite, 2015).

Destinations aim to market their food and other gastronomic features in the best and most successful way in order to become a brand. In line with this goal, destinations make various plans to promote their gastronomic elements. One of these plans is gastronomy-themed events, namely gastronomy festivals. Gastronomy festivals are events that create attraction for national or international tourists, add excitement to the destination experience, provide tourists with information about local foods and enable them to revisit and become loyal customers by satisfying them. Gastronomy festivals play a role in the perception of the destination as a brand in the minds of individuals. Destination managers who want to turn this perception into an opportunity also make plans to use gastronomy festivals as a branding tool (Lee & Arcodia, 2011; Sotiriadis, 2015; Stone et al. 2018; Yang et al. 2020).

With the help of local producers at gastronomic festivals, visitors expand their knowledge of various brands and increase their awareness of local foods. Brand-oriented perceptions occur in visitors with an increased awareness of local foods, so local food knowledge in gastronomic festivals is very important for branding (Sharples, 2008; Vukic et al. 2015). Food is becoming more and more important in the advertising and economy of tourist destinations. The reason for this situation is that today, food is effective in the marketing of destinations, being the most attractive product of the destination, attracting the attention of other tourists by highlighting the destination, creating destination loyalty and most importantly making the destination a brand (Kim et al. 2010; Smith et al. 2010; Williams et al. 2014; Folgado-Fernández et al. 2016). In his study on Lanesville Heritage Weekend, Feast of Hunter's Moon and Pierogi Festivals held in different states of India, Kalkstein-Silkes (2007) stated that food-related events and festivals play a role in creating a destination image and influencing the emotions of visitors, and that they also become symbols of local culture and therefore can be effective factors in the branding of destinations. Blichfeldt and Halkier (2013), in their study on the Mussel Festival held in Løgstør Town, Denmark, stated that the Mussel Festival has a positive effect on economic development as it generates tourist expenditure, successfully promotes the destination identity and has fulfilled an important function in the branding of the destination for years. Dimitrova and Yoveya (2014) stated that the Zimnitza Festival in Bulgaria increases the attractiveness of the region, is an important touristic event to attract tourists, and plays a role in the branding and sustainability of the destination.

Culture and Art Festivals

Cultural activities are activities that are organized in order to preserve the local characteristics that societies have accumulated for years and to leave a unique cultural heritage to the next generation, where the knowledge and support of the local people is much needed, and the personal activities of the local people, which are characterized as cultural, are also included (Allen, O'Toole & McDannell, 2005).

Unique cultural activities affect individuals socially and culturally, make the destination more attractive and a tourist attraction (Quinn, 2003; Chhabra, Healy & Sills, 2003). In touristic activities, activities are used to provide valuable experiences to visitors. Cultural and artistic activities are one of them (Brakus, Schmitt & Zarantonello, 2009). Culture and art festivals affect the reactions of visitors to their experiences in terms of design, communication and environment and their perception of brand experience (Brakus, Schmitt & Zarantonello, 2009; Barnes, Mattsson & Srensen, 2014) Cultural events are events that improve the destination image and create brand value (Echtner & Ritchie, 1991; Richards & Wilson, 2004; Pine & Gilmore, 1999). Culture and art festivals have a power that differentiates destinations and extends their recognition beyond city and country borders. These events keep destinations competitive, make the destination sustainable, provide advantages in marketing, brand the destination and bring it to the fore in the global tourism market (Aaker & Joachimsthaler, 2000; Kotler & Gertner, 2002; Connell & Meyer, 2015; Chen, King & Lee, 2018). For example, Piva et al. (2017) in their study on the perceptions of visitors participating in the Stresa Festival in Italy, stated that the festival increases the destination attractiveness, strengthens the destination image, creates attractiveness to attract tourists and is an important part of the destination branding efforts.

Music festivals are events that make destinations attractive, develop alternative activities, benefit the economy, and strengthen the image of the destination. In music festivals, visitors experience brand value components and create their perceptions of the festivals. (Frey, 1994; Leenders, 2010). Film festivals, which are one of the parts of the cultural accumulation of the destinations, contain traces of the traditions and customs, history and artistic understanding of the region where they are held. Tourists visiting these festivals also experience the natural and cultural resources of the destination and leave the festival with positive emotions. The branding and marketing power of destinations increases through film festivals. For example, the Golden Orange Film Festival and the Golden Boll Film Festival were effective in making Antalya and Adana a brand destination for film festivals (Crockett & Wood, 1999; Lee & Arcordia, 2011; Türten, 2021).

Sports Festivals

Sports events are events that create attractiveness and branding in the tourism market in order to promote touristic destinations in the international arena and attract tourists. Sports events such as the Sports tourism and events organized in this type of tourism play a role in the branding of destinations. For this reason, destinations that have the power of attractive sports activities benefit from these activities in their marketing activities and branding of their destinations (Chalip, 2004; Morgan, Pritchard, & Pride, 2010; Garcia, Gomez, & Molina, 2012; Mohammadi & Jazaei, 2016). In sports events, the participants take part in the event. For this reason, while planning such events, destinations make plans to draw participants into the event. While planning sports festivals, it focuses on elements such as aesthetics, entertainment and attractiveness in order to meet the expectations of the participants by acting from a strategic point of view. Festivals planned in this way create positive emotions in the minds of visitors (Howat & Murray, 2002; Richards & Wilson, 2006; Robertson, Chambers & Frew, 2007; Morgan, Elbe & Curiel, 2009).

Sports events are a motivational element for individuals who make touristic trips. Individuals who are motivated to the destination perceive the destination as a brand in terms of sports activity (Shipway & Fyall, 2012; Tichaawa, Bob & Swart, 2018; Hemmonsbey & Tichaawa, 2020). Sports events have an important place in the international branding of destinations and especially nations.

Therefore, events play an important role in the sustainability of destinations (Knott, Fyall & Jones, 2017; Van Niekerk, 2017). At the same time, sports festivals are events where societies share their sports activities with each other and interact. In this process, visitors from different countries have an idea about the destination that hosts the sports event (Paraschak, 1991). Thanks to activities such as the Rugby World Cup and Tokyo Olympic Games held in Japan, the brand value of Japan has increased (Hinch & Ito, 2018). In addition, Scandinavian countries have been positioned and branded as a ski resort, Netherlands as a speed skating center, and as a sumo wrestling destination in Japan (Rein & Shields, 2007; Knott & Hemmonsbey, 2015). All these reveal that sports festivals have a significant impact on destination branding and increasing recognition.

DISCUSSION, SOLUTIONS, AND RECOMMENDATIONS

When the literature review was examined, it was concluded that touristic destinations have various activities in terms of local festivals and entertainment organizations, that these activities create attraction for experiencing and have an effect on the branding of destinations. Branding of destinations ensures that the destination is preferred in a touristic sense. Destinations can increase their touristic demands by creating a strong brand and become a brand destination. In this study, the effects of the festivals organized in the destination on branding were examined. The investigations carried out theoretically prove the effect of previous research on the branding of festivals. As a matter of fact, it is seen that the activities create positive changes in people's behaviors and show the identity of the destination well. This situation has been effective in creating unique identities of destinations and branding. It has been seen in the researches how festivals have a great contribution to the brand identity created in the destinations, that the festivals play a positive role in the perception of both local people and foreign visitors and that the destination value of the festivals has improved. In order to be attractive, people need to be positively affected by a situation, and festivals are permanently in people's minds because they are events that entertain people and attract their emotions, and make the destination remembered through festivals. Therefore, festivals are events that have the ability to brand a place by staying in people's minds because they have attractive and impressive features. Tourist destinations also take advantage of these attractive features of festivals to add value to their branding, improve and develop their branding. In the literature section of this research, this situation has been proven with quite a few examples. In the background of the research, examples of these activities creating a brand are given a lot of space and a broad understanding of the subject is provided. This section tried to prove the impact of festivals on branding and the department achieved its goals. Therefore, sufficient background was provided in the literature review conducted to achieve the objectives of the department.

The brand image of destinations creates brand awareness in visitors and maintains brand sustainability. For this reason, brand managers in tourism enterprises can carry out studies on creating brand awareness in order to increase touristic demand and positively affect the behavior of consumers. A successful event organized can be remembered in the minds of the visitors and the awareness of the visitors to the destination can be increased.

Organized festivals are effective factors in destination marketing, sustainability of culture, increase in income from tourism, differentiation, being superior to competitors, providing attractiveness, changing the behavior of visitors and increasing the recognition of destinations. For this reason, festival events are important in destination branding and creating service differentiation.

Gastronomy festivals are food-themed events where local foods are exhibited, visitors and local people experience tastings, and many products with gastronomic meaning are exhibited. Food displayed at gastronomic festivals encourages visitors to experience the event at the destination. The pleasant moments that remain in the minds of the visitors after the experience can make them intend to come back to the same event. Therefore, destinations that constantly receive visitors and receive positive advice through gastronomy festivals can achieve their branding goals.

Culture and art festivals are events where a destination has products of artistic value such as literature, music, art and museums. Art festivals create emotional bonds with visitors. Studies have shown that culture and art festivals have an important role in destination branding. Since culture and art are valued at these festivals, the participation of individuals interested in this field can be quite high. Each region has its own characteristics in the formation of culture. In cultural festivals, visitors can be gained by making use of this originality. Since gaining visitors to the festival is effective in increasing recognition, culture and art festivals can be used to brand destinations.

Sports events are events where the sports activities of a single nation or different nations are carried out. Visitors who participate in sports events or want to take part as spectators can identify the destination where the event is held with the event. Therefore, such activities provide an opportunity for the destination to be recognized internationally after being identified with the destination. In the studies, it has been determined that sports festivals and events are important in destination branding. Therefore, destinations that have achieved recognition through national or international sporting events achieve success in branding. In the studies conducted, it has been seen that the effect of sportive activities on bringing people together in the branding of destinations and the effect of being permanent in people's minds by identifying with the destination and affecting branding. Ooi and Pedersen (2010) examined city branding and film festivals in their study and mentioned that the film festival increased the vitality of the destination, developed the destination and had a positive effect on branding. Lee and Arcordia (2011) stated in their study that food festivals can significantly support destination branding. Peter et al. (2013) stated in their study on the Dubai Shopping Festival and Sharjah Spring Promotion that these festivals play a role in branding the destination. Vukic et al. (2016) stated that gastronomy festivals held in Serbia can be effective in the branding of the destination. Amara (2017) in her study on El-Moez street festival and Abu Simbel Sun festival in Egypt, concluded that these festivals are associated with Egypt and are effective in promoting the country, visiting it as a touristic destination and branding it. Hemmonsbey and Tichaawa (2018) stated in their study on mega sports events in South Africa that these events have a positive effect on the development of the destination and provide an advantage in branding. In the study of İlban and Kömür (2019) on the Ayvalık International Olive Harvest Days food festival, they found that the festival increased the number of visitors to the destination, kept the destination alive and was effective in the branding of the destination. Sawrt et al. (2021) examined the destination branding of sustainable sports events in Dubai. In the study, they stated that these events were effective in positioning Dubai as a sports destination and that the diversity of the events, the attractiveness of tourism and the successful management of the event played a role in this effect. Aktürk and Akbaba (2021) stated in their study that local food festivals play an important role in destination branding. Toker (2022) stated in his study that Vegfest held in Didim is one of the promotional elements of the city. In the study of Yang et al. (2020), it is stated that food festivals play a role in the branding of the culinary destination. In the study of Laurell and Björne (2018), the relationship between dijitai festivals and destination brands is addressed and it is stated that these festivals have an impact on the branding of the destination. Blichfeldt and Halkier (2013) have shown in their study that a mussel festival held in a rural destination in Denmark is effective for the branding initiatives of the region. In their study Dimitrove and Yoveva (2014), they mentioned that rural festivals organized in Bulgaria can be a tool in branding. When the theoretical and practical contributions of previous studies are examined, the results of this study show similarities with the results of previous studies. As a matter of fact, in previous studies, the effect of music festivals or shopping festivals on branding was examined. In this study, the relationship of events with branding was mentioned, but only the role of culture and art, gastronomy and sports festivals from the festival types was mentioned and the targeted results were achieved. Research on different types of festivals has also shown to be effective in branding and has been proven to have generally positive results for the destination. Therefore, as can be understood from other studies, festivals have become events that have extremely important roles for the branding of destinations and have a superior task for attractiveness. In addition, as seen in the studies, the effect of festivals on branding is a subject that has been researched for decades and is still being researched, so it still maintains its popularity for the tourism area.

CONCLUSION AND FUTURE RESEARCH

As it is understood from the research results, festivals have a great impact on destination branding. Festivals have become an important part of today's destination visits and branding. Destination managers, who want to benefit from festivals or various events in the branding of their destinations, can add attractive activities to their plans to reveal their differences with festival organizations. For this, before the festival organization is planned in a region, the ideas of the festival experiencers about the previous festival can be learned, expectations and comments about the entertainment elements or local elements in the festival can be taken into account. In line with the tastes and expectations of the festival visitors, the elements that make the festival attractive can be increased or the presence of local elements can be strengthened by increasing the support of the local people. Within these, destination managers should follow up-to-date information about festivals that have achieved branding in international platforms. If the local people of the destination which is organized by the festivals are made aware of the effects of the festival atmosphere on the visitors, the festival environment can be organized more creatively. If the destination managers have detailed information about the modern promotion methods in online platforms in the advertisement of the festivals, by increasing the their effectiveness and memorability, the festivals role in branding the destination they are organized can be increased. The fact that the festival organizers can make the festival more attractive and increase its impressiveness can contribute to increasing the impact of the festivals by increasing the social capital made to the festival. With the help of experts who have theoretical knowledge in the organizations held at festivals, the emotional associations of the organizations in the individuals can be increased. Different plans can be made to influence the visitors who attend the festivals individually or as a group, as a matter of fact, the activities with the group can be talked about more and group psychology can be more effective. By obtaining preliminary information from the individuals participating in the festival, attractive organizations can be arranged within the framework of the festival theme according to their diseases or interests, if any. Experts or guides can be present alongside the vendors at the festival to satisfy visitors' curiosity and provide them with more information during the experience at the festival site. Special guides, they can offer foreign tourists more extensive and enlightening information about local products. At this point, if expert guides are used in the festival area, it is thought that more can be achieved in the recognition and branding of the festival area. In addition, presenting or packaging products in a quality way at the time of the festival can also play a role in branding. Research on different aspects of this issue in this area is still ongoing today. Therefore, it is thought that these suggestions can be addressed theoretically or practically in future studies.

In tourism, businesses make various investments in order to earn a certain income, to pay the wages of the employees and to carry out the activities they have planned. A certain level of income is expected as a return on their investments afterwards. With this earned income, the expenses incurred are paid and a certain gain remains for the enterprise. This situation actually shows the balance of payments of the enterprises. In other words, businesses can earn as much or more than the investment they have made (Karagöz, 2011). Investments in activities, which are an important part of generating income in tourism, are also important. When the activities are planned correctly and financial assistance is provided to an event made with good planning, good income can be obtained (Korkmaz, 2001). As it can be understood from here, festivals, which are a touristic event, can also provide qualified income if the right financial assistance is provided. It is very important that the financial support given to the festivals is managed correctly by the destination managers, because the investment made in a quality planned festival can provide a lot of positive returns and contribute to the acceleration of branding efforts. Good financial investments for a large, impressive and attractive festival organization can contribute to branding efforts, so donations to festivals can also be mentioned in future studies. In practice, if festivals are organized with the help of experts on what level of return the investments to be made in festivals will be, more contribution can be made to branding. Therefore, it is thought that a good investment in festivals for the purpose of destination branding can also be effective.

When the studies on festivals are examined, this issue is still a part of many researches and maintains its currency. The effects of festivals on the destination have been included in many studies from the past to the present. Today, these effects are still being studied and related to branding. are discussed with samples of different destinations (Gligorijević, 2022; Kádár, & Klaniczay, 2022; Hutabarat, 2022; Rahman et all. 2022). Therefore, this study remains up-to-date. Festivals play a very effective role in the development and recognition of destinations. This situation brands the destinations and codes the destinations as a brand place in the perception of individuals. In the literature, there is a wide range of examples related to the subject. In the literature review the effects of festival types on destination branding discussed in detail and these effects are illustrated with examples and resources. Therefore, the study includes a strong and detailed research in terms of examining how festivals brand a region.

In this study, a large literature review on the effect of festivals on destination branding, examples are given on the effects of festival types in detail. In previous studies, it has not been examined whether the gender of the destination has an effect on the branding of the festivals, and this includes a deficiency in the literature. In addition, in this study, the effect of the economic, cultural and local people characteristics of the destinations on the effect of the festivals on the destination and the mediating role of the value perception of the local people were not discussed, so the scope of the research only discussed branding in terms of festival types. In addition, since this study does not focus on the role of market and marketing factor, purchase intention, destination diversity, local people profile, personality characteristics, capital elements of the destination, perceived value of visitors and general image of the destination in the brand building of the festivals, this section contains deficiencies in terms of these issues. In addition, the role of the demographic characteristics of the festival participants in this branding was not mentioned. The fact that projects aimed at community-based tourism understanding play a role in festivals may have a role in the branding of destinations, but this has not been mentioned in the literature and may be mentioned in future research. In future research, the moderator effect of visitors' product knowledge and hedonic well-being in the festival area on the effect of festivals on destination branding can be examined.

REFERENCES

Aaker, E., & Joachimsthaler, E. (2000). The brand relationship spectrum: The key to the brand architecture challenge. *California Management Review*, *42*(4), 8–23. doi:10.1177/000812560004200401

Aakers, J. (1991). The negative attraction effect? A study of the attraction effect under judgment and choice. *Advances in Consumer Research. Association for Consumer Research (U. S.)*, *18*, 462–469.

Ak, M. (2011). Marka yönetimi ve imaj. (1. Baskı), İstanbul: Kültür Sanat Basım Evi.

Akdoğan, Ç. (2017). Destinasyon markalama bağlamında yavaş şehir (cittaslow) hareketi üzerine bir değerlendirme. *Uluslararası Bilimsel Araştırmalar Dergisi*, 379-390.

Akturan, O., & Oğuztimur, S. (2016). Kent markalaşması kavramının içeriği ve gelişimi: Farklı disiplinler farklı yaklaşımlar. *Planlama Dergisi*, *26*(2), 119.

Akturk, H., & Akbaba, A. (2021). Local food festivals within the scope of destination branding. In C. Cobanoglu, & V. Della Corte (Ed.), Advances in global services and retail management (pp. 1–9). USF M3 Publishing.

Alkin, K., Bulu, M., & Kaya, H. (2007). İller arası rekabet endeksi: Türkiye'deki illerin rekabetçilik seviyelerinin göreceli olarak ölçülebilmesi için bir yaklaşım. *İstanbul Ticaret Üniversitesi. Sosyal Bilimler Dergisi*, *11*, 221–235.

Allen, J., O'Toole, W., & McDonnell, I. (2005). *Festival and special event management* (3rd ed.). John Wiley and Sons Inc.

Amara, D. (2017). Events tourism: a potential to build a tourist destination branding: the case of cultural events in Egypt. *International Journal of Heritage. Tourism and Hospitality*, *11*(2), 1–12.

Ambavale, R., & Surtı, N. (2015). Is brand everythıng? Issues and challenges of brandıng in global market. *International Journal of Application or Innovation in Engineering & Management*, *4*(1), 14–18.

Atay, L., & Dülgaroğlu, O. (2017). Markalaşma açısından Cunda Adası'nın destinasyon kişiliğinin tespitine yönelik bir çalışma. *Balıkesir Üniversitesi Sosyal Bilimler Enstitüsü Dergisi*, *20*(38), 383–396. doi:10.31795/baunsobed.645146

Avcılar, M. Y., & Kara, E. (2015). Şehir markası kavramı ve marka şehir yaratma stratejilerine yönelik literatür incelemesi. *Sosyal ve Beşeri Bilimler Araştırmaları Dergisi*, *34*, 76–94.

Aydemir, B., Saylan, U., & Aydoğmuş, F. (2014). Turizmde rekabet: seyahat ve turizm (T&T) rekabet edebilirlik raporunda Türkiye'nin ve Avrupa'nın değerlendirilmesi . *Çukurova Üniversitesi İktisadi ve İdari Bilimler Fakültesi Dergisi, 18* (1), 1-15.

Bagaeen, S. (2007). Brand Dubai: The Instant City; or The Instantly Recognizable City. *International Planning Studies*, *12*(2), 173–197. doi:10.1080/13563470701486372

Baloğlu, Ş., & Mangaloğlu, M. (2001). Tourism destination images of Turkey, Egypt, Greece and Italy As Perceived by US-based tour operators and travel agents. *Tourism Management*, *22*(1), 2. doi:10.1016/S0261-5177(00)00030-3

Barnes, S. J., Mattsson, J., & Sørensen, F. (2014). Destination brand experience and visitor behavior: Testing a scale in the tourism context. *Annals of Tourism Research*, *48*, 121–139. doi:10.1016/j.annals.2014.06.002

Beerli, A., & Martin, D. (2004). Tourist Characteristics ans The Perceived İmage of Destinations: A Quantitative Analysis- a case study of Lanzorate, Spain. *Tourism Management*, *25*(5), 623–636. doi:10.1016/j.tourman.2003.06.004

Bjeljac, Ţ. (2010). *Turističke manifestacije u Srbiji* [*Tourist events in Serbia*]. Serbian Academy of Sciences and Arts.

Blichfeldt, B. S., & Halkier, H. (2013). Mussels, tourism and community development: A case study of place branding through food festivals in rural North Jutland, Denmark. *European Planning Studies*, *22*(8), 1587–1603. doi:10.1080/09654313.2013.784594

Bowdin, G. A. J., Allen, J., O'Toole, W., Harris, R., & McDonnell, I. (2012). *Events Management* (2nd ed.). Butterworth-Heinemann. doi:10.4324/9780080964317

Brakus, J. J., Schmitt, B. H., & Zarantonello, L. (2009). Brand experience: What is it? How is it measured? Does it affect loyalty? *Journal of Marketing*, *73*(3), 52–68. doi:10.1509/jmkg.73.3.052

Breiby, M. A., & Slatten, T. (2018). The role of aesthetic experiential qualities for tourist satisfaction and loyalty. *International Journal of Culture, Tourism and Hospitality Research*, *12*(1), 1–14. doi:10.1108/IJCTHR-07-2017-0082

Brown, G., & Chalip, L. Jago. L., & T. Mules, (2004). Developing Brand Australia: Examining the Role of Events. In N. Morgan, A. Pritchard, and R. Pride (Ed.), Destination Branding: Creating the Unique Destination Proposition (Amsterdam: Elsevier), (pp.279–305). A. Oxford, UK: Butterworth-Heinemann.

Brzakoviš, A., Brzakoviš, T., & Brzakoviš, P. (2019). Creating brand market value. *The Annals of the Faculty of Economics in Subotica*, *55*(42), 17–33.

Camarero, C., Garrido, M., & Vicente, E. (2010). Components of art exhibition equity for internal and external visitors. *Tourism Management*, *31*(4), 495–504. doi:10.1016/j.tourman.2009.05.011

Capriello, A., & Rotherham, I. D. (2011). Building a preliminary model of event management for rural communities. *Journal of Hospitality Marketing & Management*, *20*(3-4), 246–264. doi:10.1080/19368623.2011.562411

Cardoso, L., Dias, F., de Araújo, A. F., & Marques, M. I. A. (2019). A destination imagery processing model: Structural differences between dream and favourite destinations. *Annals of Tourism Research*, *74*, 81–94. doi:10.1016/j.annals.2018.11.001

Čavić, S., & Mandarić, M. (2021). Authenticity of gastronomic events as a function of branding a destination – *Hotel and Tourism Management,9*(1), 89-101

Chalip, L. (2004). Beyond impact: a general model for sport event leverage. In B. W. Ritchie & D. Adair (Eds.), *Sport tourism: interrelationships, impacts and issues* (pp. 226–252). Channel View Publications. doi:10.21832/9781873150672-014

Chalip, L., & Costa, C. (2005). Sport event tourism and the destination brand: Towards a general theory. *Sport in Society*, *8*(2), 218–237. doi:10.1080/17430430500108579

Chalip, L., Green, B. C., & Hill, B. (2003). Effects of sport event media on destination image and intention to visit. *Journal of Sport Management*, *17*(3), 214–234. doi:10.1123/jsm.17.3.214

Chen, Y. C., King, B., & Lee, H. (2018). Experiencing the destination brand: Behavioral intentions of arts festival tourists. *Journal of Destination Marketing & Management, 10*, 61–67. doi:10.1016/j.jdmm.2018.06.004

Chhabra, D., Erin, S., & Frederick, W. (2003). The significance of festivals to rural economies: Estimating the economic effects of Scottish Highland Games in North Carolina. *Journal of Travel Research, 41*(4), 421–427. doi:10.1177/0047287503041004012

Çiçek, E., & Pala, U. (2017). Destinasyon markalaşması üzerine bir araştırma: Tarsus örneği. *Çağ Üniversitesi Sosyal Bilimler Dergisi, 14*(2), 25-47.

Clark, C. R., Doraszelski, U., & Draganska, M. (2009). The effect of advertising on brand awareness and perceived quality: An empirical investigation using panel data. *Quantitative Marketing and Economics, 7*(2), 207–236. doi:10.100711129-009-9066-z

Çoban, M., & Süer, S. (2018). Destinasyon markalaşmasında festival turizminin rolü: Alaçatı Ot Festivali üzerine bir araştırma. *International Journal of Contemporary Tourism Research, 2*(1), 58–67.

Connell, J., & Meyer, D. (2015). Visitor attractions and events: Responding to seasonality. *Tourism Management, 46*, 283–298. doi:10.1016/j.tourman.2014.06.013

Connor, S. M. (2006). Food-related advertising on preschool television: Building brand recognition in young viewers. *Pediatrics, 118*(4), 1478–1485. doi:10.1542/peds.2005-2837 PMID:17015538

Crockett, S. R., & Wood, L. (1999). Brand western australia: A totally ıntegrated approach to destination branding. *Journal of Vacation Marketing, 5*(2), 276–289. doi:10.1177/135676679900500307

Deanna, J. R., & Robert, A. F. (1998). Overview and perspectives. *The ecology of industry: sectors and linkages*, 1.

Deng, Q., & Li, M. (2014). A model of event–destination image transfer. *Journal of Travel Research, 53*(1), 69–82. doi:10.1177/0047287513491331

Deniz, E. (2013). *Markalaşma ve reklam.* İstanbul: Kumsaati Yayın Dağıtım LTD. ŞTİ.

Dimitrova, M., & Yoveva, I. (2014). *Food and wine events in Europe - a stakeholder approach.* Routledge.

Doğan, M., Aksu, M., Çelik, Özge., & Kaymak, H. (2019). Cittaslow markasının turizme etkisi: Seferihisar'daki konaklama işletmeleri yöneticileri üzerine bir araştırma. *International Journal of Social and Economic Sciences, 4*(2), 06–11.

Domínguez-Quintero, M., González-Rodríguez, M. R., & Paddison, B. (2020). The mediating role of experience quality on authenticity and satisfaction in the context of cultural-heritage tourism. *Current Issues in Tourism, 23*(2), 248–260. doi:10.1080/13683500.2018.1502261

Durmaz, Y. (2017). The impact of psychological factors on consumer buying behavior and an empirical application in Turkey. *Asian Social Science, 10*(6), 194–204.

Echtner, C., & Ritchie, J. R. B. (1991). The meaning and measurement of destination image. *Journal of Tourism Studies, 2*(2), 2–12.

Ekinci, Y., & Hosany, S. (2006). Destination personality: An application of brand personality to tourism destinations. *Journal of Travel Research, 45*(2), 127–139. doi:10.1177/0047287506291603

Emekli, G. (2008). *Sürdürülebilir turizmde çeşitlendirmenin önemi: Çeşme-İzmir*. V. Ulusal Coğrafya Sempozyumu Bildiriler Kitabı.

Ersun, N., & Arslan, K. (2011). Turizmde destinasyon seçimini etkileyen temel unsurlar ve pazarlama stratejileri. *Marmara Üniversitesi İktisadi ve İdari Bilimler Dergisi, 31*(2), 229–248.

Eventbrite. (2015). *Millennials and the new era of food, wine and beer festival*. Eventbrite. www.eventbrite.com/academy/millennials-and-the-new-era-of-food-wine-beer-festivals/

Fathabadi, H., Nejad, M. R. O., & Alizadeh, H. (2017). An investigation of the affecting tourism destination brand equity. *Asian Journal of Social Sciences & Humanities, 6*(1), 101–113.

Ferns, B. H., & Walls, A. (2012). Enduring travel involvement, destination brand equity, and travelers' visit intentions: A structural model analysis. *Journal of Destination Marketing & Management, 1*(1–2), 27–35. doi:10.1016/j.jdmm.2012.07.002

Fırat, A., & Kömürcüoğlu, F. (2015). Muğla şehir markası ve imaj algısı: Muğla Sıtkı Koçman Üniversitesi öğrencileri üzerine bir alan araştırması. *Yönetim Bilimleri Dergisi, 13*(26), 285–304.

Folgado-Fernández, J., Hernández-Mogollón, J., & Duarte, P. (2016). Destination image and loyalty development: The impact of tourists' food experiences at gastronomic events. *Scandinavian Journal of Hospitality and Tourism*, 1–19.

Frey, B. S. (1994). The economics of music festivals. *Journal of Cultural Economics, 18*(1), 29–39. doi:10.1007/BF01207151

Fyall, A., Garrod, B., & Wang, Y. (2012). Destination collaboration: A critical review of theoretical approaches to a multi-dimensional phenomenon. *Journal of Destination Marketing & Management, 1*(1-2), 10–26. doi:10.1016/j.jdmm.2012.10.002

Garcia, J. A., Gomez, M., & Molina, A. (2012). A destination branding model: An empirical analysis based on stakeholders. *Tourism Management, 3*(3), 646–666. doi:10.1016/j.tourman.2011.07.006

Getz, D. (1997). *Event management and event tourism*. Van Nostrand Reinhold.

Getz, D. (2000). *Explore wine tourism: management, development & destinations*. Cognizant Communication Corporation.

Getz, D. (2008). Event tourism: Definition, evolution, and research. *Tourism Management, 29*(3), 403–428. doi:10.1016/j.tourman.2007.07.017

Ghosh, M. M., Ghosh, A., & Ghosh, M. (2016). Impact and influence of culture on brands in Indian market. *IOSR Journal of Business and Management, 18*(7), 54–60.

Giannopoulos, A., Piha, L., & Skourtis, G. (2021). Destination branding and co-creation: A service ecosystem perspective. *Journal of Product and Brand Management, 30*(1), 148–166. doi:10.1108/JPBM-08-2019-2504

Gligorijević, J. (2022). Practices and ideological effects of music festival branding: The case study of Serbia's two major music festivals, Exit and Gu č a. In *Musik & Marken* (pp. 117–144). Springer Fachmedien Wiesbaden. doi:10.1007/978-3-658-36472-4_6

González-Mansilla, Ó., Berenguer-Contri, G., & Serra-Cantallops, A. (2019). The impact of value co-creation on hotel brand equity and customer satisfaction. *Tourism Management, 75*, 51–65. doi:10.1016/j.tourman.2019.04.024

Güngördü, E. (2007). Türkiye'nin turizm coğrafyası. Ankara. Ankara: Asil Yayın Dağıtım.

Güripek, E. (2013). *Turizm destinasyonlarının rekabet gücünün artırılmasında stratejik destinasyon yönetimi: Çeşme Alaçatı destinasyonu üzerine bir uygulama.* Doktora Tezi, Dokuz Eylül Üniversitesi, Sosyal Bilimler Enstitüsü.

Hall, C. M. (1992). *Hallmark tourist events: impacts, management & planning*. Belhaven Press.

Hankinson, G. (2007). The management of destination brands: Five guiding principles based on recent developments in corporate branding theory. *Journal of Brand Management, 14*(3), 240–254. doi:10.1057/palgrave.bm.2550065

Hemmonsbey, J., & Tichaawa, T. M. (2018). The effects of major sport event leveraging for tourism and destination branding: The case of South Africa as an emerging destination. *Geo Journal of Tourism and Geosites, 11*(1), 213–225.

Hemmonsbey, J., & Tichaawa, T. M. (2020). Brand messages that influence the sport tourism experience: The case of South Africa. *Journal of Sport & Tourism, 24*(3), 177–194. doi:10.1080/14775085.2020.1822200

Hernández-Mogollón, J. M., Duarte, P. A., & Folgado-Fernández, J. A. (2018). The contribution of cultural events to the formation of the cognitive and affective images of a tourist destination. *Journal of Destination Marketing & Management, 8*, 170–178. doi:10.1016/j.jdmm.2017.03.004

Hinch, T., & Ito, E. (2018). Sustainable sport tourism in Japan. *Tourism Planning & Development, 15*(1), 96–101. doi:10.1080/21568316.2017.1313773

Hjalager, A. M., & Corigliano, M. A. (2000). Food for tourists-Determinants of an image. *International Journal of Tourism Research, 2*(4), 281–293. doi:10.1002/1522-1970(200007/08)2:4<281::AID-JTR228>3.0.CO;2-Y

Homburg, C., Klarmann, M., & Schmitt, C. (2010). Brand Awareness in Business Markets: When is It Related to Firm Performance? *International Journal of Research in Marketing, 27*(3), 201–212. doi:10.1016/j.ijresmar.2010.03.004

Horner, S., & John, S. (2005). Leasure marketing a global perspective, İngiltere: Elsevier.

Horng, J. S., & Tsai, C. T. (2012). Constructing indicators of culinary tourism strategy: An application of resource-based theory. *Journal of Travel & Tourism Marketing, 29*(8), 796–816. doi:10.1080/10548408.2012.730945

Howat, G., & Murray, D. (2002). The role of critical incidents to complement service quality information for a sports and leisure centre. *European Sport Management Quarterly*, *2*(1), 23–46. doi:10.1080/16184740208721910

Hutabarat, P. M. (2022). Music tourism potentials in Indonesia: Music festivals and their roles in city branding. *Journal of Indonesian Tourism and Policy Studies*, *7*(1).

İlban, M. O., & Kömür, T. (2019). Destinasyon markalaşmasında festival turizminin rolü: Ayvalık Uluslararası Zeytin Hasat Günleri Örneği. *Seyahat ve Otel İşletmeciliği Dergisi*, *16*(2), 274–295.

İlgüner, M. (2006). *Türkiye 'de marka yaratma ve yaratmanın altın kuralları*. Rota Yayınları.

Jackson, L. (2008). Residents' perceptions of the impacts of special event tourism. *Journal of Place Management and Development*, *1*(3), 240–255. doi:10.1108/17538330810911244

Janeczko, B., Trevor M., & Brent, R. (2002). *Estimating the economc impacts of festivals and events: A Research Guide*. National Library of Australia Cataloguing in Publication Data, Research Report.

Janiszewska, K., & Insch, A. (2012). Strategic importance of brand positioning in the place brand concept: Elements, structure and application capabilities. *Journal of International Students*, *5*(1), 9–19. doi:10.14254/2071-8330.2012/5-1/2

Kádár, B., & Klaniczay, J. (2022). Branding built heritage through cultural urban festivals: An instagram analysis related to sustainable co-creation, in Budapest. *Sustainability*, *14*(9), 5020. doi:10.3390u14095020

Kalkstein-Silkes, C. A. (2007). *Food and food-related festivals in rural destination branding*. [Unpublished doctoral dissertation, Purdue University, Indiana].

Karagöz, D. (2011). Etkinlik turizmi ve etkinlik turizmi bağlamında yabancı ziyaretçi harcamalarının ekonomiye etkisi: Formula 1 2005 Türkiye Grand Prix örneği [Doctoral dissertation, Anadolu University, Turkey].

Keller, K. (1998). *Strategic brand management: Building. Measuring and Managing Brand Equity*. Prentice Hall Inc.

Keller, K. L. (2003). Understanding brands, branding and brand equity. *Interactive Marketing*, *5*(1), 7–20. doi:10.1057/palgrave.im.4340213

Keskin, H. D., & Yıldız, S. (2010). Tüketicilerin marka tercihlerinde etkili olan faktörler ile marka imajının marka değeri üzerindeki etkileri: Trabzon örneği. *Dumlupınar Üniversitesi Sosyal Bilimler Dergisi*, *27*, 1–13.

Kim, S., Moon, J., & Choe, J. (2016). Comparison of destination brand equity models of competitive convention cities in East Asia. *Journal of Convention & Event Tourism*, *17*(4), 318–342. doi:10.1080/15470148.2016.1151394

Kim, S. S., Choe, J. Y. J., & Petrick, J. F. (2018). The effect of celebrity on brand awareness, perceived quality, brand image, brand loyalty, and destination attachment to a literary festival. *Journal of Destination Marketing & Management*, *9*, 320–329. doi:10.1016/j.jdmm.2018.03.006

Kim, S. S., & Morrsion, A. M. (2005). Change of images of South Korea among foreigntourists after the 2002 FIFA World Cup. *Tourism Management, 26*(2), 233–247. doi:10.1016/j.tourman.2003.11.003

Kim, Y., Suh, B., & Eves, A. (2010). The relationships between food-related personality traits, satisfaction, and loyalty among visitors attending food events and festivals. *International Journal of Hospitality Management, 29*(2), 216–226. doi:10.1016/j.ijhm.2009.10.015

Knott, B., Fyall, A., & Jones, I. (2015). The nation branding opportunities provided by a sport mega-event: South Africa and the 2010 FIFA World Cup. *Journal of Destination Marketing & Management, 4*(1), 46–56. doi:10.1016/j.jdmm.2014.09.001

Knott, B., Fyall, A., & Jones, I. (2017). Sport mega-events and nation branding: Unique characteristics of the 2010 FIFA World Cup, South Africa. *International Journal of Contemporary Hospitality Management, 29*(3), 900–923. doi:10.1108/IJCHM-09-2015-0523

Knott, B., & Hemmonsbey, J. (2015). The strategic value of sport for an African city brand. *African Journal for Physical, Health Education, Recreation, and Dance, 21*, 191–204.

Kocaman, S. (2012). *Destinasyon yönetimi kapsamında marka kimliğine etki eden faktörlerin marka imajına etkisi: Alanya Örneği.* Yayınlanmamış Doktora Tezi, Akdeniz Üniversitesi.

Korkmaz, M. (2001). Orman kaynaklarına doğa turizmi etkinliklerinin ekonomik çözümlemeleri. *Turkish Journal of Forestry, 2*(1), 111–134.

Kotler, P., & Gertner, D. (2002). Country as brand, product, and beyond: A place marketing and brand management perspective. *Journal of Brand Management, 9*(4), 249–261. doi:10.1057/palgrave.bm.2540076

Kruger, M., & Viljoen, A. (2021). Terroir wine festival visitors: Uncorking the origin of behavioural intentions. *Current Issues in Tourism, 24*(5), 1–21. doi:10.1080/13683500.2019.1667310

Laurell, C., & Björner, E. (2018). Digital festival engagement: On the interplay between festivals, place brands, and social media. *Event Management, 22*(4), 527–540. doi:10.3727/152599518X15300559276930

Lee, I., & Arcodia, C. (2011). The role of regional food festivals for destination branding. *International Journal of Tourism Research, 13*(4), 355–367. doi:10.1002/jtr.852

Lee, I., & Lee, T. J. (2009). The success factors of using events in destination branding: A case study of Woodford Folk Festival. Paper presented at the *8th Asian Pacific forum for graduate students research in tourism,* Seoul, Korea.

Lee, J. W. (2015). Unpacking brand management superiority: Examining the Interplay among brand orientation and market-linking and controlling mechanisms. [Master Thesis, Tasmanian School of Business and Economics].

Lee, W., Sung, H., Suh, E., & Zhao, J. (2017). The effects of festival attendees' experiential values and satisfaction on re-visit intention to the destination: The case of a food and wine festival. *International Journal of Contemporary Hospitality Management, 29*(3), 1005–1027. doi:10.1108/IJCHM-10-2015-0559

Leenders, M. A. A. M. (2010). The relative importance of the brand of music festivals: A customer equity perspective. *Journal of Strategic Marketing, 18*(4), 291–301. doi:10.1080/09652541003768061

Leiesen, B. (2000). Image sementation: The case of a tourism destination. *Journal of Services Marketing*, *15*(9), 51.

Lopez, C., & Leenders, M. A. A. M. (2018). Building a local identity through sellout crowds: The impact of brand popularity, brand similarity, and brand diversity of music festivals. *Journal of Strategic Marketing*, 1–16.

Lorenzini, E., Calzati, V., & Giudici, P. (2011). Territorial brands for tourism development. *Annals of Tourism Research*, *38*(2), 540–560. doi:10.1016/j.annals.2010.10.008

Mandariš, M. (2016). Strategijski brend menadžment [Strategic Brand Management]. Vrnjaţka Banja: University of Kragujevac, Faculty of Hotel Management and Tourism in Vrnjaţka Banja.

Mandariš, M., Miliševiš, S., & Sekuliš, D. (2017). Traditional values in the function of promotion of Šumadija and Pomoravlje as rural tourism destinations. *Economics of Agriculture*, *64*(2), 787–803.

Manhas, P. S., Manrai, L. A., & Manrai, A. K. (2016). Role of tourist destination development in building its brand image: A conceptual model. *Journal of Economics, Finance and Administrative Science*, *21*(40), 25–29. doi:10.1016/j.jefas.2016.01.001

Mathur, A., Moschis, G. P., & Lee, E. (2003). Life events and brand preference changes. *Journal of Consumer Behaviour*, *3*(2), 2. doi:10.1002/cb.128

Mccartney, G., & Osti, L. (2007). From Cultural Events to Sport Events: A Case Study of Cultural Authenticity in the Dragon Boat Races. *Journal of Sport & Tourism*, *12*(1), 25–40. doi:10.1080/14775080701496750

Medlik, S. (2003). Dictionary of travel. Tourism and Hospitality (3. Baskı). Oxford: Butterworth-Heinemann Publications.

Mohammadi, A., & Jazaei, M. (2016). Reviewing the sport tourism, push and pull factors in Zanjar province, Iran. *International Journal of Research in Social Sciences and Humanities*, *6*(3), 2249–4671.

Morgan, M., Elbe, J., & Curiel, E. (2009). Has the experience economy arrived? The views of destination managers in three visitor-dependent areas. *International Journal of Tourism Research*, *11*(2), 201–216. doi:10.1002/jtr.719

Morgan, N., Pritchard, A., & Pride, R. (2004). *Destination branding: Creating the unique destination proposition*. Elsevier Butterworth-Heinemann.

Morgan, N. J., Pritchard, A., & Pride, R. (2010). Destination branding. 2ndrev. (Ed.), Oxford: Elsevier.

Mossberg, L., & Getz, D. (2006). Stakeholder influences on the ownership and management of festival brands. *Scandinavian Journal of Hospitality and Tourism*, *6*(4), 308–326. doi:10.1080/15022250601003273

Oğan, O., & Yasak, Ü. (2020). Küreselleşme bağlamında mekansal kent kimliği ve markalaşmanın kent turizmine etkisi. *Türk Coğrafya Dergisi*, (74), 97–105.

Olimpia, B. (2007). Variables of the image of tourist destination. *Annals of the University of Oradea. Economic Science Series*, *7*(2), 559–564.

Ooi, C. S., & Pedersen, J. (2010). City branding and film festivals: Re-evaluating stakeholder's relations. *Place Branding and Public Diplomacy*, *6*(4), 316–332. doi:10.1057/pb.2010.30

Özaltaş Serçek, G., & Serçek, S. (2015). Destinasyon markalaşmasında gastronomi turizmin yeri ve önemi. *Journal Of Tourism Theory And Research*, *1*(1), 22–40. doi:10.24288/jttr.202819

Özdemir, G. (2014). Destinasyon yönetimi ve pazarlaması (1.Baskı). Ankara:Detay Yayıncılık.

Özilhan Özbey, D., & Başer, H. H. (2015). Kent marka imajı oluşumunda etkili unsurların önem ve yeterlilik derecelerinin Konya açısından incelenmesi. *Uluslararası Sosyal Araştırmalar Dergisi*, *8*(39), 923–932.

Paraschak, V. (1991). Sport festivals and race relations in the Northwest Territories of Canada. In G. Jarvie (Ed.), *Sport, racism, and ethnicity*. Falmer Press.

Periš, G., & Mandariš, M. (2020). Challenges in tourist destination branding in Serbia: The case of Prolom Banja. *Ekonomika Preduzeća*, *68*(5-6), 442–456. doi:10.5937/EKOPRE2006442P

Peter, S., Anandkumar, V., & Peter, S. (2013). Role of shopping festivals in destination branding: A tale of two shopping festivals in the United Arab Emirates. *Anatolia*, *24*(2), 264–267. doi:10.1080/130329 17.2013.799077

Pike, S. (2005). Tourism Destination Branding Complexity. *Journal of Product and Brand Management*, *14*(4), 258–259. doi:10.1108/10610420510609267

Pike, S., & Bianchi, C. (2016). Destination branding equity for Australia: Testing a model of CBBE in short – hauland long –haul markets. *Journal of Hospitality & Tourism Research (Washington, D.C.)*, *40*(1), 114–134. doi:10.1177/1096348013491604

Pine, B. J., & Gilmore, J. H. (1999). *The experience economy*. Harvard Business School Press.

Piva, L., Cerutti, S., Prats, L., & Raj, R. (2017). Enhancing brand image through events and cultural festivals: The perspective of the Stresa Festival's visitors. *Journal of Tourism. Culture and Territorial Development*, *8*(15), 99–116.

Quinn, B. (2003). Symbols, practices and myth-making: Cultural perspectives on the Wexford Festival Opera. *Tourism Geographies*, *5*(3), 329–349. doi:10.1080/14616680309710

Rahman, M. K., Rana, M. S., Akter, S., & Hassan, A. (2022). Technology innovation as a destination branding tool for festivals and events tourism. In *Technology Application in Tourism Fairs, Festivals and Events in Asia* (pp. 135–147). Springer Singapore. doi:10.1007/978-981-16-8070-0_8

Rein, I., & Shields, B. (2007). Place branding sports: Strategies for differentiating emerging, transitional, negatively viewed and newly industrialised nations. *Journal of Place Branding and Public Diplomacy*, *3*(1), 73–85. doi:10.1057/palgrave.pb.6000049

Richards, G., & Wilson, J. (2004). The impact of cultural events on city image: Rotterdam, cultural capital of Europe 2001. *Urban Studies (Edinburgh, Scotland)*, *41*(10), 1931–1951. doi:10.1080/0042098042000256323

Richards, G., & Wilson, J. (2006). Developing creativity in tourist experiences: A solution to the serial reproduction of culture? *Tourism Management*, *27*(6), 1209–1223. doi:10.1016/j.tourman.2005.06.002

Ritchie, B. J. R., & Ritchie, J. R. R. (1998). The branding of tourism destinations. *Annual Congress Of The International Association Of Scientific Experts in Tourism*, 1- 31.

Roberts, J. H., & Lattin, J. M. (1991). Development and testing of a model of consideration set composition. *JMR, Journal of Marketing Research*, *28*(4), 429–440. doi:10.1177/002224379102800405

Robertson, M., Chambers, D., & Frew, E. (2007). Events and festivals: Current trends and issues. *Managing Leisure*, *12*(2–3), 99–101. doi:10.1080/13606710701406485

Ruiz-Real, J. L., Uribe-Toril, J., & Gázquez-Abad, J. C. (2020). Destination branding: Opportunities and new challenges. *Journal of Destination Marketing & Management*, *17*, 100453. doi:10.1016/j.jdmm.2020.100453

Santich, B. (2004). The study of gastronomy and its relevance to hospitality education and training. *International Journal of Hospitality Management*, *23*(1), 15–24. doi:10.1016/S0278-4319(03)00069-0

Sethi, I., & Chawla, A. (2014). Influence of cultural, social and marketing factors on the buying behavior of telecom users: A comparative study of rural, semi-urban and urban areas in and around Chandigarh. *Journal of Marketing Management*, *2*(1), 97–110.

Sharples, L. (2008). Apples, cider and celebration. In C. M. Hall & L. Sharples (Eds.), *Food and Wine Festivals and Events Around the World* (pp. 134–145). Butterworth-Heinemann. doi:10.1016/B978-0-7506-8380-7.00009-9

Shi, H., Liu, Y., Kumail, T., & Pan, L. (2022). Tourism destination brand equity, brand authenticity and revisit intention: The mediating role of tourist satisfaction and the moderating role of destination familiarity. *Tourism Review*, *77*(3), 751–779. doi:10.1108/TR-08-2021-0371

Shipway, R., & Fyall, A. (2012). *International sport events: Impacts, experience and Identities*. Routledge.

Smith, S., & Costello, C. (2009). Culinary Tourism: Satisfaction With a Culinary Event Utilizing Importance-performance Grid Analysis. *Journal of Vacation Marketing*, *15*(2), 99–110. doi:10.1177/1356766708100818

Smith, S., Costello, C., & Muenchen, R. (2010). Influence of push and pull motivations on satisfaction and behavioral intentions within a culinary tourism event. *Journal of Quality Assurance in Hospitality & Tourism*, *11*(1), 17–35. doi:10.1080/15280080903520584

Song, H., & Cheung, C. (2010). Factors affecting tourist satisfaction with theatrical performances: A case study of the romance of the song dynasty in Hangzhou, China. *Journal of Travel & Tourism Marketing*, *27*(7), 708–722. doi:10.1080/10548408.2010.519675

Sotiriadis, M. D. (2015). Culinary tourism assets and events: Suggesting a strategic planning tool. *International Journal of Contemporary Hospitality Management*, *27*(6), 1214–1232. doi:10.1108/IJCHM-11-2013-0519

Spiller, K. (2012). It tastes better because consumer understandings of UK farmers market food. *Appetite, 59*(1), 100–107. doi:10.1016/j.appet.2012.04.007 PMID:22521516

Stone, M. J., Soulard, J., Migacz, S., & Wolf, E. (2018). Elements of memorable food, drink, and culinary tourism experiences. *Journal of Travel Research, 57*(8), 1121–1132. doi:10.1177/0047287517729758

Swart, K., Cazorla Milla, A., & Mataruna-Dos-Santos, L. J. (2021). From the desert to the city: How is Dubai branding itself through sustainable sport events? *Worldwide Hospitality and Tourism Themes, 13*(1), 34–48. doi:10.1108/WHATT-08-2020-0101

Tanses, G. (1999). *Türkçe-İngilizce, İngilizce-Türkçe dizinli reklam terimleri ve kavramları sözlüğü.* Adam Yayınları.

Terziš, S., & Đališ, I. (2019). Analysis of brand valuation based on a combination of valuation models. *The Annals of the Faculty of Economics in Subotica, 55*(42), 33–48.

Tichaawa, T., Bob, U., & Swart, K. (2018). Africa and sports tourism. In J. Nauright & M. Amara (Eds.), *Sport in the African world* (pp. 185–204). Routledge. doi:10.4324/9781351212755-11

Tinsley, R., & Lynch, P. (2001). Small tourism business networks and destination development. *International Journal of Hospitality Management, 20*(4), 367–378. doi:10.1016/S0278-4319(01)00024-X

Toker, A. (2022). Kent kimliği inşaasında festivallerin rolü. *Bingöl Üniversitesi Sosyal Bilimler Enstitüsü Dergisi*, (23), 121–140.

Tokol, T. (1994). *Pazarlama yönetimi*. 5. Basım Bursa: Uludağ Üniversitesi Yayınları.

Türkay, O. (2014). *Destination Management: Approaches and Tools from A Management Science Perspective*. Detail Broadcast.

Türten, B. (2021). Film festivalleri ve turizm ilişkisi: Belgesel film yönetmenleri üzerine bir çalışma. *Erciyes İletişim Dergisi, 8*(1), 205–230. doi:10.17680/erciyesiletisim.810902

Uğur, İ., Gökkaya, S., & Acar, A. (2018). Yerli turistlerin destinasyon imajına ilişkin tekrar ziyaret etme niyetleri: Safranbolu üzerine bir çalışma. *Safran Kültür ve Turizm Araştırmaları Dergisi, 1*(1), 29–40.

Unal, N. (2001). Rekabet düzeyi en yüksek sektör turizm. *Standard. Yıl, 40*(470), 27.

Visser, G. (2005). Let's be festive: Exploratory notes on festival tourism in South Africa. *Urban Forum, 16*, 155–175.

Vukic, M., Kuzmanovic, M., Martic, M., Mihajlovic, D., & Kostic-Stankovic, M. (2015). Preferences of consumers towards food festivals: Empirical study. Paper presented at *34th International Conference on Organizational Science Development*, Portoroz, Slovenija.

Vukic, M., Kuzmanovic, M., & Vukic, K. (2016). Gastronomic festivals as A form of branding Serbia as a tourist destination. In *Proceedings of the Xv International Symposium Symorg* (pp. 913-919).

Wang, C. H. (2003). Predicting tourism demand using fuzzy time series and hybrid grey theory. *Tourism Management, 25*(3), 369.

Wells, W. D. (2014). *Measuring advertising effectiveness*. Psychology Press. doi:10.4324/9781315806099

Williams, H. A., Williams, R. L. Jr, & Omar, M. (2014). Gastro-tourism as destination branding in emerging markets. *International Journal of Leisure and Tourism Marketing*, *4*(1), 1–18. doi:10.1504/IJLTM.2014.059257

Wong, I. A., Wan, Y. K. P., & Qi, S. (2015). Green events, value perceptions, and the role of consumer involvement in festival design and performance. *Journal of Sustainable Tourism*, *23*(2), 294–315. doi:10.1080/09669582.2014.953542

Wood, E. H., & Masterman, G. (2008). *Event marketing: Measuring an experience.* In *7th International Marketing Trends Congress*, Venice.

Wood, L. (2000). Brands and brand equity: Definition and management. *Management Decision*, *38*(9), 662–669. doi:10.1108/00251740010379100

Yalım, F. (2018). Yavaş şehir (citaslow) hareket ekseninde kent markalaşması ve kent iletşimi: Kırklareli "Vize" Yavaş Şehir örneği. *Trakya Üniversitesi İktsadi ve İdari Bilimler Fakültesi Dergisi*, 1- 28

Yang, F. X., Wong, I. K. A., Tan, X. S., & Wu, D. C. W. (2020). The role of food festivals in branding culinary destinations. *Tourism Management Perspectives*, *34*, 100671. doi:10.1016/j.tmp.2020.100671

Yılmaz, Y., & Çizel, B. (2000). *Türk turizminde günü kurtarmak değil marka yaratmak. Türkiye Turizm Araştırma Enstitüsü, 1.* Ulusal Türkiye Sempozyumu Tebliğ Kitapçığı.

Yükselen, C., & Güler, G. E. (2009). *Antakya marka kent Görüş Ve öneriler.* Detay Yayıncılık.

Zhang, S. N., Li, Y. Q., Liu, C. H., & Ruan, W. Q. (2021). Does live performance play a critical role in building destination brand equity – a mixed-method study of 'impression Dahongpao'. *Journal of Retailing and Consumer Services*, *59*, 102392. doi:10.1016/j.jretconser.2020.102392

KEY TERMS AND DEFINITIONS

Brand: A brand is all symbols, texts or figures that represent a business or a place.

Brand Awareness: Individuals are aware of the existence of a brand and prefer that brand more.

Culture and Arts Festival: A culture and art festival is a people's event that includes poetry, literature, music, or cultural meaning.

Destination branding: It is to differentiate the destination in the national and international arena and to ensure that it is recognized with its original elements.

Festival: It is the participation of some members of the society in a pre-planned activity at the same place, at the same time and for a specific purpose.

Gastronomy Festival: These are events whose purpose is to promote and experience food or beverages and to raise awareness of people on this issue.

Sports Festival: These are the events where people from the national or international area come together who want to compete in sports activity or watch sports activities.

Chapter 2
Antecedents of Participation in Music Festivals:
The Role of Congruence Between the Sponsor and the Festival

Sara Santos
Research Centre in Digital Services (CISeD), Polytechnic of Viseu, Portugal

Pedro Espírito Santo
Research in Management and Economics (CARME), School of Technology and Management, Portugal & Polytechnic of Leiria, Portugal

Luísa Augusto
Research Centre in Digital Services (CISeD), Polytechnic of Viseu, Portugal

ABSTRACT

In the marketing literature, the desire or need to participate in a music festival is triggered by various reasons. Some sponsorship forms have associated brands with music festivals. One of the counterparts of the sponsorship is the possibility of associating the name of the brand with the designation of the name of the festival—that is, the brand acquiring the naming right of the festival. However, this theme has aroused the academy's interest, and studies that show the importance of the festivals in connection with the sponsoring brand. Thus, this study (N = 306) was carried out through a research model, tested through the PLS-SEM methodology, which showed that cultural exploration and socialization have effects on trust and brand connection and that the congruence between the sponsoring brand and the festival's characteristics play a mediating role in this relationship. Therefore, recommendations are suggested that contribute to the theoretical-practical deepening of new forms of persuasive communication, such as sponsorship, applied to the creative and cultural industries.

DOI: 10.4018/978-1-6684-6356-7.ch002

1. INTRODUCTION

Festivals are a typology of events that have seen a substantial expansion in territories, having solid impacts on the places where they take place. Anderton (2022) considers that the places where festivals take place take on a life of their own and can function as temporary cities. The motivations for participating in a festival can be varied and interdependent, and their understanding is crucial for festival organizers to meet the needs of the public and ensure significant participation (van Vliet, 2021).

In turn, sponsorship of events, and, in particular, of festivals, is fundamental as a marketing strategy for companies (Nguyen & Vu, 2022), mainly when it is intended to create a brand image and attract the youth segment (Stejskal, 2015), thus playing an essential role in the design and operation of these festivals, having been the object of the development of some studies and research. In their study, Nguyen and Vu (2022) concluded that the congruence of the sponsor and the event positively influences the attitude towards the sponsor and, consequently, the purchase intention and may also condition participation in the festival. For Stejskal (2015), the congruence of the brand/event positively conditions the brand identity and the sympathy towards the brand.

This study intends to analyze the role of motivations in the congruence between the sponsoring brand and the festival's audience, as well as whether this sponsor and festival congruence conditions influence brand trust and attachment.

2. LITERATURE REVIEW

The events provide participants with a unique, magical, and singular environment. However, they also have substantial impacts on the places where they occur, whether in economic, touristic, social, or environmental terms, as well as projecting the image of the place where the event occurs (Kim, Uysal, and Chen, 2001). Given the multiplicity of current events, defining the concept of "festivals" in this study is interesting. "Festivals are an essential expression of human activity that contributes significantly to our cultural and social life. They are also increasingly linked to tourism to bring business activities and profits to their host communities, generating increased tourism revenue (Allen, O'Toole, McDonnell, and Harris, 2008, p. 6).

Festivals can cover various themes, including dance, film, music, arts and crafts, cultural heritage, sports, and products (Kim, Borges, and Chon, 2006). In addition to the themes, there is a great diversity of music festivals, taking into account the different musical genres, spaces, and places where they take place and provide exciting opportunities not only for artists but also for sponsors, providing a unique experience for participants and above all, to the fans (Dowd, Liddle and Nelson, 2004). The authors are unanimous in considering that festivals are an essential resource for destinations (Saayman, 2011), for which they have known remarkable growth and diversification in terms of their typology. For Kim, Uysal, and Chen (2001), events and festivals vary significantly in place, objectives, program content, and purposes, conditioned by the marketing strategy.

Hundreds of festivals are organized annually, leading to growing competition among organizers of this type of event (Kitterlin & Yoo, 2014). Thus, it is crucial to understand the motivations that drive audiences to participate in a festival to carry out strategic planning and design a more effective marketing strategy (Kitterlin & Yoo, 2014).

2.1. Motivation for Going to Festivals

Studies on the motivations that lead visitors/tourists to participate in a festival gain particular relevance, as they have a utilitarian nature for marketing professionals (Saayman, 2011). For marketers or event managers, it is essential to understand the reasons that lead people to participate in a festival, not only to plan events strategically but significantly to develop initiatives tailored to the needs and motivations of visitors (Li, Wang, and Cai, 2009; Saayman, 2011; Kim, Uysal and Chen, 2001; Thompson & Schoefield, 2009). This premise is particularly valid for boosting the first visit and promoting a repeat visit (Saayman, 2011).

Crompton and McKay (1997) distinguish three reasons that justify the relevance of understanding the motivations that lead people to participate in a festival. The first reason is that it allows event managers to design and adapt the offer and experience of the event to meet the public's expectations and needs. The second reason derives from the fact that motivation depends on post-event satisfaction on the part of the public, constituting for managers a way of evaluating the event. The third reason is the possibility for managers to understand the decision-making process of visitors.

Many studies emerged within the academy on the motivations for attending a festival (Saayman, 2011). This study aims to understand and explain the reasons for reinforcing trust and connection to the sponsoring brand in music festivals. It has practical implications for festival managers and organizers and makes an essential contribution to academia as it broadens research on motivations for going to a festival.

Festivals must meet the needs of the different audiences involved (Li, Wang, and Cai, 2009). Participating in festivals is not homogeneous but depends on the different types of festivals (Kim, Uysal, and Chen, 2001). The growth and proliferation of events, namely festivals and the motivations for going to a festival, have aroused the interest of academia, giving rise to a wide range of scientific articles (Saayman, 2011). There are several studies about the motivations of festival visitors, such as Li and Petrick (2006), Li, Wang, and Cai (2009), Crompton and McKay (1997), Thompson and Schofield (2009), Saayman (2011) and Kitterlin and Yoo (2014). It is also worth mentioning the research by Kim, Uysal, and Chen (2001) developed to understand the perception of event organizers and managers on the motivations of participants and Prentice and Andersen (2003) on the cultural consumption patterns of visitors of the festival. Kim, Borges, and Chon (2006) analyzed the motivation to go to festivals and environmental values.

For Crompton and McKay (1997, p. 425), the "decision to visit a festival is a direct action that is triggered by a desire to satisfy a need" which may result from the desire to satisfy several different needs, either at the individual level or concerning the group level. Motivation "is a dynamic process of internal psychological factors (needs/desires)" (Crompton & McKay, 1997, p. 427). The authors highlight three interpretive theoretical lines to understand tourist behavior: the Maslow hierarchy of needs, the Isto-Ahola escape/search dichotomy, and push-pull factors (Crompton & McKay, 1997; Saayman, 2011). For the authors, "a festival implies that visitors seek cultural enrichment, education, novelty, and socialization" (Crompton & McKay, 1997, p. 429).

The literature review allowed knowing the main reasons for going to the festival, listed by several authors, highlighting the need for cultural enrichment and exploration (Crompton & McKay, 1997; Lee et al., 2004; Saayman, 2011) and motives related to education (Crompton & McKay, 1997), the search for something new or novelty (Crompton & McKay, 1997; Thompson & Schofield, 2009), the need for socialization (Crompton & McKay, 1997; Thompson & Schofield, 2009) and escape (Lee et al., 2004; Saayman, 2011). Saayman (2011) adds to these reasons the attraction of the event as a motivating factor. In their study on motivation to go to festivals and environmental values, Kim, Borges, and Chon (2006, p.

959) refer that several studies have been developed that "show that the type and theme of the festival are strong determinants of motives."In the research by Thompson and Schofield (2009), sporting attractions and special local events are also highlighted. The results of the study by Saayman (2011) showed the existence of a combination of intrinsic motives (escape and cultural exploration) with extrinsic motives (attraction to the event), corroborating the ideas of Crompton and McKay (1997).

The authors unanimously consider the need to escape as a strong reason for visiting a festival. "The desire to 'escape' and 'recovery of balance' has been identified in several studies on the motivation to go to the festival" (Thompson & Schofield, 2009, p. 5). This perspective is in line with the theoretical concepts that underlie the motivations for tourism. For, also, MacCannel (1976) and Cohen (1979) consider tourism a form of escape and expression of individual freedom, a search for the experience of something new. MacCannell (1976) emphasizes that the tourist movement can be understood as a pilgrimage searching for an experience different from the usual, authentic, and genuine. Cohen (1979a, 1979b) adds that the tourist aims to escape his routine and monotonous world. With a post-modern view, Urry (2002, p. 33) considers that tourism development has boosted the transformation of real places, making them spaces of entertainment, "youth, controlled danger, anticipation, and pleasure," which provide tourists with living unique experiences. Therefore, places with specific characteristics where the service provided and the experience lived are determining factors for the tourist decision process.

In turn, Kim, Uysal, and Chen (2001) also identified five motivational dimensions based on the perceptions of event organizers about the motives of festival visitors: social/leisure, novelty, family togetherness, escape, and curiosity. Thompson and Schofield's (2009) results pointed to five dimensions of motivation for visitors who attend festivals: multi-purpose search; indifference; search for culture and sport; looking for union, socialization, and sport, looking for local events, meeting the perspectives resulting from the research of other authors. The motivational domains addressed in the literature review are related to novelty/regression (Crompton & McKay, 1997); cultural exploration (Crompton & McKay, 1997; Lee et al., 2004; Saayman, 2011; Thompson & Schofield, 2009), with familiarity (Crompton & McKay, 1997), as well as with the restoration of balance (Crompton & McKay, 1997), and the socialization of kinship or union with family and friends (Crompton & McKay, 1997; Thompson & Schofield, 2009). External interaction and socialization are also mentioned (Crompton & McKay, 1997; Thompson & Schofield, 2009) as socialization with a known group (Crompton & McKay, 1997).

2.2. Congruence Between the Sponsor's Brand and the Festival

Brand self-congruence influences brand attachment, as people feel connected to products they identify with and express who they are (Ball & Tasaki, 1992). This is demonstrated in several previous studies (Malär, Krohmer, Hoyer, & Nyffenegger, 2011; Matzler et al., 2011) in which the congruence of the self with the brand is antecedent to the attachment. Self-congruence corresponds to the consumer's perceived personality and the brand (Aaker, 1999). When there is congruence between the brand and the consumer's image, they establish emotional links and brand loyalty (Thomson et al., 2005; Park et al., 2010). Thus, consumers buy brands to express their identity (current or ideal) (Ekinci et al., 2013), also increasing their self-esteem (Escalas & Bettman, 2003).

Several marketing studies show that self-congruence results in consumer satisfaction, brand love, loyalty, and consequent WOM (Ekinci et al., 2008; Nam et al., 2011; Wallace et al., 2017). On the other hand, according to Cheng et al. (2012), when consumers develop a connection to the brand through self-congruence, they are more intent on buying and paying a premium price. While some authors consider the effects of self-congruence on the consumer direct (e.g., Wallace et al., 2017), others (e.g., Ekinci et al., 2008; Nam et al., 2011) argue that they are not direct.

Gwinner and Bennett (2008) evidenced the effectiveness of the congruence of sponsorship of sporting events, while Ko, Kim, Claussen, & Kim (2008) indicate consumer involvement in consumer responses to marketing actions. On the other hand, Koo & Lee (2018) showed that in sponsorship, when there is congruence with the sponsor, the delivery of relevant messages about the product to the consumer is more outstanding. Sponsorship can be commercial or philanthropic (Shin et al., 2018) and provide a commercial return to the sponsor (Chebli & Gharbi, 2014). In this domain, the congruence of the sponsor's brand with the festival is influenced by the motivations for which consumers participate in music festivals, so this work will test the following research hypotheses:

H1: Cultural exploration as a motivation to participate in festivals influences the perceived congruence between the sponsor and the festival

H2: When the motivation to participate in festivals is the recovery of balance, there is an influence on the perceived congruence between the sponsor and the festival.

H3: External socialization and integration with groups as a motivation to participate in festivals influence the perceived congruence between the sponsor and the festival.

2.3. Brand Trust

In the marketing literature, brand trust is a central construct in solid and long-term relationships with the brand (Delgado-Ballester & Munuera-Alemán, 2001). Chaudhuri and Holbrook (2001, p. 82) correspond to the "average consumer's willingness to trust the brand's ability to perform its declared function."In fear of opportunism, uncertainty, or minor information, trust in the brand makes consumers feel more comfortable with it (Chiu, Huang, & Yen, 2010).

To create this trust in the brand, repeated interactions and long-term relationship development are essential (Holmes, 1991; Wang & Emurian, 2005). Consequently, brand trust is reflected in the purchase process (Powers, Adlink, Austin, Graiko, & Snyder, 2012) and leads to brand loyalty. Thus, according to several authors (Chiu et al., 2010; Hong & Cho, 2011; Kim, Chung, & Lee, 2011), trust is the primary antecedent of loyalty and sponsorship (Pentina, Zhang, & Basmanova, 2013; See-To & Ho, 2014) building lasting relationships with brands (Morgan & Hunt, 1994).

Some studies have shown the relationship between brand image, satisfaction, trust, and brand loyalty (Song et al., 2019) and the importance of social media that impacts trust and loyalty (Althuwaini, 2022).

Studies by Sheinin, Varki and Ashley (2011) show that creative advertising can lead to brand trust, especially in the utility dimension. The same authors (Sheinin, Varki & Ashley, 2011) showed that trust is higher according to each person's attitude towards innovative advertising. On the other hand, communication tools such as advertising can also generate trust as long as they are consistent. In this sense, this work aims to test the following research hypothesis:

H4: The congruence of the sponsor's brand with the festival positively influences trust in the sponsor's brand.

2.4. Brand Attachment

Individuals participate in festivals that identify with their lifestyle, so the consumer and brand attachment is essential in marketing. Hiscock (2001) shows that this is the main objective of marketing: to create strong links between the consumer and the brand; however, not all consumers develop strong links with the brands (Mende et al., 2013). This emotional connection between the consumer and the brand is brand attachment (Malär et al., 2011). However, this concept's investigation is still limited to understanding its determinants and consequences (Japutra et al., 2014).

Thomson, MacInnis, and Park (2005) were the first authors to develop a brand attachment scale composed of three dimensions: affection, passion, and connection. Park et al. (2010, p. 2) define brand attachment as "the strength of the connection between the brand and the self. This bond is translated through accessible memory (or mental representation) that involves thoughts and feelings about the brand and the brand's relationship with the self". In this way, Park et al. (2010) show that the connection to the brand implies emotional and cognitive ties, which translate into the connection of the "I" to the brand (brand-self connection) and prominence.

By "brand self-connection" is meant the perception of the relevance between the brand and the consumer's "I" (Fedorikhin, Park, & Thomson, 2008), that is, the extent to which the brand is incorporated in the concept of the "I" of the consumer (Escalas, 2004). On the other hand, the concept of prominence presented by Park et al. (2010, p.2) reflects "the salience of the cognitive and affective bond that links the brand to the self."When brands are seen as part of the person, prominence is elevated because the brand's activation and the self occurs (Park et al., 2010). Brand attachment results in several consequences: the intention to recommend, buy, resistance to negative information, defend the brand (Japutra et al., 2014), loyalty, and willingness to pay a premium price (Thomson et al., 2005).

Some studies in various sectors, such as tourism (Chen et al., 2021), retail (Sheeraz et al., 2018), and luxury branding (Shetty & Fitzsimmons, 2022), demonstrate the importance of brand attachment. Chen et al. (2021) demonstrated a positive mediating effect of brand authenticity and attachment on historical nostalgia and destination brand equity. Shetty and Fitzsimmons (2022) demonstrated the effect of brand attachment on brand loyalty in the luxury context. However, recent studies about brand attachment and its relationship to music festivals have yet to be conducted.

The attachment to the brand is then assumed as a central concept in developing relationships with consumers for greater effectiveness of marketing strategies. Maintaining trust in the brand is essential as a "base for long-term customer relationships" (Burmann et al., 2009, p. 391). Thus, this study will test the following investigation hypotheses:

H5: The congruence between the sponsor and the festival influences the brand attachment.
H6: Trust in the sponsor brand has a positive influence on brand attachment.

As a result of the research hypotheses, the following research model was elaborated, as shown in figure 1:

Figure 1. Conceptual Model

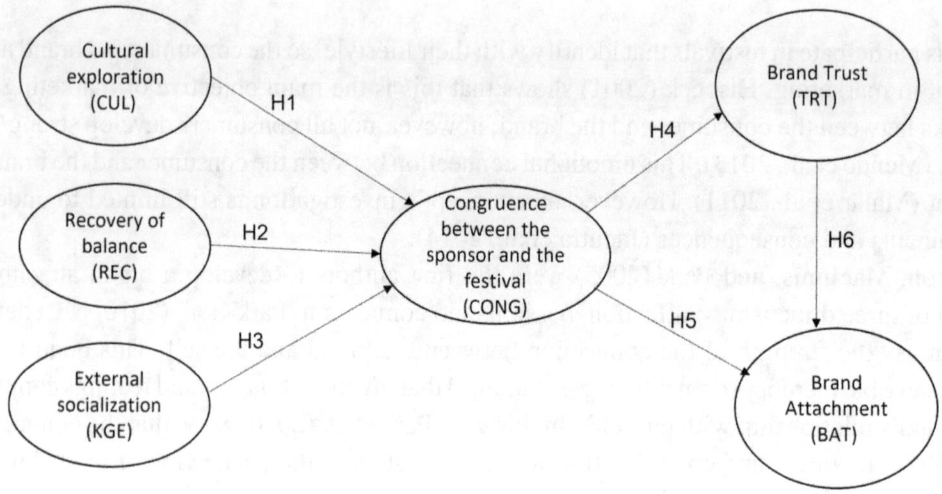

3. METHODOLOGY

The study presented assumed a conclusive-causal nature of superficial transversal characteristics, and the universe was the individuals who attended festivals recently. We collected data in 2022. The sample consisted of 306 valid responses from individuals under 30 years of age, of which 73.2% are female and 78.3% of individuals have attended music events in recent months. The festivals they attended in Portugal are RFM Somnii, MEO Sudeste, NOS Alive, and Vodafone Paredes de Coura. We analyzed these festivals because there are significant festivals in Portugal. The answers were obtained through an online questionnaire, from which responses were obtained from 5-point Likert scales tested by other authors and described in table 1.

For the sampling procedure, preliminary analyses were carried out. It was concluded that the VIF values are less than 5, which leads us to consider that there will be no multicollinearity problems. Furthermore, asymmetry values less than 3 (in absolute value) and kurtosis values less than 7 (in absolute value). Therefore, according to the literature, this data indicates that deviations from the normal distribution will not be severe (Maroco, 2014).

4. RESULTS

PLS-SEM is appropriate for small databases and when the research is exploratory and does not require data normality (Hair et al., 2016). In addition, PLS-SEM is performed in two steps. First, we analyzed the reliability and validity of the measurement model, and second, we analyzed the relationships between constructs as suggested by Hair et al. (2016). The PLS-Algorithm was run in SMART PLS v3.3.5 (Ringle et al., 2015).

From the data collection, we understood that the estimation of the conceptual model could be carried out through Partial Least Squares - Structural Equation Modeling (PLS-SEM) because it allows researchers to assess causal relationships between constructs.

Table 1. Items used in the investigation

Constructs	Items	Code	References
Cultural Exploration (CUL)	I take the opportunity to make cultural visits when I go to music festivals and concerts.	CUL01	Crompton and McKay (1997)
	I try to discover something else when I go to music festivals and concerts.	CUL02	
	Going to music festivals and concerts is to experience different customs and cultures.	CUL03	
	I try to explore more things every day	CUL04	
	I go to music festivals and concerts to increase my cultural knowledge	CUL05	
Recovery of Balance (REC)	I occasionally go to music festivals and concerts to avoid the routine.	REC01	Crompton and McKay (1997)
	I like attending music festivals and concerts to reduce anxieties and frustrations.	REC02	
	I go to music festivals and concerts to recover from my hectic pace	REC03	
	I go to music festivals and concerts to alleviate the boredom of everyday life.	REC04	
Known-Group & External Socialization (KGE)	I go to music festivals and concerts to be with my friends	KGE01	Crompton and McKay (1997)
	I like going to music festivals and concerts with a group of well-known people.	KGE02	
	I go to music festivals and concerts to be with others who enjoy the same things I do.	KGE03	
	I go to music festivals and concerts because I have fun with other people	KGE04	
	I like to meet new people when participating in music festivals and concerts.	KGE05	
Congruence (CONG)	The event sponsor's brand advertising is consistent with my way of being	CONG01	Koo and Lee (2018)
	The event sponsor's brand advertising is consistent with how others see me	CONG02	
	There is a logical connection between the music festival and its main sponsor.	CONG03	
	The event picture and the main sponsor's brand picture are similar	CONG04	
	The event and the main sponsor represent similar things.	CONG05	
	It makes perfect sense for this brand to sponsor the event	CONG06	
Brand Trust (TRT)	The event sponsor's brand is sincere	TRT01	Sheinin, Varki, & Ashley (2011)
	The event sponsor's brand is honest.	TRT02	
	Event Sponsor Branding is reliable	TRT03	
	The ads for this festival are convincing	TRT04	
Brand Attachment (BAT)	I'm emotionally attached to the event sponsor's announcements	BAT01	Park, MacInnis, Priester, Eisingerich, and Lacobucci (2010)
	The event sponsor's brand is part of me	BAT02	
	I am attracted to the event sponsor's brand advertising	BAT03	

The evaluation of the measures model shows that the measures used have factor validity, convergent validity, and discriminant validity of the constructs (Table 2).

Table 2. Measurement model results

construct	Code	Mean	Standard deviation	λ	Cronbach α (> 0,7)	Rho_A (> 0,7)	C.R. (> 0,7)	AVE (> 0,5)
Cultural Exploration (CUL)	CUL01	3,89	0,893	0,602**	0,767	0,793	0,842	0,517
	CUL02	3,11	1,016	0,750**				
	CUL03	3,76	0,912	0,793**				
	CUL04	3,49	1,066	0,748**				
	CUL05	3,75	0,893	0,688**				
Recovery of Balance (REC)	REC01	3,50	1,006	0,732**	0,851	0,874	0,899	0,690
	REC02	3,32	1,231	0,849**				
	REC03	3,24	1,270	0,870**				
	REC04	2,86	1,148	0,865**				
Known-Group & External Socialization (KGE)	KGE01	3,14	1,211	0,791**	0,800	0,807	0,855	0,544
	KGE02	3,96	1,012	0,710**				
	KGE03	4,24	0,884	0,610**				
	KGE04	3,71	0,991	0,832**				
	KGE05	3,98	0,953	0,725**				
Congruence (CONG)	CONG01	2,97	0,981	0,718**	0,841	0,843	0,883	0,558
	CONG02	2,79	1,021	0,684**				
	CONG03	3,57	1,045	0,797**				
	CONG04	3,49	0,999	0,692**				
	CONG05	3,25	1,032	0,805**				
	CONG06	3,62	0,998	0,777**				
Brand Trust (TRT)	TRT01	3,55	0,887	0,906**	0,879	0,892	0,918	0,739
	TRT02	3,54	0,857	0,928**				
	TRT03	3,73	0,843	0,878**				
	TRT04	3,90	0,835	0,709**				
Brand Attachment (BAT)	BAT01	2,67	1,122	0,887**	0,885	0,890	0,929	0,813
	BAT02	2,36	1,138	0,905**				
	BAT03	2,69	1,159	0,912**				

** $p<0.001$

The factorial validity is verified by analyzing the standardized coefficients (λ), which are more significant than 0.5 for $p<0,01$. Convergent validity exists since the composite reliability of concepts is more significant than 0.7 (C.R.>0,7), the average extracted variance is more significant than 0.5 (AVE>0,5), and Cronbach's Alpha values are more significant than 0.7 ($\alpha > 0,7$).

In addition, discriminant validity is verified by the three criteria studied:

- Fornell and Larcker criterion (1981) – to the data presented in table 3,

Table 3. Fornell and Larcker criterion (1981)

	BAT	CONG	CUL	KGE	REC	TRT
BAT-Brand Attachment	0,901					
CONG-Congruence Festival-Sponsorship	0,626	0,747				
CUL-Cultural Exploration	0,335	0,297	0,719			
KGE-Known-Group & External Socialization	0,177	0,246	0,282	0,738		
REC-Recovery of Balance	0,306	0,209	0,309	0,279	0,831	
TRT- Brand Trust	0,429	0,577	0,244	0,200	0,207	0,860

- The criterion of cross loads - by the data presented in Table 4
- Heterotrait-Monotrait (HTMT) criterion by Henseler et al. (2015) - by the data presented in Table 5.

The second step of the analysis is the test of the structural model. For this purpose, we used the bootstrap resampling technique with 5000 sub-samples to stabilize the results obtained. The accomplishment of this analysis allowed us to obtain the results that will be used to test the hypotheses under study. These results will be analyzed considering the standardized β-values and Student t-values. This analysis will validate the statistical significance of the parameters under analysis. In this sense, the results are summarized in Table 6.

Graphically, the results of the structural model are presented in figure 2:

5. DISCUSSION AND CONCLUSION

We analyzed the hypotheses understudy following the results presented to discuss the results obtained. Thus, we begin our discussion by analyzing the effects of motivations to attend a festival on the congruence of the sponsoring brand with the festival. Hypothesis H1 validated the impact of motivation for cultural exploration on the perceived congruence between the sponsor and festival. This relationship was corroborated by this study by $p<0.001$, according to the results of Koo & Lee (2018).

The research hypothesis H2 tested whether participation in festivals, based on motivations to recover individual balance, influences the perceived congruence between sponsor and festival. A possible justification for this relationship not being supported by this study ($p>0.05$) may be due to the way this motivation varies from market segment to market segment, which, in the words of Saayman (2011), also depends on the type of market. Festival visited as well as sociodemographic variables. Considering that this study is applied to individuals aged below 30 years and the festivals they attended were only music festivals, we understand that this research hypothesis, for this reason, was not supported.

Table 4 – Discriminant validity cross-loads

Item	BAT	CONG	CUL	KGE	REC	TRT
BAT01	0,887	0,518	0,356	0,142	0,286	0,371
BAT02	0,905	0,564	0,262	0,164	0,265	0,377
BAT03	0,912	0,606	0,292	0,170	0,278	0,409
CONG01	0,605	0,718	0,267	0,180	0,240	0,389
CONG02	0,611	0,684	0,200	0,100	0,234	0,339
CONG03	0,379	0,797	0,249	0,232	0,129	0,440
CONG04	0,320	0,692	0,127	0,175	0,072	0,426
CONG05	0,437	0,805	0,194	0,208	0,102	0,443
CONG06	0,393	0,777	0,268	0,213	0,125	0,548
CUL01	0,169	0,153	0,602	0,115	0,125	0,051
CUL02	0,175	0,222	0,750	0,230	0,216	0,223
CUL03	0,337	0,276	0,793	0,228	0,225	0,225
CUL04	0,245	0,219	0,748	0,205	0,220	0,171
CUL05	0,250	0,166	0,688	0,220	0,339	0,167
KGE01	0,177	0,224	0,180	0,791	0,302	0,142
KGE02	0,076	0,194	0,070	0,710	0,085	0,147
KGE03	0,093	0,046	0,202	0,610	0,236	0,046
KGE04	0,134	0,183	0,287	0,832	0,206	0,122
KGE05	0,155	0,171	0,345	0,725	0,236	0,230
REC01	0,205	0,121	0,308	0,178	0,732	0,231
REC02	0,232	0,170	0,243	0,248	0,849	0,124
REC03	0,319	0,192	0,322	0,256	0,870	0,183
REC04	0,248	0,196	0,183	0,236	0,865	0,171
TRT01	0,398	0,547	0,242	0,189	0,176	0,906
TRT02	0,420	0,520	0,233	0,183	0,182	0,928
TRT03	0,352	0,459	0,164	0,166	0,158	0,878
TRT04	0,291	0,451	0,195	0,148	0,199	0,709

Table 5. Discriminant validity Heterotrait-Monotrait (HTMT) correlation ratio by Henseler et al. (2015)

	BAT	CONG	CUL	KGE	REC	TRT
BAT-Brand Attachment						
CONG-Congruence Festival-Sponsorship	0,708					
CUL-Cultural Exploration	0,399	0,351				
KGE-Known-Group & External Socialization	0,202	0,274	0,367			
REC-Recovery of Balance	0,348	0,233	0,398	0,345		
TRT-Trust Brand	0,483	0,671	0,284	0,222	0,249	

Table 6. Results of the hypothesis test

Hypothesis	Structural Relationship	Std β	*t-Value*	Confidence Interval		p-*value*	Result
				Inferior	Superior		
H1	CUL -> CONG	0,223	3,965	0,121	0,340	0,000	supported
H2	REC -> CONG	0,096	1,711	-0,008	0,211	0,087	Not supported
H3	KGE-> CONG	0,157	2,958	0,066	0,272	0,003	supported
H4	CONG -> TRT	0,577	11,901	0,480	0,669	0,000	supported
H5	CONG -> BAT	0,568	11,573	0,469	0,662	0,000	supported
H6	TRT -> BAT	0,101	1,791	-0,011	0,209	0,073	Not supported

Figure 2. Hypothesis Test Results

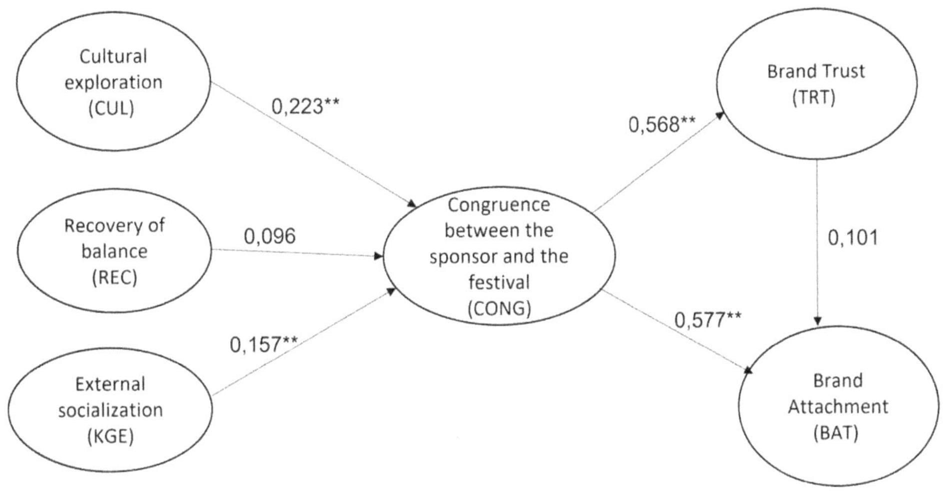

Individuals who participate in festivals to socialize perceive greater congruence between the sponsor and the festival since hypothesis H3 was supported by this study for p<0.01. Individuals seek to attend festivals with which they identify, and they find people with similar motivations in these events. This motivation is enough to attend this type of event. Since festivals are meeting places for friends, motivated individuals recognize the congruence between the festival and the sponsoring brand (Thompson & Schofield, 2009).

The perception of greater congruence between the sponsoring brand and the characteristics of the festival is influential in the trust that individuals have in the sponsoring brand since hypothesis H4 was confirmed by this study, for p<0.001. This relationship makes sense since congruent communication raises individuals' trust in brands, as stated in the paper by Sheinin, Varki and Ashley (2011).

In the analysis of hypothesis H5, it can be stated that individuals who perceive greater congruence between the sponsoring brand and the festival will be more available for a stronger attachment to the brand. Brand Attachment is supported by the existence of congruence between the sponsor's brand and the festival. This hypothesis shows that the sponsoring brand of a festival that individuals attend motivates a greater connection to the brand on the part of individuals who attend the festival.

Hypothesis H6 was not verified by this investigation (p>0.05), concluding that trust in the sponsoring brand does not influence greater brand attachment. The non-confirmation of the relationship can be justified by how the festivals are advertised and how the sponsoring brand presents itself at the event. There may be some confusion between the festival and what the brand represents outside the event. Another justification is the brands that sponsor the festivals. These brands are seen nationally as trusted (Fedorikhin, Park, & Thomson, 2008). Examples of festivals in Portugal are RFM Somnii, MEO Sudeste, NOS Alive, and Vodafone Paredes de Coura, where sponsoring brands are seen with the highest trust ranking among individuals. In this sense, trust is something already assumed by individuals. With low variability (by the data presented in the table in the measurement model), which may have caused that hypothesis, H6 was not supported by this study.

Given the hypotheses understudy, this research model is valid since the concepts show significant relationships. From these relationships, there are effects between individuals' motivations to attend festivals and the congruence perceived by individuals between brands and the festival. The congruence between the sponsor's brand and the festival will promote greater trust in the brand and a more significant connection between the individual and the brand.

The conclusions presented here are realistic since this study draws relevant conclusions for brand management. It shows that brand managers should pay special attention to the public attending festivals in their sponsorship strategies. Moreover, their motivations for these audiences are attracted to more significant brand attachment.

Although this study shows relevant conclusions in the management of events and brands, the data presented were based on a sample of young individuals who attended music festivals, which was the object of study of this work, so the generalization of the conclusions must be careful.

In this sense, it is suggested in future investigations to analyze this model in other samples and other events to understand the relationships tested here. Thus, there is work to be carried out here within the scope of the study of brand management and its relationship with communication at events.

ACKNOWLEDGMENT

This work is funded by National Funds through the FCT – Foundation for Science and Technology, I.P., within the scope of the project Ref. UIDB/05583/2020. Furthermore, we would like to thank the Research Centre in Digital Services (CISeD) and the Instituto Politécnico de Viseu for their support.

REFERENCES

Aaker, J. L. (1999). The Malleable Self: The Role of Self-Expression in Persuasion. *JMR, Journal of Marketing Research*, *36*(1), 45–57. doi:10.1177/002224379903600104

Allen, J., O'Toole, W., McDonnell, I., & Harris, R. (2008). *Organização e Gestão de Eventos* (3rd ed.). Elsevier Editora.

Althuwaini, S. (2022). The Effect of Social Media Activities on Brand Loyalty for Banks: The Role of Brand Trust. *Administrative Sciences*, *12*(4), 148. doi:10.3390/admsci12040148

Anderton, C. (2022). Festivals. In *The Bloomsbury Handbook of Popular Music, Space and Place*. Bloomsbury Publishing. doi:10.5040/9781501336317.ch-028

Ball, D., & Tasaki, L. H. (1992). The Role and Measurement of Attachment in Consumer Behavior. *Journal of Consumer Psychology, 1*(2), 155–172. doi:10.120715327663jcp0102_04

Burmann, C., Hegner, S., & Riley, N. (2009). Towards an identity-based branding. *Marketing Theory, 9*(1), 109–114. doi:10.1177/1470593108100065

Chaudhuri, A., & Holbrook, M. B. (2001). The chain of effects from brand trust and brand affect to brand performance: The role of brand loyalty. *Journal of Marketing, 65*, 81–93.

Chebli, L., & Gharbi, A. (2014). The Impact of the Effectiveness of Sponsorship on Image and Memorizing: Role of Congruence and Relational Proximity. *Procedia: Social and Behavioral Sciences, 109*, 913–924. doi:10.1016/j.sbspro.2013.12.564

Chen, X., You, S. E., Lee, T. J., & Li, X. (2021). The influence of historical nostalgia on a heritage destination's brand authenticity, brand attachment, and brand equity. *International Journal of Tourism Research, 23*(6), 1176–1190. doi:10.1002/jtr.2477

Chiu, C. M., Huang, H. Y., & Yen, C. H. (2010). Antecedents of online trust in online auctions. *Electronic Commerce Research and Applications, 9*(2), 148–159. doi:10.1016/j.elerap.2009.04.003

Cohen, E. (1979a). A phenomenology of tourist experiences. *Sociology, 13*(2), 179–201. doi:10.1177/003803857901300203

Cohen, E. (1979b). Rethinking the sociology of tourism. *Annals of Tourism Research, 6*(1), 18–35. doi:10.1016/0160-7383(79)90092-6

Crompton, J. L., & McKay, S. L. (1997). Motives of visitors attending festival events. *Annals of Tourism Research, 24*(2), 425–439. doi:10.1016/S0160-7383(97)80010-2

Delgado-Ballester, E., Manuera-Aleman, J. L., & Yague-Guillen, M. J. (2003). Development and validation of a brand trust scale. *International Journal of Market Research, 45*(1), 35–54.

Dowd, T. J., Liddle, K., & Nelson, J. (2004). Music festivals as scenes: Examples from serious music, women's music, and skatepunk. Music scenes: Local, translocal and virtual, pp. 149–67.

Ekinci, Y., Dawes, P. L., & Massey, G. R. (2008). An extended model of the antecedents and consequences of consumer satisfaction for hospitality services. *European Journal of Marketing, 42*(1), 35–68. doi:10.1108/03090560810840907

Ekinci, Y., Sirakaya-Turk, E., & Preciado, S. (2013). Symbolic consumption of tourism destination brands. *Journal of Business Research, 66*(6), 711–718. doi:10.1016/j.jbusres.2011.09.008

Escalas, J. E. (2004). Narrative processing: Building consumer connections to brands. *Journal of Consumer Psychology, 14*(1-2), 168–180. doi:10.120715327663jcp1401&2_19

Escalas, J. E., & Bettman, J. R. (2003). You Are What They Eat: The Influence of Reference Groups on Consumer Connections to Brands. *Journal of Consumer Psychology, 13*(3), 339–348. doi:10.1207/S15327663JCP1303_14

Fedorikhin, A., Park, C. W., & Thomson, M. (2008). Beyond fit and attitude: The effect of emotional attachment on consumer responses to brand extensions. *Journal of Consumer Psychology*, *18*, 281–291.

Gwinner, K., & Bennett, G. (2008). The impact of brand cohesiveness and sport identification on brand fit in a sponsorship context. *Journal of Sport Management*, *22*, 410–426.

Hair, J., Hult, G., Ringle, C., & Sarstedt, M. (2016). *A primer on partial least squares structural equation modeling (PLS-SEM)*. Sage publications.

Hiscock, J. (2001). Most Trusted Brands. Marketing, 32–33.

Holmes, J. G. (1991). trust and the appraisal process in close relationships. In W. H. Jones, & D. Perlman (Eds.), Advances in personal relationships 2 (pp. 57–104). London: Jessica Kingsley.

Hong, I. B., & Cho, H. (2011). The impact of consumer trust on attitudinal loyalty and purchase intentions in B2C e-marketplaces: Intermediary trust vs. seller trust. *International Journal of Information Management*, *31*(5), 469–479. doi:10.1016/j.ijinfomgt.2011.02.001

Japutra, A., Ekinci, Y., & Simkin, L. (2014). Exploring brand attachment, its determinants, and outcomes. *Journal of Strategic Marketing*, *22*(7), 1–15. doi:10.1080/0965254X.2014.914062

Kim, H., Borges, M. C., & Chon, J. (2006). Impacts of environmental values on tourism motivation: The case of FICA, Brazil. *Tourism Management*, *27*(5), 957–967. doi:10.1016/j.tourman.2005.09.007 PMID:32287716

Kim, K., Uysal, M., & Chen, J. S. (2001). Festival visitor motivation from the organizers' points of view. *Event Management*, *7*(2), 127–134. doi:10.3727/152599501108751533

Kim, M., Chung, N., & Lee, C. (2011). The effect of perceived trust on electronic commerce: Shopping online for tourism products and services in South Korea. *Tourism Management*, *32*(2), 256–265. doi:10.1016/j.tourman.2010.01.011

Kitterlin, M., & Yoo, M. (2014). Festival motivation and loyalty factors. *Tourism & Management Studies*, *10*(1), 119–126.

Ko, Y. J., Kim, K., Claussen, C. L., & Kim, T. H. (2008). The effects of sport involvement: Sponsor awareness and corporate image on intention to purchase sponsors' products. *International Journal of Sports Marketing & Sponsorship*, *9*(2), 79–94. doi:10.1108/IJSMS-09-02-2008-B004

Koo, J., & Lee, Y. (2018). Sponsor-event congruence effects: The moderating role of sport involvement and mediating role of sponsor attitudes. *Sport Management Review*.

Lee, C. K., Lee, Y. K., & Wicks, B. E. (2004). Segmentation of festival motivation by nationality and satisfaction. *Tourism Management*, *25*(1), 61–70. doi:10.1016/S0261-5177(03)00060-8

Li, M., Huang, Z., & Cai, L. A. (2009). Benefit segmentation of visitors to a rural community-based festival. *Journal of Travel & Tourism Marketing*, *26*(5-6), 585–598. doi:10.1080/10548400903163152

Li, X., & Petrick, J. F. (2005). A review of festival and event motivation studies. *Event Management*, *9*(4), 239–245. doi:10.3727/152599506776771526

MacCannell, D. (1976). *The tourist: A new theory of the leisure class*. Universidade da California Press.

Malär, L., Krohmer, H., Hoyer, W. D., & Nyffenegger, B. (2011). Emotional brand attachment and brand personality: The relative importance of the actual and ideal selves. *Journal of Marketing*, *75*(4), 35–52. doi:10.1509/jmkg.75.4.35

Matzler, K. E., Fuller, J., & Mooradian, T. A. (2011). Personality, Person-Brand Fit, and Brand Community: An Investigation of Individuals, Brands, and Brand Communities. *Journal of Marketing Management*, *27*(9-10), 874–890. doi:10.1080/0267257X.2010.543634

Mende, M., Bolton, R. N., & Bitner, M. J. (2013). Decoding customer-firm relationships: How attachment styles help explain customers' preferences for closeness, repurchase intentions, and changes in relationship breadth. *JMR, Journal of Marketing Research*, *50*(1), 125–142. doi:10.1509/jmr.10.0072

Morgan, R., & Hunt, S. (1994). The Commitment-Trust Theory of Relationship Marketing. *Journal of Marketing*, *58*(3), 20–38. doi:10.1177/002224299405800302

Nam, J., Ekincy, Y., & Whyatt, G. (2011). Brand equity, brand loyalty, and customer satisfaction. *Annals of Tourism Research*, *38*(3), 1009–1030. doi:10.1016/j.annals.2011.01.015

Nguyen, D. T., & Vu, H. T. (2022). Measuring Attitudes Toward Sponsor And Purchase Intention. *Cogent Business & Management*, *9*(1), 2049961. doi:10.1080/23311975.2022.2049961

Park, C. W., Eisingerich, A. B., & Park, J. W. (2013). Attachment–Aversion (A.A.) Model of Customer–Brand Relationships. *Journal of Consumer Psychology*, *23*(2), 229–248. doi:10.1016/j.jcps.2013.01.002

Park, C. W., MacInnis, D. J., & Priester, J. (2006). Beyond attitudes: Attachment and consumer behavior. *Seoul Journal of Business*, *12*, 3–35.

Park, C. W., MacInnis, D. J., Priester, J., Eisingerich, A. B., & Iacobucci, D. (2010). Brand attachment and brand attitude strength: Conceptual and empirical differentiation of two critical brand equity drivers. *Journal of Marketing*, *74*(6), 1–17. doi:10.1509/jmkg.74.6.1

Pentina, I., Zhang, L., & Basmanova, O. (2013). Antecedents and consequences of trust in a social media brand: A cross-cultural study of Twitter. *Computers in Human Behavior*, *29*(4), 1546–1555. doi:10.1016/j.chb.2013.01.045

Powers, T., Advincula, D., Austin, M. S., Graiko, S., & Snyder, J. (2012). Digital and social media in the purchase decision process: A special report from the Advertising Research Foundation. *Journal of Advertising Research*, *52*(4), 479–489. doi:10.2501/JAR-52-4-479-489

Prentice, R., & Andersen, V. (2003). Festival as a creative destination. *Annals of Tourism Research*, *30*(1), 7–30. doi:10.1016/S0160-7383(02)00034-8

Ringle, C., Wende, S. & Becker, J.-M. (2015). SmartPLS 3.

Saayman, M. (2011). Motives for attending the Cultivaria arts festival. *S.A. Journal for Research in Sport Physical Education and Recreation*, *33*(1), 109–120. doi:10.4314ajrs.v33i1.65492

See-To, E. W. K., & Ho, K. K. W. (2014). *Value co-creation and purchase intention in social network sites: The role of electronic Word-of-Mouth and trust: A theoretical analysis*. Computers in Human Behavior. doi:10.1016/j.chb.2013.10.013

Sheeraz, M., Qadeer, F., Masood, M., & Hameed, I. (2018). Self-Congruence Facets and Emotional Brand Attachment: The Role of Product Involvement and Product Type. *Pakistan Journal of Commerce and Social Sciences*, *12*(2), 598–616.

Sheinin, D. A., Varki, S., & Ashley, C. (2011). The Differential Effect of Ad Novelty and Message Usefulness on Brand Judgments. *Journal of Advertising*, *40*(3), 5–17. doi:10.2753/JOA0091-3367400301

Shetty, K., & Fitzsimmons, J. R. (2022). The effect of brand personality congruence, brand attachment, and brand love on loyalty among HENRYs in the luxury branding sector. *Journal of Fashion Marketing and Management*, *26*(1), 21–35. doi:10.1108/JFMM-09-2020-0208

Shin, H., Lee, H., & Perdue, R. R. (2018). The congruity effects of commercial brand sponsorship in a regional event. *Tourism Management*, *67*, 168–179. doi:10.1016/j.tourman.2018.01.016

Song, H., Wang, J., & Han, H. (2019). Effect of image, satisfaction, trust, love, and respect on loyalty formation for name-brand coffee shops. *International Journal of Hospitality Management*, *79*, 50–59. doi:10.1016/j.ijhm.2018.12.011

Stejskal, P. (2015). *Music Festival Sponsorship: In Investigation of the Role of Brand-Event Congruence*. [Master Thesis, Graduate School of Communication. Universiteit van Amsterdam].

Thompson, K. J., & Schofield, P. (2009). Segmenting and Profiling Visitors to the Ulaanbaatar Naadam Festival by Motivation. *Event Management*, *13*(1), 1–15. doi:10.3727/152599509789130601

Thomson, M., MacInnis, D. J., & Park, C. W. (2005). The ties that bind: Measuring the strength of consumers' emotional attachments to brands. *Journal of Consumer Psychology*, *15*(1), 77–91. doi:10.120715327663jcp1501_10

Urry, J. (2002). *Tourist gaze* (2nd ed.). Sage Publications.

Usakli, A., & Baloglu, S. (2011). Brand personality of tourist destinations: An application of self congruity theory. *Tourism Management*, *32*(1), 114–127. doi:10.1016/j.tourman.2010.06.006

van Vliet, H. (2021). Why do we go to festivals? Festivals and Experience. Amsterdam University of Applied Science. doi:, 2(19799.80805). doi:10.13140/RG

Wallace, E., Buil, I., & de Chernatony, L. (2017). Consumers' self-congruence with a "Liked" brand: Cognitive network influence and brand outcomes. *European Journal of Marketing*, *51*(2), 367–390. doi:10.1108/EJM-07-2015-0442

Wang, Y. D., & Emurian, H. H. (2005). An overview of online trust: Concepts, elements, and implications. *Computers in Human Behavior*, *21*(1), 105–125. doi:10.1016/j.chb.2003.11.008

Chapter 3
Why Do Festivals Fail?
A Case Study of Sonepur Cattle Fair Festival

Sushant Kumar
Indian Institute of Management Lucknow, India

ABSTRACT

This chapter explores the incidences and causes of failure of festivals and other threatening issues that might damage an event or the organization responsible for organizing the event. Literature often offers examples of success stories of festivals, but failures are more difficult to identify, evaluate, and document. A greater understanding of potential antecedents of failures will surely equip managers to avoid consequences and plan for contingencies. This chapter utilizes anecdotal evidence from various festivals to set the stage. This is followed by adoption of case study research design to explore the cattle fair festival organized at Sonepur, Bihar, India. Qualitative data have been collected from various sources along with field notes to analyze the reason of failure of festival. Findings are discussed in several themes that offer guidelines and framework to managers which helps in understanding and utilizing the findings in context of any festivals.

1. INTRODUCTION

Festival organizers face several challenges while organizing the festival that present the opportunities as well as threats for festivals. A crucial challenge that festivals face is the need to innovate across several areas of festival management such as strategic planning, budget planning, promotion of festivals, and human resource planning. Innovation also brings the risk of failure which if not controlled may lead to discontinuation of festivals. Festival organizers understand the dual challenges of employing innovation to improve efficiency as well as avoiding failure. The set of actions taken by festival organizers may determine the fate of festivals. Thus, it is vital that this topic is to be studied. Although challenges faced by any festival are universal, the way festival organizers respond to these challenges may set the course of the festival life cycle determining its success or failure.

DOI: 10.4018/978-1-6684-6356-7.ch003

It is widely known that festivals are often a major constituent in the economic activity of state or nations. Foreign and domestic tourists, both alike, make a substantial contribution to the economic and cultural health of a nation. Part of the services sector, tourism is travel, hotels, ecology, and many other services that support tourism, customer, and enterprise. Perhaps more than other, the tourism sector is an enormous employer. Local entrepreneurial impulses are relatively easier to channel into tourism, with or without protection.

Central to festivals is the tourists who come to experience the 'cultural heritage'. This heritage encompasses tangibility represented by the precincts of historical constructions like sites, buildings, and various structures of historic significance, as well as the intangibility of traditions, folklore, lifestyle associated with such cultures (Can, Ekinci, & Pino, 2021; Kumar, Giridhar, & Sadarangani, 2019; Naramski, Szromek, Herman, & Polok, 2022; Vecco, 2010). The tourism industry operates on local residents seeking online services like information from the internet to download information for spending their leisure time, it operates on non-local tourists who come for the experiential and emotional elements from other parts of India and elsewhere in the world (Parker, Sears, & Smith, 2012).

Festivals help in a great deal in destination marketing activities. Studies acknowledge that successful festivals strengthen the destination competitiveness, providing the competitive advantage in attracting tourists (Kumar & Dhir, 2020; Goffi, Cucculelli, & Masiero, 2019; Kubickova & Martin, 2020; Shoval & Birenboim, 2019). Further, aspiring destinations that aim to be recognized as popular destinations for national as well as international tourists are required to invest significant efforts in managing the festivals and making them successful (Fernández, Azevedo, Martin, & Martin, 2020). However, this is not always true. i.e., there were several instances of failing festivals, and some failed so miserably that they were discontinued.

Although the previous literature on tourism and hospitality has identified different factors that can contribute to destination marketing through festivals (Choi, Kang, & Kim, 2021; Lee & Jan, 2022) include branding (Xia, Vu, Law, & Li, 2019), advertisement (De Souza, Mendes-Filho, & Buhalis, 2019), green initiatives (Singjai, Winata, & Kummer, 2018), and the customization of services (Shoval & Birenboim, 2019), failure of festivals are rarely studied. There are several factors which may inhibit the growth the festivals and ultimately contribute towards the failure of festivals. This study aims to understand the reasons as why festivals fail. Thus, the research objective of this study is as follows: to explore the factors that contributes to the failure of CFF.

To address this, the study adopts a case study method approach. Here, the author study one of the largest cattle fair festivals in Asia, which is organized at Sonpur, Bihar in India. A qualitative research design is adopted where stories from several stakeholders were collected and analysed using grounded theory approach. Qualitative study uses inductive analysis which helps in development of new theories (Kumar & Sadarangani, 2018a) or conceptual model (Kumar, 2017).

Qualitative research is best suitable to explore the meaning associated with the event. Since, the Cattle Fair Festival (CFF) is organized at a grand level, there are millions of people who come to visit and see this amazing festival. Through a rigorous process, we identified the important stakeholders and collected narratives to understand as why CFF is losing its charm. We propose a framework which helps us to understand the reasons as why festivals fail.

2. HISTORICAL IMPORTANCE OF FESTIVALS

Festivals are understood as the designated period or programs of entertainment, pleasurable activities or events that include festive traits which are publicly commemorated. The public celebratory characteristic as mentioned in this definition are crucial since events have existed since very long and reflect the cultural significance. Our cultural practices created festivals as the form of wider display to public fostering collective celebration and civic rituals. Scholars (e.g. Turner, 1982) agree that these festivals are designed in all the cultures as these cultures realized the need to set aside certain time and space for collective celebration and creativity. The communal celebration of the festival dates back centuries. Primarily, these festivals were aligned with the rhythms of agrarian society. Most of these festivals were religious in nature as many of the festivals existed in Middle Ages. During renaissance times, there was wider dissemination of festivals, and the public display of civil rituals are on the rise. Due to these activities, the imperial as well as international exhibitions became part of collective responsibility as well as communal imagination in Europe in mid of 19th century. Historians have consistently emphasized that during these initial period festivals often summarize identity for the nation, state, a sense of place as well as reflect the heterogeneity among the people.

It is well known that festivals are a strong attraction for tourists and often destinations are known by the festivals that they conduct. For example, Greenwich which is recognized as the fulcrum of the earth's time zone stirred 6 festivals which lasted for a year, attracting international tourists to the city. Similarly, in 1859, the Handel Centenary festivals organized at the London Crystal palace attracted several international tourists who were informed through the process of extensive promotion that included distributing leaflets in government offices. These festivals which are transient but recurring in nature often provide a sense of collective identities to the societies hosting the festivals. When a festival fail, it endangers the local continuity and reduces the opportunity to reinforce, assert, reproduce, or contest the prevalent social norms, beliefs, and cultural values (Kumar, Giridhar, Sadarangani, & Gogoi, 2018). Thus, it is important the festivals are renewed periodically, which propagates the life stream of a community or society and provide cultural sections to its institutions (e.g. Kumar, Baishya, Sreen, Sadarangani, & Samalia, 2021).

When a festival is used as hallmark event, it serves the greater purpose by encouraging and strengthening the vision of established entity irrespective of religion. This gives the impression that festivals have a long history that encompasses the trajectory and rituals of several past events. A decline in festivals were observed from mid of 20th century onwards and again the festivals flourished in contemporary societies. The recent increase is observed by scholars globally and is associated with their potential to attract tourists. Multiple demand-driven reasons guide the growth which includes need for socialization, demand for luxury leisure, and a consumption orientation or experience orientation of tourists. On the supply side, according to urban studies literature, the wider proliferation of festivals is governed by urban restructuring processes.

For instance, scholars argue that festivals are projected as urban ephemera offering economic benefits by offering employment, increasing the sale of products, strengthening the attractiveness of destinations, related institutions, and improving the flow of capital, resources of investment along with tourists. Many states or countries use festivals as an instrument to improve the declining economies. Towards the end of the 20th century, festivals have become part of cultural strategies (e.g. Kumar, Baishya, Sadarangani, & Samalia, 2020) that helps in regenerating and orienting the economies to consumption. In contemporary consumption drive societies, festivals play a greater role in contributing holistic development of societies where luxury, leisure, tourism underpin an experience economy.

Cities which aspire to become global destinations organizing festivals are a larger part of destination marketing strategies, supported by globalization and competitions among cities. For example, Gwangju Biennale Festival organized in South Korea is used a representative of recent events where it is used to create an image of the city of art. The same image was utilized by local government to build the perception about other South Korean cities. Towards the start of the 21st century, festivals are continued to use as a strategic tool to attract tourist and improve the local economies. Increasingly technologies have been used during festivals so that tourists who may not physically attend the festival, can get the experience of the festival. The use of technologies was also boosted by COVID-19 pandemic. During COVID-19, a majority of the festivals were postponed, cancelled, or discontinued. Some festival organisers took the festivals online which boosted the need to use technologies. Increasing organizes are using technologies such as Augmented reality, virtual reality to attract the tourists.

Now, let's shift our attention to a particular festival which was already seeing a decline and hit severely by the COVID-19 crisis.

3. CASE STUDY OF CATTEL FAIR FESTIVAL (CFF)

The Cattle Fair festival is locally known by the name of Sonepur Mela. Sonepur is situated 25 KMs from the state capital of Bihar, Patna. It is inhibited by approx. 38,000 people. It is scheduled during the auspicious month of Kartik or November in Northern Bihar area Sonepur. The period of CFF is auspicious as it is held on Kartik Poornima (Full moon night). The place transforms into a sacred land where millions of devotees from all over India assemble to take a holy dip in the river Ganga and Gandak. They also visit Harihar Nath Temple to offer their prayers. That is why this fair is also known as Harihar Kshetra Mela or Chhattar Mela. Locals believe that Harihar temple was built by Lord Rama during his onwards journey to King Janak's court for marriage. Thus, it bears great significance to among the locals and tourists from pan India come to pray here.

Sonepur is a town in the Bihar state that has a rich history of holding CFF since very early as Sonepur is located near the river the Gandak and offers several natural advantages. The place where CFF is held situated at the confluence of holy river Gandak and sacred river, the Ganga. In CFF, animals such as camels, donkeys, horses, goats, birds are traded. *Haathi Bazaar* (Elephant market) until 2004 was a major attraction of this CFF. One can find the elephants to be beautifully decorated and decked for the purposes of sale.

The state run CFF has its origin in ancient history. Emperor Chandragupta Maurya used to buy animals primarily horses and elephants for his army from Sonepur and near to the Ganga region. CFF was also celebrated as the largest marketplace for elephant sale in the world. Later on, the primary buyer of animals were forest department and other businesses which need these animals for operational and logistical purposes.

Fairly unscathed by business tourism, CFF offers a glimpse into the everyday lives of rural India at the extravagant scale. The CFF is scheduled for 15 days and keeps going on for more than a month. Similar to any festivities, the first few days are very crowded. However, slowly with time, the footfall for CFF has reduced. There are certain factors which contribute to this reduced footfall which is investigated in this study.

4. RESEARCH DESIGN

To understand the failure of CFF, it is important to understand how people construct the meaning of events. This is possible by following a qualitative research paradigm where it is deciphered as how people make sense of events and activities around them (Kumar & Sadarangani, 2018; Sreen, Mukherjee, Jebarajakirthy, Kumar, & Sharma, 2022). Qualitative research has certain characteristics such as natural setting, multiple sources of data, inductive and deductive analysis, emergent design, and holistic account (Kundra, Kumar, Sreen, & Dwivedi, 2020). Few characteristics are criticized by scholars which paved the path for new approaches such as abduction. In qualitative research, interpretation and meanings are negotiated with collected data from participants as this represents the realities of participants that scholars try to reconstruct by multifaceted and iterative deduction approaches (Creswell, 2014).

In this chapter, author aims to decipher the lived experiences by collecting data from participants as well as on account of secondary information available on news articles. The chapter establishes a richer and thicker understanding of the complex concern of CFF. The case study approach would allow to have a rich and thick descriptions of the failure of CFF. Also, it would offer a framework and understanding of failure of CFF which is very limited in extant literature.

In this study, 11 individuals were contacted and requested to participate in the interview process. These individuals are associated to the CFF wither directly or indirectly. Although qualitative studies do not recommend a particular number of respondents for sampling, researchers often suggest 2 to 30 are sufficient as it depends on the theoretical saturation (Kundra et al., 2020). Theoretical saturation refers to a situation during analysis where the new data analysis does not result in the emergence of new code or categories (Sreen et al., 2022). Following other qualitative study, this study follows the convenience sampling to identify respondents. The interview followed a semi-structured approach where some of the questions were pre-conceived whereas some questions were asked to get the responses from participants. Here, probing technique is used to get the narratives from participants. Out of these 11 participants, 3 were tourists who were attending these festivals since last 10 years and aged around 52 years. Further, data were collected from 2 local entrepreneur who set up these shops during the festivals near the festival location. The study also interviewed 2 government representative who offered the perspective from organizers. Further, 4 young tourists aged around 30 years were also interviewed as these generation Y individuals have different thought process than other generations (Kumar & Sadarangani, 2018b; 2021). Few of these individuals visited the CFF only once and offered the perspective from the younger population. Pseudo names have been used to protect confidentiality.

4.1. Data Collection and Analysis

All the ethical guidelines were followed while conducting interviews. The narratives were recorded and transcribed verbatim. The combined transcribed narratives were captured in 211 pages with double spacing. Open coding processes were followed to code the data. For this, one researcher reviewed all the response, and the complete narratives were read in its entirety. In this phase of analysis, familiarization is the first step where a holistic view of each participants' narrative is done before coding. The complete narratives were multiple times to understand the greater picture. After that, each individual participants narrative was taken and read while assigning the code. In this step, a narrative fragment was taken, and a code was assigned to this fragment which represents the broader theme of the respective fragments. This process is called open coding. The similar process is followed for all the participants narrative. This step resulted in around 437 open codes. Inter-coder reliability was also examined where another scholar was asked to code, and a comparative matching was performed. The majority of codes matched offering an inter-coder reliability value of 0.8 which ensured the coding process is scientific. After open coding, these codes were merged, renamed, or transformed into other codes through the process of axial and selective coding. After that codes were arranged in a tree format to understand the themes of the data. After this process, once a list of codes and themes were confirmed, the details were sent to an external researcher to check and verify the coding scheme. Any confusion or discrepancy were discussed until it resolved through a mutual consensus.

In addition to the participants' narratives, the study also collected data from newspaper articles and various other accounts such as websites, personal account of the author's visit to CFF to triangulate the presented information. Through this, the trustworthiness in the data is ensured. Further to improve the robustness of the findings, trustworthiness and credibility were achieved by following three steps as suggested by Patton (2002). First, the coded data were kept in the electronic format to offer audit trail and shared among university scholar and academicians as well as external experts. This allowed any scholar to review given codes. Second, as mentioned previously, the consensus on data coding and themes were reached by involving other scholars who were expert in qualitative research, to confirm and ensure the appropriateness of coding structure. This improved the credibility of findings as themes and data have been reviewed without bias. In addition, added remarks were included to improve the credibility of findings and gave a tone to the theme that were revealed within the data. Lastly, the analysis approach also assigned an arithmetic code to the participants narrative in sync with the broader theme to which they belong. This step offered a quantitative profile to the narratives.

5. RESULTS AND DISCUSSION

After analysis, the failure of CFF were presented over four themes. The themes have been created around the three-dimensional metaphoric narrative inquiry space of people, place, and travel. These themes are: (1) Strategic planning failure (2) Mismanagement of festivals, (3) Regulations and legal constraints and (4) Competition from other/similar events.

5.1. Strategic Planning Failure

Respondents who have been witnessing this event for long years and have been attending the festival express their displeasure. Alka, 29 years old young lady from Hajipur, Bihar, who used to come to the festival since early childhood expresses her concern.

It's not the way it used to be. Previously we used to have a lot of events and small games. Many people used to come from across the world. It was pleasing to see few foreigners coming to our distant village. You can identify them from colour of their hair. Also, you can see all the animals such as the cow, goat, horses, elephants, and different types of birds. All are brought for sale. There were several food stalls which sold delicious street food such as Chaat and Pakoda. But things have changed nowadays. There were several events which used to happen earlier, does not happen now. That's why I have stopped going. I was regularly going for last 20 years. Now it feels that festival lost its charm. Food stalls are there but not exciting enough. A smaller number of people are interested in coming these days. You can't see the animals that one can see in previous years. Birds are also rare. I think it is because of technology. People don't find it interesting to come and visit a cattle fair festival unless you are related to cattle in some ways.

As outlined in the narratives of Alka, there are certain essential aspects of festivals which the organizers failed to continue. Figure 1 represents a picture of birds kept for sale in Sonepur Cattle Fair Festival.

Figure 1. Flocks of birds kept for sale
Source: Bihar Tourism

The respondent who worked in the organizing committee illustrated the political unwillingness to project this festival at grand level. Ram, 62 years old and retired person who used to work in the organizing committee. He expresses that

There are certain people in government who had the intention and willingness to organize the festival at grand level. But the majority of the people who worked in organizing team take their decision from senior without any justification. Often a time, it happens that the allotted budget will not meet the required expenditure for Mela. But no one questions the decision. Things will be arranged in budget after accounting for share of individuals. It means even if political willingness is there, the people may not organize it properly. You can often see there are often minimum and temporary arrangements done with poor quality of materials. One can also see several illegal activities happening during festivals. You know...this mela used to be a place for meeting of freedom fighter before independence. But after independence, insurgents used to come here, and government did not work to curb those things. Earlier, people used to fear while coming here. Now, it has reduced and faded with time.

Lack of political support in organizing the CFF is clearly seen the narratives. It is not only the political party who is in power but also the people responsible for organizing the CFF, should be taking a stand and acting upon it. People will be unwilling to go where they might anticipate some unwanted actions. People attend festivals for enjoyment and learning new things. Bureaucracy should be working for interest of people.

Another respondent highlighted that many young people are not aware about this CFF. Few who are aware don't find it exciting to come and witness the festivals. 21 years old Madhu discusses as how she surprised; she was when she got to know that her friends don't know Sonepur Mela.

Government should spend more money in promoting the Sonepur Mela. In my college there are many girls who don't even know that there is festival dedicated to cattle. I belong to a generation where Facebook, YouTube, and WhatsApp are tremendously used by people. I am aware of so many festivals organized in different countries so are my friends. But they don't know about the one of the largest cattle fair festivals which is held in our vicinity. It is not like they are completely unaware; they are born here, and they have been their parents going to this mela. We are a farmer family and so are my friends. Farmers definitely visit the Sonepur Mela as they will get a chance to which breed to cows or oxes are available for same. But they have never gone to the festival and realized the grandeur nature of this festival. The government should promote this mela in all the Indian state as well as it has to be taken to the international level. You can imagine without any promotion, so many people used to come there, thus, after promotion many more people will land here. Else, the festival will die slowly.

There are certain operational issues which if not resolved create bigger problem through bull-whip effect. Although it is well known that all festivals face operational issues, organizers can forecast them and act accordingly. Siddharth, 45 years old, illustrated how difficult it is to solve some operational issues.

I have been worked in organizing the Meal for long. It is also obvious that some problems will occur, and we are tasked to solve this problem. But sometime some problems are getting frequented every year and there is no solution. Cash flow is one such big problem. See, if we don't have money, we can't start organizing the festival. Sometimes the budget and sanction related files gets stuck in certain offices which delayed the whole process. By the time we got the money, enough time has already passed. Sonepur Meal is a very big festival. Organizing this kind of festivals takes several months, and we start working at the last hours. Sometime the issues of excessive expense happen in organizing the festival which leave very little room for other activities such as cleaning the areas post event.

There are multiple reasons for the failure of the festival. Solving operational issues and ensuring that its recurrences are avoided are certain measures that organizers must do. It has been observed that if some of these operational issues if left unchecked may create a disastrous situation damaging human and capital. In certain cases, due to these disastrous situations few festivals failed miserably and cease to exist.

5.2. Mismanagement of the Festival

The CFF attracts millions of people who come to Sonepur by several means of rail, air, and road transport. One of the respondents who was attending the festivals expresses the concern for the tourist.

If you visit Sonepur Mela, you will fine several straw huts organized for guests by Bihar Tourism. These are temporary arrangements with attached western bathrooms. I have stayed in those camps. These were very near to the festival site. But I feel the quality of arrangement can be improved. There is no security. Some people who stayed there have reported that their belongingness was missing. Tourists also have the option for staying in the bungalow where one can ask for room, the dormitory to get the hotel-types of services. However, people who are handling the services are not trained it seems. There is no courtesy or professionalism that one can find. These arrangements are not adequate. There are several instances when the incidence of theft has happened. I take caution every time I go there. Now, I don't wish to go as in my last visit I lost something valuable. There is no security and tourists suffer.

Another respondent also displeased by the festival related arrangement and decided not to attend.

There was no proper arrangement of washroom. Although there are certain temporary constructions, these are not at all adequate. If you look this kind of event from my point of view, I would like to attend those festivals where the basic necessities such as clean and secure washroom is available. It becomes really uncomfortable to go to attend nature's call. Also, one can't think of going back all the way home and come back again. This is a very crowded place and going and returning make take hours. Also, there is no arrangement of clean drinking water. Organizers must organize festivals by considering these basic necessities in mind. You know...people come from all over the globe for Sonepur Mela. Countries such as Japan, France, Switzerland, US, Portugal. What they will think of our country if we fail to provide the basic necessities of safety, security, clean washroom, and drinking water. I know that we will not change, so I had stopped going to the Sonepur mela.

Akash, a local entrepreneur also expressed his discontent.

Now festivals are not the way it used to be. In last few years, it has changed drastically. Theft was quite common. People have lost several items in Sonepur Mela. The police force is inadequate in number as well as ineffective in handling cases. Organizers should take care of this. Also, organizer should create separate facilities for different types of animals. One basic facility which is a platform structure or place where you can feed the animals or offer them water is inadequately available. There are several basic features of a Cattle fair which is missing from Sonepur Mela.

5.3. Regulations and Legal Constraints

During the COVID-19 pandemic, the festival was cancelled for year 2020. In year 2021, the state government cancelled the official fair for year 2021 as well citing the COVID-19 outbreak reasons. The tourism minister illustrated the concern of state as well as addressed the concern of people as follows:

Even if the restriction has been relaxed, big gatherings are not permitted. However, since this is an auspicious occasion, the rituals and prayers will happen as scheduled while maintaining social distance. The state is responsible to make these arrangements.

Here, despite the apprehensive of the COVID-19 outbreak, the state cancelled the fair. However, realizing the sentiments and importance of the rituals on *Karthik Purnima* (Full moon) the state relaxed the constraints when social distancing is observed.

There are certain other constraints which have affected the people's love for CFF. Sam, 30 years old male working in a multi-national company expresses his remembrance.

I remember when I was a kid, going to the Sonepur Mela to witness lots of animals was something for which I will wait for a year. Cows, goats, horses are major attraction to me. Elephants were major charm. It is very heartfelt to see a very big animal to which to we have studies only in books. My parents used to take me there…so does the parents of my friends. I used to stay there and watch elephants taking a dip in the river. Sometimes they pooled water in the trunk and throw it on people. I used to clap by seeing those actions. Elephants are singular fasciation that separated these festivals from any other south Asian festivals. Nowadays elephants don't come to Sonepur Mela. Bihar government started organizing this festival officially since 1990s and enforced Wildlife Protection Act, 1972 in 2004 which prohibited the sale of birds or animals including elephants. In festivals, there used to be market Angrezi Bazaar, which was famous for presenting and selling the rare and exotic animals. All these are history now. I don't go the Sonepur mela as I believe it has lost its charm, its major attraction.

Figure 2 represents the picture of Sonepur Cattle Fair Festival where elephants can be seen in fair diving in the holy river.

Figure 2. Devotees performing prayer rituals alongside elephant
Source: Bihar Tourism

5.4. Competition From Other/Similar Events

The failure of festival happens as there are other festivals which became more attractive agency than the previous one. There are several cattle fair festivals which are now being organized in different states of India. On account of secondary data, it was found that cattle fair festival at the Torvi village is attracting a huge crowd of sellers and buyer. Torvi is a small place in Karnataka state which is adjacent to the state of Maharashtra. The festival organizers roped in the Department of Animal Husbandry to support the treatment of animals during the festival. The arrangements have also been made to cater 100,000 farmers and visitors. Although this visitor count is less compared to the footfall at Sonepur CFF, unless some crucial decisions are taken Sonepur CFF will become an unattractive proposition for tourists. Tourists have already stopped going due to various reasons. No elephants in the Sonepur CFF as trading elephants are declared illegal. The quality of horses have also dropped. The bird market or Chiriya Bazaar is still present but requires renovation. Figure 3 represents the reasons for failure of CFF.

Figure 3. Schematic representation of failure reasons of CFF

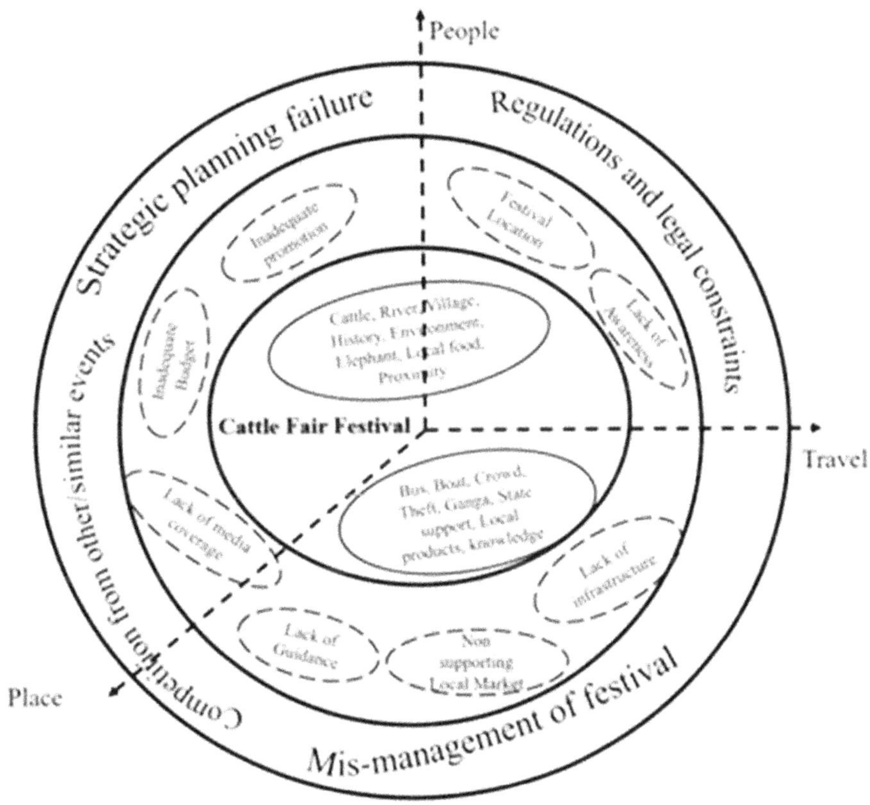

6. CONCLUSION AND RECOMMENDATIONS

The findings suggested four theme which may contribute to the failure of a festival. Festival managers are recommended to design the strategic planning as this emerged as a prominent factor of failure. This includes planning for all the regular activities as well as account for contingencies. Using adequate promotion to attract the new generation is very much required. A strong media planning is also suggested. This include having a separate public relation team for festival which will be active months before and after the event. Second prominent factor is mismanagement of festival which include the well managed infrastructure, sanitation, safety, and security. Although festival managers may not be able to alter the law and regulation, they should be strictly enforcing the rules. This offers a kind of security among tourists that this festival is legal, does not support any illegal activity, and safe. For example, since sale of elephants is ban, it should not be getting sold in the CFF or anywhere near to it. Also, presence of well-trained staff to deal with domestic or foreign tourists give confidence to the tourists.

There are many success stories of festivals. However, not all festivals succeed. There is significant proof that several festival faces challenges that bring existential crisis for organizers and some festival eventually fail. This is a global phenomenon, thus understanding the reasons for failure of festivals provide organizers with strategies and ideas to avoid failure. There are specific objectives of festivals and can be equated to types of organization for the sake of understanding. There are not for profit festivals which are organized to increase awareness and keeping the benefits of larger society in planning. There are often organized and run by volunteers and aimed not to make profit. These kinds of festivals are dependents on external sponsorship and public scrutiny is high during execution. On the other hand, there are festivals which are organized to make profit by the organizers. There are certain festivals which are organized due to its historical and cultural advantages. Most of these festivals are state run or funded by NGOs or a particular not-for profit organizing body. When looked carefully, these festivals have different success and failure factors. This chapter investigates one such festival, Cattle Fair Festival, which seeks its origin in the ancient Indian history. Through the case study of CFF, the chapter outlines the factors responsible for failure of CFF. Although it can be debated whether CFF has failed, CFF has definitely gone through a lot of transformation and on account of several narratives CFF has failed.

REFERENCES

Can, A. S., Ekinci, Y., & Pino, G. (2021). Joint brand advertising for emerging heritage sites. *Annals of Tourism Research, 91*, 103294. doi:10.1016/j.annals.2021.103294

Choi, K., Kang, H. J., & Kim, C. (2021). Evaluating the efficiency of Korean festival tourism and its determinants on efficiency change: Parametric and non-parametric approaches. *Tourism Management, 86*, 104348. doi:10.1016/j.tourman.2021.104348

Creswell, J. W. (2014). *A concise introduction to mixed methods research*. SAGE publications.

de Souza, J., Mendes-Filho, L., & Buhalis, D. (2020). Evaluating the effectiveness of tourist advertising to improve the competitiveness of destinations. *Tourism Economics, 26*(6), 1001–1020. doi:10.1177/1354816619846748

Fernández, J. A. S., Azevedo, P. S., Martin, J. M. M., & Martin, J. A. R. (2020). Determinants of tourism destination competitiveness in the countries most visited by international tourists: Proposal of a synthetic index. *Tourism Management Perspectives*, *33*, 100582. doi:10.1016/j.tmp.2019.100582

Goffi, G., Cucculelli, M., & Masiero, L. (2019). Fostering tourism destination competitiveness in developing countries: The role of sustainability. *Journal of Cleaner Production*, *209*, 101–115. doi:10.1016/j.jclepro.2018.10.208

Kubickova, M., & Martin, D. (2020). Exploring the relationship between government and destination competitiveness: The TALC model perspective. *Tourism Management*, *78*, 104040. doi:10.1016/j.tourman.2019.104040

Kumar, S. (2017). A Conceptual Paper on Dark Social–Social Cognitive and Social Capital Theory Perspective. *International Journal of Advanced Studies in Computer Science and Engineering*, *6*(9), 1–11.

Kumar, S., Baishya, K., Sadarangani, P. H., & Samalia, H. V. (2020). Cultural Influence on e-Government Development. *Electronic Journal of Information Systems Evaluation*, *23*(1), 17–33. doi:10.34190/EJISE.20.23.1.002

Kumar, S., Baishya, K., Sreen, N., Sadarangani, P. H., & Samalia, H. V. (2021). Impact of national culture on E-government development: A longitudinal study. [JGIM]. *Journal of Global Information Management*, *29*(2), 1–22. doi:10.4018/JGIM.2021030101

Kumar, S., & Dhir, A. (2020). Associations between travel and tourism competitiveness and culture. *Journal of Destination Marketing & Management*, *18*, 100501. doi:10.1016/j.jdmm.2020.100501

Kumar, S., Giridhar, V., & Sadarangani, P. (2019). A cross-national study of environmental performance and culture: Implications of the findings and strategies. *Global Business Review*, *20*(4), 1051–1068. doi:10.1177/0972150919845260

Kumar, S., Giridhar, V., Sadarangani, P. H., & Gogoi, B. J. (2018). A Cross-Cultural Study on Tourism: Strategies and Implications of Findings. In *Global Conference on Business and Economics (GLOBE 2018)* (p. 5).

Kumar, S., & Sadarangani, H. P. (2018). A Narrative Inquiry of Consumption. Shailendra Pratap Jain and Akshaya Vijayalakshmi, (eds) ACR Asia-Pacific Advances in Consumer Research 12, 79-79. Duluth, MN: Association for Consumer Research.

Kumar, S., & Sadarangani, P. (2021). An empirical study on shopping motivation among generation Y Indian. *Global Business Review*, *22*(2), 500–516. doi:10.1177/0972150918807085

Kumar, S., & Sadarangani, P. H. (2018). Study of shopping motivation and buying behaviour among generation Y in India. *ISSN*, *7832*(2362), 341.

Kundra, S., Kumar, S., Sreen, N., & Dwivedi, R. (2020). A qualitative inquiry on determinants of employee satisfaction: Evidence from it firms. [IJM]. *International Journal of Management*, *11*(8), 1207–1225.

Lee, T. H., & Jan, F. H. (2022). How does tourism image affect visitor perceptions of a festival's identity? Influence analysis of three aboriginal festivals in Taiwan. *Journal of Destination Marketing & Management*, *24*, 100704. doi:10.1016/j.jdmm.2022.100704

Naramski, M., Szromek, A. R., Herman, K., & Polok, G. (2022). Assessment of the Activities of European Cultural Heritage Tourism Sites during the COVID-19 Pandemic. *Journal of Open Innovation*, *8*(1), 55. doi:10.3390/joitmc8010055

Parker, R. D., Sears, D., & Smith, R. K. (2012). International design concepts in Internet tourism marketing: Comparing web-design practices in Atlantic Canada and New England. *Academy of Marketing Studies Journal*, *16*(2), 65.

Patton, M. Q. (2002). Two decades of developments in qualitative inquiry: A personal, experiential perspective. *Qualitative Social Work: Research and Practice*, *1*(3), 261–283. doi:10.1177/1473325002001003636

Shoval, N., & Birenboim, A. (2019). Customization and augmentation of experiences through mobile technologies: A paradigm shift in the analysis of destination competitiveness. *Tourism Economics*, *25*(5), 661–669. doi:10.1177/1354816618806428

Shoval, N., & Birenboim, A. (2019). Customization and augmentation of experiences through mobile technologies: A paradigm shift in the analysis of destination competitiveness. *Tourism Economics*, *25*(5), 661–669. doi:10.1177/1354816618806428

Singjai, K., Winata, L., & Kummer, T. F. (2018). Green initiatives and their competitive advantage for the hotel industry in developing countries. *International Journal of Hospitality Management*, *75*, 131–143. doi:10.1016/j.ijhm.2018.03.007

Sreen, N., Mukherjee, S., Jebarajakirthy, C., Kumar, S., & Sharma, H. (2022). Role of personal values on revisiting green hotels: A hierarchical value map approach. *Tourism Recreation Research*, •••, 1–21. doi:10.1080/02508281.2022.2076190

Turner, V. (1982). Introduction. In V. Turner (Ed.), *Celebration: Studies in Festivity and Ritual* (pp. 11–29). Smithsonian Institution Press.

Vecco, M. (2010). A definition of cultural heritage: From the tangible to the intangible. *Journal of Cultural Heritage*, *11*(3), 321–324. doi:10.1016/j.culher.2010.01.006

Xia, H., Vu, H. Q., Law, R., & Li, G. (2020). Evaluation of hotel brand competitiveness based on hotel features ratings. *International Journal of Hospitality Management*, *86*, 102366. doi:10.1016/j.ijhm.2019.102366

Chapter 4
The Effect of Perceived Authenticity on Satisfaction, Behavioural Intentions, and Expenditure

Ester Napolitano
University of Sassari, Italy

Antonio Salvatore Loriga
University of Sassari, Italy

Giacomo Del Chiappa
University of Sassari, Italy

ABSTRACT

The chapter investigates the influence of socio-demographics, travel-related variables, and perceived authenticity on attendee satisfaction, behavioural intentions, and expenditure level at a small-scale cultural event. A structured questionnaire was developed, and data was collected on-site during the "Autunno in Barbagia" event (Sardinia, Italy). Factor analysis was applied to items measuring authenticity, satisfaction, and behavioural intentions, and the resulting factor scores were entered in the OLS models. Findings show that objective and existential authenticity significantly and positively influence all the dependent variables. In addition, the three dependent variables were significantly influenced by certain travel-related variables and socio-demographic characteristics of attendees. Findings help deepen the scientific debate devoted to analysing the influence of socio-demographics, travel-related variables, and perceived authenticity on attendee behaviour at a small-scale event. Managerial implications are discussed and suggestions for further research are outlined.

DOI: 10.4018/978-1-6684-6356-7.ch004

INTRODUCTION

Since the 1990s, there has been a growing interest in events by academics and practitioners. Most previous studies have focused on the wide array of effects events generate onto the hosting destination, including economic, socio-cultural, environmental, physical, political, and tourism effects (e.g., Bowdin et al., 2006; Getz & Page, 2016). Several academic studies have highlighted the importance of staging events, as they are able to create high added value for both the tourists and the hosting community (Akhoondnejad, 2016), by acting as an important motivational boost for tourists (Getz, 2008). Others have stressed aspects related to the organisation and promotion of different types of events (cultural, political, entertainment, educational, sports, recreational, private, etc.) in order to help the development of destination branding strategies and actions, with positive effects on the competitiveness of the destination. According to these studies, events are useful marketing tools to further enhance the positioning of the destination identity and image, especially when considering destinations located in rural and peripheral areas. Regarding these areas, events, and in particular cultural events, often become an indispensable ingredient of the tourism offer (Getz, 2008). Thus, cultural events contribute to the transmission and the enhancement of the local identity, culture, and authenticity (Lee et al., 2008). Events exhort hospitality (Litvin & Fetter, 2006), strengthen residents' sense of belonging (Kim et al., 2006), and stimulate/facilitate networking among the various actors involved in the event (visitors, businesses, public institutions, local associations, etc.). Moreover, events have the potential to significantly support the local economy, while significantly contributing to place branding and to strengthening its link with the local identity, culture, and authenticity. In this context, authenticity has turned out to be a key attribute visitors seek when visiting destinations and attending events (Cohen, 1988; Getz & Page, 2016; Wang, 2007). In fact, nowadays, tourists are driven by the desire to enjoy an authentic experience (Lu et al., 2015), by feeling and touching the local identity and by learning about local gastronomy, lifestyles, traditions, and culture (Domínguez-Quintero et al., 2020; Kolar & Zabkar, 2010; MacCannell, 1977). This explains why existing literature (e.g., Almeyda-Ibáñez & George, 2017) suggests that a successful destination brand positioning should focus on valorising the local identity, culture, and authenticity effectively (Gilmore, 2002; Griffiths & Korstanje, 2021). In addition, in contrast to the postmodern approach, event authenticity is relevant for the place branding perspective given that this concept is primarily developed at a destination level; furthermore, it significantly influences the level of satisfaction towards the destination event (Handayani & Korstanje, 2017). Considering these elements, Getz and Page (2016) suggest that the planning and management of such events should not be left to chance. In fact, in today's highly competitive environment, it is crucial for destination managers to have a thorough understanding of the factors influencing tourist behaviour and satisfaction since these aspects determine the success or failure of the initiative over time (Del Chiappa & Atzeni, 2016). While cultural events have been increasingly regarded as critical components of destination branding, research investigating the antecedents of this kind of event experience and their impacts on tourist behaviour is lacking.

Existing studies concur that perceived authenticity positively influences tourists' satisfaction, their behavioural intentions (Robinson & Clifford, 2012), and their expenditure behaviour (Brida et al., 2013). Moreover, existing literature shows that satisfaction, behavioural intentions, and expenditure behaviour at destination are all influenced by socio-demographics (e.g., age, nationality, income, gender) and trip-related characteristics (e.g., previous experience, travel companion, length of stay, types of travel arrangement) (Valle et al., 2008; Park et al., 2019b; Park et al., 2020). However, there is still limited event-related academic research aiming at assessing whether, and the extent to which, perceived authenticity, socio-demographics (i.e., gender, age, country of residence) and trip-related variables (i.e., primary motivation to attend the event, previous participation at the event, length of stay, travel companion, type of travel arrangement) influence the level of satisfaction of participants at cultural events, their behavioural intentions, and their expenditure behaviour. This gap in the literature is even more evident with respect to small scale events in rural areas (Fotiadis & Vassiliadis, 2020). Thus, bridging this gap would be of paramount importance for both researchers and practitioners. Ultimately, a better comprehension of the antecedents of the event experience and their influence on tourist behaviour can improve, at a destination branding level, the management of these initiatives in the long term.

Thus, this study bridges this research gap by investigating empirically whether perceived authenticity, socio-demographic, and visit-related characteristics of event attendees at a small-scale cultural event are able to significantly influence their satisfaction, behavioural intentions, and expenditure behaviour. For the purposes of the study, three factor analyses and three OLS models were run on a sample of 1,639 individuals who attended the small-scale cultural event "Autunno in Barbagia" (hereafter AIB), held in a Sardinian sub-region (Italy) during the period September-December 2019.

The next section will review the literature pertaining to small-scale cultural events and will derive hypotheses, the second section will describe the methodology, and the third section will present results. The conclusions section will summarise theoretical and practical implications, while the last section will discuss limitations and implications for future research.

BACKGROUND

Authenticity

Authenticity has been regarded as an important intangible asset for destinations (Pearce & Moscardo, 1986), as it is highly effective in differentiating a certain place and in making it to be attractive for tourists (Robinson & Clifford, 2012). Based on existing literature, the search for authenticity is the result of the modern alienation and the commodification of culture (MacCannell, 1973) and, for this reason, it has been studied in relation to the tourists' motivation and experience (Wang, 1999). In the literature, many tourism studies (e.g., Brida et al., 2013a; Kolar & Zabkar, 2010) have focused on authenticity in order to investigate the meanings attached to this concept from a demand-side perspective, its underlying dimensions, and its influence on tourists' satisfaction and behavioural intention. Although a significant amount of academic studies have been devoted to authenticity, there is still a lack of consensus on how it can be defined and conceptualized. One stream of research considers authenticity as a recognisable aspect that concerns objects and cultures (Wang, 1999), where authenticity is referred to the *museum-linked usage of the authenticity of the originals* (Wang, 1999, p. 351). This type of authenticity has been named as objective authenticity since it is the original or historical artifact that gives an object legitimate power and authority. Kolar and Zabkar (2010) supported this perspective, maintaining that an authentic experience is the result of the recognition of toured objects as accurate and genuine, being made by local people according to the original custom and tradition. As Wang (1999) and Cohen (1988) suggested, the perception of authenticity can be different over time and according to the context. For instance, something initially perceived as inauthentic can become authentic over time. Moreover, the offering of artificial cultural products has led to the birth of staged experiences, specifically set up to please and appeal tourists looking for authenticity (Cohen, 1988).

Another stream of research argued that objective authenticity is insufficient to explain tourist experiences as psychologically perceived by tourists (Wang, 1999). In this case, authenticity has been regarded as existential, referring to the "*personal or inter-subjective feelings activated by the liminal process of touristic activities*" (Wang, 1999, p. 351). Existential authenticity involves both interpersonal and intrapersonal authenticity, since it is perceived from the conjunction among the visited place, the touristic behaviour, and the personal belief system. According to Kolar and Zabkar (2010), existential authenticity is related to the perceptions, feelings, and emotions of travellers when encountering and interacting with the local culture and community, and sampling local foods and beverages (Brida et al., 2013). Some researchers argue that these two approaches not just coexist, but they are also related to each other (e.g., Kolar & Zabkar, 2010; Park et al., 2019a). Similarly, Park et al. (2019a, p. 105) defined authenticity as "*the representation of a historical time and place within original or reconstructed sites and intangible characteristics as interpreted by a person and social culture*." For this reason, as Goulding (2000) claimed in her study, existential and object-based authenticity are equally important, and they should be simultaneously explored to better understand tourists' behaviour.

In recent times, some empirical studies have been carried out to investigate whether and how perceived authenticity is related to motivations and the subjective evaluation of experiences and/or activities offered at the destination. In this vein, perceived authenticity embraces both emotional and cognitive aspects of the experience (Del Chiappa & Atzeni, 2016), and was found to positively influence tourist satisfaction (Domínguez-Quintero et al., 2020) and attitudinal loyalty (Del Chiappa et al., 2019). Moreover, some studies found that the relationship between perceived authenticity and tourists' satisfaction is mediated by destination image (Lu et al., 2015). The above is probably amplified during small-scale cultural events, which allow tourists to experience the destination's true essence and uniqueness (Lu et al., 2015). Indeed, some destinations used to organize cultural events in order to offer an authentic experience to particular segments of tourists. Any failure in delivering an authentic experience in these events would compromise the uniqueness of the destination image and its related competitiveness (Park et al., 2019a). Other studies also highlighted that perceived authenticity significantly influences behavioural intentions, such as intention to return and/or to recommend to others both offline and online (Del Chiappa et al., 2019; Kolar & Zabkar, 2010), and visitors' expenditure behaviour (Brida et al., 2013; Chhabra et al., 2002). Based on some studies (e.g., Kolar & Zabkar, 2010), authenticity is a mediating construct between cultural motivation and tourist loyalty. Yet, Park et al. (2019a), in their study carried out in a cultural heritage tourism setting, showed that authenticity has a positive influence on satisfaction, revisit intentions, and recommendations to others. As far as the influence of perceived authenticity on expenditure behaviour, Brida et al. (2013) reported higher level of perceived authenticity being associated with higher level of expenditure while attending events. Similarly, Chaabra et al. (2002) found a positive relationship between the perceived object-based authenticity and the level of expenditure of cultural tourists. Based on these findings, the following hypotheses are identified:

H1: Perceived authenticity has a positive influence on visitors' satisfaction.
H2: Perceived authenticity has a positive influence on visitors' behavioural intentions.
H3: Perceived authenticity has a positive influence on visitors' expenditure behaviour.

Expenditure Behavior

When assessing the impacts of events, researchers concur that identifying and measuring the different socio-demographic factors that influence attendees' consumption behaviour is of utmost importance (Brida et al., 2013). In this regard, several studies have investigated the influence of socio-demographics and travel-related variables on the expenditure behaviour of tourists (Jang and Ham, 2009). Usually, two econometric approaches have been used to estimate visitors' expenditure behaviour (Brida & Scuderi, 2013). The most used is the one based on the analysis of Engle curve expenditure level (Marcussen, 2011), while the other one focuses on the analysis of the main determinants influencing the likelihood to purchase (Alegre & Cladera, 2010). In both of these approaches, four main categories of determinants have been usually considered, namely: budget constraints, socio-demographics, psychographic characteristics (personality, emotions, etc.), and travel-related variables (frequency of travelling, overnight stays, etc.). Estimating an OLS regression, with the expenditure level being used as the dependent variable, is the most used statistic method (Jang et al., 2004; Kozak, 2001; Marcussen, 2011). Broadly, studies pertaining to this research stream focus on studying the determinants of tourist expenditure level, understanding why consumers spend on tourism experiences, and identifying which goods or services tourists purchase at the destination (Marrocu et al., 2015). For example, Wang and Davidson (2010) examined Chinese holiday travellers' expenditure in Australia and showed that their total expenditure was influenced by their income, age, place of residence, length of stay, and travel party size. Similarly, Park et al. (2019c) showed that the level of education, the employment status, nationality, trip purposes, travel party, and information sources all significantly influence tourists' expenditure behaviour. Yet, Brida et al. (2013), found that a positive relationship exists between event attendees' length of stay and country of origin, on the one hand, and their total expenditure, on the other. Recently, Park et al. (2019b) found that first-timers at a Marathon event spent more than repeat attendees. Based on these considerations, this study aims at testing the following hypotheses in the specific context of a small cultural event in rural areas:

HP 4: Socio-demographic variables significantly influence visitors' expenditure behaviour.
HP 5: Travel-related variables significantly influence visitors' expenditure behaviour.

Satisfaction and Behavioral Intentions

Existing literature concurs that measuring customer satisfaction, expenditure behaviour, and behavioural intentions is of paramount importance for any business and tourism destination to remain competitive (Pizam & Ellis, 1999). Customer satisfaction is defined as a post-consumption judgement related to the customer "fulfilment response" that explains whether the customer likes or dislikes the service according to his/her expectations and the actual perceived performance of the experience (Oliver, 1997). Customer satisfaction has been also described as a consequence of an emotional response to a specific consumption experience, where positive or negative affect reflects satisfaction or dissatisfaction (Grappi & Montanari, 2011). When a specific setting of events is considered, past research refers to satisfaction as the overall evaluation of an attendee which encompasses both cognitive and affective aspects of his or her experience (Mason & Poggiaro, 2012). Satisfaction has been argued to be one of the most crucial elements for the success of any business and tourism destination (Kozak & Rimminmgton 2000). In fact, tourists' satisfaction drives their expenditure behaviour (Disegna & Osti, 2016) and behavioural intentions to revisit and/or to recommend the experience to others (Grappi & Montanari. 2011; Mason & Poggiaro, 2012).

It has been established that overall satisfaction during the travel could be influenced by different factors, with the socio-demographics and the behavioural ones being among the most analysed. The insights academic studies offer in this regard are sometimes contradictory. For example, Huh (2002) showed that tourist satisfaction was significantly influenced by gender, but not by age, country of origin, education level, or income. Valle et al. (2006) found that satisfaction was significantly influenced by age, educational level, and nationality, but not by gender or employment status. Regarding travel-related characteristics, Huh (2002) found that tourists' prior experience was able to predict satisfaction; however, no significant differences were found between satisfaction and travel party, length of stay, and distance to destination. Given that findings about the influence of socio-demographics and travel-related variables on satisfaction are rather contradictory, the following hypotheses are formulated:

HP 6: Socio-demographics significantly influence visitors' satisfaction.
HP 7: Travel-related characteristics significantly influence visitors' satisfaction.

Future intentions (i.e., intention to revisit and to recommend to others both offline and online) (e.g., Huang and Hsu, 2009) are important variables in tourism studies as they are generally used to measure the attitudinal dimensions of loyalty (Cole & Scott, 2004) which, in turn, contribute to shape the success and competitiveness of any business, events, and tourism destinations (Kolar & Zabkar, 2010). In this regard, the existing literature highlights both socio-demographics and travel-related variables as significant determinants of behavioural intentions (e.g., Correia et al., 2007; Valle et al., 2006); however, some inconsistencies in findings also exist here, particularly in small-scale cultural event research (Valle et al., 2008). Therefore, the following hypotheses are stated:

HP 8: Socio-demographics significantly influence visitors' behavioural intentions.
HP 9: Travel-related variables significantly influence visitors' behavioural intentions.

METHODOLOGY

The aim of this work was to analyse the influence of socio-demographics, travel-related variables, and perceived authenticity on satisfaction, behavioural intentions, and spending behaviour of attendees at "Autunno in Barbagia," a small-scale cultural event in rural areas in the island of Sardinia (Italy). This event is one of the most important cultural events at the regional level. It was launched 24 years ago in Oliena as a way to rediscover and exploit the local identity and authenticity (e.g., traditions, folklore, crafts, local cuisine, etc.) related to the local "Cortes" (courtyards of old houses). The event has grown yearly, and the number of rural villages belonging to the Nuoro Province has increased to 32 villages currently. The 2019 edition was held on weekends from mid-September to mid-December, involving around 2,500 businesses in the craft, restaurant, food and wine, and accommodation sectors, attracting more than 500,000 visitors and generating an economic impact estimated at 10 million euros (La Nuova Sardegna, Dec. 19, 2019).

For the purposes of the study, an ad-hoc survey instrument was created. Specifically, the questionnaire used for this study was developed according to a two-step structured approach. First, an in-depth literature review was conducted, focusing on studies analysing the event experience. Second, a group of professionals involved in the event organization were consulted to verify and validate the survey and to suggest potential changes, thus allowing us to adopt a theory-in-use approach (Zaltman et al., 1982).

The questionnaire included two main sections. In the first section, respondents were asked to provide some socio-demographic (i.e., age, gender, origin) and travel-related information (i.e., prior experiences with the event, primary motivation to attend the event, length of stay, travel party, type of travel arrangement, expenditure, etc.), and their average daily expenditure during the event. The second section asked respondents to assess their level of agreement or disagreement with a list of 19 statements, using a 5-point Likert scale (1=strongly disagree, 3=neither disagree nor agree, 5=strongly agree). The items were selected based on the literature review and adapted to suit the event-related setting of this study. More specifically, 9 items were adapted from Brida et al. (2013) and Robinson and Clifford (2012) to measure perceived authenticity, and 3 items to measure the overall satisfaction with the event were sourced from Babin and Griffin (1998). Moreover, 2 items to measure the respondents' willingness to revisit the area, and 5 items to measure their willingness to recommend the event to others both offline and online were adapted from Antón et al. (2018), Lee et. al (2008) and Zeithaml et al. (1996). The questionnaire was initially created in Italian and then translated into English by professional translators. To ensure the clarity of the wording, the survey was pilot tested with a small group of 20 respondents (both domestic and international visitors). Considering that no critical issues emerged from the pilot test, the two versions of the questionnaire (i.e., in Italian and in English) were finally considered for the purposes of the data collection.

Directly supervised by one of the authors, two trained interviewers collected data face-to-face by intercepting 4,500 individuals (aged 16 years old and above) attending the event during its entire duration (September-December 2019) and by using a paper-based instrument. At the end of the data collection, a convenience sample of 1,639 completed questionnaires (participation/response rate: 36.42%) was obtained. The data was then converted into an electronic format, cleaned and prepared for statistical analysis (SPSS ver. 21).

For the purposes of the study, the data were analysed in three different steps. Firstly, a descriptive statistical analysis was run on socio-demographic and travel-related variables to provide general information about the sample. Secondly, exploratory factor analyses (principal components analysis with varimax rotation) were conducted on items related to authenticity, satisfaction, and behavioural intentions to identify the factors to be included in the ordinary least square regression models (hereafter OLS; Jang et al., 2004; Kozak, 2001; Marcussen, 2011). Finally, the three OLS models were used to test our hypotheses related to the influence of socio-demographics, travel-related variables, and perceived authenticity on tourist expenditure, satisfaction, and behavioural intentions.

FINDINGS AND DISCUSSION

Descriptive Analysis

Most respondents were female (60.04%), Sardinian residents (88.41%), and belonged to the 25-34 age cohort (23.30%) (Table 1). Overall, male, middle-aged individuals (45-54 years old), and non-Sardinians residents reported a higher average expenditure when compared to their counterparts.

Table 1. Socio-demographic variables

Variables		%	Average Daily Expenditure [€]
Gender	Female	60.04%	50.27
	Male	39.96%	57.37
Age	16-18	3.54%	38.22
	18-24	16.71%	38.08
	25-34	29.35%	50.08
	35-44	17.88%	58.85
	45-54	15.43%	64.63
	55-64	12.51%	61.31
	>64	4.58%	55.28
Country of residence	Sardinia	88.41%	51.46
	Other	11.59%	65.68

Most respondents (94.75%) travelled to the destination specifically to attend the event (Table 2); these individuals also reported having a lower level of average expenditure when compared to their counterparts. A small percentage of visitors (1.22%) attended the event alone and reported having spent more (€ 76.10) than those individuals attending the event with a companion; individuals attending the event with friends reported spending the least.

Furthermore, most respondents (83.16%) reported having organised their trip without relying on any type of tourism intermediary (T.O. or Travel Agency), which resulted in higher spending compared to their counterparts (€ 46.73). Quite interestingly, 82.92% of respondents declared to have attended the events in other villages and/or to be willing to do so; however, the average expenditure was almost the same when compared to their counterparts. Finally, repeat attendees (65.77%) were reported spending less (€ 52.43) than first- timers (€ 54.41).

Attendees who stayed overnight at the destination (17.33%) spent on average € 75 during the event, € 26.5 more than daily excursionists, and around 25% of those stayed overnight visiting friends or relatives.

Factor Analysis

A series of explorative factor analysis (principal component analysis with varimax rotation) was performed in order to create the variables (i.e., authenticity, satisfaction, and intentional behaviour) to be included in the regression analysis.

The two factors associated to visitors' perceived authenticity explain 68.09% of total variance. Both the Kaiser-Myer-Olkin index (0.914) and the Bartlett's test of sphericity (chi-square = 8759.75; p-value <0.0001) confirm the goodness of the results to explain the data. Reliability was tested through Cronbach's alpha; the values were respectively 0.872 and 0.835, thus providing satisfactory levels of reliability of the extracted factors (Table 3).

"Existential authenticity" (34.65% of the total variance) was strongly related to items describing the interaction between the visitor and their experience in the destination. This factor referred to the way in which visitors come into contact with the local community, the culture and traditions of the place and the

Table 2. Travel-related variables

Variables		%	Expenditure [€]
Purpose of the visit	Autunno in Barbagia	94.75%	52.37
	Other	5.25%	66.56
Travel companions	Alone	1.22%	76.10
	Wife/Husband	43.38%	60.36
	Children	11.35%	71.33
	Other relatives	15.74%	64.44
	Friends	62.84%	48.46
	Other	3.11%	49.35
Number of travel companions	0	1.16%	78.53
	1	17.21%	50.9
	2-5	52.35%	52.38
	6-10	18.67%	54.29
	>10	10.62%	55.43
Arrangement with TO or AV	No	83.16%	54.4
	Yes	16.84%	46.73
Participation in other stages of the event	No	17.08%	53.25
	Yes	82.92%	53.08
Overnight in the destination	No	82.67%	48.50
	Yes	17.33%	75,00
Previous experience in AIB	Repeat time visitors	65.77%	52.43
	First time visitors	34.23%	54.41
	1-2 times	17.51%	48.77
	3-5 times	24.89%	54.89
	6-10 times	8.91%	49.03
	>10 times	14.46%	54.74

existential experience they had. "Objective authenticity" (33.44% of total variance) is strongly associated with items defining the genuineness and uniqueness in term of its ability to properly represent the Sardinian culture and traditions, thus making Sardinia a perfect destination for cultural and authentic tourism.

Secondly, an exploratory factor analysis was run on satisfaction items and a single factor was extracted (Table 4). This factor was identified with 73.85% of the total variance. The goodness of this factor was confirmed by the Kaiser-Myer-Olkin index (0.710) and the Bartlett's test of sphericity (chi-square = 1,845.80; p-value <0.0001). In addition, the Cronbach's alpha was calculated to test the reliability of the extracted factor (0.815).

Thirdly, an exploratory factor analysis (principal component analysis and varimax rotation) was applied to items related to behavioural intentions (Table 5). One single factor emerged (57.16% of total variance) including items related to intention to revisit the area, to participate in the event again, and to recommend it (and the hosting area) to others. The goodness of the factor was tested through the analy-

Table 3. Factor analysis on authenticity items

	Loadings	Eigenvalue	% Var. expl.	% var cum	Cron bach's α
Factor 1: Existential authenticity		3.12	34.65%	34.65%	0.872
This event gave me the possibility to interact with the local community	0.628				
This event gave me the possibility to learn something about Sardinian customs and traditions	0.648				
During the event I had the chance to enjoy the authenticity of local food	0.825				
During the event I had the chance to enjoy the authenticity of local beverages	0.863				
During the event it was easy for me to find Sardinian handcrafts	0.637				
Factor 2: Objective authenticity		3.01	33.44%	68.09%	0.835
In its genre, this event is unique	0.688				
This event represents the local Sardinian culture appropriately	0.844				
The event represents a Sardinian historical tradition	0.829				
This event makes Sardinia a destination for cultural and authentic tourism	0.619				
Goodness of fit: Chi-square = 8759.76 d.f. = 36 Sig = 0.000 – KMO = 0.914					

Table 4. Factor analysis on satisfaction items

	Loadings	Eigenvalue	% var expl	% var cum	Cron-bach's α
Factor 1: Satisfaction		2.21	73.85%	73.85%	0.815
I am satisfied with the decision to attend this event	0.868				
In its genre, this event is one of the best I have ever attended	0.828				
I don't regret attending this event	0.881				
Goodness of fit: Chi-square = 1845.81 d.f. = 3 Sig = 0.000 – KMO = 0.710					

sis of the Kaiser-Myer-Olkin index (0.842) and the Bartlett's test of sphericity (chi-square = 6,552.03; p-value <0.0001). Reliability was confirmed by Cronbach's alpha (0.861).

Regression Analysis

To identify the most suitable sets of variables to consider as determinants of visitor expenditure, a preliminary OLS regression was run on the sample including socio-demographics (e.g., gender, age, and place of residence) and travel-related variables (e.g., travel party, information sources, accommodation facility, prior experience at the event). In addition, the scores of the two authenticity factors obtained from the factor analysis (i.e., objective, and existential authenticity) were added to the model as independent variables. According to the model, visitor expenditure (EXP_k) was related to socio-demographics (S_k, i.e.,

Table 5. Factor analysis on intentional behaviour items

	Loadings	Eigenvalue	Var. expl. %	Var. cum. %	Cronbach's α
Factor 1: Behavioural intentions		4.00	57.16%	57.16%	0.861
I will say positive things about this event to other people	0.825				
I will say positive things about this event to other people	0.821				
Considering this event overall, I think I will write a positive post on my social network profiles (e.g. my Facebook Page, my Twitter/Instagram profile, etc.)	0.683				
After visiting this area, because of the event, I would recommend others to visit it	0.799				
Considering this event overall, I think I will write a positive comment/ review on a travel review website (e.g. Tripadvisor, Zoover, etc.)	0.671				
In the future I will attend to this event again	0.764				
After visiting this area, because of the event, I would like to come back on holiday	0.713				
Goodness of fit: Chi-square = 6552.3 d.f. = 21 Sig = 0.000 – KMO = 0.842					

gender, age, and country of residence), travel-related (TR_k, i.e., event-related motivation, prior experience with the event, number of times at AIB events, travel party, travel arrangement with tour operator or travel agency, type of accommodation, overnight stays), objective authenticity (OA_k) and existential authenticity (EA_k). The first model had the following form:

$$EXP_k = f(S_k, TR_k, OA_k, EA_k) \tag{1}$$

Specifically, the overall model used was (equation 2):

$$EXP_k = \alpha x + \varepsilon_k \tag{2}$$

where x was the vector which included the determining factors, namely: α was the vector of coefficients, which defined the marginal effect of the variable on the expenditure level; ε_k was the error term independent of the other determinant, normally distributed and with zero expected value.

To carry out the other two regressions, the scores of the two factors extracted (i.e., satisfaction, ST_k, and intentional behaviour, IB_k) were used as dependent variables, while the independent variables were the same as in the previous analysis. Therefore, the other two models can be described as follows (equation 3 and 4):

$$ST_k = f(S_k, TR_k, OA_k, EA_k) \tag{3}$$

$$IB_k = f(S_k, TR_k, OA_k, EA_k) \tag{4}$$

And the overall models used were (equation 5 and 6):

$$ST_k = \beta x + \tau_k \quad (5)$$

$$ST_k = \phi x + \sigma_k \quad (6)$$

In which, β *and* ϕ were the vector of coefficients; τ_k *and* σ_k were error terms independent of the other determinant, normally distributed and with zero expected value.

The scores obtained of the four factors were used to test the previously stated hypotheses and OLS models. Specifically, objective and experiential authenticity scores were used as independent variables in order to examine their effect on expenditure, satisfaction, and behavioural intention. In addition, satisfaction and behavioural intention scores were entered in two different models as dependent variables in order to identify their determinants.

Using perceived authenticity factors, visitors' socio-demographic and travel-related characteristics as independent variables, three econometric analyses were performed to explore the main contributing factors of expenditure behaviour, satisfaction, and behavioural intentions (dependent variables). Table 6 reports the results from the three OLS regression models tested.

Table 6. OLS results: determinants of expenditure, satisfaction, and behavioural intentions

	OLS Model 1		OLS Model 2		OLS Model 3	
	Expenditure		Satisfaction		Behavioural intentions	
Independent Variables	Coefficients		Coefficients		Coefficients	
Constant	27.24**	(11.93)	-0.14	(0.18)	-0.26**	(0.11)
Country of Residence	2.1	(5.12)	0.01	(0.07)	0.04	(0.06)
Gender	4.57	(3.22)	-0.01	(0.04)	0.02	(0.04)
Age	4.26***	(1.09)	0.02	(0.01)	0.04***	(0.01)
First-time visit	8.22*	(4.54)	0.01	(0.05)	0.02	(0.05)
Number of times in AIB	0.56	(0.5)	0.01	(0.01)	0.01	(0.01)
Purpose of the visit	-6.49	(5.47)	-0.16*	(0.09)	-0.13	(0.08)
Alone	4.95	(13.5)	0.32*	(0.17)	0.24	(0.16)
With wife or husband	-1.05	(5.37)	0.07	(0.05)	0.03	(0.04)
With Children	31.00***	(9.08)	0.00	(0.06)	0.10*	(0.06)
With other relatives	15.62***	(5.82)	0.04	(0.05)	0.05	(0.05)
With friends	0.22	(4.9)	0.07	(0.05)	-0.03	(0.05)
Number of travel companions	0.05	(0.1)	0.00	(0.01)	0.01	(0.01)
Arrangement with TO or AV	-8.05**	(4.08)	-0.13**	(0.06)	-0.08	(0.05)
Participation in other AIB stages	1.33	(4.9)	0.13**	(0.05)	0.16***	(0.05)
Overnight at the destination	33.62***	(5.43)	0.03	(0.06)	0.02	(0.05)
Hosted by friends or relatives	-0.56	(3.8)	0.03	(0.10)	0.16	(0.10)
Existential authenticity	5.53***	(1.67)	0.47***	(0.02)	0.5***	(0.02)
Objective authenticity	2.84*	(1.56)	0.45***	(0.02)	0.46***	(0.02)
*Standard errors in brackets * p<0.10, **p<0.05, *** p<0.01*						

First, the effect of socio-demographic variables (S_k, i.e., gender, age, and place of residence), travel-related variables (TR_k, i.e., event-related motivation, prior experience with the event, number of times at the AIB events, travel party, travel arrangement with tour operator or travel agency, type of accommodation, overnight stays), objective authenticity (OA_k) and existential authenticity (EA_k) on expenditure was supported by the data. Coefficients show that only some of the independent variables are able to significantly influence the expenditure. Particularly, the expenditure was found to be higher for attendees reporting higher level of perceived authenticity, thus confirming HP1; in fact, both coefficients for objective and experiential authenticity were positive with a p-value lower than 5%. HP4 and HP5 were only partially supported by the empirical findings. Specifically, a significant relationship was found between spending behaviour and the following socio-demographic and trip-related variables: age, attending the event with children or relatives, first-time visit, overnight at the destination, and having organized the visit with tour operators or travel agency. Specifically, results suggest that older attendees visiting the area with children and relatives, spending at least one night at the hosting destination and being first timers, tend to spend more than their counterparts. On the contrary, expenditure was lower for those who organise their visit with a tour operator or agency. Finally, results show that gender, place of residence, and prior experience at the event did not influence the expenditure level.

The independent variables were then included in the second model to study their effect on satisfaction. Findings show that H2 is supported. More specifically, those individuals who perceived higher levels of objective and existential authenticity reported as being more satisfied. Regarding HP6 and HP7, findings show that satisfaction is influenced by only some of the independent variables, namely: event-related motivation, travel party, participation in other stages of the same event edition, and having organized the visit with a travel agency or a tour operator. In particular, those attending the event alone and those who decided to participate in other stages reported higher levels of satisfaction. By contrast, satisfaction levels were lower when the visit was organized by a tour operator or a travel agency and when the purpose of the visit was not visiting the area with the main purpose of attending the event. No significant relationships were found with the other independent variables.

Thirdly, the effect of the independent variables on behavioural intention was tested. Results showed that perceived authenticity had a positive influence on visitors' expenditure, thus supporting HP3. On the other one hand, HP8 and HP9 were just partially supported. Specifically, behavioural intentions were positively and significantly influenced only by age, travelling with children, and participating in other stages of the event. By contrast, none of the other variables were found to significantly influence attendees' expenditure behaviour.

SOLUTIONS AND RECOMMENDATIONS

From a managerial perspective, this study offers useful recommendations to policy makers, destination marketers, and event organisers attempting to effectively plan and execute marketing strategies to attract attendees and make them satisfied, prone to spend, and willing to return and to engage in traditional and electronic word-of-mouth behaviour.

Firstly, findings suggest that the event "Autunno in Barbagia" should be planned, organised, and promoted as an authentic experience. In this direction, the study suggests involving the local community and local stakeholders in all stages of the event. Meeting and interacting with locals are perceived by the visitor as a strong authentic element associated with satisfaction, behavioural intentions, and expenditure behaviour. To achieve this aim, it would be certainly useful for policy makers, destination marketers, and event organisers to do their best to support the local community in hosting the event and building a sense of pride for it (Kim et al., 2006; Litvin & Fetter, 2006).

Finally, results also highlight that practitioners should tailor their marketing strategies based on the socio-demographics and travel-related variables. For example, if the goal is to increase visitor spending and/or encourage positive behavioural intentions, rural event organisers should focus on providing an authentic experience (and to further stress this aspect when running event promotion) to older visitors and participants attending the event with children. However, if the goal is to have satisfied visitors, more focus should be on customising aspects of authenticity to target those who attend the event alone.

FUTURE RESEARCH DIRECTIONS

While this study makes several contributions to theory and practice, some limitations remain, generating opportunities for further research.

Firstly, this study is based on a convenience sample collected during just one specific small-scale event in rural areas and in just one edition of it; all these circumstances render the findings hardly generalizable. Future studies might consider repeating the study in order to survey a representative sample of respondents, at different editions of the event (i.e., longitudinal studies), and/or at similar small-scale cultural events staged in other tourism destinations. Moreover, in this research, only some socio-demographic and trip-related variables have been considered to test their influence on attendees' behaviour. Future research should consider a wider array of variables. For example, regarding travel-related characteristics, and according to relatively recent studies (e.g., Park et al., 2020), it would be useful to collect data about the informational sources used by individuals when planning their event participation, to understand whether and the extent to which different information sources might significantly influence attendees' expenditure, their satisfaction, and behavioural intentions (Park et al., 2019). Finally, according to recent studies carried out in the specific context of museums (Antón et al., 2018), it would be also useful to consider whether and the extent to which socio-demographics, travel-related variables, and the perceived authenticity influence not only the online behaviour of attendees, but also their intensification (i.e., the interest of attendees in retrieving more information about event and the destination, post-participation).

CONCLUSION

This empirical study was carried out to deepen the scientific debate around the key determinants of expenditure behaviour, satisfaction, and behavioural intentions of attendees of small-scale cultural events staged in rural areas. Specifically, the purpose of this research was to identify and test the influence of perceived authenticity, socio-demographic aspects, and trip-related characteristics on visitors' spending behaviour, satisfaction and intention to return and to recommend the destination to others, both online and offline. With this purpose in mind, descriptive statistics, factor analyses and OLS models were performed on a sample of 1,639 attendees participating in a cultural event, called "Autunno in Barbagia," hosted in rural Sardinia during the 2019 autumn season.

The results of this study confirm that both dimensions of perceived authenticity (i.e., objective and existential) significantly influence not only attendees' satisfaction and their behavioural intentions

(Domínguez-Quintero et al., 2020; Park et al., 2019a), but also their expenditure levels (e.g., Chhabra et al., 2002; Brida et al., 2013; Robinson & Clifford, 2012). Moreover, this study also showed that attendees' expenditure behaviour was influenced to a greater extent by existential authenticity. Similarly, existential authenticity was reported to predict attendees' behavioural intentions more that objective authenticity did. Moreover, the OLS results also show that certain socio-demographic and travel-related characteristics significantly influence the three dependent variables (e.g., Huh, 2002; Park et al., 2019b; Park et al., 2020; Valle et al., 2008; Wang & Davidson, 2010). Specifically, expenditure behaviour was influenced by only one socio-demographic aspect (i.e., age), and four travel-related variables (participation with children or other relatives, first-time visit, overnight stay in the destination, and arrangement with tour operators or travel agency). On the other hand, visitors' satisfaction levels were influenced only by travel-related characteristics (i.e., event-related motivation, travel companions, participation in other stages of the same event edition, and tour arranged with travel agency or tour operators). In fact, neither age, gender, nor country of residence were significantly correlated to attendees' satisfaction. Finally, results confirm that visitors' willingness to recommend the event to others and to return to the destination of the event were influenced by age and by only two travel-related variables (i.e., travelling with children and participating in other stages of the event).

Our findings are useful for both researchers and practitioners. On the one hand, they contribute to further deepen the current scientific understanding about the attendees' experience and behaviour in small-scale cultural events organised in rural areas. Particularly, the findings shed light on the crucial role that perceived authenticity, socio-demographics, and travel-related variables exert in shaping visitors' expenditure, satisfaction, and their behavioural intentions. In general, this study only partially confirms existing studies on tourism and event tourism, thus further emphasizing that an inconsistent relationship between the event experience, the expenditure behaviour, and behavioural intentions might also characterise event-related studies (Del Chiappa et al., 2017). Moreover, findings attest that perceived authenticity is crucial to understanding tourist satisfaction and behaviour (Park et al., 2019a) in cultural destinations. Another theoretical contribution of this study is highlighting the potential role of perceived authenticity in stimulating online word-of-mouth behaviour in the specific context of events. In this regard, results are consistent with non-event studies, which found that individuals who reported higher levels of perceived authenticity tended to be more prone to posting comments and review online (e.g., Del Chiappa et al., 2019).

ACKNOWLEDGMENT

This study is based on the research project "*Il ruolo degli eventi nella valorizzazione dell'autenticità e dell'offerta turistica delle aree rurali: un'analisi sui partecipanti ad Autunno in Barbagia*" [The role of events in exploiting the authenticity and the tourism offer of rural areas: an empirical study on participants at Autunno in Barbagia], financially supported by the Nuoro Chamber of Commerce.

REFERENCES

Akhoondnejad, A. (2016). Tourist loyalty to a local cultural event: The case of Turkmen handicrafts festival. *Tourism Management*, *52*, 468–477. doi:10.1016/j.tourman.2015.06.027

Alegre, J., & Cladera, M. (2010). Tourist expenditure and quality: Why repeat tourists can spend less than first-timers. *Tourism Economics*, *16*(3), 517–533. doi:10.5367/000000010792278419

Almeyda-Ibáñez, M., & George, B. P. (2017). The evolution of destination branding: A review of branding literature in tourism. *Journal of Tourism* [JTHSM]. *Heritage & Services Marketing*, *3*(1), 9–17.

Antón, C., Camarero, C., & Garrido, M. J. (2019). What to Do After Visiting a Museum? From Post-consumption Evaluation to Intensification and Online Content Generation. *Journal of Travel Research*, *58*(6), 1052–1063. doi:10.1177/0047287518793040

Babin, B. J., & Griffin, M. (1998). The nature of satisfaction: An updated examination and analysis. *Journal of Business Research*, *41*(2), 127–136. doi:10.1016/S0148-2963(97)00001-5

Bowdin, G., O'Toole, W., Allen, J., Harris, R., & McDonnell, I. (2006). *Events Management* (2nd ed.). Elsevier. doi:10.4324/9780080457154

Brida, J. G., Disegna, M., & Osti, L. (2013). The effect of authenticity on visitors' expenditure at cultural events. *Current Issues in Tourism*, *16*(3), 266–285. doi:10.1080/13683500.2012.674105

Brida, J. G., & Scuderi, R. (2013). Determinants of tourist expenditure: A review of microeconometric models. *Tourism Management Perspectives*, *6*, 28–40. doi:10.1016/j.tmp.2012.10.006

Chhabra, D., Sills, E., & Rea, P. (2002). Tourist expenditures at heritage estivals. *Event Management*, *7*(4), 221–230. doi:10.3727/152599502108751613

Cohen, E. (1988). Authenticity and commoditization in tourism. *Annals of Tourism Research*, *15*(3), 371–386. doi:10.1016/0160-7383(88)90028-X

Cole, S. T., & Scott, D. (2004). Examining the mediating role of experience quality in a model of tourist experiences. *Journal of Travel & Tourism Marketing*, *16*(1), 79–90. doi:10.1300/J073v16n01_08

Correia, A., Barros, C. P., & Silvestre, A. L. (2007). Golf tourism repeat choice behaviour in the Algarve: A mixed logit approach. *Tourism Economics*, *13*(1), 111–127. doi:10.5367/000000007779784489

Del Chiappa, G., & Atzeni, M. (2016). The role of emotions in a sport event experience. *Tourismos*, *11*(5).

Del Chiappa, G., Melis, G., & Atzeni, M. (2017). Driving first-time and repeat spectators to cultural events. In A. Correia, M. Kozak, J. Gnoth, & A. Fyall (Eds.), *Co-creation and wellbeing in tourism* (pp. 47–59). Springer. doi:10.1007/978-3-319-44108-5_4

Del Chiappa, G., Napolitano, E., & Atzeni, M. (2019). Perceived Authenticity, Satisfaction and Behavioural Intentions at Wineries. *Micro and Macro Marketing*, *1*, 119–138.

Disegna, M., & Osti, L. (2016). Tourists' expenditure behaviour: The influence of satisfaction and the dependence of spending categories. *Tourism Economics*, *22*(1), 5–30. doi:10.5367/te.2014.0410

Domínguez-Quintero, A. M., González-Rodríguez, M. R., & Paddison, B. (2020). The mediating role of experience quality on authenticity and satisfaction in the context of cultural-heritage tourism. *Current Issues in Tourism*, *23*(2), 248–260. doi:10.1080/13683500.2018.1502261

Fotiadis, A., & Vassiliadis, C. (Eds.). (2020). *Principles and Practices of Small-scale Sport Event Management*. IGI Global. doi:10.4018/978-1-7998-4757-1

Getz, D. (2008). Event tourism: Definition, evolution, and research. *Tourism Management, 29*(3), 403–428. doi:10.1016/j.tourman.2007.07.017

Getz, D., & Page, S. J. (2016). Progress and prospects for event tourism research. *Tourism Management, 52*, 593–631. doi:10.1016/j.tourman.2015.03.007

Gilmore, F. (2002). A country—can it be repositioned? Spain—the success story of country branding. *Journal of Brand Management, 9*(4/5), 281–293. doi:10.1057/palgrave.bm.2540078

Goulding, C. (2000). The commodification of the past, postmodern pastiche, and the search for authentic experiences at contemporary heritage attractions. *European Journal of Marketing, 34*(7), 835–853. doi:10.1108/03090560010331298

Grappi, S., & Montanari, F. (2011). The role of social identification and hedonism in affecting tourist re-patronizing behaviours: The case of an Italian festival. *Tourism Management, 32*(5), 1128–1140. doi:10.1016/j.tourman.2010.10.001

Grayson, K., & Martinec, R. (2004). Consumer perceptions of iconicity and indexicality and their influence on assessments of authentic market offerings. *The Journal of Consumer Research, 31*(2), 296–312. doi:10.1086/422109

Griffiths, M., & Korstanje, M. E. (2021). Issues of authenticity in religious and spiritual tourism. In D. H. Holsen & D. J. Timothy (Eds.), *The Routledge Handbook of Religious and Spiritual Tourism* (pp. 276–287). Routledge. doi:10.4324/9780429201011-23

Handayani, B., & E Korstanje, M. (2017). Place Brand Authenticity in Social Media Interaction: A Postmodern Perspective. *Études caribéennes*, (37-38).

Huang, S., & Hsu, C. H. (2009). Effects of travel motivation, past experience, perceived constraint, and attitude on revisit intention. *Journal of Travel Research, 48*(1), 29–44. doi:10.1177/0047287508328793

Huh, J. (2002). *Tourist satisfaction with cultural/heritage sites: The Virginia Historic Triangle* [Thesis, Virginia Polytechnic Institute and State University, Blacksburg, VA].

Jang, S. S., Bai, B., Hong, G. S., & O'Leary, J. T. (2004). Understanding travel expenditure patterns: A study of Japanese pleasure travelers to the United States by income level. *Tourism Management, 25*(3), 331–341. doi:10.1016/S0261-5177(03)00141-9

Jang, S. S., & Ham, S. (2009). A double-hurdle analysis of travel expenditure: Baby boomer seniors versus older seniors. *Tourism Management, 30*(3), 372–380. doi:10.1016/j.tourman.2008.08.005

Kim, H., Borges, M. C., & Chon, J. (2006). Impacts of environmental values on tourism motivation: The case of FICA, Brazil. *Tourism Management, 27*(5), 957–967. doi:10.1016/j.tourman.2005.09.007 PMID:32287716

Kolar, T., & Zabkar, V. (2010). A consumer-based model of authenticity: An oxymoron or the foundation of cultural heritage marketing? *Tourism Management, 31*(5), 652–664. doi:10.1016/j.tourman.2009.07.010

Kozak, M. (2001). Repeaters' behavior at two distinct destinations. *Annals of Tourism Research*, *28*(3), 784–807. doi:10.1016/S0160-7383(00)00078-5

Kozak, M., & Rimmington, M. (2000). Tourist satisfaction with Mallorca, Spain, as an off-season holiday destination. *Journal of Travel Research*, *38*(3), 260–269. doi:10.1177/004728750003800308

Lee, Y. K., Lee, C. K., Lee, S. K., & Babin, B. J. (2008). Festivalscapes and patrons' emotions, satisfaction, and loyalty. *Journal of Business Research*, *61*(1), 56–64. doi:10.1016/j.jbusres.2006.05.009

Litvin, S. W., & Fetter, E. (2006). Can a festival be too successful? A review of Spoleto, USA. *International Journal of Contemporary Hospitality Management*, *18*(1), 41–49. doi:10.1108/09596110610641966

Lu, L., Chi, C. G., & Liu, Y. (2015). Authenticity, involvement, and image: Evaluating tourist experiences at historic districts. *Tourism Management*, *50*, 85–96. doi:10.1016/j.tourman.2015.01.026

MacCannell, D. (1973). Staged authenticity: Arrangements of social space in tourist settings. *American Journal of Sociology*, *79*(3), 589–603. doi:10.1086/225585

MacCannell, D. (1977). *The Tourist*. Schockon.

Marcussen, C. H. (2011). Determinants of tourist spending in cross-sectional studies and at Danish destinations. *Tourism Economics*, *17*(4), 833–855. doi:10.5367/te.2011.0068

Marrocu, E., Paci, R., & Zara, A. (2015). Micro-economic determinants of tourist expenditure: A quantile regression approach. *Tourism Management*, *50*, 13–30. doi:10.1016/j.tourman.2015.01.006

Mason, M. C., & Paggiaro, A. (2012). Investigating the role of festivalscape in culinary tourism: The case of food and wine events. *Tourism Management*, *33*(6), 1329–1336. doi:10.1016/j.tourman.2011.12.016

Oliver, R. L. (1997). *Satisfaction: A behavioral perspective on the consumer*. McGrawHill.

Park, E., Choi, B. K., & Lee, T. J. (2019a). The role and dimensions of authenticity in heritage tourism. *Tourism Management*, *74*, 99–109. doi:10.1016/j.tourman.2019.03.001

Park, K., Park, J. Y., & Back, R. M. (2019b). Determinants of Marathoners' event expenditures: An in-depth exploration of past experience. *Journal of Hospitality and Tourism Insights*, *2*(2), 110–120. doi:10.1108/JHTI-09-2018-0062

Park, S., Woo, M., & Nicolau, J. L. (2020). Determinant factors of tourist expenses. *Journal of Travel Research*, *59*(2), 267–280. doi:10.1177/0047287519829257

Pearce, P. L., & Moscardo, G. M. (1986). The concept of authenticity in tourist experiences. *The Australian and New Zealand Journal of Sociology*, *22*(1), 121–132. doi:10.1177/144078338602200107

Pizam, A., & Ellis, T. (1999). Customer satisfaction and its measurement in hospitality enterprises. *International Journal of Contemporary Hospitality Management*, *11*(7), 326–339. doi:10.1108/09596119910293231

Robinson, R. N., & Clifford, C. (2012). Authenticity and festival foodservice experiences. *Annals of Tourism Research*, *39*(2), 571–600. doi:10.1016/j.annals.2011.06.007

Valle, P. O., Correia, A., & Rebelo, E. (2008). Determinants of tourism return behaviour. *Tourism and Hospitality Research*, *8*(3), 205–219. doi:10.1057/thr.2008.19

Valle, P. O., Silva, J. A., Mendes, J., & Guerreiro, M. (2006). Tourist satisfaction and destination loyalty intention: A structural and categorical analysis. [IJBSAM]. *International Journal of Business Science and Applied Management*, *1*(1), 25–44.

Wang, N. (1999). Rethinking authenticity in tourism experience. *Annals of Tourism Research*, *26*(2), 349–370. doi:10.1016/S0160-7383(98)00103-0

Wang, Y. (2007). Customized authenticity begins at home. *Annals of Tourism Research*, *34*(3), 789–804. doi:10.1016/j.annals.2007.03.008

Wang, Y., & Davidson, M. C. (2010). A review of micro-analyses of tourist expenditure. *Current Issues in Tourism*, *13*(6), 507–524. doi:10.1080/13683500903406359

Warnick, R. B., Bojanic, D. C., & Xu, F. (2015). Using a trade market analysis technique to refine measurements for economic impact analysis of special events. *Journal of Travel Research*, *54*(1), 52–65. doi:10.1177/0047287513513160

Zaltman, G., LeMasters, K., & Heffring, M. (1982). *Theory construction in marketing: Some thoughts on thinking*. John Wiley & Sons.

Zeithaml, V. A., Berry, L., & Parasuraman, A. (1996). The behavioural consequences of service quality. *Journal of Marketing*, *60*(2), 31–46. doi:10.1177/002224299606000203

Chapter 5
Managing Festivals With Smart Technologies and Their Contribution Towards Destination Promotion

Sahil Sharma
Himachal Pradesh University, India

Jeet Dogra
Indian Institute of Tourism and Travel Management (IITTM), Noida, India

ABSTRACT

This paper presents a scholarly synthesis of contemporary smart event technologies under a clear three phase event classification. The constructed conceptualisation is designed as an ideal synthesis of tech basics, utilisation benefits, and real examples. In the pre-event phase, uses of 3D event diagramming software are discussed. The actual-event phase provides a comprehensive discourse on - beacon technology, geo fencing, radio frequency identification (RFID), near field communication (NFC), wrist band, mobile app, chat bots, virtual reality (VR), and augmented reality (AR) - using some real examples from iconic international events/festivals such as Coachella, Tomorrowland, Glastonbury, etc. Lastly, in the post-event phase basics and uses of smart event data is explained.

INTRODUCTION

Festivals and Cultural events have always occupied a special place in societies (Piva et al., 2017). Authors describe an event as something which happens at a specific place and time each year (Jayswal, 2008; Pırnar et al., 2019). However, sometimes a festival does not need to happen every year, for instance, the Kumbh festival (the largest religious gathering on earth) in India is celebrated every 12th year by shuffling among four different locations. So, the condition that a festival happens "every year" at the same place and time may not be a necessity to be called a festival.

DOI: 10.4018/978-1-6684-6356-7.ch005

Festivals are the authentic source of culture. A lucrative opportunity for one to get deeply immersed in local culture. Festivals are also an excellent resource for cultural tourism. For ages, festivals are known to attract visitors and guests to the destination or the location of the festival. Today festivals are being recognised as the most distinctive resource of any tourism destination that differentiates it from its competitor destination. Because generally the culture is different at every new tourism destination and geographical features stay the same across large areas. For example, all-mountain destination looks the same, all beach locations look the same, etc. But the culture at any destination is usually unique. Hence, cultural products like festivals can be the greatest factor to differentiate one tourist destination from another.

The cultural events are currently being utilised for destination branding, image enhancement and tourism development (Piva et al., 2017). Events/festivals earn free publicity, as they are covered by media and the information about the festival and the destination is automatically marketed to a larger audience (Jayswal, 2008). Therefore, festivals are increasingly recognised as an excellent element of destination marketing and branding (Jago et al., 2003; Pırnar et al., 2019).

For destination marketers, the main agenda is to attract more visitors as the - visitors are the main variable of interest for whom the branding and marketing campaign are being laid out. Tourism is a service industry; consumers mainly seek experiences in exchange for money spent on tourism products/services. Satisfaction in tourist experiences reflects the quality of tourism products/services. Positive experience results in positive word-of-mouth (WOM) and vice-versa. So, particular focus should be given to the enhancement of positive attendee experience while they are at the festival. Digital technologies today are easing human life in many spheres. Similarly, there is a growing trend among iconic international festivals to innovatively use smart technologies for overall management and experience enhancement. Such trends need to be simplified and explained through academic contributions so that every festival can benefit from the use of smart technologies.

Event tourism has seen a significant shift due to development in technologies such as 5G, the Internet of Things (IoT), big data, decision support systems, and artificial intelligence (AI). And, a growing number of events are integrating technology (Cimbaljević et al., 2021; Oestergaard, 2019; Simeng et al., 2020; Tom Dieck et al., 2018). Tourists are also growingly becoming tech-savvy and expect service delivery through smart systems. Aware of the increasing fact, event managers are utilising technologies to effectively meet guest expectations and to enhance the experience (Türkmendağ & Türkmendağ, 2022, p. 104). Osman (2022) writes that today technology has a main role to play in the successful management of large festivals (p. 301). So, smart technologies are being leveraged for the effective and efficient management of mega-events, and at the same time, festivalgoers also seek tech-savvy environments. Therefore, the success of festivals or mega-events can significantly be improved by the proper utilisation of digital technology, because it effectively manages and takes care of each visitor.

Today's technologies such as beacons, geofencing, augmented reality, radio frequency identification, smart passes, face detection, heat maps, smart devices, wireless sensors, real-time visitor feedback, and many others are improving festival management and visitor experience to a significant extent. So, through this paper, the researcher aims to postulate a scholarly synthesis of all current technological developments and trends revolutionising the event and festival industry. The results can serve as a guide for everyone from policymakers to festival organisers to comprehensively understand and implement technology in the management of festivals/events. Besides, new areas for future research.

LITERATURE BACKGROUND

Festivals, Event Tourism and Smart Technology

Since ancient times festivals have remained an integral part of the culture and religion of prehistoric tribes, and with time, festivals grew into more diversified forms (Cudny, 2016). And till today, festivals are crucial components of human culture and are studied under disciplines such as anthropology, sociology and human geography (Cudny, 2013). Festivals are researched as an important sub-field of event studies as well (Getz, 2010). According to Getz (2008), the events sector was unknown as a separate field of study during the 1960s-70s, and only after the 1980s event tourism have grown rapidly as a separate field of study. Festivals can be understood as organised events, showcasing diverse flavours of culture, during which humans meet irrespective of their work (Cudny, 2013).

Event tourism is both a sub-academic field and an area of destination management application, therefore, event tourism-related studies require drawing implications for both the academic community and destination managers (Getz, 2008). Today destination competitiveness is an important topic and festivals are increasingly becoming a key component in the differentiation strategy of tourism destinations (Van Niekerk, 2017). Because festivals generate tourist traffic, they significantly influence the surrounding space (Cudny, 2013). Branding based on events and festivals attracts new customers and ensures repeat visitors to the tourism destination (Van Niekerk, 2017). Moreover, tourism is not the only beneficiary, events play important role in community-building, urban renewal, cultural development, and fostering national identities as well (Getz, 2008).

Customer experience has become a key area of research and implementation today. Event experiences are never the same, every time they are unique, so to get a unique experience a person has to be there, if not, it's a lost opportunity (Getz, 2008). Many people attend festivals in a quest for extraordinary experiences that enrich their lives, and though not emotionally attached for traditional reasons, people want to visit again and again (Cudny, 2016). Thus, the visitor's experience is playing a key role in the success of an event/festival. Even though the visitor's experience is personal, it is largely influenced by the environment created by the organisers such as apps, intelligent information access, recommendation systems, support systems, etc. (Cimbaljević et al., 2021). Integration of technology in the overall infrastructure of destinations is one aspect that leads us to the concept of smart tourism where the main focus lies on increasing the quality of life for residents and enhancing the experience of visitors (Sharma, 2022). The smart festival concept is defined by Cimbaljević et al.- (2021) as an "Event that is part of cultural life and traditions of the city/destination, where smart technologies are embedded to co-create, facilitate and manage visitor experience; co-operate and organise all actors involved; and favour tourism development, destination competitiveness and local business economy". There are various festivals around the globe being organised and managed using modern technology (e.g., Tomorrowland, Coachella, Glastonbury, etc.). Hence, it is becoming increasingly important to understand smart technologies and how they can be implemented in festivals and events in order to make festivals smart to enhance visitor experience, strategize branding and marketing, or to increase returns on investments (ROI).

CLASSIFICATION OF SMART TECHNOLOGIES IN EVENTS

Pre-Event Phase

The pre-event phase is generally devoted to planning and preparations of events/festivals. This phase usually contains tasks like determining the objectives, finding a suitable venue, looking for chief guests, budget planning, marketing strategy etc. A large part of these tasks can be simplified and catalysed using technological tools. Technology provides all sorts of tools and techniques for making the event planning process faster, more collaborative, and easier to manage (Eventbrite, 2019). *Event Diagramming Software (EDS)* are very useful for designing and planning the event in the earlier phase. Event Diagramming Software (EDS) is a user-friendly computer application that allows event organisers to make a visual replica of the event venue in 2D and 3D formats. The digital twin of the event venue can be used to showcase the virtually tangible idea to investors, clients, and policymakers (Figure 1).

Figure 1. A 3D Diagram of the Event Venue created by Event Diagramming Software (EDS)
Source: *www.eventdraw.com*

EDS helps in collaboration among planners, teammates and other stakeholders, all of whom can propose changes and amend master plan according to their needs in real-time; and the stunning 3D tours allows planners to go deep down to the finest details to ensure that nothing is missed or inappropriate (Cvent, 2022). With EDS planners can change the position of chairs, tables, mic, buffet, etc., which helps in ensuring that everything is well-managed. While the current visual example (Figure 1) is of an indoor small-scale event, the same software can be utilised for mega-event planning in outdoor locations.

Often the EDS are equipped with social media, marketing, and ticketing tools that allow planners to promote the event more effectively. Technological tools also let potential attendees visually see the event before registration. Using technological tools in event planning saves a huge time, reduces cost, increases attendance, and increases return on investment (ROI) (Mcintosh, 2019). Technological tools can be utilised for sustainability and responsible behaviours at the event. Some popular service providers in this space are Cvent (formerly known as Social Tables), Event Draw, Smart Draw, All Seated, etc., and most of them provide free features for trials that event planners can experiment with.

Actual Event Phase

The actual event phase is often a roller coaster ride for organisers, someone is coming, others are going, something is left here, someone is demanding there, - crowd, sound, and everything else - makes it a fully vibrant environment. Often it is likely for an event organiser or managers or the host team to get lost and lose their calm in this environment, however, thanks to the technology today that everything can be managed properly and effectively in a smart manner without losing calm and control. New technologies enhance the visitor experience at the festivals and provide data-driven insights to organisers that were previously not possible (Hookings, 2019). *Glastonbury* and *Tomorrowland* are prime examples which show an application of new innovative technologies used in the management of festivals to revolutionise the attendee experience (Arkoudis, 2022). Statistical studies show that event technology can increase attendance by 20%, productivity by 27% and decrease costs by 20-30% (Enterprise Event Marketing, 2016). Once considered gimmicky technologies such as facial recognition, radio frequency identification (RFID) and virtual reality (VR) are revolutionising the immersive quality experience for festival goers (Thompson, 2016).

For this section, all identified technologies will be introduced to the readers one by one under separate sub-headings.

Beacon Technology

Beacon technology was first introduced by Apple in 2013, which now has diversified and become revolutionary for many industries (PlotProjects, 2022b). Beacons is a technology that transforms a small-scale area into a smart environment where internet of things (IoT) devices seamlessly interact with each other establishing a robust communication and service delivery atmosphere between host and guest/consumer. Beacons work on Bluetooth signals to transmit information and on a small scale, they provide more accurate location info than a GPS. Figure 2 provides three pics of beacon technology: the first explains the beacon process in the retail sector, the second shows the internal components of a beacon device, and the third shows beacon placement in the festival's pole.

In a festival or event, the beacons are used for several functions such as registration, guiding attendees, gamification for guests, social networking, proximity marketing, and for tracking the behaviour of attendees. On arrival, the attendees can scan a QR code or provide basic information through a mobile app, upon successful registration the beacon will let him/her in and direct will direct them further (Shaikh, 2021). After check-in, the beacon can guide the guests about the venue like where is food stall, bathroom, etc. The proximity function of beacon technology allows organisers to track guests' behaviour and communicate directly with attendees' mobile devices by sending them push notifications (Eventbrite, 2016). Using proximity event planners can generate floor maps, guide attendees about food and beverages, establish specialised communication, and keep attendees engaged using gamification (Shaikh, 2021).

Reeperbahn festival's innovative logbook function goes beyond push notification, created using beacon technology the logbook allows attendees to effortlessly record their journey throughout the festival (Ciurkot, 2016; Thompson, 2016). At the end of the day the logbook shows the journey in a Facebook

Figure 2. Beacon Technology
Source: *Pic 1-2. www.businessinsider.in, Pic 3. www.kontakt.io*

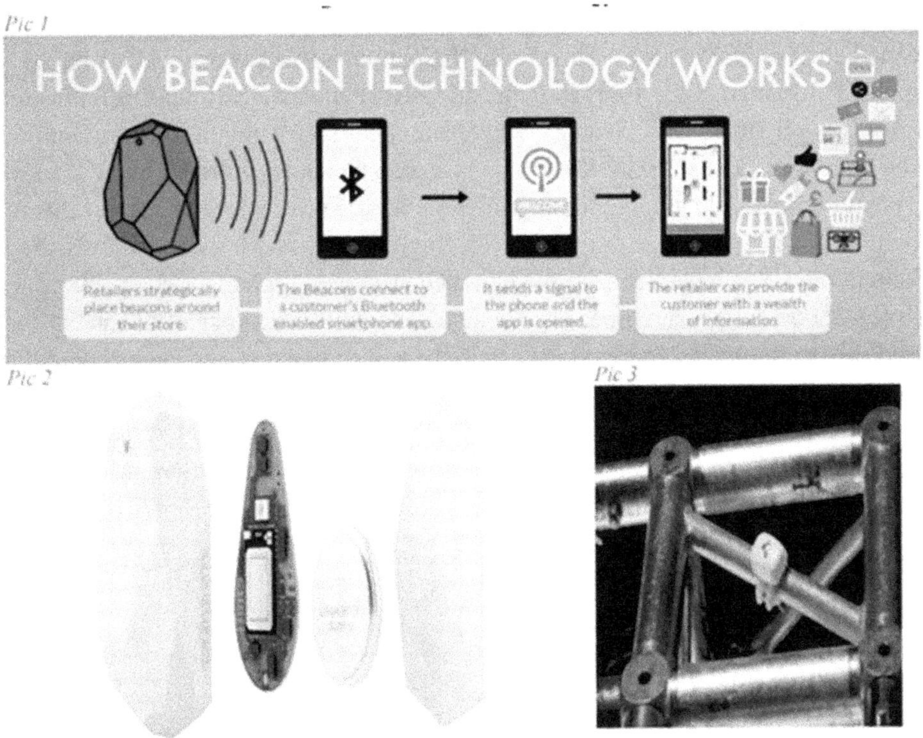

like feed containing all the information related to the places they went, which artist and song they heard, the photos, videos, etc., the logbook even populates attendee's Spotify playlist automatically using this information. Also, beacon technology is used for building social networks at events. For example, the 2014 Cannes lion film festival used beacons to develop social and professional networks using the "Around Me" function in the event app (Shaikh, 2021). The *Around Me* function allowed attendees to see who is currently present at the event including details such as name, designation, employer and a quick LinkedIn search provided enough information to strike up a conversation. Hence, beacon technology has greater potential to provide multiple functions and aspects to ordinary events giving new shape to ordinary events towards the more sustainable and smarter side.

Geo-Fencing

Beacons and Geo-Fencing are close associates that support each other in proper functioning. While the beacons are used for micro-locations, geo-fencing supports macro-location services (Goicochea, 2017). Geo-fencing as the name suggests creates a virtual fence/boundary using GPS and RFID technologies, it is used in festivals/events to track people's movements like entry, exit and how people roam around, geofence also supports direct digital communication with attendees based on their particular location in an event (Fujisan Marketing, 2018; PlotProjects, 2022c). In events geo-fencing helps in managing the crowd and avoiding lines; it can guide attendees around the venue and can tell what's happening at a certain corner of the event; people can use it for branding, marketing and sending promotional offers; altogether it helps in engagement, and immersion of a guest at an event (PlotProjects, 2022a). Geo-fencing has become very popular in many areas. Festivals such as Cannes lion international festival, the Bonnaroo music festival, and the Reeperbahn festival are well known for their innovative use of location-based technologies (Goicochea, 2017). Bonnaroo is a 4-day long music festival attended by 85,000 people enjoying music from 10-plus stages spread across 700 acres, where attendees stay using the tent accommodation. For a few years, the organisers are using Geofencing and Beacon technologies to communicate with the guest and to understand how people are moving in that large area. Ford motors have used this technology to spread marketing campaigns across the region. Talking about marketing campaigns using geofencing. Montana State in the US is a perfect example here because they implemented an ambitious large-scale geofencing project covering each corner of Montana. The whole territory of Montana was geo fenced and they used this technology to promote ski tourism and to enhance the experience of visitors who came here to ski. Using geofencing Montana tracked who's coming inside the borders of Montana and are going towards for the snowbound regions for ski, they then used a targeted marketing campaign on those people to show them better areas where the snow is richer. Furthermore, Montana also assessed the effectiveness such campaigns by comparing the people who they have shown the advertisement and those who have actually visited the place.

Figure 3. Geofencing at a market shop
Source: *Adapted from PlotProjects (2022c)*

Geo-fencing has been greatly used in events as well, suppose a group of friends visiting an event. As they enter they get a push notification to install an app that is meant to guide them throughout the event, as they come across mainstage they get a notification about the info of the performing band and about the upcoming artists, and they get a notification to visit a booth, participate in a poll, share their stories on social media, and after leaving they get a notification to rate their experience of the event (Fujisan Marketing, 2018). For location-based, direct, and digital marketing events, conventions, conferences, trade shows, concerts, fairs, music festivals, and cultural events are the perfect opportunities to showcase products and services using geofencing technology (Seminatore, 2021). Geofencing supports real-time customer engagement using geolocation-specific marketing and advertising content, this also provides enriching insights on consumer behaviour as well, for example, in a supermarket store data on consumer behaviour can tell how the customer is navigating the store, how much time they spend watching certain products or at a certain section in the store, which areas in-store section are in high demand and others (Fujisan Marketing, 2018). Hence, the application of geofencing (figure 3) and beacon technology opens many doors for innovative marketing and management at festivals.

RFID, NFC, and Wrist Bands

Radio frequency identification (RFID), Near field communication (NFC) and Wrist band technology enhance attendee experience to a greater extent. It lets attendees roam around the festival without the need to carry any entry pass, money or phone; with these technologies, attendees can show ID, do registration, pay money digitally, and even let a person connect with a friend on social media whom they just met in the festival/event. RFID popular at hotel-door entry and highway toll tax can take the festival experience to another level, it lets you speed up the registration process, allows cashless payments, reduces fraud, and helps organisers show ROI to the sponsors (Eventbrite, 2022a). With the convenience of festival attendees, RFID technology allows organisers to collect data about customer profiles, general behaviours, social media activity, and festival spending in real-time, these insights are capable of a continuous enhancement of attendee experience (Thompson, 2016).

Figure 4. Wrist Band scanning at a festival
Source: *Adapted from Eventbrite (2022a)*

Wrist bands have become very popular among festivals, Tomorrowland festival-goers use them to scan their way around the festival and to connect with friends on Facebook (Thompson, 2016). These wristbands contain email linking that allows attendees to connect with friends on Facebook and share experiences, these wristbands even allow one-tap payment making a seamless experience for festival goers (Arkoudis, 2022). Safety and Security is also a major concern for both organisers and the festival goers, therefore, some festivals have started storing information related to friends, parent, and guardian, so in a time of need someone close to the festival-goer can be contacted immediately by the support staff (Hookings, 2019). NFC tags are another useful smart technique to spread information at festivals. NFC allows event goers to tap on product and service booths so that information related to the product can be given to the smartphone of the attendee, using NFC embedded smart posters readable data can be transferred to a smartphone, NFC also allows check-in, and access to event schedules (Bezborodkina, 2019).

NFC is a sub-set of RFID technology, while the earlier requires a close touch with a sensor the latter works from a multi-meter distance. Example: credit/debit card swipe works on NFC and automatic toll tax works on RFID system. Many festivals have gone cashless using NFC wrist bands such as Coachella, Toronto Music Festival, Isle of wight festival, Liverpool sound city festival, Tomorrowland, etc. (NFC-direct, 2019). NFC technology is used for cashless payments, check-in/-out, information sharing, and buying food-beer-merchandise can be done using NFC Wrist Bands (Globalpayments, 2020; Ussmartcards, 2019). Therefore, RFID, NFC, and Wrist Bands (figure 4) are changing the way festivals/events used to happen and these elements are much needed today for any successful and memorable event.

Mobile Apps and Chatbots

The digital economy is growing rapidly today, and disruptive transformations are occurring well before time, which is further accelerated by pandemics and emergencies like COVID-19. In such a scenario, mobile apps are becoming a basic need for consumers to ensure ease of access to digital information sharing and transactions. Mobile apps are the most convenient form of information sharing, which adds to the experience of festival attendees by offering facilities according to their needs like the multilinguistic nature of apps, in addition, can provide enriching data to festival organisers (Eventbrite, 2022b). At Smukfest in 2019, IBM created an app for the festival that asked several questions to attendees determining their archetypes (e.g., hipster, metalhead, hippie, etc.) that allowed organisers to frame programs accordingly (Oestergaard, 2019). The app determined the success rate of programs automatically by comparing the actual attendance of people.. Finally, Smukfest has given a consolidated package of festival experiences (fun facts, pictures, archetypes, etc.) to the attendees which will add to overall experience. And, Smukfest is capable of improving this experience further in future events.

Mobile apps add to the experience of attendees by allowing them to quickly scan a barcode, get registered, queue less entry, and automatic creation of activity logbook. For example, Greencopper created a mobile logbook for the Reeperbahn festival's attendees which resulted in Facebook like feeds: pictures, videos and automated Spotify playlists (Ciurkot, 2016). Reeperbahn is a large music festival consisting of 600 concert-like events, workshops, and conferences held across 70 different locations. Due to the diverse festival line-up, it was challenging for attendees to remember the events and conferences they have visited. Aware of the issue, the organisers along with Greencopper (a mobile solutions company) developed an app feature that will chronologically track each activity of the attendee. Reeperbahn festival did not use beacons for sending push notifications rather they used this technology to track users' journeys around the festival and to show location-specific content (Ciurkot, 2016). Using mobile apps chatbots feature automatically based on artificial intelligence (AI) can answer many queries of festival attendees. For example, IBM Watson assistant powered two chatbots for Smukfest 2019, first to answer the queries of volunteers and the second to assist the attendees: Scout Bot and Walther Bot respectively (Oestergaard, 2019; Shah, 2020). While the Scout bot has answered queries from volunteers such as

"Where do I find the nearest dumpster?", "How can I swap shifts with my colleague", etc.; Walther bot has answered festival attendees' queries such as *"Where is the nearest toilet?"* and *"Who is performing on stage", etc.* (Oestergaard, 2019).

For crowd management in mega-events like Kumbh Mela (in India), research studies highlight the importance of mobile apps and technology for accurate counting of attendees, better monitoring, multilingual information sharing, and safety (Tiwari & Chowdhary, 2022). Therefore, the festival organiser must maintain a mobile application which can provide information about the festival, registration, on-site activities, and travel preparation, and educate attendees on how to use smart technology at a festival (Cimbaljević et al., 2021).

Virtual Reality (VR) and Augmented Reality (AR)

Virtual reality (VR) and Augmented reality (AR) are revolutionising the events industry today. VR using digital images, videos and sounds are letting people attend and enjoy the same festival from many parts of the world. And, AR is providing new dimensions or layers of experience to traditional festivals. Now people are becoming more tech-savvy and seek a personalised experience using VR headsets, control sets, and sensory devices people's experience could be enhanced (Eventbrite, 2022b). Exposure to these tech-driven immersive dimensions can have a long-lasting impact on the experience of festival attendees which works as a motivation for revisits. In the growing future, technology will even reduce people's need to physically attend the event. Similarly with the growing popularity and acceptance of metaverse people are less likely to seek ordinary festival experiences. A study on *"Future of Events: Glastonbury 2050"* predicts that by 2050 people going to the Glastonbury festival will purchase a weekend ticket at 600 euro and VR ticket at 1200 euros (Brooks, 2020). In the year 2050, while virtually attending an event is popular and more preferred among the general public, physical attendance is sought by people who crave real social interactions and resist technology. So, such predictions show how might the event industry is going to evolve in the future. Despite the uncertainty of the future, the omnipresence of technology is certain, which highlights the need today to establish familiarity with the growing trends.

Virtual reality (VR) is something which entirely exists in a virtual world and Augmented reality (AR) is something which augments/integrates real/live situations/objects/places with some digital objects/information. VR has been adopted by some world-renowned festivals such as Tomorrowland, Coachella, and Glastonbury. The Tomorrowland music festival is held every year in Belgium. Due to its popularity and love from across the world has invented a new concept *"UNITE with Tomorrowland"*. Since last few years, Tomorrowland establishes a live connection using satellite connection with several parties occurring at various locations on earth (Figure 5). For example, the 2016 UNITE with Tomorrowland was celebrated in real-time across 8 different countries - Mexico, Colombia, Germany, Israel, South Africa, India and Japan - using a live satellite connection from Tomorrowland mainstage in Belgium (Tomorrowland, 2016).

In 2020, the Glastonbury festival was cancelled due to COVID-19, so Glastonbury's Shangri-La team launches a new VR Lost Horizon festival which has been described as "a real festival in a virtual world" (Moore, 2020). A multi-stage fully interactive VR festival was open to explore using PC, VR, Mobile App, Live Streaming and directly available at anywhere on the planet. Furthermore, as part of 50th anniversary the BBC launched an pop-up channel to air the festival (Skinner, 2020). Similarly, utilising the potential of VR technology, US's most popular festival Coachella launched an VR app and all ticket holder were given a google cardboard VR headset for viewing performances and exploring a 360° view

Figure 5. UNITE with Tomorrowland 2016 Stage
Source: experience.tomorrowland.com/

of the festival ground (Thompson, 2016). Therefore, VR technology is already there and expected to be adopted by a growing number of festivals in coming years.

While Virtual reality (VR) operates in a simulated world, Augmented reality (AR) offers a real-time enhanced experience by integrating what the audience can already see (Eventbrite, 2022b). The beauty of AR is the non-requirement of experience or complicated hardware, VR can be experienced using Mobile phones or using on-site equipment, AR provides super-imposition of virtual elements (image, sound, graphic, or information) onto an already existing physical environment using a Camera (Eventbrite, 2022b). This has the ability to diversify what actually a place has to offer by adding new layers of dimensions to the existing festival. AR is now increasingly adopted by organisers to add to the entertainment of festival audiences (Hookings, 2019). Coachella festival is continuously experimenting with AR technology to provide its attendees with a uniquely futuristic and unforgettable experience. The first large-scale experiment with AR was done in 2018 at the U2 Concert where Eminem and Marron 5 performed using AR effects, since then the era of AR tech in festivals has begun (Dragani, 2019). In 2019, the Sahara tent of the Coachella festival was themed on AR space objects (Figure 6). Continue experimenting with AR, the Flume Coachella 2022 was themed around wildlife and nature: during the performances, the live stage was integrated with AR objects like trees, birds, dragons, etc (Chow, 2022). Thus, virtual reality, augmented reality and metaverse are the new future of the music festival industry.

Figure 6. Augmented reality (AR) driven Space theme at Coachella 2019
Source: *Image Credit to Goldenvoice, adapted from Melnick (2019)*

Post Event Phase

Post-event phase is a time to reflect back and do some analysis. After using smart devices for the operations and management of the festival the organisers will be possessing a lot of digital data which can be utilised for developing insights to further improve the festival experience next time. Several tools such as

Event registration/management software, RFID/NFC devices, chatbots, social media, and other on-site systems can be used for event data collection (Splash, 2022).

Mr Ben a CEO of *Splash* (a service provider for event data and marketing) explains event data comes in three categories (Figure 7) such as attendee data, intent data and value data (Splash, 2022). This data is available on each attendee that visits the event. The attendee data type contains basic information such as name, age, email, profession, and other preferences on food, music, products, etc. The attendee data is collected at the first step when a person registers for the event, and at the time of check-in. This can be done by scanning QR codes, wristbands, etc. After attendee data comes the turn for the intent data collection which was collected during the event. Intent data can be collected using tech devices such as sensors, geofencing, beacons, RFID, NFC, face detection, eye detection, etc., and this data can explain the behavioural characteristics of every person present in the event. For example, time spent at a certain stall, activity, product, stage, or artist; patterns of eye movements on branding initiatives; in-event response pop-up surveys; engagement; clicks etc. Finally, the intent data is analysed for obtaining valuable information which then is called value data. The value data describe who the attendees actually are, what they like and did not like, how much they spend, and who could be the potential guest for future events, which artists, stages or sections were more popular during the event. Therefore, the knowledge, segment of attendees, the behaviour of individuals, etc., developed in the value dataset can help organisers determine the success of the event, and return on investment (ROI), guide future marketing endeavours, and improve the overall festival experience for the next time.

Figure 7. Three types of data collected on each individual at a smart event
Source: *Author based on the articulations of CEO Ben (Splash, 2022)*

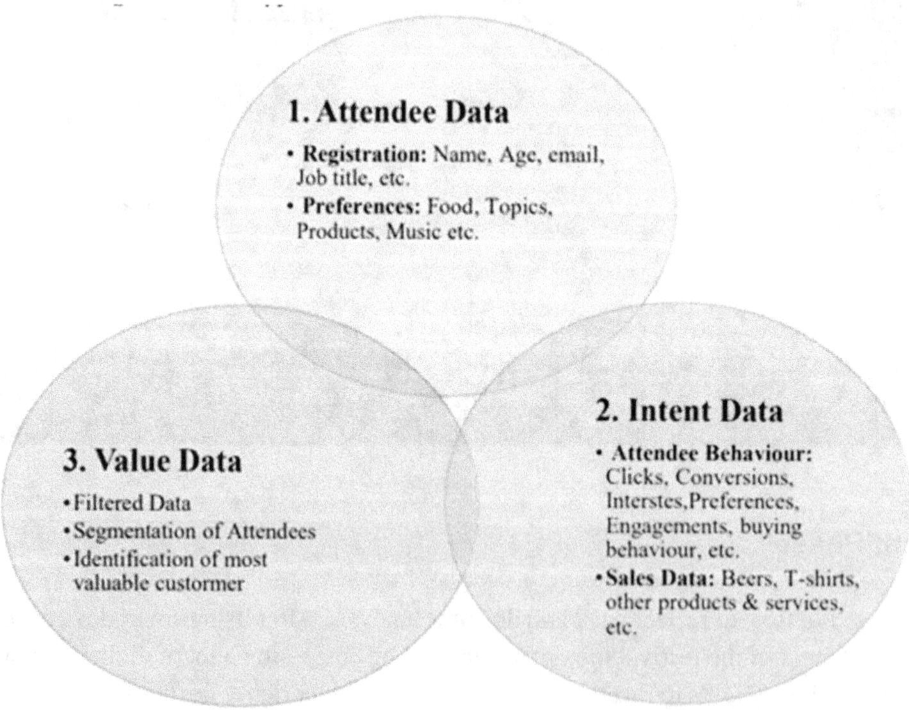

After a full analysis of the event data, organisers have rich details on attendee statistics and their behaviour, interests, contact information, ideas on what they (organisers) did best, and what might have not gone as expected. Using the filtered data (value dataset) and event management software (EMS) other important functions are executed such as sending *thank you!* emails and feedback form link to all the attendees. Some popular EMS service providers such as Eventbrite, Cvent, Splash, etc., even provide these features for free in initial trials. While renting the services of these companies can improve small and medium events to great success, it would not be a suitable option for larger events/festivals. Therefore, for the big and recurring events/festivals, building their own software is advisable through the services of programmers. Using a specialised software for the management of event/festival facilitates central information exchange (and storage), marketing, promotion, and other automated functions making each attendee feel their presence valued by the organisers. Furthermore, to flourish social connections linked with the event, a social network of all attendees can be leveraged using digital technology because sometimes one main outcome required from events is to develop connections. So, developing a social network of the event is fruitful for organisers and attendees both: helping the earlier to market, promote, brand effectively and making it easier for latter to develop social and professional connections. Moreover, multimedia posts, quizzes, rewards, and opinion polls can be shared on social media to keep the attendees engaged and motivating them for next visit.

DISCUSSION, SUMMARY, AND CONCLUSION

This paper has postulated a detailed discourse on modern technological trends and smart devices which are pioneeringly being used for the management of events/festivals by some of the iconic international festivals. These modern technologies have significant potential to improve management of festivals/event at all levels (small, medium and large events). Technology integration in festivals could fulfil one component to make festivals smart. Smart festivals enhance attendee experience and sustainability efforts. Therefore, in this study, the authors have attempted to scholarly synthesise and conceptualise all key arguments, facts, figures, and case examples of smart technologies in festivals under a clear three stage classifications of an event such as pre-event, actual-event, and post-event phase.

In the pre-event phase, software technology which specialises in making 3D diagrams of the event has been discussed. Implementation this technology in initial phase of the event increases effectiveness in event planning vis-à-vis enhances prospects of higher funding by investors. Event planning software, decrease the likelihood of making basic mistakes such as insufficient spacing, wrong seating plan, insufficient food, wrong placement of stalls, etc. In other words, an Event Diagramming Software allows event organisers to construct a digital twin of event in the pre-event phase. Using the digital twin planners and organisers can see, arrange and make changes about how the chairs will be placed, where the food stalls will be located, how would the venue look like with full of peoples, etc., all to experience digitally before the event actually occurs. This reduces the likelihood of making mistakes and significantly enhances the overall event experience. Also, showing such planning ideas in a digitally tangible format in front of investors, government officials, and policy makers increases the likelihood of getting investments and support for the event. Therefore, starting the planning of a festival/event by making its 3D digital twin is the very first step in making the overall festival smart.

For the actual-event phase, the chapter have discussed figure-head and fundamental technologies such as beacons and geofencing; RFID, NFC and wrist bands; use of mobile apps and chat bots; and finally, some case examples on the use of virtual reality (VR) and augmented reality (AR) in festivals.

- The beacons and geofencing technology allow real-time, location based, tracking, communication, and service facilitation with the attendees of a festival. While the geofencing is used in macro-dimension; the beacons are used in a micro-environment such as festivals/events. And, together they could create wonders for both attendees and organisers. Beacons and geofencing opens door for location specific direct marketing and branding. At festivals, it helps in issues such as crowd management, waste management, illegitimate entries, guiding attendees, etc. Overall, beacons and geofencing establishes stronger control over the event and helps in processes such as registration, catering and guiding each attendee, branding, direct marketing, and in promotion of sustainable practices, etc.
- The RFID, NFC and wrist bands are growingly being used around various festivals globally. It allows certain function such as automatic registration, entry, cashless payments, making friends on social media, safety-cum-security, and collection of data on each attendee. Using this technology, the attendee can roam freely around the festival the need to carry an ID, money, or the ticket. By tapping their NFC equipped wrist bands they can enter (register) an event, purchase (beer, food, t-shirts, etc.), and connect with friends on social media. Furthermore, the wrist bands can be equipped contact information related to friends and relative so in-case of any emergencies the supporting staff can immediately contact them. Moreover, the data collection done using RFID, NFC and wrist bands are instrumental in bringing insights on attendee behaviour about who the attendee actually, their likes/dislikes, buying behaviour, preferred brands, time spent at a certain artist, personality type, etc., such information is utilised for better knowing the attendees and their preferences, success of an event, popularity of artists/services/products/stalls, return on investment (ROI), etc., and for marketing. This is also done in real-time at many festivals companies are coming to brand, market and sell their products. For example, Ford motors, hunter garments, etc. Hence, RFID, NFC, and wrist bands are another technology combo that helps in making a certain festival/event smart.
- The mobile apps and chat bots are increasingly becoming basic need for businesses, organisations, institutions or any other services or establishments that has a public facing windows. For events and festivals, the mobile apps facilitate in registration, marketing, augmented reality, virtual reality, payments, and for automated tracking-cum-recording of an attendee's whole journey at a festival. Chat bots are becoming a must have feature for everyplace where there are a lot of people asking questions to a single entity. Similar is the scenario in festival where many people are visiting and they have many frequently asked questions (FAQ) like where to register, eat, sleep, toilet, put trash, and buy medicine, etc. So, a chatbot based on artificial intelligence (AI) equipped in the mobile app of a festival is smart way to answer all FAQ ask1ed by attendees. Chat bots will never let any attendee go un-heard and misguided which significantly can improve their experience at the festival. Therefore, mobile apps and chatbots are the must have component of any smart event.
- The virtual reality (VR) and augmented reality (AR) technologies are giving festivals new dimension and in the coming future (2050) people are going to prefer more expansive virtual tickets then the ordinary physical tickets. Main difference between VR and AR is that the earlier exists entirely in a digital/virtual environment while the latter augments/integrates the real situation with digital

objects such as animation, images and sound. Tomorrowland music festival started utilising VR since 2015-16 when the mainstage DJ in Belgium was connected with other DJ sets across 8 countries in real-time. So, using VR technology and satellite connections the Tomorrowland (UNITE with Tomorrowland) is celebrated every year across multiple countries at the same time and with the same DJs and music. Another example of VR implementation in festivals is Glastonbury, due to COVID-19 the Shangri-La team celebrated the festival in virtual environment which was even telecasted by BBC. In terms of augmented reality (AR), Coachella the biggest music festival in America, is experimenting with this technology since few years. The culture of AR in music festivals started when Eminem performed at U2 concert with AR effects. The Coachella festival continued to experiment with AR technology themed the music festival around space objects. The festival attendee near Sahara tent were able to space objects flying over them using the app camera. In the next year, Coachella amazed its attendee with wildlife related AR objects being integrated and shown during the festival. Hence, AR and VR are revolutionising festivals, giving new dimension to attendee to experience and these experiences have long lasting impact on the attendee. Thus, enhancing revisits and increasing the success of events.

Finally, in the post-event phase, this study discusses potential of event data, importance of sending *thank you!* Emails and feedback forms, and event management software. When a festival is embedded with smart devices for management of event and facilitation of guests. In the post-event stage, there will be a lot of data that has been collection such data usually is called big data and require data science methods for processing and analysis. The big data of smart event can be categorised into three types attendee data, intent data and value data. Attendee data is the very first data a festival/event receives at the time of registration and check-in. Now whatever attendee do at the event, their patterns of movement, buying behaviour, time spending at certain sections, eye movements, preferred brands, likes/dislikes, etc., are gathered under Intent data section. Finally, the Value data is something which comes after sophisticated analysis and tells information like attendee statistics, segmentation of attendees, ROI, most valuable attendees, and insights about the overall success of a festival. Therefore, post-event is also an important phase of an event where time must be spent on analysing the event data and keeping in touch with the attendees which could be easily done by an event management software and technology related data experts like data scientists.

Conclusively, this paper has conceptualised contemporary use of smart devices in festival or mega-events and promotes them for better sustainable future. This study comprehensively explains what, how, and why of smart technology integration in festivals/events. Hence, the knowledge created through this chapter must guide the event planners and organisers towards implementation of smart technologies in festivals to create amazing experience and promote sustainable practices.

REFERENCES

Arkoudis, S. (2022). How technology amplifies music experiences in festivals. *Tuned Global*. https://blog.tunedglobal.com/how-technology-music-experiences-festivals

Bezborodkina, N. (2019). IoT for Event Management: What's in for Event Planners. *KeepGo*. https://business.keepgo.com/blogs/articles/iot-for-event-management-what-s-in-for-event-planners

Brooks, R. (2020). *The Future of Events: Glastonbury 2050.* UKF. https://ukf.com/words/the-future-of-events-glastonbury-2050/27005

Chow, A. R. (2022). Coachella Gave Fans a Wild AR Experience. It Could Be the Future of Concerts. *Time.* https://time.com/6168688/coachella-enters-the-metaverse/

Cimbaljević, M., Stankov, U., Demirović, D., & Pavluković, V. (2021). Nice and smart: Creating a smarter festival – the study of EXIT (Novi Sad, Serbia). *Asia Pacific Journal of Tourism Research, 26*(4), 415–427. doi:10.1080/10941665.2019.1596139

Ciurkot, A. (2016). How Beacons Helped the Reeperbahn Festival Boost In-App Music Streaming by 86%. *Kontakt.* https://kontakt.io/blog/beacons-at-music-festivals/

Cudny, W. (2013). Festival Tourism - The Concept, Key Functions and Dysfunctions in the context of Tourism Geography studies. *The Geographical Journal, 65*(2), 105–118. https://www.sav.sk/journals/uploads/03101218Cudny.pdf

Cudny, W. (2016). The Concept, Origins and Types of Festivals. In W. Cudny (Ed.), *Festivalisation of Urban Spaces: Factors, Processes and Effects* (pp. 11–42). Springer International Publishing., doi:10.1007/978-3-319-31997-1_2

Cvent. (2022). *Cvent Event Diagramming®: Better events by design.* Cvent. https://www.Cvent.com/en/supplier-venue/event-diagramming-software

Dragani, R. (2019). *Augmented reality will change concerts forever.* Verizon. https://www.verizon.com/about/our-company/fourth-industrial-revolution/augmented-reality-will-change-concerts-forever

Enterprise Event Marketing. (2016). *Technology - Key experiential marketing stats for 2016.* EEM. https://www.slideshare.net/BEcauseExperientialMarketing/technology-key-experiential-marketing-stats-for-2016

Eventbrite. (2016). *The Buzz on Beacons: Introducing the Latest Festival Technology.* Eventbrite. https://www.Eventbrite.com/blog/the-buzz-on-beacons-introducing-the-latest-festival-technology-ds00/

Eventbrite. (2019). *Event Life Cycle: Before, During & After an Event.* Eventbrite. Eventbrite. https://www.Eventbrite.com/blog/event-life-cycle-ds00/

Eventbrite. (2022a). *5 Questions to Ask When Choosing an RFID Vendor for Your Event.* Eventbrite. https://www.Eventbrite.com/blog/choosing-rfid-vendor-ds0c/

Eventbrite. (2022b). *Exciting Event Technology You Can Use to Enrich Your Attraction.* Eventbrite. https://www.Eventbrite.com/blog/exciting-event-technology-you-can-use-to-enrich-your-attraction-ds00/

Fujisan Marketing. (2018). Geofencing: What Is It, And How Is It Used? *Medium.* https://medium.com/@fujisanmarketing/geofencing-what-is-it-and-how-is-it-used-1b2754499316

Getz, D. (2008). Event tourism: Definition, evolution, and research. *Tourism Management, 29*(3), 403–428. doi:10.1016/j.tourman.2007.07.017

Getz, D. (2010). The Nature and Scope of Festival Studies. *International Journal of Event Management Research, 5*(1), 1–47.

Globalpayments. (2020). *What is the Difference Between NFC and RFID?* Global Payments Integrated. https://www.Globalpaymentsintegrated.com/en-us/blog/2020/04/21/what-is-the-difference-between-nfc-and-rfid

Goicochea, E. (2017). *3 Music Festivals that use Location-based Marketing to create amazing mobile experiences*. MOCA. https://www.mocaplatform.com/blog/3-music-festivals-use-proximity-marketing-create-amazing-experiences

Hookings, M. (2019). Guest Blog: How technology is enhancing festivals. Event Industry News. https://www.eventindustrynews.com/news/guest-blog-how-technology-is-enhancing-festivals

Jago, L., Chalip, L., Brown, G., Mules, T., & Ali, S. (2003). Building Events Into Destination Branding: Insights From Experts. *Event Management*, *8*(1), 3–14. doi:10.3727/152599503108751658

Jayswal, T. (2008). Events Tourism: Potential to build a brand destination. *Conference on Tourism in India – Challenges Ahead*, (pp. 251–262). D-Space. http://dspace.iimk.ac.in/xmlui/handle/2259/178

Mcintosh, S. (2019). How Event Diagramming Is Shaping The Event World. *Purple Pass*. https://www.purplepass.com/blog/how-event-diagramming-is-shaping-the-event-world/

Melnick, K. (2019). *Coachella Debuts Its First Interactive AR Stage*. VR Scout. https://vrscout.com/news/coachella-ar-interactive-stage/

Moore, S. (2020). *Glastonbury's Shangri-La team launch new VR festival Lost Horizon*. NME. https://www.nme.com/news/music/glastonbury-shangri-la-team-launch-new-vr-festival-lost-horizon-2683835

NFCdirect. (2019). *What are the uses of NFC at festivals?* NRF Direct. https://www.NFCdirect.co.uk/blog/what-are-the-uses-of-nfc-at-festivals.htm

Oestergaard, F. (2019, July 19). How Data Rocks at Smukfest 2019. *IBM Nordic Blog*. Www.Ibm.Com. https://www.ibm.com/blogs/nordic-msp/smukfest-2019/

Osman, S. (2022). Technology Application in Tourism Fairs, Festivals and Events in Malaysia. In A. Hassan (Ed.), *Technology Application in Tourism Fairs, Festivals and Events in Asia* (pp. 301–311). Springer Singapore., doi:10.1007/978-981-16-8070-0_18

Pırnar, I., Kurtural, S., & Tütüncüoğlu, M. (2019). Festivals and destination marketing: An application from Izmir City. *Journal of Tourism. Heritage & Services Marketing*, *5*(1), 9–14. doi:10.5281/zenodo.2640987

Piva, E., Cerutti, S., Prats, L., & Raj, R. (2017). Enhancing Brand Image through Events and Cultural Festivals: the Perspective of the Stresa Festival's Visitors. *Almatourism - Journal of Tourism, Culture and Territorial Development*, *8*(15 SE-Articles and reports), 99–116. doi:10.6092/issn.2036-5195/6888

PlotProjects. (2022a). *10 Ways to Use Geofencing and Beacons for Events, Conferences and Festivals to Increase Attendee Engagement*. PlotProjects. https://www.plotprojects.com/blog/10-ways-use-geofencing-beacons-events-conferences-festivals/

PlotProjects. (2022b). *Beacon Technology – What it is & Why You Need it | 2022*. PlotProjects. https://www.plotprojects.com/blog/beacon-technology-why-the-market-is-booming/

PlotProjects. (2022c). *Geofencing Software – How to Set up Geofencing in 2022 | Free Trial*. PlotProjects. https://www.plotprojects.com/blog/how-to-set-up-geofencing/

Seminatore, C. (2021). *Geofencing for Events*. Get GeoFencing. https://www.getgeofencing.com/geofencing-for-events-for-targeted-audiences/

Shah, H. (2020). *Powering Smukfest with IBM Data and AI*. Linkdin. https://www.linkedin.com/pulse/powering-smukfest-ibm-data-ai-hetu-shah/

Shaikh, N. (2021). How to use beacons to enhance in-event experience? *Peer Bits*. https://www.peerbits.com/blog/how-to-beacons-to-enhance-in-event-experience.html

Sharma, S. (2022). Smart Tourism-A Solution for Tourism Challenges in Himachal. *Journal of Smart Tourism*, *2*(2), 21–32. doi:10.52255/SMARTTOURISM.2022.2.2.4

Simeng, Z., Yizhou, Z., & Raj, V. (2020). Explore the Improvement of the Management of China's International Film Festivals Based on Artificial Intelligence. *AICSconf '20: Proceedings of the 2020 Artificial Intelligence and Complex Systems Conference*, (pp. 63–67). ACM. 10.1145/3407703.3407716

Skinner, T. (2020). *BBC announce pop-up Glastonbury channel as part of 50th anniversary coverage plans*. NME. https://www.nme.com/news/music/bbc-announce-glastonbury-50th-anniversary-celebration-plans-2676162

Splash. (2022). Event Data and Technology. *Splash*. https://Splashthat.com/resources/event-marketing-guide/event-data-and-technology

Thompson, N. (2016). *How technology is giving more life to festivals*. UK Tech. https://www.uktech.news/news/technology-trends-give-festivals-life-beyond-physical-now-20160727

Tiwari, P., & Chowdhary, N. (2022). Technology and Crowd Management at Events: a Case Study of Kumbh Festival in India. In CABI Books. CABI Books. doi:10.1079/9781789248685.0015

Tom Dieck, M. C., Jung, T. H., & Rauschnabel, P. A. (2018). Determining visitor engagement through augmented reality at science festivals: An experience economy perspective. *Computers in Human Behavior*, *82*, 44–53. doi:10.1016/j.chb.2017.12.043

Tomorrowland. (2016). *Tomorrowland 2016, UNITE Trailer*. [Video]. Youtube. https://youtu.be/Mdit-0COM0ZM

Türkmendağ, T., & Türkmendağ, Z. (2022). Technology Use and Innovation Strategies in Event Tourism. In C. M. Q. Ramos, S. Quinteiro, & A. R. Gonçalves (Eds.), *ICT as Innovator Between Tourism and Culture* (pp. 104–121). IGI Global. doi:10.4018/978-1-7998-8165-0.ch007

Ussmartcards. (2019). *How can NFC be used at festivals?* US Smart Cards. https://www.usmartcards.com/2019/05/29/how-can-nfc-be-used-at-festivals/

Van Niekerk, M. (2017). Contemporary issues in events, festivals and destination management. *International Journal of Contemporary Hospitality Management*, *29*(3), 842–847. doi:10.1108/IJCHM-01-2017-0031

Chapter 6
The Role of Culinary Festivals in Building Destination Brand Equity

Davut Kodas
Mardin Artuklu University, Turkey

ABSTRACT

The current chapter theoretically aims to reveal the relationship between culinary festivals and destination branding, focusing on studies investigating aspects of culinary festivals such as festival experience, quality, and motivation in the relationship between the culinary festivals and destination brand equity. In this chapter, the role and importance of culinary festivals in developing destination brand equity are theoretically discussed through reviewing hospitality and tourism literature. Theoretical results of the present chapter indicate that some topics of culinary festival studies such as food experience, food quality, and food festival motivation from a tourist perspective have played an essential role in how increasing core constructs of destination brand equity which included destination brand image, destination brand awareness, perceived quality, and brand loyalty.

INTRODUCTION

Today in the competitive market, all destination wants to have a strong brand by developing destination image, destination quality, awareness, and loyalty which are widely used in measuring destination brand equity. To do this, the tourism destinations are using local attractions of destination that are regarded as push and pull motivational factors in attracting visitors to destination (Kodaş & Sekizkardeş, 2022). Events and festivals have huge potential to be a core constituent of the growing hospitality and tourism industries (Getz, 2008; Getz & Page, 2016) as they are one of the major tourist attractions (Felsenstein, & Fleischer, 2003). Hence, all tourism destination has started to organize events or festivals. By organizing kinds of festivals, destinations make significant contributions to their destination familiarity and build a strong image in tourist mind about the destination (Jago Chalip, Brown, Mules, and Ali, 2003). Festivals provide the festival attendees with some kinds of experiences such as social, leisure, cultural

DOI: 10.4018/978-1-6684-6356-7.ch006

and culinary experiences (Axelsen & Swan, 2010; Harb, Fowler, Chang, Blum, & Alakaleek, 2019). Festivals are considered as one of the effective destination strategical destination promotional tools as they are a valuable and significant source for destination development. Festivals are very important and special events which could introduce regional culture to their visitors and build a destination brand identity. Culinary festivals are also regarded as a special event which focuses on local food and beverage and assists in enhancing the destination brand strength.

As culinary or gastronomic attractions of destination may be the primary motivation for visiting a different country as suggested by Quan and Wang, (2004) and Mkono, Makwell and Wilson, (2013), today majority of tourism destination in the different region has been using local foods and beverage in their festival which focused culinary or gastronomy. So, food festivals play a key role in determining choice among domestic or international tourists who are interested in different cuisine (Wan & Chan, 2013). Gastronomy or culinary festivals are designated as special events enhancing gastronomy tourism opportunities for the destination and building the image, awareness, and loyalty of the destination (Axelsen & Swan, 2010). A culinary festival is one of the efficient ways to bring people together and provide an opportunity to meet local people and tourists and achieve a better cultural exchange between local people, local producers, and tourists (Organ Koenig-Lewis, Palmer, and Probert, 2015). Besides, culinary festivals allow visitors to share each other's food experiences (Silkes, 2007). In this book chapter, the comprehensive literature review will be conducted to understand how events and festivals, in particular culinary festivals, can contribute to destination brand equity which has dominant dimensions including brand awareness, brand image, perceived quality, and brand loyalty. This chapter contributes to augment the literature by discussing conceptually the role of culinary festivals in destination brand equity based on studies in relevant literature which linkages between some aspects of culinary festivals studies and destination branding.

THEORETICAL BACKGROUND

Culinary/Gastronomy Festivals

A festival is a "collective experience" where festival goers can be highly influenced by other people who are there with the same purpose and people who shared their experiences (Maisello, Bonetti and Izzo, 2019). Thenceforth the onset of recorded history, kinds of festivals have played a significant role in the communities (Shone & Parry, 2004). Throughout the history, festivals associated with the local community have increased and in time turned into commercial products, and by attracting the attention of tourism destinations, it has been organized as small, medium, and large-scale festivals in many tourism destinations. Festivals have grown considerably over the last two decades owing to their positive economic, social and cultural impacts on tourist destinations and societies (Choo and Park, 2020). Events and festivals are also a great way to promote a destination by creating a destination image and assisting in improving the destination (Brown, Chalip, Jago, & Mules, 2004; Li & Vogelsong, 2006; Getz & Page, 2016).

A festival which is held in a particular destination can attract many visitors to the destination and increased festival expenditure of visitors contributes to the economic well-being of the local community (Felsenstein, & Fleischer, 2003; Getz & Page, 2016). Litvin & Fetter (2006) stated that there are several positive contributions of festival tourism to the development of tourism destinations. The authors argue that festivals are one of the key factors of tourism attractiveness by emphasizing the extension of the tourism season, accelerating the development of cities, diversifying tourism in the destination, increasing awareness of the destination, encouraging new investments, and improving the destination image (Litvin & Fetter, 2006). In this context, tourism destinations are trying to organize kinds of festivals, especially traditional festivals at certain times every year. For example, some well-known festivals related to food and beverage are organized in different countries or cities such as the Bordeaux Wine Festival in France, Taste of Amsterdam in the Netherlands, Barcelona Beer Festival in Spain, Taste Florence Food Festival in Italy, Munich Oktoberfest, Taste of Adana Festival in Turkey. Culinary or gastronomy festivals which are of valuable festivals organized in a destination which are familiar with their gastronomic attractions is a significant component of destination promotional tools and are considered important events for developing destination branding (López-Guzmán, Lotero, Pérez Gálvez and Rivera, 2017; Keskin, 2018).

As aforementioned in previous studies, gastronomy attractions of destinations are being the primary motivation for some tourists to visit tourist destinations (Mitchell & Hall, 2004; Quan & Wang, 2004). These tourists are looking for new tastes, unique food experiences and they tend to explore different cultures (Kodaş & Aksoy, 2022). In addition to tasting local flavours at the destination visited and attending culinary festivals, they experience different food experiences. It can be stated that culinary or gastronomy festivals provide an environment for the tourist to experience the destination's food cultures. Therefore, local/traditional food and beverages are valuable sources for gastronomy festivals due to its some important contributions to destination building and developing of strong and lasting bonds between the tourist and the host community (Hernández-Mogollón, Folgado-Fernández, & Duarte, 2014; Hermann, Boshoff, & Ncala, 2019). For this reason, the importance of local food for destination and gastronomy festivals needs to be discussed.

Local food which is a meaningful part of regional brands enables to provide tourists with the regional culture and local history of a district and provides tourists with a playful means for becoming aware of the destination (Gordin & Trabskaya, 2013). Local food is a gastronomy product that destination regarded as an important of destination marketing tools, have been one of the basic elements for developing destination improvement. In addition, local food/regional foods are seen as valuable resources for tourism destinations to differentiate themselves from other destinations.

Gastronomy tourism products reflect the cultural identity of a destination and create a perfect symbiosis for tourists who wants to pursue different taste, products, and experiences. (Pineda et al., 2019). Because tourists are increasingly travelling to destinations in search of gastronomy experiences, and tasting new and authentic foods (Wijaya et al., 2013). To do this, besides local foods of the destination must be offered on the menu of food-beverage establishments or hotel menus, the destination offers local foods to the tourist as a tourism product for attracting more visitors to their region by organizing different food-theme festivals. This can be helpful for marketers and developers of tourist destination who wants to organize gastronomy festivals at certain times of the year by offering special, local foods to enhance destination images, awareness, loyalty, and perceived quality.

According to Du Rand, Hearth and Albert (2003: 97-99), the involvement and curiosity of tourists towards local foods make the regional cuisines an important destination attractiveness. Therefore, the food-related festival is considered to be significant event and contribute to the development of tourist attractions (Park, Reisinger, & Kang, 2008). As tourist is searching for novel and unique food experience, it can be stated that tasting the local foods is one of the best ways to new experience for the tourist. Travelling to destinations to experience local food and wine is turned into a serious leisure experience (Hall & Mitchell, 2005; Park, Reisinger, & Kang, 2008). For this reason, destination stakeholders have realized the growing importance of local food in destination marketing and have been trying to organize a food-related festival to advertise their local products and destination and also satisfy visitors. Because food-related festival is not only a way to taste new experience but also an event that brings people together by offering fun and entertainment for them. Culinary/gastronomy festivals are important events which provide the tourist with many elements such as presentation, production, tasting, recipes, cultural food rituals and ceremonies, culinary cultures and traditions (Organ et al., 2015; Davras & Özperçin, 2021, p 1).

Consumer-Based Destination Brand Equity

Destination branding is crucial for tourism destinations to have competitive power with rival destinations (Hall, 1999; Wong, Wu and Cheng, 2018). Destination branding has been growing attention in the tourism and destination marketing discipline since the 1990s. Until today, parallels to this advancement, many studies were conducted to reveal dimensions of destination brand equity from a tourist perspective in various destinations. In addition, destination marketers and managers are cognizant of the growing importance of brand equity in enhancing destination familiarity and promoting their destination (Ferns & Walls, 2012). Destination branding can be defined as:

Destination branding is the set of marketing activities that (1) support the creation of a name, symbol, logo, word mark or other graphic that readily identifies and differentiates a destination; that (2) consistently convey the expectation of a memorable travel experience that is uniquely associated with the destination; that (3) serve to consolidate and reinforce the emotional connection between the visitor and the destination; and that (4) reduce consumer search costs and perceived risk. Collectively, these activities serve to create a destination image that positively influences consumer destination choice'(Blain, Levy and Ritche, 2005: p. 337).

In studies related to destination branding, destination brand equity has been one of the essential issues in the marketing and management discipline. Destination brand equity is defined as 'it is the combination of key constructs that can be described as the overall utility that tourists place in the destination brand when compared to its competitors (Ferns & Walls, 2012, p. 27). It has been a crucial factor in understanding the perception and evaluations of visitors about a destination (Boo, Busser, & Baloglu, 2009; Horng, Liu, Chiu, & Tsai, 2012a; 2012b).

In destination branding studies, many scholars have tried to evaluate customer-based brand equity using the brand equity dimensions of Aaker (1991) and Keller (1993) (Konecnik & Gartner, 2007; Boo, Busser, & Baloglu, 2009; Pike 2010; Myagmarsuren & Chen, 2011; Gomez, Lopez, & Molina, 2015; Pike & Bianchi, 2016; Fathabadi, Nejad, & Alizadeh, 2017; Chekalina, Fuchs, & Lexhagen, 2018; Liu, 2020; Kodaş, 2021). Moreover, some special tourism destinations such as wine, and culinary have adopted the consumer-based destination brand equity dimensions in various studies (Horng et al, 2012a; Horng et al., 2012b; Gomez Lopez and Molina, 2015; Jiang et al., 2017). In addition to key elements of destination brand equity, different dimensions such as perceived value have been investigated as constructs of brand equity (Boo, Busser, & Baloglu, 2009; Gomez, Lopez, & Molina, 2015). To illustrate, Gomez et al., (2015), examined the effects of destination brand image and destination image on wine tourism destination brand equity which includes destination brand awareness, destination perceived value and destination brand loyalty in Spain. Besides, many researchers have widely conducted to reveal antecedents of destination brand equity in the various types of destinations such as medical, ski, and culinary (Horng et al., 2012a:2012b; Das & Mukherjee, 2016; Kodas, 2018; Chekalina, Fuchs & Lexhagen, 2018). For example, in the study of Kodaş (2018), the relationship between gastronomy experience, motivation, destination satisfaction and culinary destination brand equity was examined in the Gaziantep culinary destination located in Turkey. The results of this study indicated that gastronomy experience, gastronomy motivation and satisfaction have positively influenced the destination brand equity in the context of a culinary destination. It can be asserted that even if dominant dimensions of destination brand equity have been conducted in several studies, different dimensions of destination brand equity were explored in different destination contexts.

MAIN FOCUS OF THE CHAPTER

This chapter is focusing on understanding the relationship between culinary festival and destination branding. To achieve this, comprehensive literature review was conducted to understand how the culinary festivals are increasing the destination branding in the framework of destination brand equity, in particularly gastronomy destination. Thus, this chapter evaluated the relationship culinary festivals and destination brand equity elements. By doing this, the impact of some construct of culinary festival on destination brand equity elements was reviewed. For example, culinary motivation is considered as a significant driver of destination brand equity dimension such as destination brand image, destination brand loyalty etc. (Deng and Li, 2014; Folgado-Fernandez, Hernández-Mogollón, & Duarte, 2017). From methodological point, this chapter was conducted by reviewing a comprehensive literature review. In this line, this study investigated the relevant literature carefully by reviewing the studies dealing with the relationship between culinary festivals and destination brand equity and its components.

Culinary Festivals as a Key Enhancer of Destination Brand Equity

The relevant literature shows that food-related festivals have been frequently used in destination branding for culinary tourism development (Horng & Tsai, 2010; Lee & Arcodia, 2011; Sotiriadis, 2015; Yang, Won, Tan and Wu, 2020; Davras & Özperçin, 2021). In order to improve destination competitiveness, destination marketers and managers always generate some positive solutions to enhance their brand equity focusing on perceived destination image, perceived quality, destination awareness and destination brand loyalty which are dominant dimensions of brand equity (Im, Kim, Elliot, & Han, 2012; Chi, Huang & Nguyen, 2020). Festivals are one of the essential elements for destinations in building solid brands (Kodaş, 2021). Marzano and Scott (2006, p. 199) stress that "the challenge for events planners and managers is to consistently fit the events into the interrelationships among the elements of the marketing mix and therefore enhancing the destination brand equity". Local festivals which are special events are organized in a location to promote the destination and introduce the region's culture to the visitors (Wan & Chan, 2013). Gastronomy/Culinary festivals which are defined as special event is the best way to build a destination brand. For this reason, a large number of food-related festivals are held in various tourist destinations, but it should not be forgotten that locality is required when organizing a local festival which helps to promote the destination. Food-themed festival, which is considered cultural events, makes contributions to destination branding as a way of enhancing destination image, destination awareness, and destination loyalty which are essential elements of destination brand equity.

In branding studies, local foods are seen as a valuable attraction and indicator of destination image, perceived quality, brand awareness and loyalty (Kodaş & Özel, 2022). In the underpinning literature, if tourist visits a specific destination to consume local food, their food experience will be conveyed peak experience which leaves an unforgettable memory for the tourist (Quan & Wang, 2004). So, with a post-Fordist consumption approach, the need for searching for more special, different, and novel food experiences increases the significance of food tourism for the destination and tourism firm (Everett & Aitchison, 2008). In this context, local or traditional food is seen as the primary motivation source for tourists and a way to get a memorable food experience (Davras & Özperçin, 2021). Local food or traditional food has been highlighted by many researchers as a significant tourist attractiveness (Björk & Kauppinen-Räisänen, 2014; Henderson, 2009; Kim, Eves, and Scarles, 2009; Kodaş, 2021). Food experience also is the main motive for visiting different destinations for tourists who search for authenticity and a unique experience. Therefore, destinations aim to strengthen perceived destination brand awareness, quality, loyalty, and destination brand image of tourists by prospering tourist local food experiences by offering their traditional and local foods, especially in the gastronomy festivals held in the destinations. A destination's competitiveness depends on its ability to increase visitors' expenditure and arrivals by ensuring them satisfying and memorable experiences and it is vital in destination branding, especially focusing on customer-based destination brand equity (Wong, Wu and Cheng, 2018). By organizing food-related festivals and offering different food experiences, destinations make an effort to reinforce their brand strengths. Hence, food-related festival experiences are valuable sources for destination branding.

In the existing tourism literature, evaluations of the visual perspective of tourism were not enough to understand tourist behavior and preference (Urry, 1990). Due to changed tourist behavior, many tourism organizations had to shape their product structure. Parallel o this evolution, destination stakeholders have felt themselves to adapt to the new tourist who searches for variety. As it was stressed that tourists are looking for unique, new experiences, memorable experiences, tourism organizations and destination marketers and managers should pay attention to these tourist requisitions (Pine & Gilmore, 1999). At this point, the experimental value of the touristic product is getting more important for tourist who wants a different experience to satisfy their psychological needs (Schmitt, 1999). It is stressed in the hospitality and gastronomy literature that gastronomy/food experience is an important indicator of destination brand equity. Relevant literature shows that food experience or experience related to gastronomy festivals influenced the perceived dimensions of destination brand equity in the context of a tourist destination.

Festival quality is generally used as a predictor of tourist satisfaction and behavioral intentions of tourists (Wu et al., 2014; Culha, 2020). The food quality of festivals is one of the important antecedents of destination branding (Lee & Arcodia, 2011). Because a good quality of food affects positively destination image (Davras & Özperçin, 2021) and visitors' satisfaction (Wong, Wu, & Cheng, 2015). Festival satisfaction is all about the feeling of fulfilment that a participant receives from a festival (Kim et al., 2010). If it is a high level of satisfaction with the food offered at the festival, it will be positive outcomes such as recommending others to the food festival or revisiting the festival (Wong, Wu, & Cheng, 2015). Consequently, tourist destinations will have a chance to enhance destination image and loyalty via food-themed festivals.

It can be stated that the quality and experiences of food-themed festivals can trigger the destination brand image and destination brand loyalty (Deng and Li, 2014; Folgado-Fernandez et al., 2017; Yayla, 2017; Yang et al., 2020). Because visitors who have a good experience and are satisfied with the food-related festival can recommend others to participate, and experience food and beverage in the food-themed festivals and it will transition to a favourable destination image and enhanced destination loyalty (Yoon,

Lee, & Lee, 2010; Hui, Wan, and Ho, 2007; Yang et al., 2020; Aksoy & Kodaş, 2021). For example, Yoon et al., (2010) stated that there is a positive correlation between gastronomy events image and tourists' behavioural intentions and Folgado-Fernandez et al., (2017) revealed that gastronomic experiences related to gastronomic events have a positive effect on destination image and loyalty.

Motivations positively influence the perceptions of destination image (Baloglu & McCleary, 1999) and destination loyalty (Yoon & Uysal, 2005), if visitors are satisfied. Festival motivation is one of important of predictor of culinary destination image (Bigné, Sánchez & Sánchez, 2001; Yang et al., 2020) and destination awareness (Lee, Lee, & Lee, 2005). In the context of the event and destination branding, understanding or exploring dimensions of motivations can provide remarkable information for building effective branding effectiveness (Albayrak & Caber, 2018). Thus, it can be suggested that high events attendee food-related festival motivations led to increase the perceived destination brand equity. Because tourist who are pursuing the authentic, novel, different food experience is highly motivated to experience regional foods served in the destination visited (Björk and Kauppinen-Räisänen 2014). Gastronomy festivals is considered one of valuable source to meet these demands of tourist. So, it can be stated that the gastronomy/culinary festivals may trigger and enhance the motives of tourist local food consumptions (Chang & Yuan, 2011; Kim, Savinovic, & Brown, 2013; Krajíčková & Šauer, 2018; Kılıçhan & Köşker, 2020;) that result in higher perceived destination image and loyalty.

FUTURE RESEARCH DIRECTIONS

Based on the reviewed literature, through culinary festivals, destinations can promote their local/traditional food, build powerful food and destination images, and enhance the level of visitor awareness. Meanwhile, to develop destination brand loyalty, the destination can increase visitors' satisfaction with festival by focusing on successful and fascinating culinary festivals. In this chapter, it is discussed theoretically that food experience, food motivation and food quality with culinary festivals might be significant indicators of destination brand equity. Besides other tourist attractions through culinary festivals, destinations can increase perceived destination brand image, awareness, perceived quality, and brand loyalty of tourists if they want to build a successful destination brand for developing and achieving destination competitive advantage. Finally, it can be stated that culinary festivals help destinations to improve strategies for branding, hence they should dedicate more efforts to organizing culinary festivals as it is a powerful tool to enhance the destination brand equity.

This chapter has without limitations and have some implications for further research. First, aim of this study to reveal theoretically the effects of some various research perspectives of culinary festival such as festival motivation, experience, and food quality on tourist's perception towards the consumer-based destination brand equity. For this reason, some other antecedents of destination brand equity in the context of culinary festival such as festival involvement, festival identity, festival familiarity can be investigated using qualitative and quantitate approach. Second, the views of destination stakeholder could be investigated to create a successful culinary festival organization that help to enhance the perceptions of domestic and international tourist regarding destination brand.

CONCLUSION

On the global scale, we have witnessed a dramatic increase in the organization of culinary festivals as it is stated that culinary festival is a major attraction of destinations. Therefore, destinations among others, try to attract more visitors by organizing culinary festivals on a small- medium and large scale and they are considering culinary festivals as a way of promoting destination by focusing on locality. Culinary festivals, in particular, play an important role in building destination image, awareness, and destination loyalty. The literature demonstrates that culinary festivals are one of the effective ways to promote the destinations, increase destination familiarity and enhance tourist awareness of the destination. According to the results of this theoretical paper, if a destination wants to make build a successful destination brand via culinary festivals, they must focus on the quality of festivals, festival motivations, visitors' festival satisfaction, and festival experience. Because to build a successful destination branding strategies, all destination stakeholders should not ignore these elements. Through the culinary festival, festival goers discover new tastes, have a chance to socialize, enhance their level of cultural knowledge, and finally have strong connections with the destination. With these acquirements, destination image and loyalty degrees of festival attendees are increasing. Consequently, the destination builds a powerful destination brand and gains competitiveness. A growing body of literature shows that festival is an important driver of destination branding (Dimanche, 2002; Lee & Arcodia, 2011; Yang et al., 2020; Davras & Özperçin, 2021). Culinary festivals especially are paid attention from festival and event organizers in the destination context, since tourists are searching for different, unique, and exciting experiences.

REFERENCES

Aaker, D. A. (1991). *Managing Brand Equity: Capitalizing on the Value of a Brand Name*. The Free Press.

Aksoy, Z., & Kodaş, B. (2021). Determining the food experiences of tourists visiting Istanbul through online reviews. In Cihan Cobanoglu, Ebru Gunlu Kucukaltan, Muharrem Tuna, Alaattin Basoda & Seden Dogan (Eds.), Advances in Managing Tourism Across Continents (pp. 623–636). USA: USF M3 Publishing

Albayrak, T., & Caber, M. (2018). Examining the relationship between tourist motivation and satisfaction by two competing methods. *Tourism Management*, *69*, 201–213. doi:10.1016/j.tourman.2018.06.015

Axelsen, M., & Swan, T. (2010). Designing festival experiences to influence visitor perceptions: The case of a wine and food festival. *Journal of Travel Research*, *49*(4), 436–450. doi:10.1177/0047287509346796

Baloglu, S., & McCleary, K. W. (1999). A model of destination image formation. *Annals of Tourism Research*, *26*(4), 868–897. doi:10.1016/S0160-7383(99)00030-4

Bigné, J. E., Sánchez, M. I., & Sánchez, J. (2001). Tourism image, evaluation variables and after purchase behaviour: Inter-relationship. *Tourism Management*, *22*(6), 607–616. doi:10.1016/S0261-5177(01)00035-8

Björk, P., & Kauppinen-Räisänen, H. (2014). Culinary-gastronomic tourism–a search for local food experiences. *Nutrition & Food Science*, *44*(4), 294–309. doi:10.1108/NFS-12-2013-0142

Björk, P., & Kauppinen-Räisänen, H. (2016). Local food: A source for destination attraction. *International Journal of Contemporary Hospitality Management, 28*(1), 177–194. doi:10.1108/IJCHM-05-2014-0214

Blain, C., Levy, S. E., & Ritchie, J. B. (2005). Destination branding: Insights and practices from destination management organizations. *Journal of Travel Research, 43*(4), 328–338. doi:10.1177/0047287505274646

Boo, S., Busser, J., & Baloglu, S. (2009). A model of customer-based brand equity and its application to multiple destinations. *Tourism Management, 30*(2), 219–231. doi:10.1016/j.tourman.2008.06.003

Brown, G., Chalip, L., Jago, L., & Mules, T. (2004). Developing brand Australia: Examining the role of events. In N. Morgan, A. Pritchard, & R. Pride (Vol. Eds.), Destination branding: Creating the unique destination proposition. Vol. 2. Destination branding: Creating the unique destination proposition (pp. 279–305). Oxford: Elsevier Butterworth-Heinemann.

Chang, W., & Yuan, J. (2011). A taste of tourism: Visitors' motivations to attend a food festival. *Event Management, 15*(1), 13–23. doi:10.3727/152599511X12990855575024

Chekalina, T., Fuchs, M., & Lexhagen, M. (2018). Destination brand promise: The core of customer-based brand equity modeling. *Tourism Analysis, 23*(1), 93–107. doi:10.3727/108354218X15143857349800

Chi, H. K., Huang, K. C., & Nguyen, H. M. (2020). Elements of destination brand equity and destination familiarity regarding travel intention. *Journal of Retailing and Consumer Services, 52*, 101728. doi:10.1016/j.jretconser.2018.12.012

Choo, H., & Park, D. B. (2020). Comparison between local and non-local visitors for local food festivals. *Asia Pacific Journal of Tourism Research, 25*(6), 692–705. doi:10.1080/10941665.2020.1760905

Culha, O. (2020). The effect of food festival quality on place attachment and destination recommendation intention through festival experience and festival satisfaction: The case of the Didim International Olive Festival. *Journal of Convention & Event Tourism, 21*(5), 387–416. doi:10.1080/15470148.2020.1775743

Das, G., & Mukherjee, S. (2016). A measure of medical tourism destination brand equity. *International Journal of Pharmaceutical and Healthcare Marketing, 10*(1), 104–128. doi:10.1108/IJPHM-04-2015-0015

Davras, Ö., & Özperçin, İ. (2021). The relationships of motivation, service quality, and behavioral intentions for gastronomy festival: The mediating role of destination image. *Journal of Policy Research in Tourism, Leisure & Events*, 1–14. doi:10.1080/19407963.2021.1968881

Deng, Q., & Li, M. (2014). A model of event–destination image transfer. *Journal of Travel Research, 53*(1), 69–82. doi:10.1177/0047287513491331

Dimanche, F. (2002). The contribution of special events to destination brand equity. In K. W. Wöber (Ed.) *City Tourism 2002: Proceedings of European Cities Tourism's International Conference* (pp. 73-80). Springer. 10.1007/978-3-7091-6125-8_8

Du Rand, G. E., Heath, E., & Alberts, N. (2003). The role of local and regional food in destination marketing: A South African situation analysis. *Journal of Travel & Tourism Marketing, 14*(3/4), 97–112. doi:10.1300/J073v14n03_06

Everett, S., & Aitchison, C. (2008). The role of food tourism in sustaining regional identity: A case study of Cornwall, Southwest England. *Journal of Sustainable Tourism*, *16*(2), 150–167. doi:10.2167/jost696.0

Fathabadi, H., Nejad, M. R. O., & Alizadeh, H. (2017). An investigation of the factors affecting tourism destination brand equity. *Asian Journal of Social Sciences & Humanities*, *6*(1), 101–113.

Felsenstein, D., & Fleischer, A. (2003). Local festivals and tourism promotion: The role of public assistance and visitor expenditure. *Journal of Travel Research*, *41*(4), 385–392. doi:10.1177/0047287503041004007

Ferns, B. H., & Walls, A. (2012). Enduring travel involvement, destination brand equity, and travelers' visit intentions: A structural model analysis. *Journal of Destination Marketing & Management*, *1*(1-2), 27–35. doi:10.1016/j.jdmm.2012.07.002

Folgado-Fernández, J. A., Hernández-Mogollón, J. M., & Duarte, P. (2017). Destination image and loyalty development: The impact of tourists' food experiences at gastronomic events. *Scandinavian Journal of Hospitality and Tourism*, *17*(1), 92–110. doi:10.1080/15022250.2016.1221181

Getz, D. (2008). Event tourism: Definition, evolution, and research. *Tourism Management*, *29*(3), 403–428. doi:10.1016/j.tourman.2007.07.017

Getz, D., & Page, S. J. (2016). Progress and prospects for event tourism research. *Tourism Management*, *52*, 593–631. doi:10.1016/j.tourman.2015.03.007

Gómez, M., Lopez, C., & Molina, A. (2015). A model of tourism destination brand equity: The case of wine tourism destinations in Spain. *Tourism Management*, *51*, 210–222. doi:10.1016/j.tourman.2015.05.019

Gordin, V., & Trabskaya, J. (2013). The role of gastronomic brands in tourist destination promotion: The case of St. Petersburg. *Place Branding and Public Diplomacy*, *9*(3), 189–201. doi:10.1057/pb.2013.23

Hall, C. M., & Mitchell, R. (2005). Gastronomic tourism: Comparing food and wine tourism experiences. In M. Novelli (Ed.), *Niche tourism: Contemporary issues, trends and cases* (pp. 73–88). Butterworth-Heinemann. doi:10.1016/B978-0-7506-6133-1.50014-7

Hall, D. (1999). Destination branding, niche marketing and national image projection in Central and Eastern Europe. *Journal of Vacation Marketing*, *5*(3), 227–237. doi:10.1177/135676679900500303

Harb, A. A., Fowler, D., Chang, H. J. J., Blum, S. C., & Alakaleek, W. (2019). Social media as a marketing tool for events. *Journal of Hospitality and Tourism Technology*, *10*(1), 28–44. doi:10.1108/JHTT-03-2017-0027

Henderson, J. C. (2009). Food tourism reviewed. *British Food Journal*, *111*(4), 317–326. doi:10.1108/00070700910951470

Hermann, U. P., Boshoff, L., & Ncala, T. T. (2019). Understanding beer festival attendee motivations to a craft beer festival in South Africa. *Journal for New Generation Sciences*, *17*(2), 18–30.

Hernández-Mogollón, J., Folgado-Fernández, J., & Duarte, P. (2014). Event tourism analysis and state of the art. *European Journal of Tourism. Hospitality and Recreation*, *5*(2), 83–102.

Holcomb, B. (1999). Marketing cities for tourism. In D. Judd & S. Fainstein (Eds.), *The Tourist City* (pp. 54–70).

Horng, J. S., Liu, C. H., Chiu, H. Y., & Tsai, C. Y. (2012a). The role of international tourist perceptions of brand equity and travel intention in culinary tourism. *Service Industries Journal, 32*(16), 2607–2621. doi:10.1080/02642069.2011.594879

Horng, J. S., Liu, C. H., Chou, H. Y., & Tsai, C. Y. (2012b). Understanding the impact of culinary brand equity and destination familiarity on travel intentions. *Tourism Management, 33*(4), 815–824. doi:10.1016/j.tourman.2011.09.004

Horng, J. S., & Tsai, C. T. S. (2010). Government websites for promoting East Asian culinary tourism: A cross-national analysis. *Tourism Management, 31*(1), 74–85. doi:10.1016/j.tourman.2009.01.009

Hui, T. K., Wan, D., & Ho, A. (2007). Tourists' satisfaction, recommendation and revisiting Singapore. *Tourism Management, 28*(4), 965–975. doi:10.1016/j.tourman.2006.08.008

Im, H. H., Kim, S. S., Elliot, S., & Han, H. (2012). Conceptualizing destination brand equity dimensions from a consumer-based brand equity perspective. *Journal of Travel & Tourism Marketing, 29*(4), 385–403. doi:10.1080/10548408.2012.674884

Jago, L., Chalip, L., Brown, G., Mules, T., & Ali, S. (2003). Building events into destination branding: Insights from experts. *Event Management, 8*(1), 3–14. doi:10.3727/152599503108751658

Jiang, W. H., Li, Y. Q., Liu, C. H., & Chang, Y. P. (2017). Validating a multidimensional perspective of brand equity on motivation, expectation, and behavioural intention: A practical examination of culinary tourism. *Asia Pacific Journal of Tourism Research, 22*(5), 524–539. doi:10.1080/10941665.2017.1287106

Kalkstein-Silkes, C. A. (2007). *Food and food related festivals in rural destination branding*. [Unpublished Ph. D. Thesis, Purdue University, Indiana.]

Keller, K. L. (1993). Conceptualizing, measuring, and managing customer-based brand equity. *Journal of Marketing, 57*(1), 1–22. doi:10.1177/002224299305700101

Keskin, S. N. (2018). *Katılımcılarının Motivasyonlarına Göre Pazar Bölümlendirmesi: Alaçatı Ot Festivali Örneği. (Yüksek Lisans Tezi)*. Mersin Üniversitesi Sosyal Bilimler Enstitüsü.

Kılıçhan, R., & Köşker, H. (2020). Participation Motivation of Visitors of Local Food-Themed Gastronomy Festival. *Journal of Tourism and Gastronomy Studies, 8*(1), 78–91. doi:10.21325/jotags.2020.537

Kim, S., Savinovic, A., & Brown, S. (2013). Visitors' Motivations in Attending an Ethnic Minority Cultural Festival: A Case Study of the Feŝta Croatian Food and Wine Festival, South Australia. *Event Management, 17*(4), 349–359. doi:10.3727/152599513X13769392444828

Kim, Y. G., Suh, B. W., & Eves, A. (2010). The relationships between food-related personality traits, satisfaction, and loyalty among visitors attending food events and festivals. *International Journal of Hospitality Management, 29*(2), 216–226. doi:10.1016/j.ijhm.2009.10.015

Kodaş, B. (2021). Web Revolution and Events: Development and Progress. In K. Birdir, S. Birdir, A. Dalgıç, & D. Toksöz (Eds.), *Impact of ICTs on Event Management and Marketing* (pp. 18–32). IGI Global. doi:10.4018/978-1-7998-4954-4.ch002

Kodaş, B., & Aksoy, Z. (2022). Understanding of tourists' memorable local food experiences: A netnography study. *Journal of Economy Culture and Society*, *66*, 379–396.

Kodaş, B., & Sekizkardeş, M. (2022). Mardin'e Gelen Yerli Ziyaretçilerin Gastronomi Deneyimlerinin Demografik. Özelliklere Göre İncelenmesi. *Journal of Tourism and Gastronomy Studies*, *10*(4), 2862–2880.

Kodaş, D. (2018). *The relationship between gastronomy experience, gastronomy motivation, destination satisfaction and destination brand equity: case of Gaziantep*. [Doctoral dissertation, Anadolu University, Institute of Social Sciences, Eskisehir.]

Kodaş, D. (2021). Kodaş, D. (2021). Understanding Food Consumption Motivations: An Interpretivist Approach Using User-Generated Content. In Cihan Cobanoglu, Ebru Gunlu Kucukaltan, Muharrem Tuna, Alaattin Basoda & Seden Dogan (Eds.), Advances in Managing Tourism Across Continents (pp. 623–636). USA: USF M3 Publishing.

Kodaş, D., & Özel, Ç. H. (2022). (Article in Press). Antecedents of Gastronomy Destination Brand Equity: An Examination of Gastronomy Experience, Motivation, and Destination Satisfaction. [AHTR]. *Advances in Hospitality and Tourism Research*.

Konecnik, M., & Gartner, W. C. (2007). Customer-based brand equity for a destination. *Annals of Tourism Research*, *34*(2), 400–421. doi:10.1016/j.annals.2006.10.005

Krajíčková, A., & Šauer, M. (2018). Differences in motivation of food festivals visitors: A view from the Czech Republic. *Geographica Pannonica*, *22*(3), 189–200. doi:10.5937/gp22-17050

Lee, C.-K., Lee, Y.-K., & Lee, B. (2005). Korea's destination image formed by the 2002 World Cup. *Annals of Tourism Research*, *32*(4), 839–858. doi:10.1016/j.annals.2004.11.006

Lee, I., & Arcodia, C. (2011). The role of regional food festivals for destination branding. *International Journal of Tourism Research*, *13*(4), 355–367. doi:10.1002/jtr.852

Li, X. R., & Vogelsong, H. (2006). Comparing methods of measuring image change: A case study of a small-scale community festival. *Tourism Analysis*, *10*(4), 349–360. doi:10.3727/108354206776162769

Litvin, S., & Fetter, E. (2006). Can A Festival Be Too Successful? A Review of Spoleto, USA. *International Journal of Contemporary Hospitality Management*, *18*(1), 41–49. doi:10.1108/09596110610641966

Liu, C. H. (2020). Integration of different perspectives of culinary brand equity. *Journal of Hospitality and Tourism Management*, *45*, 152–161. doi:10.1016/j.jhtm.2020.08.007

López-Guzmán, T., Lotero, C. P. U., Gálvez, J. C. P., & Rivera, I. R. (2017). Gastronomic festivals: Attitude, motivation and satisfaction of the tourist. *British Food Journal*, *119*(2), 267–283. doi:10.1108/BFJ-06-2016-0246

Marzano, G., & Scott, N. R. (2006). Consistency in Destination Branding: The Impact of Events. *Global Events Congress and Event Educators Forum*, Brisbane, Australia.

Masiello, B., Bonetti, E., & Izzo, F. (2020). Multiple identities of a festival: Intended, communicated and perceived brand personality in the social media environment. *International Journal of Contemporary Hospitality Management*, *32*(2), 749–768. doi:10.1108/IJCHM-11-2018-0937

Mitchell, R., & Hall, C. M. (2004). Consuming tourists: Food tourism consumer behaviour. In *Food tourism around the world* (pp. 72–92). Routledge.

Mkono, M., Markwell, K., & Wilson, E. (2013). Applying Quan and Wang's structural model of the tourist experience: A Zimbabwean netnography of food tourism. *Tourism Management Perspectives*, *5*, 68–74. doi:10.1016/j.tmp.2012.10.007

Myagmarsuren, O., & Chen, C. F. (2011). Exploring relationships between destination brand equity, satisfaction, and destination loyalty: A case study of Mongolia. *Journal of Tourism. Hospitality & Culinary Arts*, *3*(2), 81–94.

Organ, K., Koenig-Lewis, N., Palmer, A., & Probert, J. (2015). Festivals as agents for behaviour change: A study of food festival engagement and subsequent food choices. *Tourism Management*, *48*, 84–99. doi:10.1016/j.tourman.2014.10.021

Park, K. S., Reisinger, Y., & Kang, H. J. (2008). Visitors' motivation for attending the South Beach wine and food festival, Miami Beach, Florida. *Journal of Travel & Tourism Marketing*, *25*(2), 161–181. doi:10.1080/10548400802402883

Pike, S. (2010). Destination branding case study: Tracking brand equity for an emerging destination between 2003 and 2007. *Journal of Hospitality & Tourism Research (Washington, D.C.)*, *34*(1), 124–139. doi:10.1177/1096348009349820

Pike, S., & Bianchi, C. (2016). Destination brand equity for Australia: Testing a model of CBBE in short-haul and long-haul markets. *Journal of Hospitality & Tourism Research (Washington, D.C.)*, *40*(1), 114–134. doi:10.1177/1096348013491604

Pine, B. J., & Gilmore, J. H. (1999). *The experience economy: Work is theatre and every business a stage*. Harvard Business Press.

Piva, L., Cerutti, S., Prats, L., & Raj, R. (2017). Enhancing brand image through events and cultural festivals: The perspective of the Stresa Festival's visitors. *Journal of Tourism. Culture and Territorial Development*, *8*(15), 99–116.

Quan, S., & Wang, N. (2004). Towards a structural model of the tourist experience: An illustration from food experiences in tourism. *Tourism Management*, *25*(3), 297–305. doi:10.1016/S0261-5177(03)00130-4

Schmitt, B. H. (1999). *Experiential marketing: How to get customers to sense, feel, think, act and relate to your company and brands*. The Free Press.

Selby, M. (2004). *Understanding Urban Tourism: Image, Culture and Experience*. IB Tauris.

Shone, A., & Parry, B. (2004). *Successful Event Management. A Practical Handbook* (2nd ed.). Thompson.

Sotiriadis, M. D. (2015). Culinary tourism assets and events: Suggesting a strategic planning tool. *International Journal of Contemporary Hospitality Management*, *27*(6), 1214–1232. doi:10.1108/IJCHM-11-2013-0519

Urry, J. (1990). *The tourist gaze: Leisure and travel in contemporary societies*. Sage Publications.

Wan, Y. K. P., & Chan, S. H. J. (2013). Factors that affect the levels of tourists' satisfaction and loyalty towards food festivals: A case study of Macau. *International Journal of Tourism Research, 15*(3), 226–240. doi:10.1002/jtr.1863

Wijaya, S., King, B., Nguyen, T. H., & Morrison, A. (2013). International visitor dining experiences: A conceptual framework. *Journal of Hospitality and Tourism Management, 20*, 34–42. doi:10.1016/j.jhtm.2013.07.001

Wong, J., Wu, H. C., & Cheng, C. C. (2015). An empirical analysis of synthesizing the effects of festival quality, emotion, festival image and festival satisfaction on festival loyalty: A case study of Macau Food Festival. *International Journal of Tourism Research, 17*(6), 521–536. doi:10.1002/jtr.2011

Wong, P. P. W. (2018). Role of components of destination competitiveness in the relationship between customer-based brand equity and destination loyalty. *Current Issues in Tourism, 21*(5), 504–528. doi:10.1080/13683500.2015.1092949

Wu, H. C., Wong, J. W. C., & Cheng, C. C. (2014). An empirical study of behavioral intentions in the food festival: The case of Macau. *Asia Pacific Journal of Tourism Research, 19*(11), 1278–1305. doi:10.1080/10941665.2013.844182

Yang, F. X., Wong, I. A., Tan, X. S., & Wu, D. C. W. (2020). The role of food festivals in branding culinary destinations. *Tourism Management Perspectives, 34*, 100671. doi:10.1016/j.tmp.2020.100671

Yayla, Ş. (2017). *Yöresel yemek festivaline katılım motivasyonu ile festivallerin destinasyon imajına ve davranışsal niyetlere etkisi* [Master's thesis, ESOGÜ, Sosyal Bilimler Enstitüsü].

Yoon, Y., & Uysal, M. (2005). An examination of the effects of motivation and satisfaction on destination loyalty: A structural model. *Tourism Management, 26*(1), 45–56. doi:10.1016/j.tourman.2003.08.016

Yoon, Y. S., Lee, J. S., & Lee, C. K. (2010). Measuring festival quality and value affecting visitors' satisfaction and loyalty using a structural approach. *International Journal of Hospitality Management, 29*(2), 335–342. doi:10.1016/j.ijhm.2009.10.002

ADDITIONAL READING

del Pilar Leal Londoño, M., Georgescu-Paquin, A., & Arcos-Pumarola, J. (2022). Food festivals as builders of the image of a tourist destination. *Journal of Policy Research in Tourism, Leisure & Events, 14*(3), 262–278. doi:10.1080/19407963.2022.2076690

Dimitrova, M., & Yoveva, I. (2014). Food festivals as purpose-created tourism attractions and their impact on destination branding. *Food and wine events in Europe: A stakeholder approach*, 172-181.

Hsu, F. C., Park, S. H., & Miller, J. C. (2023). Segmenting food festivalgoers: Experiential value, emotional state and loyalty. *British Food Journal, 125*(1), 29–48. doi:10.1108/BFJ-05-2021-0549

Kompaniets, O. R. (2023). The role of wine festivals in Swedish wine tourism marketing. In *Routledge Handbook of Wine Tourism* (pp. 347–358). Routledge.

Pirnar, I., Kurtural, S., & Tutuncuoglu, M. (2019). Festivals and destination marketing: An application from Izmir City. *Journal of Tourism. Heritage & Services Marketing*, *5*(1), 9–14.

Shi, H., Liu, Y., Kumail, T., & Pan, L. (2022). Tourism destination brand equity, brand authenticity and revisit intention: The mediating role of tourist satisfaction and the moderating role of destination familiarity. *Tourism Review*, *77*(3), 751–779. doi:10.1108/TR-08-2021-0371

Tran, A. T., Le, A. D., Bui, L. A. P., Le, V. V., & Vu, L. T. (2022). (Article in Press). Food festival research review in contemporary tourism. *International Journal of Tourism Cities*. doi:10.1108/IJTC-01-2022-0009

KEY TERMS AND DEFINITIONS

Brand equity: It is defined as 'a set of brand assets and liabilities linked to a brand, its name, and symbol, which add to or subtract from the value provided by a producer, by a product or service to a firm and/or to that firm's customers' (Aaker, 1991).

Culinary festival: It can be defined as a specific event that offer local food and beverages of destination to the festival-attendees.

Destination Brand Equity: It can be described as an adaption of Aaker's (1991) definition of brand equity model to the tourism destination. This brand equity model has four dimensions: awareness, associations/image, perceived quality, loyalty Which are widely accepted in measuring destination brand equity.

Festival: A festival is a "collective experience" where festival goers can be highly influenced by other people who are there with the same purpose and people who shared their experiences (Maisello, Bonetti and Izzo, 2019).

Chapter 7
Festivals as a Resource of Heritage Tourism:
A Bibliometric Analysis Using the Vosviewer

Ruchika Kulshrestha
Amity University, Gurgaon, India

Surabhi Srivastava
Om Kothari Institute of Research and Management, Kota, India

ABSTRACT

Rich and colorful festivals are a witness to our diversified arts and heritage around the world. Festivals are increasingly finding a place in every segment with integrity, where tangible and intangible heritage treasures are combined. To reveal the potential for tourism through promoting festivals as a resource of heritage, this research uses bibliometric analysis to carry out a systematic analysis of the literature. From the Web of Science data over the last 10 years (2012–2022), 199 English-language publications were gathered for analysis using VOS viewer. The most cited articles, contributions to heritage festivals research by various authors, nations, organizations, journals, and so on for innovations in the tourism sector, provided useful insights. This research entails a thorough examination of recent advancements in festival tourism, as well as a discussion of the researchers' future plans.

INTRODUCTION

Since ancient times, people have expressed their cultural celebrations and ceremonies to express their achievement, happiness, and delight. Around the whole world, various events are held cyclically on a community, provincial, nationwide, and international level. Festivals are becoming the key players in the development and promotion strategies of several destinations since they serve as imperative tourism motivators.

DOI: 10.4018/978-1-6684-6356-7.ch007

Festivals have been an integral component of the intangible cultural inheritance that communities have created through their qualities. It is more difficult to safeguard tangible cultural legacy than intangible cultural heritage, nevertheless. A community's traditions, customs, folklore, beliefs, dialect, music, dance, phrases, stories, and poetry are anthologies of intangible and invisible elements. One can only simply preserve festivals by frequently commemorating it, and by doing so, we may pass on our cultural heritage to subsequent generations (Yenipinar & Yieldiz, 2016).

In terms of possibility and connotation, the word "festival" positions not without its limitations. The term "festival" has the potential to symbolize everything, particularly enormous geopolitical graphics. Most people agree that a "festival" can be distinguished from other cultural events by either its frequency or the fact that it cannot be organised over so many different locations, making year-round celebrations unsustainable (Perry, Ager, & Sitas, 2020). These are global phenomenon and cultural activities with many different forms and expressions.

Festivals are extensively studied as social rituals that shape crucial, sometimes competing, patterns of human behaviour, according to sociology, anthropology, and other literature. Because festivals "hold a special place in practically all societies," Among these disciplines, the historical analysis of festivals is well-established (Getz, 2010). They are tangible traditions that have been used by early human cultures to organise time, behaviour, and the pursuit of pleasure.

Festivals are very important in the development of exciting municipal programs because they can promote a joyous and festive environment (Richards & King, 2022). Festivals' cultural components are crucial in establishing a connection between the host community and guests. To promote their cultures to tourists, several nations, regions, and towns have devised and designed festival programs. Precisely, a festival is a way to preserve the past of communities that have undergone urban development (Richards & Leal Londoño, 2022).

In addition to its social and cultural significance, festivals are becoming more and more popular as tourism attractions. They are treasured for their potential to bring life into towns, hotels, and destinations and are regarded as instruments in place development and destination designing. Therefore, cultural and social decision-makers as well as place promotion and tourism organisations are interested in how much longer they can serve as attractions while still upholding their social and cultural duties at the regional level (Andersson & Getz, 2008).

Festivals and events are a particularly effective instrument for development because they are adaptable and can be designed to address both local and global requirements. Due to their capacity to foster a euphoric, celebratory atmosphere, cultural festivals are particularly significant in the creation of eventful urban programming (Richards & King, 2022). Additionally, festivals are in a strong position to support the socio-cultural and economic aspects of the host and national communities. Festivals are events that uphold and promote heritage while creating destination branding, generating economic advantages, combating seasonality, and supporting local communities and regional businesses (Sinha & Murdia, 2018). Visitors learn about the local community's customs, traditions, values, and celebrations of their culture, and the host community shares its way of life with them by exchanging knowledge and experience through such religious and regional festivals. It also provides tourists with an opportunity to interact with the local community to meet their cultural needs, which fosters greater understanding and brotherhood (Sinha & Murdia, 2018).

Festivals are the hub that brings together various facets of living heritage, raising majorly two issues: Whether festivals reflect the regional and trans-regional legacy? How do these two domains balance out? These questions are bound to rethink the phrase "Unity in diversity" which can be stated as a commemoration of the intangible heritage expressions that remain local in nature yet celebrated globally.

This raises another question about the role of customs & traditional values as a creative process and the disposition of heritage as a "product," highlighting the role of the individual as a cultural actor as well as challenges with authenticity and identity. How is it possible for a single person, or even a small group, to claim to speak for everyone?

Bibliometrics is a statistical approach that, via mathematical methods, might quantitatively analyse one particular subject in the research articles involved (Chen et al., 2017). The foundation of bibliometric analysis is the systematic interpretation of a few of the metrics, which helps create an actual perspective of the researched data and makes it testable and verifiable (Celebi, Pirnar, and Eris, 2020). The quality of studies, the analysis of major study topics as well as the direction of future studies may potentially be predicted. Almost every major research article, which also contains integrated analytical tools to create representative numbers, is included in the Web of Science databases (WOS). Moreover, for additional examinations such as VOS viewer, search results from WOS might be transferred to software.

However, there is no publication have not been done on bibliometric analyses of Festival Tourism papers. Further knowledge from this reference should be derived, it is critically necessary for the bibliometric study. Our study was thus carried out in time to give a wide knowledge of Festival Tourism and future prospects for research.

This study applies bibliometric analysis to synthesize existing knowledge from 218 articles published on Web of Science between 2012 and 2022 as the study is only related to past 10 years only. To recognize contributions from chosen authors, organizations, and nations, the research looked at keywords, co-authorship and citations, as well as publishing networks.

As a result, the current study fills up this knowledge gap by responding to two research questions (RQs):

RQ1. Who are the Festival's main contributors as a source for heritage tourism research?
RQ2. What are the most significant future research areas that scholars investigating this issue should consider?

The present study creates fragmented the literature related to the Festival as a resource of Heritage tourism that response to the RQs

Festivals and Cultural Heritage

In recent years, several tourist locations have placed a high value on cultural heritage events in order to advance their economies and cultures. The growth of cultural tourism is significantly aided by festivals. MICE tourism planners are putting more stress on illustrating historical and cultural themes to craft cultural events as a mega showcase of traditions and customs of communities to pull maximum tourists and brand the artistic, cultural reflection of the host destinations by holding and celebrating a festival at the regional level. These occasions or celebrations are not targeted at any specific tourist group. The objective of organizing such events is to preserve & promote the growth of ethnicity of local heritage which has significant economic and cultural advantages. It is evident that the host community played a major role in promoting, sustaining and organizing festivals in an authentic way.

These customs and cultural heritage are culturally and socially significant and delicate in nature and they were created and advertised as tourist attractions. (Duran, 2013). Festivals give the community a reason to celebrate by sharing their traditions, experiences, and knowledge.

Festivals of culture and art are occasions that would significantly contribute to the sustainability of cultural values. The promotion and development of city image values are now achieved through culture and art festivals, which have evolved into living cultural legacy (Sazak & Eyübolu, 2009). According to Sazak and Eyübolu (2009), towns that acknowledge the value of festivals and create suitable strategies benefit in terms of culture, society, economy, and space. Local festivals also significantly benefit destinations by prolonging the tourist season, reviving the area in the off-season, building a brand, making the product selection more enticing, and preserving and promoting the region's natural heritage, customs, and traditions. (Şengül & Genç, 2016, Yenipinar and Yieldiz, 2016)

Festivals are being used to showcase and rediscover "local values" that had fallen into obscurity. Values expand beyond the bounds of customs, allowing for the development of authentic cuisines that reflect regional tastes and the blending of various flavour profiles. Festivals motivate individuals to travel to new places by igniting their interest about other cultures and ways of life, which also strengthens the bonds between communities. Such occasions spread a pleasant attitude and an attitude of acceptance toward various places and communities. Yuan and Chong (2007) Cultural heritage festivals offer a destination to commemorate the occasion with the cultural activities, as well as exhibiting the spirit to enlighten the cultural significance among the people (Kızılırmak, 2006). The modern festivals have placed a strong focus on the magical rituals' motivations, which sought to explain and discover the causes of the summer and winter days, oceans, waterfalls, forests and the sky (İmirgi, 2005). As a result, festivals are now thought of as activities conceded with seasons, history and have particular meanings. When these activities are connected with communities, their worth and authenticity increases. In a nutshell, local support is essential to the endurance and success of festivals (Şengül & Genç, 2016). According to Şengül and Genç (2016), all stakeholders, especially locals, must provide moral and physical support for tourism initiatives and the growth of the destination's branding. Festivals propagate cultures, and their enduring existence preserves and exercise the traditions and customs that enhance the pride among communities (İmirgi, 2005). In simple terms, festivals cannot continue to exist and be successful unless communities accept them and incorporate them into their social lives. Festival facilitators are aware about the fact of the significance of community involvement, so they always consider the local artist as assets. In order to rejoice and feel proud of one's country, as well as to demonstrate happiness and well-being as an entire nation, people get together to celebrate fairs and festival days.

In summary, Festivals can indeed be thought of as holistic locations where the concrete and intangible elements of cultural history come together and interact, with both beneficial and detrimental results (Perry, Ager, and Sitas, 2020). By integrative sites, we refer to occasions where context and content interact dynamically and where space, meaning, and cultural expression are all intertwined. Additionally, while defying rigorous descriptive variations, the rationale of holistic places suggests multiple levels of significance within festivals themselves. Festivals are symbolic, contingent, and contextual collections of events and understandings that are typically only understandable in context. They are geographically and chronologically restricted composites of several realities.

The festivals honor and recognize various manifestations of intangible heritage, which is defined as a variety of cultural expressions, forms, meanings, histories, and identities. Festivals offer platforms for surfacing and valuing many historical meanings in the pursuit of more sustainable urban transitions. In addition to their impact on regional economics and tourism initiatives, festivals can be studied for their varying qualitative implications of place and heritage throughout history. Indeed, the festivals that were looked at her offer intriguing ways to incorporate a holistic view of cultural heritage into theories and methods of built-up progress (Perry, Ager, and Sitas, 2020).

Celebrations and gatherings of a particular region are increasingly becoming discussion platforms where individuals can discuss more general cultural, economic, and political problems. Due to industrialization and development, discussions frequently become divisive between those who want variation and individuals who want to maintain "traditional" or "local" culture (Crespi-Vallbona & Richards, 2007).

RESEARCH METHODOLOGY

Data Collection

Benckendorff and Zehrer (2013) state that Web of Science was used to collect the data because it is widely regarded as the most in-depth abstract idea available for study. The current chapter explores "Heritage Festivals," "Heritage," and "Festivals," which are listed in the Web of Science under the title and include 229 publications, in accordance with standard operating procedure (Mukherjee et al., 2020). Tavakoli and Wijesinghe (2019) contend that they have little influence on the expansion of general knowledge. The data was excluded before the year 2012. This resulted in 224 publications but later on, the papers in English were only considered. Therefore the total number of articles included for review was 199 which consisted of 193 articles, 7 early access documents, 4 editorial material, 1 record review, and 1 review article from 2012 to 2022 from all the citation topics in Web of Science database. The data was then derived for the final bibliometric evaluation using Vosviwer.

Data Analysis

For bibliometrics, a variety of software tools are available. A statistical technique called bibliometrics uses mathematical techniques to quantitatively measure research papers on a given subject. Additionally, it could assess the study's quality, examine the main research areas, and predict the course of future research. However, almost all of the important research publications are available in the Web of Science (WOS) online database, which also has constructed analytical capabilities for generating representative statistics. Furthermore, WOS search results might be transferred to a programme for additional study, such as VOSviewer.

Additionally, the WOS collection database was searched for any global literature about heritage festivals published between 2012 and 2022. The search words used to find the publication with the highest similarity were heritage festivals which were used as the keyword in the title.

The information needed to complete the necessary documents contained complete file and stated references which incorporate the elements consisting of year of publication, language, journal, title, author, affiliation, keywords, document type, abstract and counts of citations which had been exported into different document format after which to tab (delimited win) format. The retrieval occurred on 24th August 2022. VOSviewer (version 1.6.18). This tool allows for the biometric presentation of an in-depth summary of a subject or journal (Aria & Cuccurullo, 2020). Two common weight attributes are used, and their definitions are as follows: "Links attribute" and "Total link strength attribute".

ANALYSIS AND DISCUSSION

Bibliometric Analysis of all the Keywords

The final analysis incorporated keywords supplied by the authors as long as they appeared at least five times in the primary WOS database. 35 out of the 1,150 keywords satisfied the criteria. The overall strength of concurrent links with other keywords was computed for each of the 35 keywords. Tourism, festivals, festivals, heritage, cultural heritage, identity, and intangible cultural heritage were among the confirmed handful of the chosen keywords. The two keywords with the highest frequency were "tourist" and "festival," which were strongly related to "heritage" and "cultural heritage" (Figure 1). The figure shows the 35 keywords with 4 clusters each having 13, 10, and 8 terms.

Figure 1. Bibliometric analysis of the key words

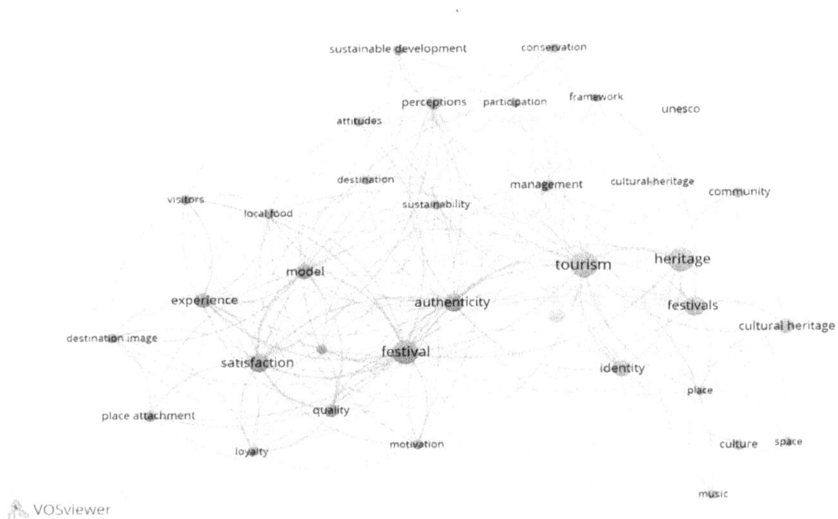

Bibliometric Analysis of the Author Keywords

The author chooses keywords that accurately describe the text. The bare minimum of keywords in Figure 2 was five. Eight of the 873 keywords pass the test. The overall strength of the co-occurrence links with other phrases was computed for each of the eight keywords. Tourism and festivals, which were mentioned the most (approximately 14 times each), were followed by "heritage" and "cultural heritage," which were mentioned 12 times.

Figure 2. Bibliometric analysis of the author key words

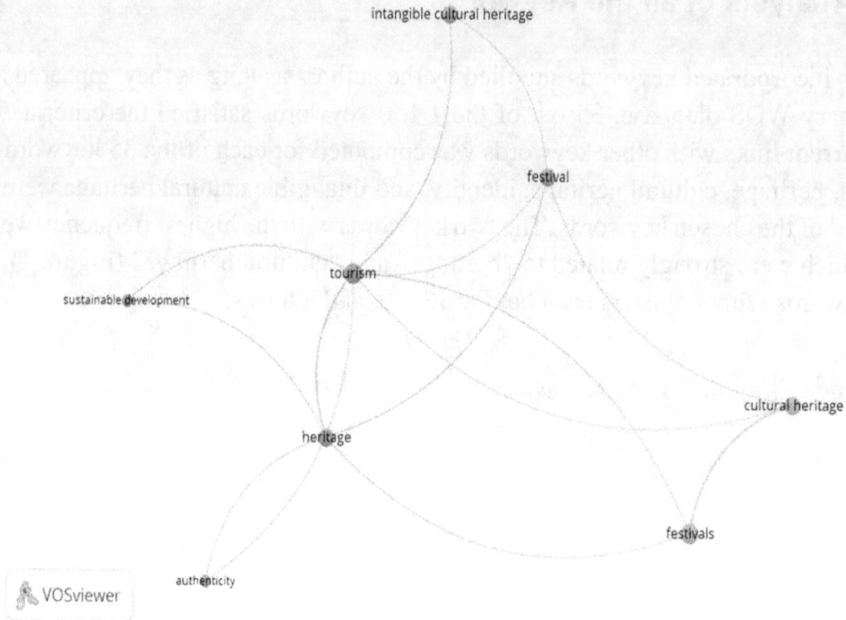

Bibliometric Analysis of the Co-Authorship

Co-Authorship Bibliometric Analysis

18 researchers out of 518 who contributed to the creation of articles about Heritage Festivals meet the requirement. The overall strong point of the co-authorship link with other writers was considered for each of the 18 authors. The researcher with the most powerful links overall was chosen, and the author with the lowest number of documents was 2, also chosen. Woosnam, Kyle Maurice from University of Georgia, Athens, GA, USA, has four document and 63 citations. Aleshinloye, Kayode D. from University of Central Florida USA has 156 citations with 3 documents.

Co-Authorship and Organizations Bibliometric Analysis

Through a review of the literature on the web of science, it is discovered that 322 organizations have published articles on heritage festivals, with 4 meeting the threshold criteria. University of Surrey, United Kingdom (UK)has published 4 documents, which have received 123 citations for a total link strength of 4. The University of Central Florida is following with the same number of documents but with 128 citations and the total link strength is 3.

Figure 3. Co-authorship bibliometric analysis

Figure 4. Co-authorship and organisations bibliometric analysis

Co-Authorship and Countries Bibliometric Analysis

In Figure 4 of the 60 countries, 15 meet the threshold in which the highest number of authors are from the United states of America. USA has published 33 documents with 286 citations and the total link strength is 20. The authors from England have 24 publications with 214 citations with a total link strength of 13. Almost as remarkable, the People Republic of China has 30 documents with 217 citations which has a total link strength of 13. and contributes 91 citations with 5 documents. New Zealand, Taiwan, India and Poland also worked with other nations to share expertise on the approach to heritage festivals.

Figure 5. Co-authorship and countries bibliometric analysis

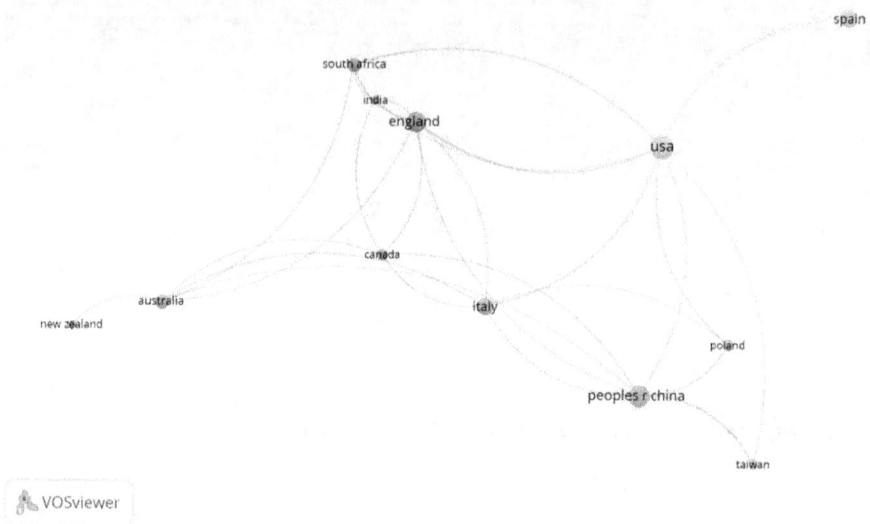

Citation Report Analysis

Time cited and Publications over Time

The articles were cited roughly 1,132 times across 199 publications. The results set's total items have, on average, 6.11 referencing articles. The total number of citations is 1,216. A collection of publications that have been sorted by the quantity of Times Cited is used to determine the h-index value. For the publications listed in the article presented in the above table, the h index is 19.

BIBLIOGRAPHIC COUPLING

Bibliographic coupling takes place when two publications quote the same third source in respective bibliographies. It suggests that there's a chance the two works are discussing the same thing. Two documents are said to be bibliographically related if they cite the same one or more other documents.

Table 1. The top 10 publications by number of citations

Rank	Title	Journal	Article Type	Country of corresponding author	Publication Year	Citations
1.	Residents' Emotional Solidarity with Tourists: Explaining Perceived Impacts of a Cultural Heritage Festival	Journal of Hospitality and Tourism Research	Article	South Africa	2018	51
2.	Overtourism as a perceived threat to cultural heritage in Europe	Current Issues in Tourism	Article	UK	2019	38
3.	The Influence of Event Authenticity and Quality Attributes on Behavioral Intentions	Journal of Hospitality and Tourism Research	Research Article	Italy	2016	34
4.	Analyzing the effects of an urban food festival: A place theory approach	Annals of Tourism Research	Article	Hong Kong	2019	32
5.	Celebration of heritage: linkages between historical re-enactment festival attributes and attendees' value perception	Journal of Travel and Tourism Marketing	Article	U.S.A.	2017	26
6.	Trip Cultural Activities and Tourism Expenditure in Emerging Urban-cultural destinations.	International Journal of Tourism Research	Research Article	UK	2020	23
7.	The roles of cultural worldview and authenticity in tourists' decision-making process in a heritage tourism destination using a model of goal-directed behavior.	Journal of Destination Marketing & Management	Research Article	Spain	2020	21
8.	Intangible Heritage as a Festival; or, Folklorization Revisited	Journal of American Folklore	Research Article	Iceland	2018	20
9.	Celebration of heritage: linkages between historical re-enactment festival attributes and attendees' value perception	Journal of Travel & Tourism Marketing	Research Article	USA	2018	20
10.	Cultural Heritage entanglements: festivals as integrative sites for sustainable urban development	International Journal of Heritage Studies	Research Article	England	2020	19

Bibliographic Coupling with Documents

Figure 7 depicts the bibliographic coupling map of documents. A document's minimum number of citations was one. Of the 199 documents, 199 meet the threshold. The research identified ten clusters. Cluster 1 contains 23 items, with a total link strength 25 (shown in red). Cluster two also has 17 items followed by cluster 3 with 12 items. While cluster 4 and cluster 5 has 10 and 9 items respectively which discuss the influence of heritage festivals.

Figure 6. Time cited and publications over the time

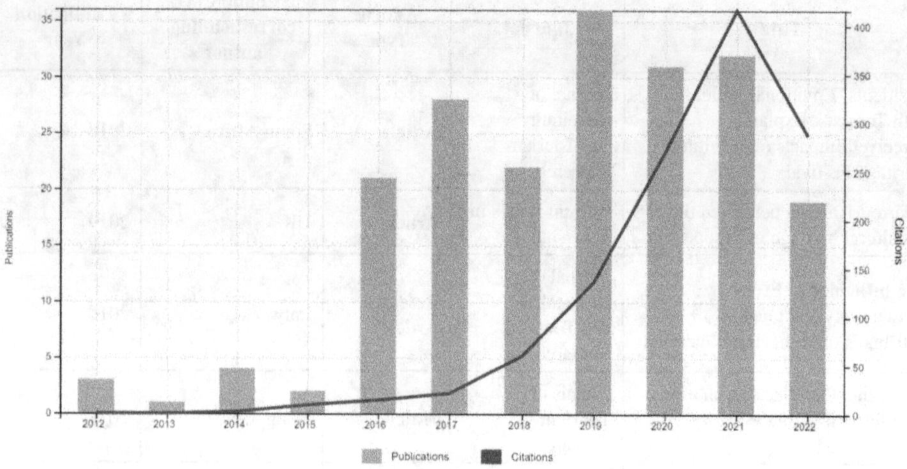

Figure 7. Bibliographic coupling with documents

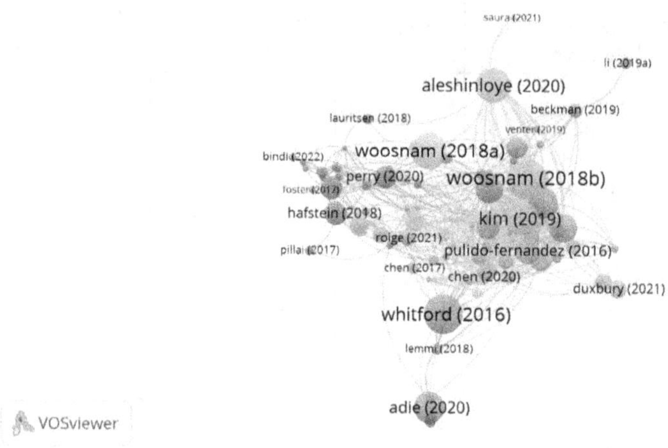

Bibliographic Coupling with Authors

Figure 8 shows the source coupling map with the authors. Of the 518 authors, 18 meet the thresholds. The bare minimum for an author was one document. The number of authors to be selected were 518. In the 88 items, there were 28 clusters and the total link strength is 8030 which has a link about 49716. Cluster 1 has 83 items (in red color) which has 102 links with total link strength 136 .

Figure 8. Bibliographic coupling with authors

CO-CITATION ANALYSIS

Co-citation analysis is a novel approach to investigating the cognitive structure of science. Tracking pairs of papers that are cited together in the source articles is what co-citation analysis is all about. When multiple authors cite the same pair of papers, research clusters emerge. The papers that are co-cited in these clusters usually have a common theme. Co-citation analysis generates paradigms/clusters to reveal interdisciplinary research trends within institutions. It is one of the most effective methods for tracking the evolution and structure of science.

Co-Citations and Cited References

The minimum number of a citations of a cited reference was selected 5. Of the 9621 cited references, 3 meet the threshold.For each of 39 cited references, the total strength of co-citations link with other cited references was calculated .The 4 clusters were formed with red, blue, green and yellow color .Cluster 1-3 has 11 items each while cluster 4 has 6 items.

Co-Citations and Cited Sources

The minimum number of citation of a source is 20.Of the 6000 sources, 27 meet the threshold. The Tourism management has highest number of citations i.e. 410 with the total link strength of 14592. While the annals of tourism research has 277 citations with total link strength 9692.The figure has 5 clusters in which cluster 1 (red color) has 12 items, cluster 2 (green color) has 11 items, cluster 3 (blue color) has 7 items, cluster 4 (light green color) has 4 items, cluster 5 (purple color) has 3 items.

Figure 9. Co-citations and Cited references

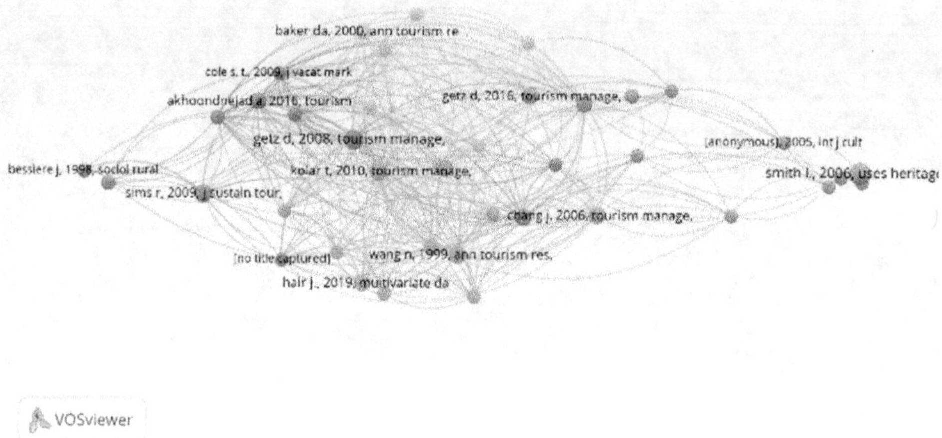

Figure 10. Co-citations and Cited sources

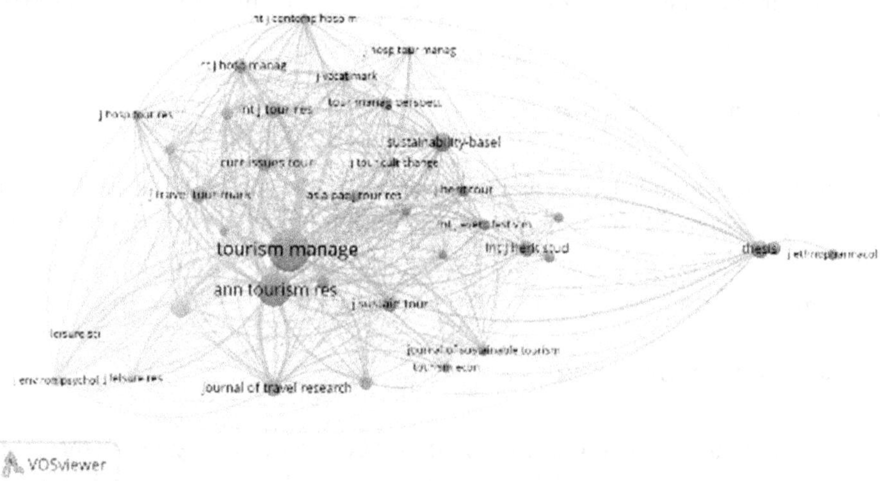

Figure 11. Co-citations and Cited authors

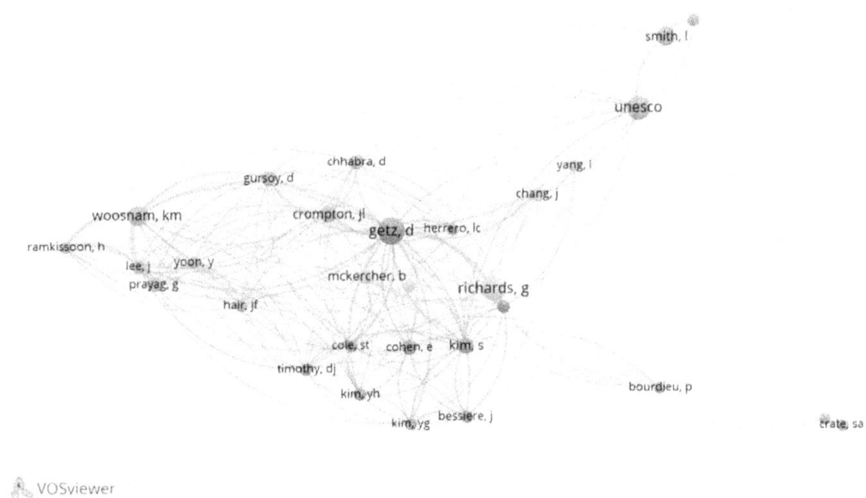

Co-Citations and Cited Authors

The minimum number of citations of an author selected were 10 . Of the 7518 authors, 32 meet the threshold. With a total 30 items, it has 6 clusters. Getz D. from Canada has the highest number of citations which is 49 and the total link strength is 365.

THEORETICAL IMPLICATIONS

From a scholarly standpoint, the bibliometric study helps by offering some answers to important concerns that academics should take into account while investigating festivals, heritage, and festivals as heritage. This study can aid researchers in recognising and understanding fresh developments in the field, as well as the lingo, publications that are devoted to these inquiries, and field reference studies. It gives insight into the significance of the topic and, in reality, might be applied to the creation of new tourism-related goods. It is the first study to employ a bibliometric approach to assist close the gap, identifying the resources, journals, and most importantly the conceptual subdomains, such heritage, festivals, intangible cultural heritage, and cultural identity, that will be crucial in future studies.

This chapter examines the intent, impact, and repercussions of heritage festivals critically. The research reveals how intangible cultural heritage is mobilised within events to particular agency, particularly in the creation and consumption of intrinsic and instrumental benefits for tourists, local communities, and performers.

This study serves as a reminder to planners of the three challenges that researchers regularly explore. First, identifying resources, safeguarding cultural heritage as festival, and maintaining the integrity of heritage festivals are the first steps in the resource planning and sustainability process. Second, planners must take into account how tourism development will affect stakeholders, including performers, inheritors, locals, the government, and Intangible cultural heritage itself. Thirdly, festival as product is a fresh approach to forging a distinctive image of a location at a time when it is crucial for planners to understand how to gauge, forecast, and market visitor behaviour.

The findings emphasized the significance of historical re-enactment festivals, social interactions, event design, and physical aspects. Each element has a unique impact on how valuable a heritage festival was researched by scholars. This chapter reflects the connections between heritage festivals and tourism promotion that provide a diverse path for the upcoming research.

CONCLUSION

Instead of just being a way to create a sense of place, festivals should be valued and seen as intrinsically important. The goal of this chapter is to undertake a bibliometric analysis of the academic literature on festivals and heritage tourism. Through bibliometric analysis, we were able to put the documents on the topic in perspective and draw attention to the key sections and research gaps, the top authors, the nations with the most publications and most favourable evaluations, the most illustrious journals, and the most frequently cited articles.

The complete number of publications on the topic is not recorded from the WOS database, highlighting the study's limitations. We hope that more researchers in other linguistic domains will duplicate the study since it was limited to English-language publications and did not investigate papers in other languages. Since there are more publications on the topic, it is important to assess the integrity of the papers offered in order to acquire the most pertinent data. The study's findings regarding festivals as a cultural legacy were restricted to works published between 2012 and 2021. Besides, bibliometric analysis some more qualitative research can be done using content analysis which can provide more knowledge to its contribution in the festivals as heritage (Lu & Nepal, 2009).

This study has many ramifications for heritage and festivals. First, it establishes a conceptual framework that clarifies the many ideas explored in the article, illuminating future researchers who could be interested in writing about festivals as a cultural heritage (Singh et al., 2021). Second, by providing information about the most productive authors, nations, and institutions, researchers may be able to gain a general knowledge of the partnership structure.

There is no question that there is still a long way to go and that advancement in science is necessary to address the issues with the study. Finally, in order to better comprehend the relevance of festivals as a legacy, this study contributes to enhancing knowledge transfer from research to the tourism industry.

REFERENCES

Ali, F., Park, E., Kwon, J., & Chae, B. (2019). 30 Years of contemporary hospitality management: Uncovering the bibliometrics and topical trends. *International Journal of Contemporary Hospitality Management, 31*(7), 2641–2665. doi:10.1108/IJCHM-10-2018-0832

Andersson, T. D., & Getz, D. (2008, September). Stakeholder management strategies of festivals. [). Taylor & Francis Group.]. *Journal of Convention & Event Tourism, 9*(3), 199–220. doi:10.1080/15470140802323801

Aria, M., & Cuccurullo, C. (2017). Bibliometrix: An R-tool for comprehensive science mapping analysis. *Journal of Informetrics, 11*(4), 959–975. doi:10.1016/j.joi.2017.08.007

Bashir, A., Singh, R., & Ps, S. (2022). Mapping the foundation and trends in tourism research: A bibliometric review of literature in tourism: An international interdisciplinary journal. *Tourism: An International Interdisciplinary Journal, 70*(3), 480–492. doi:10.37741/t.70.3.10

Benckendorff, P., & Zehrer, A. (2013). A network analysis of tourism research. *Annals of Tourism Research, 43*, 121–149. doi:10.1016/j.annals.2013.04.005

Celebi, D., Pirnar, I., & Eris, E. D. (2020). Bibliometric analysis of social entrepreneurship in gastronomy tourism. Tourism. *An International Interdisciplinary Journal, 68*(1), 58–67. doi:10.37741/t.68.1.5

Chen, H., Jiang, W., Yang, Y., Yang, Y., & Man, X. (2017). State of the art on food waste research: A bibliometrics study from 1997 to 2014. *Journal of Cleaner Production, 140*, 840–846. doi:10.1016/j.jclepro.2015.11.085

Crespi-Vallbona, M., & Richards, G. (2007). The meaning of cultural festivals: Stakeholder perspectives in Catalunya. *International Journal of Cultural Policy, 13*(1), 103–122. doi:10.1080/10286630701201830

Duran, E. (2013). A swot analysis on sustainability of festivals: The case of International Troia Festival. *Journal of International Social Research, 6*(28).

Fahimnia, B., Sarkis, J., & Davarzani, H. (2015). Green supply chain management: A review and bibliometric analysis. *International Journal of Production Economics, 162*, 101–114. doi:10.1016/j.ijpe.2015.01.003

Getz, D. (2010). The Nature and Scope of Festival Studies. *International Journal of Event ManagementResearch, 5*, 1-47. doi:10.1108/17852951011029298

İmirgi, A. (2005). Festival kavramıüzerinedüşünceler. *Milli Folklor, 17*(65), 29–36.

KIZILIRMAK, İ. (2006). Türkiye'de düzenlenen yerel etkinliklerin turistik çekicilik olarak kullanilmasina yönelik bir inceleme. [An investigation on the use of local events organized in Turkey as a tourist attraction.]. *Manas Üniversitesi Sosyal Bilimler Dergisi, 8*(15), 181–196.

Kockel, U., Craith, M. N., Clopot, C., & Tjarve, B. (2019). Heritages, identities and Europe: Exploring cultural forms and expressions. In *Heritage and Festivals in Europe*. Taylor & Francis. doi:10.4324/9780429202964-1

Lu, J., & Nepal, S. K. (2009). Sustainable tourism research: An analysis of papers published in the Journal of Sustainable Tourism. *Journal of Sustainable Tourism, 17*(1), 5–16. doi:10.1080/09669580802582480

Perry, B., Ager, L., & Sitas, R. (2020). Cultural heritage entanglements: Festivals as integrative sites for sustainable urban development. *International Journal of Heritage Studies, 26*(6), 603–618. doi:10.1080/13527258.2019.1578987

Richards, G., & King, B. (2022). The experience of cultural festivals: Evidence from Hong Kong. *Journal of Policy Research in Tourism, Leisure & Events, 14*(3), 296–309. doi:10.1080/19407963.2022.2033249

Richards, G., & Leal Londoño, M. D. P. (2022). Festival cities and tourism: Challenges and prospects. *Journal of Policy Research in Tourism, Leisure & Events, 14*(3), 219–228. doi:10.1080/19407963.2022.2087664

Sazak, Ş., & Eyüboğlu Erşen, A. (2009). Bir Kültür Endüstrisi Olarak "Kırkpınar Güreş Festivali". ["Kırkpınar Wrestling Festival" as a Cultural Industry.]

Şengül, S., & Genç, K. (2016). Festival Turizmi Kapsaminda Yöresel Mutfak Kültürünün Destekleyici Ürün Olarak Kullanilmasi: Mudurnu İpekyolu Kültür Sanat ve Turizm Festivali Örneği. *Pamukkale Üniversitesi Sosyal Bilimler Enstitüsü Dergisi*, (23), 79–89.

Sinha, M. R., & Murdia, M. (2018). Local Residents notion on the contribution of cultural tourism in Murshidabad, West Bengal. *IJRAR–International Journal of Research and Analytical Reviews, 5*, 1430–1438.

Tavakoli, R., & Wijesinghe, S. N. (2019). The evolution of the web and netnography in tourism: A systematic review. *Tourism Management Perspectives, 29*, 48–55. doi:10.1016/j.tmp.2018.10.008

Yenipinar, U., & Yildiz, E. (2016). Festivals as cultural heritage: the Mesir festival of Manisa. R. Efe, Global Issues and Trends in Tourism, 323.

ADDITIONAL READING

Bhattacharyya, J., & Dash, M. K. (2021). *What do we know about customer churn behaviour in the telecommunication industry? A bibliometric analysis of research trends, 1985–2019*. FIIB Business Review. doi:10.1177/23197145211062687

Dundes, A. (Ed.). (2005). *Folklore. Critical Concepts in Literary and Cultural Studies (I-IV): I. From Definition to Discipline. II. The Founders of Folklore. III. The Genres of Folklore. IV. Folkloristics: Theories and Methods*. Routledge.

Ellegaard, O., & Wallin, J. A. (2015). The bibliometric analysis of scholarly production: How great is the impact? *Scientometrics, 105*(3), 1809–1831. doi:10.100711192-015-1645-z PMID:26594073

Littell, J. H., Corcoran, J., & Pillai, V. (2008). *Systematic reviews and meta-analysis*. Oxford University Press. doi:10.1093/acprof:oso/9780195326543.001.0001

Chapter 8
Gastronomic Festival Tours in Turkey:
Motivations and Expectations of Tourists

Suat Akyürek
Gümüşhane University, Turkey

ABSTRACT

The demand for gastronomy tourism has been increasing in recent years, and gastronomy tours and festivals are marketed as an important gastronomic product in meeting this demand. However, research on the motivation and expectations of tourists participating in gastronomy festivals are still not sufficient in the literature. In this study, it is aimed to determine the motivation and expectations of tourists who participated in the tours organized under the name of gastronomic festival tours. Face to face interviews were carried out with 23 tourists who participated in three different gastronomic festival tours arranged in Turkey. In the light of the findings, the motivations of tourists participating in gastronomic festival tours were evaluated within the scope of push and pull factors. Expectations of tourists from gastronomic festival tours were classified into five categories as general services, relations with other people, environmental conditions and internal expectations, which were especially for products/activities that are the focal point of the tours.

INTRODUCTION

One of the most important travel motivations since primitive ages can be considered as the desire to find food and experience different tastes (Korstanje, 2009). Therefore, interest in researches on tourist motivation in the gastronomy tourism literature is increasing day by day. The gastronomic values of destinations can increase the travel motivation of tourists towards that destination (Boniface, 2003; Hall, & Sharples, 2003; Quan, & Wang, 2004; Kivela, & Crotts, 2005) and can function as a determinant for the selection of destination (Boniface, 2003; Hsu, Tsai, & Wu, 2009; Kim, Goh, & Yuan, 2010). Besides, some authors (Tikkanen, 2007; Hsu et al., 2009; Cruz et al., 2019) emphasize that gastronomy is one of the most important travel motivations for destinations. In the research conducted by International

DOI: 10.4018/978-1-6684-6356-7.ch008

Tourism Trade Fair-FITUR (2018), it was determined that the gastronomy of the destination is among the main motivations of 15% of the tourists in choosing the tourism destinations they will travel (cited in Berbel-Pineda et al., 2019).

Since the beginning of tourism research, researchers have tried to determine travel motivations (Lee, 2015). Motivation is a process that explains the basis of forces that mobilize tourists and the reasons behind tourist behavior (Crompton, 1979). Goossens (2000) emphasizes the importance of marketers understanding the factors that lead to consumer decisions and consumption behaviors in order to market tourism services and destinations well. Consumer researchers argue that "experiential" aspects of consumption such as consumption dreams, emotions and entertainment play an important role in consumer choice behavior. In this context, in order to perceive consumer behaviors; it is critical to understand the needs and the motives and expectations that trigger them (Koç, 2012).

Motivation is a key element in the design and creation of gastronomy tourism, down to physiological, safety and social needs. This includes belonging and personal need motivation, need for prestige (gastronomy and luxury), status or self-realization. Therefore, motivation represents many desires and needs (Horne, & Manzenreiter, 2006; Ellis et al., 2018). Gastronomy can allow tourists to discover the cultural origins of the region they go to and thus can be one of the important motivations in destination selection (Cruz et al., 2019). Otto and Ritchie (1996) suggested that the concepts of tourism experience and motivation are highly correlated. In fact, determining the motivations of individuals planning to travel will shed light on what kind of experiences they want to have. Lee (2015) suggests that travel motivations may be effective in the formation of experiences. Fields (2002) asserted that developing experiences with tourist motivations will make gastronomy tourism even more popular in the future. According to the research conducted by Berbel-Pineda et al. (2019) in Seville (Spain), it was suggested that the travel motivation of tourists affects their gastronomic experiences. Gastronomic experiences gained increase the satisfaction of the destination. Destination satisfaction and gastronomic experiences constitute tourist loyalty. According to Park et al. (2008), tourists behave differently as they want to meet their different needs. Understanding the travel purposes of tourists, designing and tailoring the presentations for specific target markets is an essential prerequisite for creating appropriate experiences and satisfaction for these markets. Gastronomic supply sources can create the most intense experiences for tourists. This is especially prominent when gastronomy tourism is the main travel motivation of tourists such as gastronomy tours, gastronomy themed festivals, wine tourism (Zamparini, 2018). In this context, it is important to understand the motivations of tourists to participate in gastronomic festival tours and their expectations from these tours. In this sense, the main purpose of this study is to determine the reasons and expectations of tourists participating in gastronomic festival tours. Thus, it is aimed to provide some suggestions to the relevant field and practitioners.

As a result, the demand for gastronomy tourism has been increasing in recent years and gastronomy festivals are marketed as an important gastronomic product in meeting this demand. However, researches on the motivation and expectations of tourists participating in gastronomy festivals are still not sufficient in the literature. In this study, it is aimed to determine the motivation and expectations of tourists who participated in the tours organized under the name of gastronomic festival tours. Face to face interviews were carried out with 23 tourists who participated in three different gastronomic/festival tours arranged in Turkey. I also participated as a participant observer in these tours organized within the scope of gastronomic festivals. In the light of the findings, the motivations of tourists participating in gastronomic festival tours were evaluated within the scope of push and pull factors.

1. GASTRONOMIC FESTIVAL TOURS

In the tourism sector, there is an increasing demand for gastronomy tourism and the variety of gastronomic products is increasing day by day. Gastronomy tourism includes a series of activities such as participating in any gastronomic festival, events and activities, visiting gastronomic production areas, visiting gastronomy museums, having conversations about gastronomy history, product, harvest etc., watching gastronomic competitions and shows. The gastronomy tourism specifically consists of tours organized to places where food and beverage are produced, as well as gourmet tours that include testing food products. Gastronomy tour is defined as a touristic product prepared by a tour agency, where participants experience the local food culture and food traditions besides the culture and heritage of the destination (Ko et al., 2018). Hjalager (2002) evaluated gastronomy tours as the tours carried out in different regions for the purpose of learning the production processes and history of the products belonging to the food and beverage culture, and experiencing their preparation and taste.

Nowadays, gastronomy tours can be done on foot, by bicycle, car, bus, train or boat. Walking tours are usually short (up to 6 hours) and an authentic way for visitors to experience the culinary journey. The main purpose here is to provide the right amount of exercise with training and culinary experience. Kemeraltı flavor tour in Turkey is the best example of the walking tour. Bicycle gastronomic tours can be performed in Europe. Especially in rural areas, it is an important source of attraction for tourists who want to experience both the nature and local food-beverages on site (Stanley, & Stanley, 2015). In addition, thanks to applications such as EatWith (https://www.eatwith.com/); visitors can go to local residents' houses nowadays and have a meal with the family, and participate in special gastronomy tours in the region.

Shenoy (2005) defined gastronomy tours as cooking tours, dinner tours in restaurants and tours for food production areas. Many specific gastronomy tours are organized worldwide such as local food tours, cooking course tours, tasting tours, gourmet tours, wine tours, restaurant tours, cheese tours, olive and olive oil tours, tea tours, coffee tours, mushroom tours, chocolate tours, crop harvest tours and gastronomic themed festival tours. In addition to these, many gastronomic tours can be arranged such as tours for a special meal (molecular, fusion, street food etc.), tours to eat or watch a show made by a famous chef, wine (vintage) tours, gastronomic museum tours, farm tours, food production factories, vegan tours and romantic dinner tours. However, gastronomy tours for gastronomy festivals, especially in Turkey, are also very popular in recent years. Table 1 includes examples of some gastronomy tours organized in different regions of the world.

2. MOTIVATIONS FOR PARTICIPATING IN GASTRONOMY TOURISM

Gastronomy tourism is a sector that can create important and loyal customers in terms of its economic returns (Kivela, & Crotts, 2005). So much so that food is among the most important features of destinations (du Rand ve Heath, 2006) and can significantly affect the travel motivation of tourists (Tikkanen, 2007). Food and beverage is the main motivation source for 93% of travelers (World Food Travel Association, 2019). In a 2012 survey by Visa and PATA, it was determined that food and beverages were the third main reason for participants to visit Asia Pacific countries and the first reason participants would revisit a destination (Wijaya, 2014).

Table 1. Gastronomy tours organized in different regions of the world

Name of Tour	Place	Theme	Agency's Site
1- Wine and Tapas Tour	Spain	Wine and Tapas	https://www.tours4fun.com/valencia-wine-and-tapas-tour.html https://www.withlocals.com/experience/the-ultimate-tapas-win e-tour-a7485cfb/?adults=2&children=0
2- Gourmet Cheese Tours	France	Cheese	https://www.exclusive-france-tours.com/en/gourmet-cheese-tou r-franc/
3- Food Discovery of Sri Lanka	Sri Lanka	Food and drink	https://www.intrepidtravel.com/
4- China Tea Tour	China	Tea	https://www.easytourchina.com/tour-c8-china-tea-tours
5- Mexican Real Food Adventure	Mexico	Food and drink	https://www.intrepidtravel.com/
6- Morocco Real Food Adventure	Morocco	Food and drink	https://www.intrepidtravel.com/
7- Vegan Flavors Feast in Delhi	India	Vegan foods	https://www.intrepidtravel.com/
8- Mysterious Wine Tour	Italy	Wine	https://www.secretfoodtours.com/rome/secret-wine-tastings/
9- Street Food Tour	Thailand	Street Foods	https://www.bangkokfoodtours.com/bangkok-best-eats-midnight-food-tour-by-tuk-tuk/
10- Small-Group Tour: Historical Pub Walking Tour of London	England	Historical Pubs	https://www.tripadvisor.com/AttractionProductReview-g186338-d11451643-Small_Group_Tour_Historical_Pub_Walking_Tour_of_Lo ndon-London_England.html
11- Small Group Hanoi Street Food Tour with a Real Foodie	Vietnam	Street Foods	https://www.tripadvisor.com/AttractionProductReview-g293924-d12878612-Small_Group_Hanoi_Street_Food_Tour_with_a_Real_Foo die-Hanoi.html
12- Kentucky Bourbon tour	USA	Bourbon	https://kybourbontrail.com/plan-your-trip/
13- Chocolate Tour	Switzerland	Chocolate	https://www.viator.com/Switzerland-tours/Chocolate-Tours/d69 -g6-c32020
14- Sunset Wine & Cheese tasting canal boat tour	Netherlands	Food and drink	https://www.eatwith.com/events/49204
15- 10 Tastings With Locals: Athens Favorite Food Tour	Greece	Food and drink	https://www.withlocals.com/passions/food/greece/athens/
16- Berlin: Vegan Food Tour	Germany	Food and drink	https://www.getyourguide.com.tr/activity/berlin-l17/berlin-v egan-yemek-turu-t170481?utm_force=0 https://www.withlocals.com/experience/the-ultimate-vegan-ber lin-tour-38ed2629/?adults=2&children=0
17- Urla Artichoke Festival Tour	Turkey	Artichoke dishes	https://www.vesturizm.com.tr/urla-enginar-festivali-turu
18- Izmir Kemeraltı Taste Tour	Turkey	Food and drink	http://italtur.com.tr/tr/italtur_gastro/detay/10/kemeralti-g astronomi-turu
19- Tea Gathering Tour	Turkey	Tea	https://www.ritur.com.tr/cay-toplama-turu-cay-bahcesi-gezisi -rize-cay-uretimi/
20- Istanbul on Food - Culinary Tours	Turkey	Food and drink	https://www.tripadvisor.com.tr/Attraction_Review-g293974-d58 49400-Reviews-Istanbul_on_Food_Culinary_Tours-Istanbul.html

Source: Table was created by author

Table 2. Tourist motivations for participating in gastronomy tourism

Motivations for Participating in Gastronomy Tourism	Resear-chers
1. Physical motivators: Gastronomy offers physical attractions that appeal to the five senses. **2. Cultural motivators:** Gastronomy reflects the culture of a region. **3. Interpersonal motivators:** Gastronomy is a means of socialization. **4. Status and prestige motivators:** As described by Veblen (1899), meals show outstanding forms of entertainment and class (Ramshaw, 2016).	Fields, 2002
1. Anxiety: It is the desire to show courage and celebrate food and drink on holiday. It is the pursuit of nostalgia with the need for comfort and escape with traditional, non-routine foods. **2. Need to show distinction, affluence and individualism:** It includes the desire to consume special foods and beverages that act as media. It is the desire to consume food and beverages representing a "common" transferable language between daily life and vacation. **3. Curiosity and desire for knowledge and exploration:** It is the desire to consume food and drink to show others his/her education and superiority. **4. The need to move away from the impact of globalization:** Investigate production areas, rural areas, farms, etc. to see how food is obtained. **5. A need for tactile and sensory pleasure and experience:** It is the desire to consume food and drink to obtain different experiences.	Boniface, 2003
1. High interest: The main motivation for the travel destination is visiting a restaurant, market or winery or participating in food-related activities. (Gourmet, gastronomic and culinary tourism). **2. Moderate interest:** Visiting local markets, festivals, restaurants or wineries, as well as many activities at the destination. (Culinary tourism). **3. Low interest:** Visiting local markets, festivals or wineries in the destination because they are different. (Rural / urban tourism). **4. Low interest / no interest:** It is going to restaurants that are familiar to have to eat.	Hall, & Sharples, 2003
1. Primary motivation: Refers to a tourist choosing an exact destination to try the local cuisine. **2. Secondary motivation:** It focuses on causes other than gastronomy, but gastronomy is also a powerful factor in deciding whether to visit a destination or not.	Quan, & Wang, 2004
1. Exciting experience: It involves the search for excitement by experiencing gastronomic elements that tourists have not seen and tasted before. **2. Escape from routine:** It refers to going out of daily routine nutrition. **3. Health concern:** It includes the search for factors such as nutritional, freshness and naturalness. **4. Learning knowledge:** It includes learning the food culture of different cultures. **5. Authentic experience:** Refers to focus on local food and drink for its uniqueness and originality. **6. Togetherness:** It refers to getting together with relatives and increasing relationships. **7. Prestige:** Expresses that food and beverages improve the self and increase the personality value in the eyes of others. **8. Sensory appeal:** It refers to the driving force of the smell, taste and visual appearance of the food. **9. Physical environment:** It includes elements such as the cleanliness, atmosphere and order of the restaurants.	Kim, Eves, & Scarles, 2009
1. Push factors: It consists of 3 factors: "knowledge and learning", "entertainment and new experiences", "relaxation with the family". **2. Pull factors:** It consists of 3 factors: "field quality and value", "activity quality" and "food variety".	Kim, Goh, & Yuan, 2010
1. Cultural experience: It includes experiencing the taste of a local meal, getting to know the local culture through the cuisine, discovering new things and seeking authenticity. **2. Excitement seeking:** It is the desire to feel excitement, relax, escape from routine. **3. Interpersonal relationship:** It includes elements such as the desire to talk to others about local dining experiences, the desire to develop kinship and friendship relations, and the desire to spend pleasant time with their relatives. **4. Sensory appeal:** It involves the pursuit of pleasant fragrance, flavor and presentation. **5. Health concern:** It includes the search for elements such as nutritional, freshness and naturalness.	Kim, Eves, & Scarles, 2013
1. To get a break from your day-to-day environment **2.** To relax and relieve stress **3.** To create lasting memories **4.** To enrich your relationship with your family **5.** To see or do something new different **6.** To have a life with no fixed schedule **7.** To enrich your perspective on life **8.** To gain knowledge of history, other cultures or other places **9.** To stimulate your mind/be intellectually challenged **10.** To renew personal connections with people **11.** To be pampered	Sohn, & Yuan, 2013
1. Relaxation **2.** Socialization **3.** Getting new experiences **4.** Culture **5.** Food experience	Björk, & Kauppinen-Räisänen, 2014

Source: Table was created by author

The literature on tourism behavior has shown that tourism patterns have changed for various reasons and that travelers are not a homogeneous cluster (Turnbull, & Uysal, 1995). Again, Smith, & Costello (2009) examined the relevant literature of the motivation for participating in festivals and determined that the motivations vary according to the theme of the event and the type of visitor. Fields (2002) draws attention to the difficulty of classifying tourists into homogeneous groups according to their motivations, stating that the motivating factors in tourism are as many as the number of tourists. Likewise, Cruz et al. (2019) show that tourists' gastronomic motivations are heterogeneous and conditioned by their attitudes and/or interests in gastronomy in their travels. Despite this, classifications about tourist motivation have been made by different researchers in the literature. Table 2 includes some studies related to the motivation to participate in gastronomy tourism.

Table 2 includes classifications or reasons made by different authors, which are related to the motivation for participating in gastronomy tourism. As can be seen from the table, the reasons for participating in gastronomy tourism have been put forward in different ways by each author. In particular, the desire to experience unfamiliarity and difference and to learn about the culture of others is an important reason to visit a destination (McKercher, & So-Ming, 2001). As a result, the basis of tourists' travel and tourism motivations has the aim of entertainment, enjoyment, renewal and experience and knowledge. Also the main motivation for tourists participating in gastronomy tourism is to get to know and try the local food culture of the region or district. Again, the concept of curiosity is at the center of motivation and drives people to learn, do, experiment, explore and experience. Curiosity therefore plays a central role in shaping the tourist experience (Tse, & Crotts, 2005).

Food and beverages can be a supportive or main travel motivation for travel and is an activity that tourists will not avoid spending money (Chen, 2018). Goeldner, & Ritchie (2006) state that food and drink cannot be ignored by tourists, and they cannot visit a country without consuming food and drink. It has been proven by many authors (Hjalager, 2002; Boniface, 2003; Hall, & Sharples, 2003; Quan, & Wang, 2004; Kivela, & Crotts, 2005; Hsu et al., 2009; Björk, & Kauppinen-Räisänen, 2014) that gastronomy is the primary/main or secondary/supporting source of attraction that triggers travel motivation.

3. EXPECTATIONS FROM GASTRONOMIC FESTIVAL TOURS

Expectations consist of individuals' general knowledge and past experiences (Oliver, 1977). Expectations are examined to help explain the quality of experience (Dickson, & Hall, 2006). Gnoth (1997) emphasizes that it is important to manage the expectations of tourists because it affects the selection process and destination experience perceptions of tourists. According to Zeithaml, Bitner, & Gremler (2017), expectations are defined as customer requests or needs, which are believed to be offered especially by service providers. However, Du Rand, & Heath (2006) emphasize the importance of understanding the desired benefits from the eyes of tourists for marketing gastronomy tourism, which is a niche tourism type. Similarly, Richards (2002) points out that in order to develop strategies for marketing gastronomy tourism, it is important to know the unique expectations in this market segment, to find value-added features and to increase positive tourist experiences. In this context, the participants were asked questions about their expectations (before travel) from the tour they participated in and whether their expectations were met (after the trip).

4. METHODOLOGY

Qualitative research method was preferred in the study. Qualitative methods offer the opportunity to work on deep and detailed issues (Patton, 2001). Carlsson (2001) points out that qualitative research offers important opportunities in understanding people's experiences with their environment and their relations with this environment. In this context, qualitative method was preferred in the study. Semi-structured interview and participant observation were preferred to collect data. In qualitative research, interview is the basic technique in the field of research to understand the phenomenon. The main purpose of interviews is to collect private information. Interviews are essential to learn about behaviors, experiences, ideas, perceptions, emotions that we cannot observe, and how people express the world around them (Patton, 2001; Merriam, 2009). Berg, & Lune (2017) defined the interview as a simple conversation with a purpose. Merriam (2009) suggested that interviews should be conducted in a semi-structured manner with open-ended questions in qualitative studies. Because semi-structured interviews can enable the participant to express the perceived world with their own thoughts. Kim, Eves, & Scarles (2009) refer to the necessity of semi-structured and open-ended questions to encourage the free expression of the opinions and feelings of the interviewees. Kastenholz, & Lima (2011) state that semi-structured individual interview is a suitable technique for data collection in investigating emotions, meanings, other personal opinions and behavioral options. However, qualitative research requires going to a real environment and working there personally (Creswell, 2017). Therefore, direct participation and observation of the phenomenon is required to fully understand the complexity of many situations (Patton, 2001). In particular, the researcher participated in tours/festivals as a tourist and interacted with other people in the tour group to better understand the participants' thoughts, motivations and expectations.

Sampling selection in qualitative research is generally not random, it is aimed at a specific purpose (Miles & Huberman, 2016; Merriam, 2009; Berg, & Lune, 2017). Creswell (2017) states that in the sample selection required for the research, it is important to select the people who have experience in the researched phenomenon. Therefore, purposeful sampling method was used in the study. As a result, the sample of the study consists of 23 people who participate in gastronomy tours and consider themselves a gastro tourist. The interviews were carried out with 10 people who participated in Alaçatı Herb Festival Tour on April 6, 2019, 6 people who participated in the Yeşil Üzümlü Morel Mushroom and Dastar Festival Tour on April 7, 2019, and 7 people who participated in Bodrum Bitter Herb Festival Tour on March 14, 2020. When the number of interviewees reached 23 people, the answers received from the participants began to be repeated and the interviews were ended. Inclusion in all the tours were realized as participant observer.

Basit (2003) states that there is no standard list of things to do in data analysis of qualitative research. Thus, the necessity of data collection and data analysis to be a simultaneous process in qualitative research is supported by many authors (Basit, 2003; Merriam, 2009; Patton, 2001; Berg, & Lune, 2017). The data collected in this context were immediately converted to text, and codes and themes were started to be drawn. According to Basit (2003), the way to analyse is determined depending on the size of the project, available funds and duration, and the researcher's tendency and expertise. Patton (2001) points out that there is no exact formulation or standard serial steps in the analysis of qualitative data. In fact, in qualitative data analysis, computer programs do not analyze the qualitative data. Again, it is the researcher himself who will do everything with the analysis. The researcher decides what kind of a structure to create, what the code and the theme will be (Saldaña, 2013; Çelik et al., 2020). Patton (2001) argues that although computer programs can facilitate analysis, they cannot present the creativity and intelligence that make qualitative analysis unique. Chametzky (2016) suggests manually coding because it does not have the fineness we make. Miles, & Huberman (2016) state that the most important measurement tool in a research is the researcher himself. Many authors (Saldaña, 2013; Patton, 2001; Merriam, 2009; Berg, & Lune, 2017) emphasize the importance of repeated reading in the analysis of qualitative data. In this context, the interview records and observation notes were repeatedly read by the researcher, notes were taken, and as a result, the analysis of the data was carried out by the researcher. All interviews and analyzes were conducted in Turkish and all quotations were translated into English by the author.

5. FINDINGS

16 of the people participating in the interview are women and 7 of them are men. Although the education level of the participants varies, participation at the undergraduate level is predominant. The majority of the participants are married and generally participated with their spouse or friends.

5.1 The Reasons for Participating in Tours for the Tourists

Fields (2002) revealed that understanding consumer behavior related to gastronomy in tourism requires empirical evidence about the motivation of tourists. Gastro tourists can have many motivations for their visit, and therefore, real motivations for visiting the attraction can only be discovered by asking the gastro tourists themselves (Tikkanen, 2007). In this context, participants were asked why they participated in gastronomy tours. According to the answers given, it was seen that the motivations for participation in gastronomy tourism emerged within the scope of push and pull factors of Crompton (1979). While the factors that make people need touristic goods and services are accepted as driving factors, the factors that cause them to prefer the touristic supply source are accepted as attractive factors (Dann, 1981). Traditionally, reasons for push are useful for explaining the desire to go on vacation, while reasons for pull help explain the choice of destination (Goossens, 2000). In this context, factors that push tourists to participate in gastronomic festival tours are the desire to escape from the daily routine, the desire to seek different/new flavors, the desire to be informed in terms of culture and gastronomy, the desire to reach healthy/natural products, the need to socialize, to the pursuit of excitement, the desire to relieve stress, curiosity, adventure, the desire to gain prestige and the pursuit of entertainment. Below are some of the direct answers given by the participants.

While I and my wife were on vacation, we wanted to do something different and we learned that there was such a tour. The fact that it was a day trip also attracted us and we participated to make a difference (P3).

We joined this tour for fun. We tought to have a change of air with my wife. We said let's eat, drink, have fun, make a difference at the weekend... (P16).

In general, we can access natural products more easily at such festivals and learn how to make them... Also, we were curious about this festival. We were especially curious about the products made with bitter herbs. That's why we said not to participate when we had time and money (P22).

Table 3. General information about participants

Participants	Tour	Gender	Age	Education	Marital status	Companion	Tour days
Participant 1	Alaçatı Herb Festival	Female	53	High school	Married	Daughter	1
Participant 2	Alaçatı Herb Festival	Female	22	Bachelor	Single	Mother	1
Participant 3	Alaçatı Herb Festival	Male	30	Bachelor	Married	Wife	2
Participant 4	Alaçatı Herb Festival	Female	29	Bachelor	Married	Husband	2
Participant 5	Alaçatı Herb Festival	Female	28	Bachelor	Married	Husband	1
Participant 6	Alaçatı Herb Festival	Female	55	Bachelor	Married	Husband	3
Participant 7	Alaçatı Herb Festival	Male	62	Bachelor	Married	Wife	3
Participant 8	Alaçatı Herb Festival	Female	33	Bachelor	Single	Friends	1
Participant 9	Alaçatı Herb Festival	Female	61	Bachelor	Married	Sister	1
Participant 10	Alaçatı Herb Festival	Male	71	High school	Married	Wife	1
Participant 11	Morel Mushroom and Dastar Festival	Female	26	Bachelor	Single	Friends	1
Participant 12	Morel Mushroom and Dastar Festival	Female	56	High school	Married	Husband	1
Participant 13	Morel Mushroom and Dastar Festival	Female	-	Bachelor	Married	Friends	1
Participant 14	Morel Mushroom and Dastar Festival	Male	32	Postgraduate	Married	Friends	1
Participant 15	Morel Mushroom and Dastar Festival	Male	39	Bachelor	Married	Wife	1
Participant 16	Morel Mushroom and Dastar Festival	Female	-	High school	Married	Husband	1
Participant 17	Bitter Herb Festival	Female	55	High school	Married	Friends	1
Participant 18	Bitter Herb Festival	Female	34	Bachelor	Single	Friends	1
Participant 19	Bitter Herb Festival	Female	45	Bachelor	Married	Friends	1
Participant 20	Bitter Herb Festival	Male	24	Associate degree	Single	Friends	1
Participant 21	Bitter Herb Festival	Male	57	Bachelor	Married	Wife	1
Participant 22	Bitter Herb Festival	Female	46	Associate degree	Married	Husband	1
Participant 23	Bitter Herb Festival	Female	40	Bachelor	Married	Husband	7

The factors that attract tourists to participate in gastronomic festival tours were usually determined as the food and drinks specific to the region in particular which are the main theme of the tour, the gastronomic festivals held in the region, the recognition/popularity of the region/festival/cuisine, the presence of chefs-gourmet-experts in the region and/or participating in the festival, the places/museums reflecting the gastronomic culture of the region, gastronomic diversities/possibilities and the presentation of natural/organic products. Below are some of the direct answers given by the participants.

The herb festival was effective in my participation in this tour. So you come normally to Alaçatı and travel, but this is different. Now you see a lot of things. There is an enthusiasm in people. You see them. You get different experiences... You taste flavors... You learn things you do not know (P1).

Alaçatı is already a popular destination. Especially here, the products made with gum mastic are quite a lot. And the festival is already a well-known one. That's why we wanted to experience it and choose this tour (P9).

My reason for joining this tour is to taste different flavors that I did not know before. Visiting the local places where these flavors are made... I also like to follow famous chefs on social media. For example, Master Chef Mehmet Yalçınkaya (A famous cuisine chef in Turkey) came here. I came to meet him, to take a photo... (P19).

Figure 1. Keşkek (Traditional Turkish food) being beaten/prepared in front of the participants at the Bitter Herb Festival (left), a view from Alaçatı Herb Festival (right)
Source: *Photos were taken by the researcher.*

5.2 Expectations of Tourists from the Tours

According to the answers given by the tourists, their expectations from the gastronomy tours they participated were classified in five categories. These are the expectations for the products/activities which are the focal point of the tour, expectations for the general services of the tour, expectations for relations with other people, expectations for general environmental conditions and internal expectations.

Expectations for the products/activities which are the focal point of the tour; to taste new products, to get to know the gastronomic products specific to the region, to learn the historical texture and process of the gastronomic products offered, to see the gastronomic areas, to participate in the harvest of the products grown in the region are the expectations of the tourists from gastronomic festival tours. The answers of some participants are given below.

It was to know and learn everything about herbs. No, it didn't meet my expectations. It was aimed at the promotion and sale of homemade products rather than the herb festival. We saw similar products. It was not aimed at promoting the local products of the region. The introduction of herbs was insufficient (P5).

Actually, I didn't have a great expectation... I recommend them to focus on whatever their focus is. If the focus is on mushrooms, let them be mushrooms. For example, there was a Rumi stone print (a min-

iature gift) in a place that sells gifts, and I thought what would be the relevance between. But it would be better if they had morel stone print. I would also buy that. It was the mushroom what I was looking for there, what I wanted. Okay, it may not be the morel. But for example, it may introduce another type of mushroom. But I think the basis must be the mushroom (P11).

I cannot say that my expectations were fully met. It was overcrowded and believe me we had difficulty walking… The stands were very complicated. Everyone was selling something to their liking. Officially everywhere is under smoke. We had trouble breathing. Stands could be set up in a more organized and pre-planned manner in this regard... (P16).

Expectations for the general services offered on the tour; are towards basic needs such as transportation, accommodation, guide, food and beverage. Especially during the tours, expectations about the guide come to the fore. The guide is expected to be friendly and sincere, to provide information about the gastronomy culture of the region and the festival, to focus his talks on gastronomy, to offer suggestions about where to choose and eat and drink, and to provide information about the recipes of the products tasted.

The things he (tour guide) told in the minibus were very nice, but not relevant to our tour. I can already search the internet if I wanted to investigate the history or past history of Turkey. Also, if I wanted to know the history of Van or Göbekli Tepe, I would go there. But I'm coming here because I want to know about Alaçatı and Çeşme. (Observation Note: Information on general Anatolian history was given during the tour. Alaçatı-Çeşme region was not mentioned much. Not much information was given about eating and drinking). The tour guide could of course tell about İzmir and its surroundings, but instead of talking about the Central Anatolia, the East or Urfa, he could talk about the place where we live. Also, the theme of this Alaçatı Herb Festival in its 10th year was chard. I searched this on the internet. For example, he never mentioned it (P2).

The name of the tour was a gastronomy tour, but even the breakfast was terrible… I think something more local could be offered… In general, the festival was good, but the tour did not have much appeal in terms of eating and drinking (P21).

Considering the expectations for relationships with other people; elements have emerged such as having fun/chatting with other participants on the tour, fraternizing with local people, meeting famous chefs, and obtaining information from local service providers. The answers by P9 is given below.

Today my expectations were not met. It wasn't much fun. The group was very detached from each other, we could not go around and taste it together… It was too crowded. Of course, it was effective in the rain, in this case (P9).

Expectations for general environmental conditions consist of elements such as weather, traffic, cleaning / hygiene, basic need areas (washbasin, sitting areas, etc.), infrastructure and superstructure facilities. P22 says the following on the subject.

I also liked the festival area very much. They designed it very well. The atmosphere is very pleasant. The stands are very well arranged. It is very reassuring in terms of hygiene. I can say that this was the most attention-grabbing place. Almost everyone was wearing gloves and their heads were covered. They stretched the food. This was a beautiful thing. I can say that my expectations have been met (P22).

The inner expectations expected from the gastronomic festival tours are the desire to taste new things, the desire to learn new things, the search for difference and the desire to experience different cultures, and the reasons for getting rid of routine/stress.

My expectation was to be in a different atmosphere, to live a different day... We saw and experienced new things... I can say that it met my expectation but I think it could have been better... (P14).

Figure 2. Expectations from gastronomic festival tours
Source: *Figure was created by author*

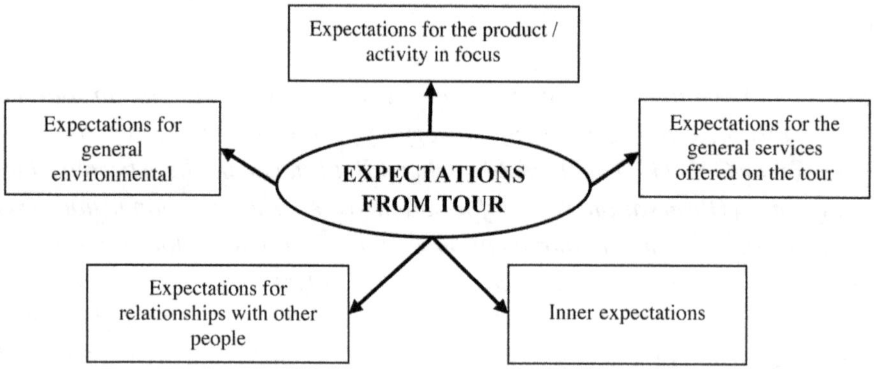

CONCLUSION

With the growing interest in gastronomy tourism, an increase is also observed in the variety of gastronomic products offered at destinations. Gastronomy tours are considered an important tourism product within the scope of gastronomy tourism in recent years, too. When we examine worldwide, there are wine tours in countries such as Italy and France, street food tours in countries such as South Korea, China and Thailand, and coffee tours in countries such as Colombia and Costa Rica. In Turkey also, there is an increasing interest in recent years for gastronomic tours (especially gastronomic festivals) which are organized under different themes. As a matter of fact, it is an important issue to determine the motivation and expectations of tourists for the development of gastronomy tourism and providing better service to tourists interested in such products. In this context, it was also aimed in this study to determine the motivation and expectations of tourists participating in gastronomic festival tours. According to the findings obtained, the motivations of tourists participating in gastronomy tours were revealed within the scope of push and pull factors of Crompton (1979). Factors that push tourists to participate in gastronomy tours are the desire to escape from the daily routine, the desire to seek different/new flavors, the desire to be informed in terms of culture and gastronomy, the desire to reach healthy/natural products, the need to socialize, the pursuit of excitement, the desire to relieve stress, curiosity, adventure, the desire to gain prestige and the pursuit of entertainment. The factors that attract tourists to participate in gastronomy tours were usually determined as the food and drinks specific to the region in particular which are the main theme of the tour, the gastronomic festivals held in the region, the recognition/popularity of the region/festival/cuisine, the presence of chefs-gourmet-experts in the region and/or participating in the festival, the places/museums reflecting the gastronomic culture of the region, gastronomic diversities/ possibilities and the presentation of natural/organic products. In short, each of the gastronomic products that the destination or the tour can offer are attractive factors that enable the tourists to buy the tour or to choose the destination. In addition, gastronomy can play a role as a primary or secondary motivation source as Quan and Wang (2004) stated within the scope of the reasons for the participation of gastro tourists. In the study of Kim, Goh and Yuan (2010), they identified three factors that push tourists to participate in gastronomy tourism as knowledge and learning, entertainment and new experiences, and relaxation with family. They identified the attractive factors in three factors again, namely field quality and value, activity quality and food variety. An evaluation of the findings obtained in the study overlaps with the study of Kim et al. As a matter of fact, it is seen that gastronomic values come to the fore more when they are considered within the scope of gastronomy tours. However, it is seen that the results obtained are similar to previous studies (Björk, & Kauppinen-Räisänen, 2014; Sohn, & Yuan, 2013; Kim, Eves, & Scarles, 2013; Kim, Eves, & Scarles, 2009; Fields, 2002; Boniface, 2003) which examine the reasons for tourists to participate in gastronomy tourism.

Expectations of tourists participating in gastronomic festival tours were evaluated in five categories. These are; expectations for the products/activities which are focal point of the tour, expectations for the general services of the tour, expectations for relations with other people, expectations for general environmental conditions and internal expectations. Expectations of tourists for the products/activities which are focal point of the tour; to taste new products, to get to know the gastronomic products specific to the region, to learn the historical texture and production processes of the gastronomic products offered, to see the gastronomic areas and to participate in the harvest of products which are grown specific to the region. Expectations for general services offered on tours are aimed at basic needs such as transportation, accommodation, guide, food and beverage. Especially during the tours, expectations about the guide come to the fore. The guide who is friendly and sincere while accompanying the tours, providing information about the gastronomy culture of the region and the festival, concentrating his/her conversations on gastronomy, offering suggestions about where to shop, eat and drink, and giving information about the recipes of the tasted products are the elements which the tourists expect from the guide. For tourists who participate in gastronomic festival tours, their relations with other people are also important. In this context, tourists expect to have fun/chat with other participants in the tour, fraternize with local people, meet famous chefs, and obtain information from local service providers. The expectations for general environmental conditions on the tours are also remarkable. Expectations were expressed regarding elements such as weather, traffic, cleanliness/hygiene, areas for basic needs (washbasin, sitting areas, etc.), infrastructure and superstructure facilities. The reasons such as the desire to taste new things, the desire to learn new things, the search for difference and the desire to experience different cultures, getting rid of routine/relieving stress are gathered in the category of internal expectations. Consequently, the motivations and expectations of tourists participating in gastronomic festival tours are quite heterogeneous. Pine, & Gilmore (1999) draw attention that presenting the unexpected is rather the way to exceed expectations. Oliver (1977) states that customer expectations consist of four factor which are general knowledge or biases about products or services, past experiences, experiences of others and marketing initiatives of the company. Expectations of tourists regarding the dining experience in their travels may depend on previous dining experiences in previously visited destinations, food-related cultural values of the destination, suggestions from mass media, and restaurant/destination marketing initiatives (Chang et al., 2011). Below are some suggestions for practitioners in order to organize gastronomic festival tours that will motivate tourists to participate in gastronomy-themed tours and meet their expectations.

Participation motivation of tourists participating in gastronomic festival tours emerges as pushing (internal motivation elements) and pulling (factors that make the destination different and preferred) factors. In this context, it is important to highlight these push and pull factors in the marketing of gastronomic festival tours, and to include attractive factors that will mobilize tourists in the tours. According to the findings, it has been determined that tourists participate in gastronomic festival tours to get out of the holiday routine even during the holiday. In this context, it is important to increase such tours in order to increase the variety of products especially in touristic destinations and to ensure that tourists leave the destination by accumulating satisfied and memorable experiences. Increasing such touristic products can increase the awareness of the destination, re-visit and recommendation to others.

Considering the expectations of tourists from gastronomic festival tours, it is understood that just eating and drinking is not enough. Tourists' expectations vary considerably in such tours. In this context, it is important that such tours are carried out by experts in their field within a detailed program prepared before. At first, it is important to meet the expectations for the product which is the main theme of the tour (wine, coffee, tea, street food, festival etc.). Besides, arrangements should be made regarding the general services offered on tours, general environmental conditions and the relationship with other people whom to be interacted during the tour. Especially the travel agency and guide organizing the tour here have important duties. The guide must be an expert and experienced in the field of gastronomy. The fact that travel agency and/or the guide establish the background to meet the expectations before the tour, make deals with the relevant businesses (food and beverage businesses, food production facilities, gastronomy museum, herbalist or a shop assistant, etc.) and/or areas (such as tea/coffee gardens, orchards) are key factors for the success of the tour. As a result, the emphasis on gastronomic resources and the diversification of such resources are effective in the success of the tour.

The increasing importance of gastronomic tourism and festivals throughout the world reveals different types of gastronomy tourism and the presentation of gastronomic products. As a matter of fact, it is necessary to understand the gastro tourist in the best manner to be able to perform a better gastronomy tourism and festival. The study is limited to three gastronomy festivals. In addition, the study was limited to the motivation and expectations of Turkish tourists. In this context, for future studies, it may be suggested to researchers to conduct similar studies for tourists participating in different types of gastronomic festival tours. For example, similar studies can be done in specific areas such as wine festivals, tea and coffee festivals, olive and olive oil festivals. Furthermore, examining the motivation and expectations of tourists participating in gastronomic festival tours in different countries may present different perspectives.

NOTE

This study was prepared on the basis of the doctoral thesis titled "Components of Experience of Gastro Tourists: A Research in the Scope of Gastronomy Tours" prepared by the author (2020).

REFERENCES

Basit, T. N. (2003). Manual or Electronic? The Role of Coding in Qualitative Data Analysis. *Educational Research*, *45*(2), 143–154. doi:10.1080/0013188032000133548

Berbel-Pineda, J. M., Palacios-Florencio, B., Ramírez-Hurtado, J. M., & Santos-Roldán, L. (2019). Gastronomic Experience as a Factor of Motivation in the Tourist Movements. *International Journal of Gastronomy and Food Science*, *18*, 1–10. doi:10.1016/j.ijgfs.2019.100171

Berg, B. L., & Lune, H. (2017). *Qualitative Research Methods for the Social Sciences*. Pearson.

Björk, P., & Kauppinen-Räisänen, H. (2014). Culinary-Gastronomic Tourism - A Search for Local Food Experiences. *Nutrition & Food Science*, *44*(4), 294–309. doi:10.1108/NFS-12-2013-0142

Boniface, P. (2003). *Tasting Tourism: Traveling for Food and Drink*. Ashgate Publishing.

Carlsson, B. (2001). Depicting Experiences. *Scandinavian Journal of Educational Research*, *45*(2), 125–143. doi:10.1080/00313830120052723

Çelik, H., Başer Baykal, N., & Kılıç Memur, H. N. (2020). Qualitative Data Analysis and Fundamental Principles. *Journal of Qualitative Research in Education*, *8*(1), 379–406. doi:10.14689/issn.2148-2624.1.8c.1s.16m

Chametzky, B. (2016). Coding in Classic Grounded Theory: I've Done an Interview; Now What? *Sociology Mind*, *6*(04), 163–172. doi:10.4236m.2016.64014

Chang, R. C., Kivela, J., & Mak, H. A. (2011). Attributes that Influence the Evaluation of Travel Dining Experience: When East Meets West. *Tourism Management*, *32*(2), 307–316. doi:10.1016/j.tourman.2010.02.009

Chen, Z. (2018). A Pilot Study of the Co-creation Experience in Traditional Cantonese Teahouses in Hong Kong. *Journal of Heritage Tourism*, *13*(6), 506–527. doi:10.1080/1743873X.2018.1444045

Creswell, J. W. (2017). *Nitel Araştırmacılar İçin 30 Temel Beceri* (H. Özcan, Trans.). Anı Yayıncılık.

Crompton, J. L. (1979). Motivations for Pleasure Vacation. *Annals of Tourism Research*, *6*(4), 408–424. doi:10.1016/0160-7383(79)90004-5

Cruz, F. G. S., Tito, J. C., Perez-Galvez, J. C., & Medina-Viruel, M. J. (2019). Gastronomic Experiences of Foreign Tourists in Developing Countries. The Case in the City of Oruro (Bolivia). *Heliyon*, *5*, 1–8. PMID:31312731

Dann, G. (1981). Tourism Motivations: An Appraisal. *Annals of Tourism Research*, *8*(2), 189–219. doi:10.1016/0160-7383(81)90082-7

Dickson, S., & Hall, T. E. (2006). An Examination of Whitewater Boaters' Expectations: Are Pretrip and Post-Trip Measures Consistent? *Leisure Sciences: An Interdisciplinary Journal*, *28*(1), 1–16. doi:10.1080/01490400590962452

Du Rand, G. E., & Heath, E. (2006). Towards a Framework for Food Tourism as an Element of Destination Marketing. *Current Issues in Tourism*, *9*(3), 206–234. doi:10.2167/cit/226.0

Ellis, A., Park, E., Kim, S., & Yeoman, I. (2018). What is Food Tourism? *Tourism Management*, *68*, 250–263. doi:10.1016/j.tourman.2018.03.025

Fields, K. (2002). Demand for the Gastronomy Tourism Product: Motivational Factors. In A-M. Hjalager ve G. Richards (Eds.). Tourism and Gastronomy, 36–50. London & New York: Routledge.

Gnoth, J. (1997). Tourism Motivation and Expectation Formation. *Annals of Tourism Research*, *24*(2), 283–304. doi:10.1016/S0160-7383(97)80002-3

Goeldner, C. R., & Ritchie, J. R. B. (2006). *Tourism: Principles, Practices and Philosophies*. John Wiley and Sons, Inc.

Goossens, C. (2000). Tourism Information and Pleasure Motivation. *Annals of Tourism Research*, *27*(2), 301–321. doi:10.1016/S0160-7383(99)00067-5

Hall, C. M., & Sharples, L. (2003). The Consumption of Experiences or the Experiences of Comsumption? An Introduction to the Tourism of Taste. In M. C. Hall, L. Sharples, R. Mitchell, N. Macionis ve B. Cambourne (Eds.). Food Tourism Around the World: Development, Management and Markets. Oxford: Butterworth Heinemann, 1–24.

Hjalager, A. M. (2002). A Typology of Gastronomy Tourism. In A. M. Hjalager ve G. Richards (Eds.). Tourism and Gastronomy. Londra ve New York: Routledge, 21–35.

Horne, J., & Manzenreiter, W. (2006). An introduction to the sociology of sports mega-events. *The Sociological Review*, *54*(2, suppl), 1–24. doi:10.1111/j.1467-954X.2006.00650.x

Hsu, T.-K., Tsai, Y.-F., & Wu, H.-H. (2009). The Preference Analysis for Tourist Choice of Destination: A Case Study of Taiwan. *Tourism Management*, *30*(2), 288–297. doi:10.1016/j.tourman.2008.07.011

Kastenholz, E., & Lima, J. (2011). The Integral Rural Tourism Experience from the Tourist's Point of View – A Qualitative Analysis of Its Nature and Meaning. *Tourism & Management Studies*, *7*, 62–74.

Kim, Y. G., Eves, A., & Scarles, C. (2009). Building a Model of Local Food Consumption on Trips and Holidays: A Grounded Theory Approach. *International Journal of Hospitality Management*, *28*(3), 423–431. doi:10.1016/j.ijhm.2008.11.005

Kim, Y. G., Eves, A., & Scarles, C. (2013). Empirical Verification of a Conceptual Model of Local Consumption at a Tourist Destination. *International Journal of Hospitality Management*, *33*, 484–489. doi:10.1016/j.ijhm.2012.06.005

Kim, Y. H., Goh, B. K., & Yuan, J. (2010). Development of a Multi-Dimensional Scale for Measuring Food Tourist Motivations. *Journal of Quality Assurance in Hospitality & Tourism*, *11*(1), 56–71. doi:10.1080/15280080903520568

Kivela, J., & Crotts, J. C. (2005). Gastronomy Tourism: A Meaningful Travel Market Segment. *Journal of Culinary Science & Technology*, *4*(2/3), 39–55. doi:10.1300/J385v04n02_03

Ko, S., Kang, S., Kang, H., & Lee, M. J. (2018). An Exploration of Foreign Tourists' Perceptions of Korean Food Tour: A Factor-Cluster Segmentation Approach. *Asia Pacific Journal of Tourism Research*, *23*(8), 833–846. doi:10.1080/10941665.2018.1494613

Koç, E. (2012). *Tüketici Davranışı ve Pazarlama Stratejileri Global ve Yerel Yaklaşım*. Seçkin Yayıncılık.

Korstanje, M. E. (2009). Reconsidering the roots of event management: Leisure in ancient Rome. *Event Management, 13*(3), 197–203. doi:10.3727/152599509790029783

Lee, Y.-J. (2015). Creating Memorable Experiences in a Reuse Heritage Site. *Annals of Tourism Research, 55*, 155–170. doi:10.1016/j.annals.2015.09.009

McKercher, B., & So-Ming, B. C. (2001). Cultural Distance and Participation in Cultural Tourism. *Pacific Tourism Review, 5*(1), 23–32.

Merriam, S. B. (2009). *Qualitative Research: A Guide to Design and Implementation*. Jossey-Bass.

Miles, M. B., & Huberman, A. M. (2016). *Genişletilmiş Bir Kaynak Kitap: Nitel Veri Analizi* (S. Akbaba Altun & A. Ersoy, Trans.). Pegem Akademi.

Oliver, R. L. (1977). Effect of Expectation and Disconfirmation on Postexposure Product Evaluations: An Alternative Interpretation. *The Journal of Applied Psychology, 62*(4), 480–486. doi:10.1037/0021-9010.62.4.480

Otto, J. E., & Ritchie, J. R. B. (1996). The Service Experience in Tourism. *Tourism Management, 17*(3), 165–174. doi:10.1016/0261-5177(96)00003-9

Park, K. S., Reisinger, Y., & Kang, H. J. (2008). Visitors' Motivation for Attending the South Beach Wine and Food Festival, Miami Beach, Florida. *Journal of Travel & Tourism Marketing, 25*(2), 161–181. doi:10.1080/10548400802402883

Patton, M. Q. (2001). *Qualitative Research & Evaluation Method*. Sage Publications.

Pine, B. J., & Gilmore, J. H. (1999). *The Experience Economy: Work is Theatre and Every Business a Stage*. Harvard Business School Press.

Quan, S., & Wang, N. (2004). Towards a Structural Model of the Tourist Experience: An Illustration from Food Experiences in Tourism. *Tourism Management, 25*(3), 297–305. doi:10.1016/S0261-5177(03)00130-4

Ramshaw, G. (2016). Food, Heritage and Nationalism. In D. J. Timothy (Ed.), *Heritage Cuisines Traditions, Identities and Tourism* (pp. 53–64). Routledge.

Richards, G. (2002). Gastronomy: An Essential Ingredient in Tourism Production and Consumption? In A. M. Hjalager ve G. Richards (Eds.). Tourism and Gastronomy. London: Routledge, 3–20.

Saldaña, J. (2013). *The Coding Manual for Qualitative Researchers*. Sage.

Shenoy, S. S. (2005). Food Tourism and the Culinary Tourist. [Unpublished Doctoral Thesis, Graduate School of Clemson University, ABD.]

Smith, S., & Costello, C. (2009). Segmenting Visitors to a Culinary Event: Motivations, Travel Behavior, and Expenditures. *Journal of Hospitality Marketing & Management, 18*(1), 44–67. doi:10.1080/19368620801989022

Sohn, E., & Yuan, J. (2013). Who are the Culinary Tourists? An Observation at a Food and Wine Festival. *International Journal of Culture, Tourism and Hospitality Research*, *7*(2), 118–131. doi:10.1108/IJCTHR-04-2013-0019

Stanley, J., & Stanley, L. (2015). *Food Tourism a Practical Marketing Guide*. CABI. doi:10.1079/9781780645018.0000

Tikkanen, I. (2007). Maslow's Hierarchy and Food Tourism in Finland: Five Cases. *British Food Journal*, *109*(9), 721–734. doi:10.1108/00070700710780698

Tse, P., & Crotts, J. C. (2005). Antecedents of Novelty Seeking: International Visitor's Propensity to Experiment Across Hong Kong's Culinary Traditions. *Tourism Management*, *26*(6), 965–968. doi:10.1016/j.tourman.2004.07.002

Turnbull, D., & Uysal, M. (1995). An Exploratory Study of German Visitors to the Caribbean: Push and Pull Motivations. *Journal of Travel & Tourism Marketing*, *4*(2), 85–92. doi:10.1300/J073v04n02_07

Wijaya, S. (2014). *Encounters with Local Food: The Culinary Experiences of International Visitors in Indonesia*. [Unpublished Doctoral Thesis, College of Business Victoria University, Melbourne, Australia].

World Food Travel Association. (2019). *What Is Food Tourism?* WFTA. https://www.worldfoodtravel.org/cpages/what-is-food-tourism

Zamparini, L. (2018). The Role of Gastronomy and Typical Foods in the Tourism Experience. In A. Isoni, M. Troisi ve M. Pierri (Eds.). Food Diversity Between Rights, Duties and Autonomies, 191–202. sSpringer. doi:10.1007/978-3-319-75196-2_11

Zeithaml, V. A., Bitner, M. J., & Gremler, D. D. (2017). *Services Marketing: Integrating Customer Focus Across the Firm*. McGraw-Hill Education.

Chapter 9
Local Fairs:
Festivals and Sustainable Tourism Development in Rajasthan

Anukrati Sharma
University of Kota, India

Shruti Arora
University of Kota, India

Debendra Nath Dash
Mahatma Gandhi National Council of Rural Education, India

ABSTRACT

Festivals and fairs abound throughout India. More festivals are observed in this nation than everywhere else in the globe. Social and cultural festivals and fairs have become a tool for promoting regional economies, extending the tourist season, and developing the tourism industry. The effects of these fairs and festivals matter for the local economy as well as the tourism industry. The culture, traditions, and festivals provide an explanation of the "way of life of a place" and aid in our understanding of a certain locale. To discover more about the customs and traditions of the host town, many tourists go to attend a variety of festivals. The chapter has therefore concentrated on different regional fairs and festivals of Rajasthan that have not yet been explored with the intention of enhancing the local economy, and such chances occasionally go unused. The recommendation would therefore be to invest in festival development and community involvement while providing tourists with distinctive experiences.

INTRODUCTION

National parks, wildlife parks, historic sites, museums, cultural and heritage sites, art galleries, amusement parks, sports facilities, transportation facilities, casinos, shopping centers, and other attractions are included in a nation's assets. There are also numerous temporary attractions, such as fairs and festivals. Festivals, fairs, and events are some of the tourism sectors that are expanding the quickest worldwide.

DOI: 10.4018/978-1-6684-6356-7.ch009

With the aid of traditions and culture, various festivals are being created all over the world in both large cities and small villages in an effort to draw tourists and improve the reputation of the location. These celebrations are a crucial component of cultural tourism.

The promotion of many locations' desire to share their culture, customs, environment and social values with tourists. Festivals act as stress relievers and assist us in maintaining emotional equilibrium. In India, Rajasthan is a colorful, bright and vivacious place. It is at the top of the list in terms of celebrations, customs, and culture. It is sometimes referred to as a Jewel in the Desert because it shines more brightly during its fairs and festivals. Festivals have also been a significant component of Rajasthani culture because of their celebratory tone. The lyrics to the well-known folk song "Padharo Mhare Desh" is "Welcome to my country." The hospitality of Rajasthan is renowned around the world. The fairs and festivals of this nation give major contributions to the image of Incredible India. The domestic and international markets have shown a tremendous deal of interest in local festival celebrations. The major stakeholders feel that with the right festival marketing or positioning, this specific category may prove to be India's biggest tourist draw. This new trend is developing into a tremendous advantage for the Indian tourism industry.

REVIEW OF LITERATURE

The term 'sustainability' refers to the efforts to meet the needs of present and future generations or to reduce poverty and preserve life support systems (Kates, 2011; Walker, 2017). Sustainability has a spiritual and scientific dimension which means fulfilling the needs and paying attention to the aspirations of future generations. In the context of tourism, sustainability means meeting the needs of present tourists and host regions while protecting and enhancing opportunities for the future. It is envisaged as leading to the management of all resources in such a way that economic, social and aesthetic needs can be fulfilled while maintaining cultural integrity, essential ecological processes, biological diversity, and life support systems. Meanwhile, fairs and festivals are an important part of social activities. Fairs and festivals, which showcase a variety of goods and activities, are a staple of Indian culture. Many products created by regional artists are displayed at some of the fairs, some of which are arranged for religious or seasonal purposes (Jain 2016, 2017b). They are arranged all over the world in a variety of ways. Some celebrations are specific to certain areas and some are celebrated by different groups, and communities in certain villages, cities, or even states or countries. The local festivals are naturally influential in enhancing domestic tourism and making a major contribution to the economy of the region. Moreover, festivals are a key strategy in promoting those places which have suffered from underinvestment and have been essentially ignored (Bairwa & Patawari, 2021). The studies on the sustainability of fairs and festivals cannot be separated from the concept of sustainable tourism. Local festivals provide many contributions to the public not only economically but also socially and psychologically (Gözgeç Mutlu, Avcıkurt & Filiz, 2020). This gives raise to fairs and festivals with their own local flavors. The word "festival" is derived from the Latin word "festivus", taken from Anglo- Norman French (Srikanth & Ram, 2013). From a cultural and anthropological point of view, festivals are sacred and/or religious celebrations. There are various expressions in this sense in the literature (Choi, Kang & Kim, 2021). Festivals are a kind of themed special events with a formal program, open to the public, rooted in history, anthropology, or natural aspects (Getz & Page, 2016), that provide participants with celebration and enjoyment feelings (Ma & Lew, 2012). Sinha & Murdia (2018) mentioned that festivals, being a significant factor of cultural tourism have an essential task to carry out for the growth of cultural tourism in the host communities. Festivals are been considered to contribute fundamentally to the advancement of cultural tourism, likewise, festivals are utilized as an instrument to boost the regional economy. The investigation of the effect of festivals and occasions turned into a productive region to advance cultural tourism. Enormous quantities of tourists travel to go to numerous festivals so as to have some understanding of the way of life of the host network. Panda et al. (2017) expressed that festivals are significant throughout the entire existence of the person. Festivals and occasions have prospered in ongoing decades and enthusiasm for understanding their centrality in the tourism institute has additionally risen as needs be. Lately, festival tourism is another type of tourism production the economic growth of a country and drawing the attention of policymakers across the world. Sharma (2017) proposed that India has a scope of fairs and festivals to offer and the USP (Unique Selling Proposition) of this tourism product is that a similar celebration has an absolutely new flavor in an alternate district that is the Dussehra at Kullu and the Dussehra at Mysore. The Indian display of fairs and festivals is loaded with such models. These fairs and festivals, to date, have been utilized as a promoting device for pulling in foreign tourists alone, however now with the place light being shared by the domestic tourist itis the need of great importance to make tourism systems that are not just comprehensive of residential tourism, rather some should concentrate solely on local tourism.Recognizing the potentiality that this segment holds, Subhash Goyal (2014) President, of the Indian Association of Tour Operators (IATO) said in the article, India Developing Cultural Tourism through Festivals" (Travel Trends Today) that "Festivals have the potential to extend tourist seasons, peak seasons and introduce a "new season" for a destination. Events such as festivals not only attract tourists but also help to develop or maintain a community or regional identity. Festivals constitute one of the most exciting and fastest growing forms of leisure, business, and tourism- related phenomena." In addition, there is a relationship between the growth of festivals and the increase in tourist demand. In current usage, the concept of festival tourism is similar to the term well-developed event tourism. When festivals are used in this sense, they refer to consciously created marketing tools by considering city branding/place positioning/destinations (Quinn, 2006).

Kulkarni and Bhopatkar (2013) argued that local festivals promote tourism activities in the area where they are held and create the cultural environment in the area. They concluded that these festivals have a significant economic impact too. Local events, which provide key opportunities for extending tourism activities over a 12-month period and diversifying touristic products, also help the region promote itself, create an image, and increase its tourism share (Baysal, Özgüre & Çeken, 2016). According to Raina and Agarwal (2004), India has vast tourist attractions being rich in fairs and festivals. He also pointed out that Rajasthan along with Kerala and Goa has the capacity to attract around thirty percent of international tourism. In addition to the festivals celebrated by the Hindus, Muslims and others, there are also traditional fairs in Rajasthan like animal fairs, religious fairs and fairs to mark the changing season. Sustainable tourism is defined as the use of tourism resources without being consumed, polluted, destroyed, and used in a way that will enable them to be used by future generations.

The World Tourism Organization defines sustainable tourism as providing the continuity of cultural integrity, basic ecological processes, biological diversity, and life support systems while meeting the needs of existing tourists and the host region (Avcıkurt, 2009).

FAIRS AND FESTIVALS

All throughout the year and around the globe, events, fairs, and festivals of all shapes and sizes take place, from local business gatherings to the biggest music festivals, athletic events, and conferences. The majority of people like and celebrate fairs and festivals as normal parts of life. A fair is a location where a lot of people come together at a certain time to buy and sell things. A festival is a time when the community comes together to celebrate a special religious or cultural event with family, friends, and other members of the community. The role of fairs and festivals is crucial for improving the economic situation of local and neighboring populations. They are particularly beneficial in reducing economic inequality, creating jobs, and reinforcing cultural values. The fairs and festivals can help heal many contemporary social ills brought on by materialism (Shee, 2020).

The following are fairs' and festivals' primary characteristics:

Table 1. Characteristics of fairs and festivals (compiled by authors)

Uniqueness	Distinctive features of festivals like clothing, food and beverage.
Festival spirit	Values and belonging through ritual, fantasy, magic
Hospitality	Willingness of community to host tourist
Strengthening relationships	Bonds among public and private organizations, government, and neighborhood groups, staff, volunteers and interested residents.
Seasonality	Harvesting season, Change in season

Festivals are defined by Ma and Lew (2012) in four distinct contexts that result from the historical (or chronological) and geographic (or spatial) circumstances of unique events.

Local heritage festivals, regional modern festivals, national heritage festivals, and international modern festivals fall under this category. Local heritage festivals are defined as events that are molded by the promotion and evaluation of problems related to authenticity and local identity. The fact that local heritage festivals are rooted in both location and history increase the likelihood that they will support authenticity. These events represent the traditional definition of festivals and have symbolic value and significance pertaining to the local sense of place. Local communities always develop new methods to celebrate life through local modern festivals. Even if there is a greater emphasis on entertainment than on faithful recreations of traditions, these events nonetheless contribute to the local identity.

These gatherings are more enjoyable and less solemn than local heritage celebrations (Ma & Lew, 2012). Local festivals are therefore an intricate cultural phenomenon that is more than just a collection of cultural artifacts on the show; they are also a cultural good in and of itself, a cultural expression, and a cultural process wherein culture is ingested and passed down through generations (Del Barrio, Devesa & Herrero, 2012).

NEED AND SCOPE OF THE STUDY

Currently, an important and fruitful area of tourist research analysis is the study of festivals and fairs. Particularly after COVID-19, tourists will be more active in market segments for the country's heritage, culture, and rural locations rather than traveling abroad. As an immediate result, the local travel market will grow significantly in comparison to the global tourist sector, relieving the community's tension from being confined at home. In order to satisfy the tourist's thought process while providing financial benefits to the destination and local community, fairs and festivals must be organized in this fashion.

Objective of the Study

The study's objective is to investigate several regional festivals in Rajasthan that are currently undiscovered but have the potential to enhance the area's reputation and economy.

The study is focused on the role of the local community in protecting and promoting local festivals and fairs. The present paper highlights the main initiatives taken by the Government of Rajasthan in promoting the traditional and indigenous festivals and fairs of the state.

METHODOLOGY

The chapter is built on descriptive research. In order to achieve the above objective, secondary data was gathered from a variety of research papers, journals, books, newspaper articles, and publications on tourism by the Indian Ministry of Tourism and the Government of Rajasthan from 2011 to the 2022.

Regional Fairs and Festivals of Rajasthan

Rajasthan is a happy state that hosts a variety of fairs and festivals throughout the year. The opportunity for tourism to discover the art, culture, and customs that are exquisitely intertwined with the state's regal history is provided by these sparkling fairs and festivals. The inhabitants of Rajasthan believe in celebrating life, therefore a variety of events and competitions are held throughout the year, including turban-tying contests, puppet plays, camel racing, dance performances, bullfights, and camel and cow trading. Rajasthan is known for its many festivals, events, and traditions, all of which serve as top tourist draws and stand out for their grandeur, joy, and majesty like Mewar Festival, Teej Festival, Jaipur Literature Festival and so on. But still, there are a few local and regional festivals that are yet still to be tapped at national and international levels to attract more tourists like.

Local Fairs: Festivals and Sustainable Tourism

Local fairs and festivals are more frequently employed as tools to promote tourism and strengthen the local economy. When seen from a worldwide perspective, many tourist sites employ fairs or festivals as a significant tourist draw, making them a crucial component of the overall tourism offering. Thus, the appeal of the festival and the location of its hosting are inextricably linked. Fairs and festivals are more frequently associated with tourism in order to boost local economies and local residents' earnings. Festivals, according to Yeoman (2004), can extend peak season, extend tourist seasons, or accompany a "new

Table 2. Important fairs and festivals of Rajasthan Source – (Source - data compiled by authors)

S.No.	Regional Fairs and Festivals
1	*Chaksu Fair or Basoda, Chakshu*: Chakshu fair is celebrated as a tribute to Sheetla Mata, a local deity who wards off epidemic diseases like measles and chicken pox as well as skin conditions in the family as well in the society. A temporary market opens up during the fair. The locals can be seen trading in goods including footwear, apparel, food, kitchenware, agricultural equipment, etc.
2	*Baneshwar Festival, Dungarpur:* It is one of the largest and most unique tribal fairs in the nation. It highlights its religious significance while reflecting on their habits and traditions. During this festival, Lord Shiva is prayed.The fair's main draws are "Tribal Sports," "Religious processions," "Tribal Bazar," and performances of the "Gair Dance" by men dressed in vibrant red ghagras.The sale of handmade or handcrafted goods produced by indigenous tribes is the focus on stalls and businesses.
3	*Gogamedi Fair, Hanumangarh*: The Gogamedi fair, which is held in honor of the Serpent God, starts on Goga Navmi. It is commonly held that anyone with a snakebite can be cured by chanting the name of Goga Ji Maharaja. Every year, the festival is typically held in the months of August and September. Additionally, Goga Ji Maharaja is presented with delicacies made of honey, sugar, and coconut. Rajasthan's Gogamedi is a well-known fair and is regarded as the main animal fair in the area. Even Gogamedi panoramic view is breathtaking and mind-blowing, and it's a terrific place to take pictures.
4	*Matsya Festival, Alwar*: The "Pride of Alwar" celebration, which lasts for two days, is held in November.The vivid and colorful Matsya Festival in Alwar is well-known for its stunning art exhibitions, diverse sporting events, and cultural performances. At the location of the festival, an archaeological exhibition called the Dharohar is staged. In Sanskrit, Matsya means "fish". Matsya is regarded as a symbol and one of Lord Vishnu's avatars in Hindu mythology. The Matsya Festival honors some of the most well-known regional traditional games. The Rumal Jhapatta is the most well-known. It's a sport that calls for extraordinary quickness and agility. The Rassa Kasi, or tug of war, is another well-liked activity in this area. The tug-of-war game is played while sporting the traditional ghagra choli outfit in the women's competition. Numerous water activities can be enjoyed during the festival at the well-known Siliserh Lake.
5	*Kapil Muni Fair, Bikaner*: The largest fair in the Bikaner area, the Kapil Muni Fair, is held on Kartik Poornima at Kolayat, also recognized as Kapilayatan and named for the sage Kapil, who is thought to have meditated here for the benefit of humanity. The fair's location is at Kolayat. It has long been believed that visiting Kolayat is an extremely significant pilgrimage, or "tirtha," and that one day spent there have the same positive effects as ten years spent at any other holy site. In the holy lake, people are seen floating lit lights. Devotees passionately sing songs that glorify Kapil Muni's qualities. This Kapil Muni Fair includes a cattle fair. Lots of livestock, horses, camels, and buffaloes are sold. The fair awards certificates and prizes to the top breeders.
6	*Chandrabhaga Fair,Jhalrapatan*: The Chandrabhaga Fair, a unique fusion of commerce, culture, and musical events, showcases Rajasthan's genuine customs and old-world charm, organized in the Hindu calendar's Kartik month (October/November).The Chandrabhaga Fair serves as a significant camel trading hub. This fair hosts a variety of entertaining events and competitions, such as the Matka race, Tug of war, turban tying, Rangoli creation, and mustache competitions. Additionally, the Fair offers visitors a chance to mingle with residents and become familiar with their traditions.
7	*Sitabari Fair, Baran:*The Sitabari Fair is hosted in the town of Sitabari, which is located close to the Rajasthani village of Kelwara Kasba and 45 kilometers from the city of Baran. This location has a sizable tribal fair on Jyeshta Amavasya, which occurs between May and June. People frequently come to this location, to take holy baths and dips in the tanks and kunds. The location has a noteworthy characteristic, according to which Lakshman shot an arrow into the ground to get the water that supplies the location's significant water. Different varieties of cattle are sold by cattle breeders. People from all around the state of Rajasthan and neighboring Madhya Pradesh attend this fair, where the Sahariya tribe's Swayamvar, a rite for choosing grooms, is held. When a Sahariya male offers a Sahariya girl a handkerchief at a fair, if the Sahariya girl accepts, it is said that they have decided to become married. After receiving their parents' blessings, the bride and groom complete the seven rounds of the Hindu marriage ceremony performed under the Barnawa tree.
8	*Kabir Yatra, Bikaner:* One of the events honoring Rajasthani folk music in the greatest way is held in the month of November. Most importantly, it is a delight for tourists who want to listen to the folk music of the state while they are there. It is the fact that it is a mobile music festival that travels around and around Bikaner for six days and serves as a platform for regional artists and vocalists to showcase their talents to the fullest.
9	*Abhaneri Festival:*The village of Abhaneri in the Dausa district, around 90 kilometers from Jaipur on the Agra road, is the inspiration for the name. The popularity of this two-day festival among travellers from around the world has grown significantly because of local folk performances such Kachhi Ghori, Kalbeliya, Ghoomar, and Bhawai, as well as various Rajasthani performances. The festival, which Rajasthan Tourism started in 2008, is very important to Rajasthan. The name Abha Nagri, which means "city of light," was initially given to the town of Abhaneri. The Chand Baori-step well, one of the biggest step wells constructed more than a thousand years ago, is a famous landmark in the area.
10	*Ranakpur Jawai Bandh Festival:*The city of Ranakpur, which is in the district of Pali, transforms into a charming location for one of Rajasthan's most well-known festivities during the Ranakpur Jawai Bandh festival.This event, which is put on by the tourism department, gives visitors a fascinating look into the history and culture of the area.The Ranakpur Festival is a must-attend event with enjoyable activities like yoga, nature hikes through the Aravalli foothills, trips to the Ranakpur Jain temple, hot air balloon rides, intriguing events like tug of war, lovely decorations, cultural programmes, appealing folk and classical performances presented every evening at the Sun Temple, and much more. Every year on December 21 and 22, this event welcomes visitors from all over the world to immerse them in a vibrant celebration of cultural diversity.
11	*Saras Mela*: The Rajasthan Grameen Aajeevika Vikas Parishad (Rajeevika) is the organization behind the Saras Ajeevika Mela, which showcases an exquisite fusion of India's rich and wealthy cultures, traditions, and folk arts.Rural SHG craftswomen from all states and UTs curate an abundance of aesthetically pleasing handicrafts, handloom, and organic food goods for each Saras Mela. Cultural activities and delectable Indian cuisine provide a pleasant flavor, heightening the joy and celebration.

Figure 1. Fairs & Festivals
Source: Authors

season" in a community. These regional celebrations assist to foster or preserve a sense of community or regional identity in addition to drawing tourists. Through the expansion of fairs and festivals, small towns and villages are becoming more open to sharing their culture, heritage, customs, and natural surroundings with visitors. Rajasthan is renowned for its warm welcome and kind residents. People frequently choose to visit during fairs and festivals because they enjoy learning about the area from the perspective of the locals. Every day, more people visit fairs and festivals. They also give guests a chance to escape from their regular daily lives. In the communities where they are hosted, they support the preservation and enhancement of local culture. Additionally, they aid towns economically, advance regional tourism, and lengthen the tourist season (Huang, Li & Cai, 2010).Festivals can offer opportunities for the host community, such as new infrastructure development and employment prospects, in addition to enabling a community to rejoice. Festivals are advantageous for both social advancement and the sustainability of tourism, Gunsoy & Hannam (2013). Additionally, governments support festivals because they boost the local economy (Arcodia & Whitford, 2006).Some festivals serve as tourist draws by showcasing the nation's cultural values or historical landmarks (Zdemir & Ulha, 2009).

Federation of Indian Chamber of Commerce and Industry (FICCI) and Rajasthan Tourism

FICCI is a non-profit, non-governmental organisations that represents business and industry in India. The FICCI-Yes Bank Tourism Study states that by 2029, "Indian tourism will exceed 35 trillion." Additionally, it adds that in order to realize tourism's full potential, tourism destinations must be grown, maintained, and developed sustainably, with an emphasis on creating new tourism-related opportunities.

For the past three decades, FICCI Tourism has been collaborating closely with the Ministry of Tourism, the Government of India, and the Tourism Departments of various State Governments to promote tourism and investment in tourism infrastructure to the states through the use of a variety of platforms, special initiatives, and addressing important issues that serve to catalyze policy change on a national and international scale. The platforms for tourism offer coverage for a variety of niche markets, including inbound tourism, domestic tourism, investments in tourism infrastructure, digital and travel technology,

M.I.C.E., wedding tourism, film tourism, etc. Every platform has a knowledge report that serves as the foundation for presenting trends and an actionable agenda to the government.

Community Development Through Local Fairs and Festivals of Rajasthan

Festivals are the key to showcasing our heritage and culture. Local fairs and festivals in Rajasthan have a natural impact on boosting domestic travel and significantly boosting the local economy. Moreover, festivals play a significant role in promoting locations that have received little attention and under-investment. The opportunity for members of many communities and villages to spend quality time together is made possible by these holiday celebrations. Better comprehension and a sense of belonging among the masses are produced by social engagement. No matter where the tourists are from, what their preferred languages are, etc., the festivals and fairs are so overwhelming that the rhythm, noises, and atmosphere never fail to awe them. Nonetheless, despite the enormous potential of local fairs and festivals, the area does not profit from the potential benefits of and support for displaying the undiscovered and secret festivals to the general public. One must realize that these regional fairs and festivals serve as the foundation for a region's development and rebranding as well as its source of income. As the regional song "Kesaria Baalam Aavoni Padharo Mhare Desh" embodies, Rajasthani hospitality to visitors is second to none. Fairs and festivals provide several direct and indirect economic and social benefits to the local communities. Caste diversity and the existence of numerous communities are expected in a state as big as Rajasthan. Together with larger populations, Rajasthan also has numerous tribal groups that are an important part of its culture and tradition. Some significant communities are Rajput, Brahmins, Jat, Gurjars, Bhils, Kalvi, Garasia, and Kanjars, which are further divided into many castes and tribal groups.

According to a report in the Indian Express from December 4th, 2022, Rajasthan has 22 different types of folk festivals that the state government plans to organize at all levels, from the panchayat to the state, to promote tourism, preserve the arts and cultures of the various regions of the state, and work to uplift artists who struggled during the pandemic with a budget of INR 27 crore.

POST COVID–19: TOURISM BACK ON TRACK

Business in the tourism sector was severely impacted by COVID-19, and it completely stopped. Prior to the Covid-19 pandemic, tourism was one of the most significant economic sectors in the world, accounting for around 10% of the global GDP and employing millions of people across the globe. Rajasthan attempted to restart its tourism industry by distributing vaccinations and booster doses, as well as by implementing basic safety precautions, by viewing COVID- 19 as a challenge and an opportunity through its regional fairs and festivals. Every festival is distinctive in some manner. Through these fairs and festivals, the tourists learn about the diversity of each location. It makes the state more appealing to tourists. These fairs and festivals give visitors the chance to travel to the area during the festival season and offer a special chance to interact with the local population, giving them the chance to learn about or engage with local customs and culture. Local festivals are increasingly being used as tools to promote tourism and strengthen the community's economy. Every fair and festival has a distinct goal, such as a religious one, a social one, an amusement goal, a goal for national integration, etc.

Following COVID-19, the local government and tourism boards use these local or regional festivals and fairs as a platform to present and market their art, culture, and tourism. Rajasthan topped the charts, with 10.96% of travellers preferring it over all other Indian destinations, according to trends identified by Thrillophilia in the final quarter of 2021 (The Print, 13th Jan. 2022). 3.07% of domestic visitors intend to go to Kashmir. 4.78 percent preferred the North East for their vacation, while 3.54 percent preferred Kerala. 7.89% of respondents who intended to travel said they would go to Himachal Pradesh, while 6.88% said they would go to Uttrakhand. In addition to the hill states, 6.5% of respondents intended to travel to Goa, and more than 8.6% wished to visit Rajasthan. People in the country's remaining 58.71% decided to take vacations in other states. The dramatic increase in tourism in Rajasthan was largely a result of the state's Covid regulations, which aimed to impose the fewest restrictions and provide the greatest level of safety.

CONCLUSION

Fairs and festivals play a significant role in cultural tourism, which uses cultural events to attract more visitors. They are a constantly expanding sector of the global tourist and entertainment industries. They have a large impact on the host or destination community's economy, socio-culture, and politics. It is impossible to determine to what extent fairs and festivals contribute to local economic prosperity. The most noticeable impact, however, is the rise in local revenue and job prospects brought on by more visitors. Although the success of a festival depends in part on the organizer, if a festival cannot be organized successfully the destination not only loses income but also loses the destination's reputation. Fairs and festivals can enhance its image and be positioned as a cultural destination.

Suggestions

As with all things, there are other aspects of fairs and festivals that may be enhanced in order to make a bigger impact on the growth of the tourism industry. One of the most popular and watched inbound and outbound visitor marketplaces in the world is Rajasthan. Festival planning has a lot of possibilities, and researchers have offered several ideas for it:

- People of all backgrounds must experience fairs and festivals in order to appreciate Rajasthan's customs and culture.
- To pique visitors' interest in the local culture, cultural events such as road shows and Heritage walks should be conducted in an excellent way.
- To create a brand name for the celebration, a brand ambassador must be chosen.
- The area that already has some hotels is unable to maintain standards and quality, so the government and local stakeholders should take action in this regard. Basic infrastructure, such as improved road connectivity, electricity, and water supply, as well as multi-cuisine restaurants, must be established immediately in order to increase the tourist flow.
- A better display of their traditional art and culture should be made with the art and craft cuisine.
- The frequency of transportation services must be enhanced throughout the event.
- Appropriate PR and marketing strategy should be used by fair and festival organizers.

ACKNOWLEDGEMENT

We want to acknowledge this chapter under the major research project entitled "Rural Community Development and Engagement through Rural Tourism: Strategies of Strengthening the Capacity of Rajasthan's Villages" under ***Mahatma Gandhi National Council of Rural Education (MGNCRE), Department of Higher Education, Ministry of Education, Government of India.***

REFERENCES

Arcodia, C., & Whitford, M. (2006). Festival attendance and the development of social capital. *Journal of Convention & Event Tourism*, *8*(2), 1–18. doi:10.1300/J452v08n02_01

Avcıkurt, C. (2009). *Turizm sosyolojisi genel ve yapısal yaklaşım*. Detay Yayıncılık.

Bairwa, M., & Patawari, S. (2021). Festivals for Sustainable Tourism Development: A Case Study of Marwar Region of Rajasthan, India. In LOCAL FOOD AND COMMUNITY EMPOWERMENT THROUGH TOURISM (pp.82-96) Eureka Publications.

Baysal, K., Özgürel, G., & Çeken, H. (2016). Aydın yöresindeki yerel etkinliklerin kırsal turizm açısından değerlendirilmesi. *International Journal of Social and Economic Sciences*, *6*(1), 100–108.

Choi, K., Kang, H. J., & Kim, C. (2021). Evaluating the efficiency of Korean festival tourism and its determinants on efficiency change: Parametric and non-parametric approaches. [online]. *Tourism Management*, *86*, 1–20. doi:10.1016/j.tourman.2021.104348

Del Barrio, M. J., Devesa, M., & Herrero, L. C. (2012). Evaluating intangible cultural heritage: The case of cultural festivals. City. *Cultura e Scuola*, *3*(4), 235–244. doi:10.1016/j.ccs.2012.09.002

Getz, D., & Page, S. J. (2016). Progress and prospects for event tourism research. *Tourism Management*, *52*, 593–631. doi:10.1016/j.tourman.2015.03.007

Gözgeç Mutlu, H., Avcıkurt, C., & Filiz, A. (2020). Festival katılımcılarının motivasyonları ve turistik deneyimleri. *International Journal of Contemporary Tourism Research*, *4*(2), 219–233.

Gunsoy, E., & Hannam, K. (2013). Festivals, community development and sustainable tourism in the Karpaz region of Northern Cyprus. *Journal of Policy Research in Tourism, Leisure & Events*, *5*(1), 81–94. doi:10.1080/19407963.2013.774201

Huang, J. Z., Li, M., & Cai, L. A. (2010). A model of community-based festival image. *International Journal of Hospitality Management*, *29*(2), 254–260. doi:10.1016/j.ijhm.2009.10.010

Indian Express 4th Dec. 2022. Rajasthan govt to organise folk art festivals from panchayat to state level to boost tourism. Retreived from https://indianexpress.com/article/lifestyle/destination-of-the- week/rajasthan-govt-to-organise-folk-art-festivals-from-panchayat-to-state-level-to-boost-tourism- 8305324/

Jain, V. (2016). Candy of Cannabis sativa L. - A traditional intoxicant in Udaipur, Rajasthan. *J Trad Folk Pract*, *4*, 129–131.

Jain, V. (2017b). Ethno-biology of exhibits in a mega handicraft fair of Rajasthan. *Indian Journal of Traditional Knowledge*, *16*(3), 524–532.

Kates, R. W. (2011). What kind of a science is sustainability science? *Proceedings of the National Academy of Sciences of the United States of America*, *108*(49), 19449–19450. doi:10.1073/pnas.1116097108 PMID:22114189

Kulkarni, S., & Bhopatkar, A. (2015). *The Impact of Festivals in promoting cultural tourism: A case study of Ganesh Utsav in Pune City, Maharashtra*. RTD. http://rtd7.org/wp-content/uploads/2013/09/Alka-Bhopatkar-paper.pdf

Ma, L., & Lew, A. A. (2012). Historical and geographical context in festival tourism development. *Journal of Heritage Tourism*, *7*(1), 13–31. doi:10.1080/1743873X.2011.611595

Özdemir, G., & Çulha, O. (2009). Satisfaction and loyalty of festival visitors. *Anatolia*, *20*(2), 359–373. doi:10.1080/13032917.2009.10518914

Panda S.N., Panda P. & Majhi R. (2017): An empirical analysis of socio-economic growth and development by festival tourism: open-air theatre dhanuyatra, *International Journal of Physical and Social Science, 7*(5).

Quinn, B. (2006). Problematising "festival tourism": Arts festivals and sustainable development in Ireland. *Journal of Sustainable Tourism*, *14*(3), 288–306. doi:10.1080/09669580608669060

Raina, A. K., & Agarwal, S.K. (2004). The essence of tourism development. India: Sarup & Sons.

Sharma, P. M. (2017). *Fairs & Festivals: A Pull Factor for Domestic Tourist*. Utilizing the Social Media Tool.

Shee, P. (2020): Impact of fairs and festivals on local economy: A case study on tarakeswar town in Hugli district of West Bengal. *International Journal of Scientific and Technology Research, 9*(3),

Sinha, R., & Murdia, M. (2018): Heritage Festivals: An important factor to promote cultural tourism (A study of Murshidabad, West Bengal). AACH, 3(2).

Srikanth, P., & Ram, M. R. (2013). Economic impact of festivals: Evidence from Diwali effect on Indian stock market. *Researchers World: Journal of Arts, Science and Commerce*, *IV*(2), 27–37.

Walker, P. (2017). Sustainability: A discipline and a political agenda? *Environmental Hazards*, *16*(2), 93–98. doi:10.1080/17477891.2017.1296810

Chapter 10

Karaga Jatre of Karnataka: A Tryst With History, Anthropology, Belief, Culture, and Tourism

Shaheed Khan
Dharthi NGO, India

Freeda Maria Swarna M.
Dharthi NGO, India

ABSTRACT

Karaga Jatre/procession is observed in Thigalarapete in Bangalore, India with religious fervour by a community known to be warriors and gardeners who have part of history for more than four hundred years. The scale is huge, with close to half a million people attending the eleven-day festival in the heart of the city of Cosmopolitan Bangalore, now known as Bengaluru. It is unique because in the canvas of India, the facets of communal harmony come to the core through this festival and so many others. However, with the changes in time and modernity appearing with new generations coming to the fore in greater numbers, will we as a people be able to hold on to the cultural ethos of a time that has gotten us to freeze in history and heritage? The researchers ask the primordial question regarding the future of this festival, and if it will be in the same shape and form. How should the federal and provincial governments intervene to ensure the festival, the Jatre moves ahead, binding us as humanity.

INTRODUCTION

The Karaga Jatre (Yatra/Procession)

Spread over eleven days, the Karaga *Jatre* originates from a ritual dedicated to Draupadi, who is known locally as *Droupadamma* and has its origins in the Mahabharata (fight between Draupadi and demon *Timirasura* whilst the Pandavas were ascending heaven) and signifies the emergence of Adishakti Draupadi as a symbol of strength and feminity; and the festival of Karaga is the symbolism of Draupadi coming to earth once every year for ten-days. Karaga is the decorated earthen pot which is several feet

DOI: 10.4018/978-1-6684-6356-7.ch010

in height, made of unbaked clay and carries water, as a representation of Goddess Ganga (Gange *Puje/* Pooja*)*. Accompanied by dancers from the *Thigala* (*Vaniya Kula Kshtriyas*) (originally from the Tamil country) community (oldest social grouping of the city who are agriculturists and grew vegetables and flowers) of Bangalore (now Bengaluru) and the *Veerakumaras* dressed in white trousers (earlier *dhotis*), with turbans and being bare chested for the *Jatre*, carry swords (*Katti*) and beat them on their chests whilst reciting *'Govinda'........'dik-di dik-di,'* (Lord Govinda/Krishna.....show us the way) as the pot carrier (*Karaga*) dressed as female in the form of Draupadi moves around the city of Old Bengaluru. The carrier of the pot, the priest is identified six-months prior to the *Jatre* that takes place during March/ April breaks off from the family, wherein the wife lives the life of a widow, and priest now living as a woman (Draupadi) in the Dharmaraya Swamy temple, built by the *Thigala* community over eight hundreds ago, where besides Draupadi, Krishna, Dharmaraya (Yudhishtir), Bhima, Arjuna, Nakula and Sahadeva are worshipped. It is the presence of the priest for six-month, which enables the him to usher in the characteristics of the Goddess in himself.

THE THIGALA COMMUNITY, UNITED IN KARAGA

The Thigalas call themselves by various names, which essentially are sub-groups (Select Venue, 2021); *Vahni Kuladavaru*, following the Lunar calendar (hence all festivities take place after moonrise), *Shambhu Kula Kshatriya*, and *Agnikula Kshatriya*, follow the Solar calendar (hence celebrate all festivities linked to Karaga during the daytime). Most of them even consider themselves to be the descendants of Agni Banniraya. The community is found in Bangalore, Mysore, Dharwad, but today, one can see them on account of education, position in the society, migrating to other parts of the world as well. There traditional occupations have been to manage the Kitchen and the flower gardens. Though at a stage in life many moved to agriculture, white and blue collared jobs as well; besides Government service. They speak Kannada (Language of Karnataka) and Tamil (Language of Tamil Nadu) as their mother tongue. Govindarajan (2014) reports that, "....some even believe that they spoke a mix of Tamil, Kannada and Telugu languages, at a point in time would have had their own script......"

The Thigalas are known to be the devotees of Dharmaraya, the first Pandava; but amongst the Thigalas, there are both Shaivas and Vaishnavas. Besides, Dharmaraya, the Thigalas also worship, Maramma, Kollapuramma, Yellamma and Siddedevaru. *Ulli Thigala* (Karnataka) and *Arava Thigala* or the Dharmarayana Vokkalu are two endogamous divisions in the Thigala community. Though there are many exogamous divisions within the major systems that have evolved into castes. Karaga festival is the one that units them all, irrespective of caste or any other division structure. In a way the Karaga festival and other festivities that are part of the eleven-day event, brings all of the Thigalas together, who celebrate the same with all religious fervor and belief that has held them together for hundreds of years now. True to their origins, the Thigalas are a warrior community, who were protected Draupadi and they consider themselves as the offshoots of *Veerakumars*, hence, the warrior status for the community (Govindarajan, 2014).

Natarajan (2017) mentions of the fact that, in all likelihood the Thigalas moved to Bengaluru from Tamil Nadu and Andhra Pradesh and became part of the workforce of Nada Prabhu Kempegowda I (1510-1569) who was instrument in building the core city of Bengaluru. Then on, in the eighteen century, Hyder Ali and Tipu Sultan who from 1742 to 1799 had displaced the Wodeyar Dynasty of Mysore encouraged the Thigalas to join them for their skills in horticulture. Whilst on his wars in Tamil Nadu,

Hyder Ali had come across the Thigala community and came to know of their trait, as not only warriors, but also as gardeners (Govindarajan, 2014). True, Hyder and Tipu contributed to the construction in the city of Bengaluru. No doubt there are ample references to the Thigala community who were gardeners help Hyder and his son, to layout the Lal Bagh in Bengaluru which was commissioned in 1760 (Arena Flowers, 2021). Prasad (2014) notes that the Thigala who supported Hyder and Tipu in the Lal Bagh mission, settled in the Siddapura-Lal Bagh area and built nurseries, which made Queen Elizabeth when she visited Bangalore in the late 1950s refer the area as, 'Little Holland.' So much for the Thigalas being called the Gardeners. Prasad (2014) further notes that the plants and seedlings business were centered around Lal Bagh in 1930s and 1940s and it was in 1964 that Dr. MH Marigowda, considered to be the father of Horticulture in India who set up a society that ensured unity in the unorganized Thigala community, to continue their trade of gardening. Clearly if Bangalore received the sobriquet of 'Garden City,' we will reassure to ourselves, whose hand it was, the Thigala community. Jayaram (2021) whose deep interest in Lal Bagh was triggered by the fact that he himself was a Thigala, refers to the community as 'gardeners of Lalbagh' and further mentions that, they extended their expertise well beyond the public gardens, but they ensured the greening of Bengaluru.

DHARMARAYA SWAMY TEMPLE

Locally known as the Dodda Dharmaraja Devastana (Latitude/Longtitude: 12.9654° N, 77.5834° E), it is the principal centre of all the rituals during the Karaga festival. No doubt all over the India, there are temples dedicated to Lord Shiva, Vishnu, Mother Lakshmi and other manifestations; there are also temples dedicated to Hanuman, Ganesh, Karthikeya (Murugan), and the multitude of Gods and Goddess, big and small temples which attract people and their devotion. But having a temple dedicated to

Figure 1. The Dharmaraya Temple, Thigalarapete (Picture in Centre depicts the Pandavas along with Draupadi and Lord Krishna)
Source: Authors own pictures

Dharmaraya, i.e., Yudhistra, the eldest of the Pandavas of the Mahabharata is something unique. Besides the Dharmaraya Swamy temple there are other temples in Bengaluru viz., Chikka Dharmaraja temple, Dharmaraja and Panchali temple at Kalasipalya and the Ekambareshwara Dharmaraja temple at Shivajinagar. All of the temples have festivals in honor of the presiding deity which have the nomenclature of Karaga. But it is the *pete* (old/market town) Karaga, that is organized at Dharmaraya Swamy temple in Thigalarapete, also known as the Bengaluru Karaga which is considered as the primary civic ritual.

DRAUPADI, A LEGEND NO MORE... SHE LIVES ON... SHE IS ADI SHAKTI

The Karaga festival and accompanied festivities are a reminder of the power of the woman, the Adi Shakti, to fight out and ensure for oneself and society. Draupadi too; having been married off to five brother, the Pandavas, then being humiliated in the palace by the Kauravas, then proceeding to exile and the trials and tribulations of exile and the final move to heaven with the Pandavas and Lord Krishna; demonstrates the life line of destiny that a human would have go through, and what Draupadi went through went would be an inspiration to all of society; the ability to face the trials and tribulations, the issues, standing for oneself, protecting oneself, becoming a model for others. In a lighter vein, when we all as children had our own, protectors, our Superheroes- Spiderman, Superman, He-man, Giant Robot (and the other evolutions) Batman, wonder woman and the female genre; Draupadi surely is a Super human being, a Wonder woman, who withstood it all, thousands of years ago and came out successful. She demonstrated the *Naari-Shakti* that is innately present in every woman, should help us and the society that we are part of to be in good stead.

Figure 2. The Priest gets ready to dress up as Draupadi for the procession
Source: Select Venue (2021)

At a point in time, after the exile, when Lord Krishna, the Pandavas and Draupadi were on the way to heaven; and Draupadi had lagged behind. Demon Timarasura, who had a boon from Lord Shiva that in a battlefield, from every single droplet of blood fell from Timarasura, would arise thousands of Timarasuras called out to Draupadi and even teased her to fight. Upon seeing that the Pandavas and Lord Krishna were ahead, knowing fully well, she had to fight the demon with superpowers all by herself; she assumed the primeval form of *Adi Shakti* and the moment she shook the cloth covering her breast, sprang warriors like *Durga Pujari, Gante Pujari* (bell priest), *Ganachari* and the *Veerakumaras* (hero protectors of the Goddess). The battle between the two was long and a terrible; and as per the boon granted by Lord Shiva to Timarasura, more demons sprang out from the blood that fell on earth. Thence, Draupadi in the form of the Goddess struck that one blow that brought down the demon, and with the powers that she possessed, licked clean the blood that flowed, ensuring no more demons came about, and even swallowed the demon, to ensure a complete victory over evil. Post the battle, Draupadi regained the original form and promised to visit her warrior sons once every year and no doubt, the Karaga is the place where she meets with her children, the *Veerakumaras*, who not only protect the person carrying the Karaga with swords, who is now a representation of Draupadi herself, but in case of the pot falling (it's never happened), have to put an end to the procession by killing the person. But then, from historical times, we have no record to prove that that *Veerakumaras* had used the swords in a gory way; for that is the belief, and considering that the person carrying the Karaga has taken the form of the Goddess and is in a state of trans, nothing wrong is bound to happen; and no wrong will be done. It is a lesson to the people, the society, the believers, to have a belief, to carry on…

THE *JATRE*, THE FESTIVAL ROUTE

The carrier of the pot, the contents of which are unknown to the common human for centuries, is in a trance and dances along with the *Veerakumaras* as the procession with tens and thousands following and leading, and thousands more thronging the streets of Bengaluru to pay obeisance and witness the glory of culture, religion, and a belief.

Figure 3. The Karaga Carrier surrounded by the Veerakumaras
Source: Select Venue (2021)

A number of musical instruments viz., Thavi, Nadaswaram, Muni, Udukka and Pamba besides others accompany the dancing pot carrier who moves to various temples across the city dedicated to Annamma, Mariamma, Patalamma, Muthyalamma, Yellamma, and Gadagamma; all of whom appear as the Siblings of Draupadi/*Droupadamma*.

Besides the Amman temples, a temple dedicated to Lord Hanuman is also visited besides, fort (*kote*), ponds (*kere*), gardens (*tota*) during the *Jatre*. The Karaga also makes a stopover at the Dargah (mausoleum) of eighteenth-century Sufi Saint, Hazrat Tawakkal Mastan Baba, where the Muslim community welcomes Draupadi with all fervor and a belief in the good; which demonstrates the secular fabric of the *Jatre* and the alliance of communities, in this case the Muslims and the Hindus. Every year, the Karaga procession makes a customary halt at the dargah to pay obeisance to Baba Tawakkal Mastan. The tradition stems from the belief that once, during the procession Sufi saint hurt himself and he cursed that the

Figure 4. The Karaga Route Map followed in 2019 as released by the Deputy Commissioner of Police, Traffic West, Bangalore City
Source: https://t.co/KmA5FpaJfw

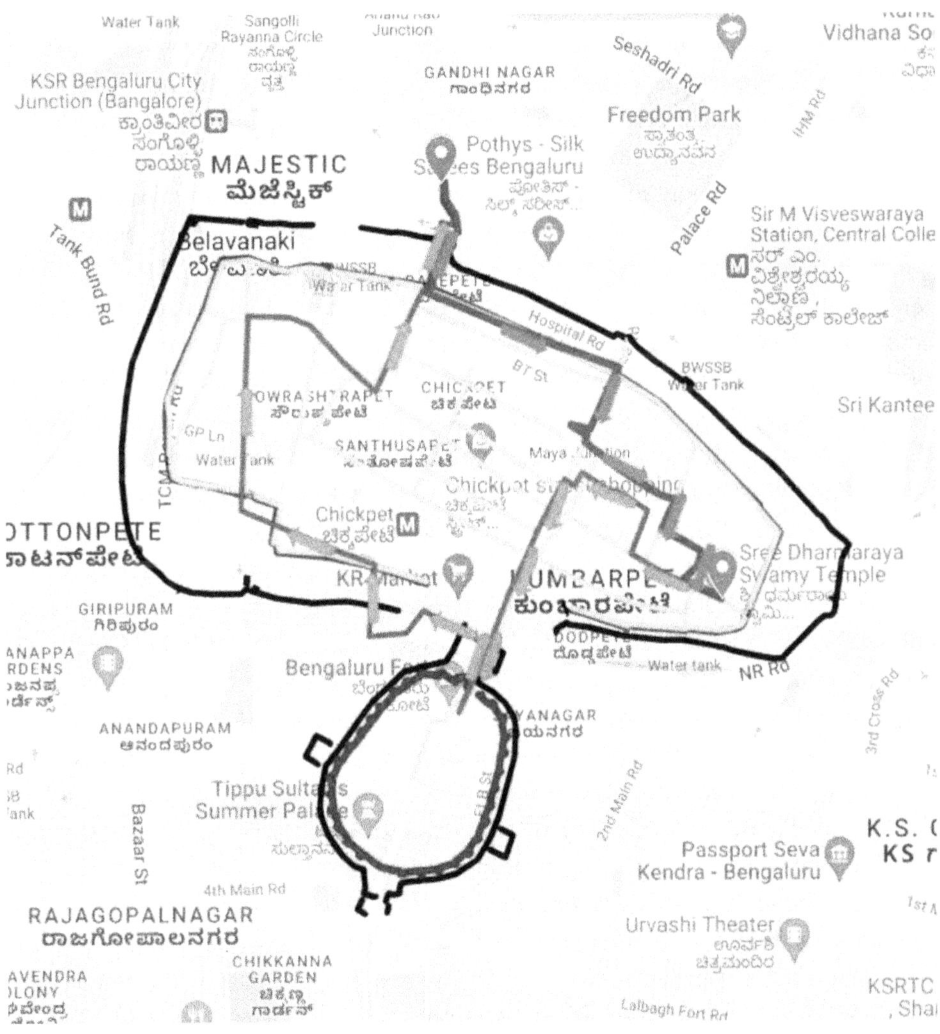

Karaga will not move forward until turmeric from the temple was applied to his wound to pacify him. Ever since, the Karaga visits the Dargah to mark this event (News Nine, 2022).

On the eleventh day, the night-long procession, which is highlight of the festival starts on a full moon night around midnight from the Dharmarayaswamy temple in Tigalarapet. The Karaga then proceeds through the streets of Ulsoorpet, Nagarthpet, Cubbonpet, Ganigarapet, Avenue Road, Doddapet, Aralepet, Balepet, Kumbarpet and Gollarpet (Rakshitha, 2019) before returning back to the Dharmaraya Temple in the early hours of the morning.

THE DESTINATION, THE FESTIVAL, THE WAY FORWARD

The Karaga has become one of the many festivals of Karnataka, however, it is localized to the city of Bengaluru and a few other locations in Karnataka, where the festival is celebrated with a lot of religious fervor. No doubt thousands of locals who gather for the various events of the festival over eleven days; visitors from other parts of Karnataka and all the South India States visit the festival area, move across the identified temples in the route of the Karaga. But it is on the day of the march of people along with Priest carrying the Karaga across the city that ushers in a maximum number of visitors from far and wide. There is vibrancy, a sense of disbelief, a wanting to propitiate the Gods and be blessed in all those who are part of *Jatre*. As researchers and visitors to various events of the festivals, we would see many foreign tourists being part of the celebrations; but in the recent years, the numbers have dwindled for a multitude of reasons. This clearly showcases that the branding of the festival or even the destination/ destinations mentioned throughout the route of the march, is something to be desired.

Though the tourism department of Karnataka provides for information and inputs about the festival; much more can be done in the form of a build up to the March/April event. First and foremost, most of the monuments that the march covers are not included in the day tour packages that are on offer, or even as special tours both the domestic and international visitors. In our discussions with the planners and some of the locals who have been watching the festival and its evolution for two to three decades are of the view, that organizing walks to the places where the events unfold, considering that they are part of 'old Bangalore,' it will become a unique feature for the city and the necessary build up can be provided for the future visitations by tourists. It will also mean that besides promoting the various tourist destinations of the city of Bangalore that have been included as normal; the lore of the 'mud-fort' of Bangalore, the various gates to the metropolis of today; which still stand a testimony to the glorious past (though some have been demolished) and the temples and the bazaars that dotted the places of the Karaga event, will usher in a new understanding of the city of Bangalore and the way the *Thigala* community which front ends the festivals are also known as the 'Gardeners of Bangalore' and true to the sobriquet of 'garden city of Bangalore,' the festival which is trilingual (Kannada, Tamil and Telugu) is a cynosure of happenings and the opportunities that provides for creating a niche in the dynamics of tourism and evolving a product that will leave a memory in the mind of the visitor. Though a Researcher and author says, "….in some ways, the Karaga festival is to Bengaluru what the Mardi Gras is to New Orleans……," the *Jatre* showcases the intangible heritage of the cosmopolitan city, which demonstrates the alliance of the communities of Bengaluru, their aspirations.

Karaga an urban festival in a city that is beaming in cosmopolitan culture, the mores, the morals of the alliance of communities, Karaga has its own story to render to the past, present and the future and is as famous as the Mysuru *Dasara* festival to anyone from the city of *Bendakaluru* (city of boiled beans),

Bangalore, now Bengaluru. Kamath (1983) reiterates, "..whatever said, the Karaga is a celebration in honour of Shakti, which is a core-concept in the *Shakta* philosophy, which has wide acceptance not only in Karnataka, but in Assam, Bengal and Kashmir….in the *Samkhya* dualism between the feminine *prakriti* and the masculine *purusha*, its assumption of the proliferation of the productive powers of *prakriti*, of the auxiliary role of the mail divinities, of the representation of the union of male and female powers, in water symbolized by *kalasha*…are doctrines that are basic to these cultic practices…"

It is here that one needs to go back to the Mahabharata, when, Draupadi, the wife of the five Pandavas created several *Veera Kumaras* (brave sons) to form a mini army and defeat demon Timirasura. At the time of her death, these *Veera Kumaras* begged Draupadi not to leave them. Draupadi instead promised to return each year during the full moon day of Chaitra masa (month). Karaga festival is celebrated each year to mark the return of Draupadi in the form of Adishakti, what is also the community deity of the Vanihikula Kshtriyas (Yamini, 2022), invariably falling in the months of March and April (Karnataka Tourism, n.d.) every year, or what is referred as the Chaitra month according to the Hindu calendar (Yamini, 2022).

As has seen from the past, the festival attracts lakhs of not only devotees, but believers and tourists, from different parts of the country. Organized for a total of eleven days, the festival celebrates the rituals and traditions that have an origin in the Mahabarata as we are aware of. The stripping of Draupadi by the Kauravas, the loss in the game between the cousins, the exile of the Pandavas, the sons of Draupadi, whom she raises to fight the evil asura; but who die at the hands of Ashwathama are episodes from the Mahabharata that are remembered and celebrated to eulogize the power of Draupadi, the woman. The festival is a celebration of the trials and tribulations of the events that Draupadi went through during her time, and the way she stood strong, throughout; Draupadi emerges as a symbol of strength and ideal womanhood, the Goddess of Strength (*Shakti*).

THE KARAGA *JATRE*, THE DHARMARAYA TEMPLE, *VOX POPULI*

Many aspects have been stated about the Karaga *Jatre*. the temple in Thigalarapet, the story woven around Mahabharat and the episodes of Draupadi, the researchers felt, apart from the available literature about the temple and the events that happen on a yearly basis, what is that the common people think of the destination. So, we got interviewing a few who were aware of the temple complex and the events that occur in and around.

Balaji (Balaji, 2022) a businessman in large scale printing, born and brought up on Bangalore, now Bengaluru, mentions that, "……from childhood I was aware that there was a festival which involved Draupadi, and that a man would dress as a woman and parade throughout the old city of Bengaluru……. but as I grew up I realized the importance of the festival, rather a *Jatre,* that was steeped in history and had its own cultural ethos built into." Balaji adds further, "…..I was an adult when I realized that the festival was of the people, conduct by the people and was meant for the people…….and hence, all came forward to participate with vigor and ensure that they are blessed by Draupadi and the clan……it was the belief that has kept the festival going…and I am sure, it will last a long-time, I have seen it for fifty years, it will surely last…" Balaji goes on add, …it was the night out, the excitement, and seeing the militant aspect of the warrior clan, the *thigalas* who would have naked swords prancing towards the Karaga carrier, and some would even have flexi swords which they would swing on to the ground, creating sparks… but most important of all, the fear that, if the Karaga falls, the carriers head would be beheaded…that fear was always there…but then the divine, the divinity surrounding the event…the belief, ensured, and I am given to understand, never in the history…the Karaga has fallen…" Balaji provided another element in the discussion, which the researchers acknowledge as may be a turning point. Balaji mentions, …earlier the area where the Dharmaraya Temple from where the Karaga *Jatre* would start has people belonging to the Thigalas, the Telugus, the Kannadigas…most of them have sold their properties and moved to other parts of Bengaluru or even moved out to other countries…lot of properties are occupied by the Marwari community from Rajasthan, who have accultured themselves to the milieu…not sure, if the *Jatre* will continue in the future, as has been happening over the years…" Balaji, who considered the Karaga *Jatre* as important as the Dasara habba and the procession that happened in Mysore (Mysuru); should not become part of a lore in the future, the generations to come should know and see the *Jatre,* the festivities, which will ensure the culture, folklore, the mores continue to exist…otherwise they will only be available on YouTube and similar…"

Srinivas (Srinivas, 2022), a local, who currently works as a parking attendant in the area close to temple, observed, "…as a child, we would go to our nearby village to be part of the festival, it was only when I was about nineteen, when an elder in the area mentioned, 'you go to the festival in your village which is good, have you even realized the *mahime* of the temple dedicated to the Pandavas and the *Jatre* that occurs. Not for nothing the temple has come up in the place. The place is a *punya Bhoomi*, people who visit the place, whoever it is, poor, the rich, the haves, the have-nots… will be blessed." Srinivas says, "…after I heard about the place from the elders, Dharmraya Temple and the *Jatre* became the centre point of my life…" Srinivas even went on to state, "…the mere fact that you are writing something about the place itself, if your Karma, and the destiny…that has made you to visit the *punya Bhoomi*…otherwise there are so many in Bangalore, who know about the temple, the festival, the procession, but have never been a witness to anything…" Srinivas even mentions that the elders in the temple who speak with the community have said, ".that according to the *Sthalapurana* (the local tradition) the temple is believed to have been constructed by the Ganga Arasu (King) [Ganga Dynasty of Talakadu a place near Mysore]…"

Saravana Kumar, a businessman who seen lived in an area nearby as a toddler and grew up with the people of the area, mentions,

Karaga festival, the jatre is the pride of the Thigala community. True there is lot of religious fervour attached to the festival, but in the current times, the new generation will always look at it as something different. The population composition in the area of the Dharmaraya temple has changed over the years; more and more the community have moved out and new caste groupings have come in; besides, lot of traditions are being lost on account of the movement from joint family to nuclear family structures. In joint families, the community deity, was a critical element to bound the family together; as nuclear families each have a choice to participate or not to. Hence, one sees the reduction in footfalls as well, besides, the factor of time and speed at which the world moves.

Saravana comes to an interesting point in the discussion with the researchers, and says,

every area in Bangalore has a festival, we see the Maria Matha (Feast of St. Mary Basilica) (Latitude/ Longitude: 12.9844° N, 77.6044° E) festival at Shivajinagar, which is localized and come what may the festival will remain and the fervour in the people will continue. So, to with the Karaga Jatre; the Jatre will continue, the fervour will remain. The new generations may have a disconnect to the Jatre, the temple; but the mobile generation who are more used to touch screen will surely envisage something different, as they are not in touch with reality, and will surely loose out understanding the culture, the ethos...not blaming the new generation for anything...but that seems to be the reality.

Going back to the discussion on the composition of the Thigalapete (area of the Thigalas) who dominated the space, Saravana mentions,

the Sociological change today it is the Chowdhury, Rajputs and Purohits (CRP), with a almost five decade gap that was dominated by the Jain Marwaris, who are moving out in small numbers...earlier to the Jain Marwaris, it was the Thigala community that dominated the area...but with the composition changing, is it feasible to retain the fervour, the grandeur, and the very ethos the festival/Jatre stood for, once upon a time.

Saravana laments that, "we are at the cusp of a society, that wants everything aspect of life to be done with quickly…and many do not even have the time…religion and philosophy sound esoteric to many… in this environment, is there is a feasibility of maintaining the elaborate festivals of the kind that the Karaga is known for?" A pertinent question was brought about in the discussion with a few, though the researchers wanted to connect to as many.

When the researchers spoke with a Priest at the Dharmaraya Temple on the condition of anonymity, as he was preparing the various sanctums in the temple premises for the *pooja* and other offerings on a normal day as part of his routine; he mentioned,

the people of today, the younger generation just do not have the time - everything should be done fast and quick. The more we delay, we will have people moving out and watching the celebrations, and the pooja over YouTube and other channels that are available in a plenty. Somehow the older generations are still continuing the traditions with more of the new generation moving to new age jobs and migrating to other cities and even countries, what will happen of the festival, the jatre time only will tell.

The Priest who was in his mid-thirties, smiled to the questions of the researchers and said, "…time is the essence, and time will tell us, whether such events will continue or not…so many cultural symbols have died…because of apathy and non-participation of the people…." The Priest with twinkle in the eye and smile, as he was cleaning up the space where an image of Lord Ganesh is located mentioned, "…don't worry, my generation and another one will surely continue the efficacy of the festival and ensure it is done…I am sure the traditions will continue, but to what extent may be questioned…we as commoners should help retain the traditions, the Government of the day, should help promote events of this kind, as it encourages people to witness a historical – cultural event which has a bearing on society and considering the pressures we go through in our daily life and chores…these events are not 'speed breakers,' but add value to our lives and make us believe in the oneness of the Supreme Power to be… thereby helping us emotionally and ensuring our way of life…"

Rohit Hangal (Hangal, 2022), who is the Director of Sphere Travel Media (https://www.spheretravelmedia.com/) an entity that supports in tourism promotion mentions, "…Karaga is the biggest festival of Bangalore…the cities culture is on display…it is bit of Bangalore or Bengaluru or Bendakaaluru that each one of us would love to share with others…" Rohit goes on to add, "the festival is historic in its own way considering the history the festival possess, and if given an opportunity, Karaga, should become the state festival along with biggest of events in Karnataka that is organized as State (*naadu*) festival, the Mysore Dasara…it is a festival of the people, like the Dasara, there is communal harmony, people of all walks of life get involved…it is a festival wherein the city comes alive…and what more to promote, tourism, bring in the local, the domestic and even the International to cherish the culture, the heritage, the visual arts, the dance, drama, the Mahabarat with all its connotations, that the festival has a link with…it surely is a splendour…and none need to miss it…" Rohit, describes further, "…the festival is just part of a brochure, we should bring it out and make it a 'calendar event' like the Ganesh Chaturti in Mumbai, where visitors throng from far and wide to get a glimpse of the phenomenal occasion of religious fervor, of beliefs, of or culture, that people follow…" On be questioned by the researchers on the role of the Government, Rohit mentions, "….the Government is aware of the catalytic role tourism plays….how on account of tourism…people take an interest, in the art form; make an effort to retain it for the people, for the future… usher in respectability, recognition, livelihood options for those involved… this is catalytic, media writes about the same, all come to know about it…in the end, we have preserved a culture, an ethos for posterity……the way Kathakali in Kerala and Yakshanagana today have become part of any tourism event and the visitors are enthralled to events…all on account of the initial Government intervention…the forms are today vibrant and live to see another generation…" We need to do the same with Karaga, the Government has been supportive, but making it a State Festival, the importance given will enable the festival, the *Jatre* to move and ensure its efficacy for the future generation..."

Providing another perspective, is Kishore Shivagupta (Shivagupta, 2022) (Trip Kraft India) (https://www.facebook.com/TripKraft/), a Custom Travel Delight Planner and a Travel Junkie himself, mentions, "…I have customized, promoted and sold tour packages globally and within in Indian to our people…they have accepted visiting places of festivals, *Jatre* and similar occasions at destinations… but it is sad to speak…that no one really has asked for Karaga…even when I market global packages to exotic nations, travelers ask for specific festivals and processions and for me organize visits and ensure they participate…but it is an irony, that though I do a little inbound into India…and I do Bangalore and Karnataka destinations with elan…none have asked for Karaga." Kishore, whilst interacting, raises a question, "….where is the lag….it is the promotion of the festival, the destination, the temple, the story of Mahabharata, the story of the trials and tribulations of Draupadi…none of it has been done." Kishore mentions with a gloom, "…even the tourism brochure on Bangalore, has all but one or two sentences on the Dharamaraya Temple and the festival…" Kishore also mentions that, "I can sell the destination and the festival to the prospective visitors….but a buzz has to be created…..in the mind of the visitors.and that job can be done through Roadshows, with the other tour operators and travel agents knowing about the destination…and guess who can do the same, the Government of Karnataka…..they should, like what they do with the State Festival of Dasara in Mysore, promote Karaga as another festival, which is known for culture, heritage, mind you, the colour, the pomp, the glory, the esoteric world of religion, the belief system…and the total devotion..." "If this is done," Kishore adds, "...we can promote the destination and ensure visitation by the travelers, who are bound to enjoy what's on offer…"

Coming to the views of an expert, who had adorned the space of History, Culture, Sociology and Tourism, CN Shaik Mastan (Mastan, 2022), Professor of History, Bangalore University, states that,

whatever it is, the vibrant festival of Karaga is an enigma in itself, it is a legacy from the historical, socio-cultural perspective and I am confident the future generation will keep it alive....As we all are aware, the festival is associated with the worship of Draupadi Cult as a mark of seeking war weapons secured in the banni tree. True, in course of time and space, it developed as a syncretic and accommodative historic and cultural legacy. It became a festival of the people, when Nada Hiriya Prabhu Kempe

Gowda encouraged one and all to participate. Karaga will remain the pride and identify of Bangalore for year to come.

ROLE OF GOVERNMENT IN PROMOTION OF KARAGA

Constitutionally speaks, the States (Provinces) in India have all the powers to ensure quality tourism promotion. The Federal (Central) Government in New Delhi can provide for directions, programs, and funds to implement schemes that will help in assuaging quality tourism and also support in destination development. The Indian Federal Government in 2019, through the Ministry of Tourism, Government of India, (MoT, GoI), had come out with schemes that had one of the best intents of quality tourism promotion; besides in-bound the focus was on Domestic Tourism, which had been the focus for many years. The Government had introduced some of the finest of programmes, viz., *Dekho Apna Desh*, *Swadesh Darshan Scheme*, Pilgrimage Rejuvenation and Spiritual Heritage Augmentation Drive (PRASHAD) under the Incredible India 2.0 campaign; which not only focused on 'footfalls', but spoke of enhancing tourist experience and employment opportunities in a sustainable fashion. These programmes that came into existence in 2019, the Indian Travel, Tourism and Hospitality (TTH) eco-system was looking for a phenomenal opportunity, to ensure an increase in domestic tourism, which primarily focuses on pilgrimage tourism, and yes these days, also ventures out for ecotourism, beach tourism, and the younger generation has made adventure tourism, come to a turning point that the stakeholders are identifying more circuits and destinations to satiate the palate of the domestic tourists. The Foreign Tourist Arrivals by and large focus on the culture, heritage besides beach tourism. In such a scenario the Federal Government has made efforts to focus on Religious and pilgrimage tourism, with augurs well for the system (Khan & Swarna, 2022).

According to the Karnataka Tourism Vision Group (KTVG) which presented a series of recommendations to the Government of Karnataka (the Government of the State of Karnataka) in January 2014 (KTVG, 2014), the committee created a sample guide to usher in tourist destination classification, wherein they got to mention the Bangalore Karaga, as an excellent (Illustration 1) event to promote considering the fact of i) number of visitors, ii) composition of visitors-local, domestic and global, iii) Management

Figure 5. Grab from the Karnataka Tourism Vision Group document
Source: KTVG (2014)

Criterion	Exceptional	Excellent	Good	Potential
#Number of Visitors #Composition of Visitors (Local / National / International) Management Maturity Calibre of talents & Depth of Intellectual pursuit	Mysore Dasara Kannada Sahitya Sammelana BIFFES Rangashankara Theatre Festival Suchitra Short Film Festival	Hampi Festival Karnataka Janapada Sammelana Bangalore Literature Festival Academy of Music Festival at the Chowdiah Memorial Hall Lalbagh Flower Show Kadalekai Parishe Bangalore Karaga	Pattadakal Badami Aihole Navraspur Sree Rama Seva Mandali - Fort High School St. Marys Festival Lalbagh Mango Festival	Vasantha Habba Bangalore Habba Bangalore Roof Top Film Festival Gudiya Sambrana

Maturity, iv) event calibre, considering the talent and the depth of intellectual pursuit that one can have through the festival. It is clear the Karaga festival is on the top of the Government agenda, but then it is what the common people can do, to promote and propagate the festival to 'live on' generation after generation and enthrall the viewer, from all aspects, religious and cultural.

The researchers whilst speaking with one of the Central Tourism officials in the who mans the counter in the airport at Bangalore; mentioned, they do not get many queries on Karaga or any other festival from the visitors who arrive at the airport. The researchers even spoke to an officer of the India Tourism Office in the city of Bangalore, who stated on all counts of anonymity, "...that no one has asked about the festival of Karaga from the other Tourist Offices in India or overseas or even checked with the same..." The officer did mention that, "....we do get queries about the Mahamastabhisheka in the Jain Destination of Sravanabelagola...but no one has asked about Karaga per se..." The officer even mentioned that, "....we do not have specific information on Karaga, but the inputs provided through the brochure on Bangalore does mentions a line or two about the same...." However, Pradeep who has been working a Tour Guide with the India Tourism Office in Bangalore had this to say about the Festival and its promotion, "...the Karaga is festival which is unique not only to Bangalore but a few other places that promote the same, by calling it *Hasi Karaga*....all those visit Bangalore as domestic and International travelers rarely get to know about the Karaga in Bangalore, though some of us guides do make a passing mention to the same; but the visitors are more interested in the monuments, the parks, the shopping etc...even the brochures produced both by the Central Government in India and the State of Karnataka, mention Karaga in one if not two sentences...and that's the end of it..." Pradeep adds, "...from a festival perspective, we celebrate the world renowned Mysore Dasara with lot of elan...every roadshow of Karnataka has something about the festival and the procession....but then the it is not the same about Karaga..." Pradeep arguing out for the cause of Karaga, notes, "....even the forest department of Karnataka and Karnataka Tourism, promote the Bird sighting Festival (*Hakki Habba*/Bird Festival) that happen once in a year, in one location of Karnataka with a lot of importance and many visitors travel to the destinations...these festivals have helped in the preservation and conservation of the ecosystem in the area, no doubt...the Government should also focus on the finer aspects of Culture and promote festivals with a heritage like Karaga... one unique festival of the cosmopolitan city of Bangalore...if the Government promotes the festivals, there will be a multitude of business opportunities that will come about, needless to say...walk tours of the city, Cuisine Tour in the old and new city, walk tours of temples, churches, mosques, gardens... the colonial building that still adorn...the Bungalows of the Raj, that are found in the city...Bus tours, tours by bikes...the list is endless....but the Government of the day has to ensure a start; give opportunity to the business to build up on the momentum...but alas, all this does not happen...the festival of Karaga goes on..." True, the words of Pradeep the Tour Guide of Bangalore is to be taken seriously; if there is a momentum to promote the festival, then nothing will stop the juggernaut from rolling and creating more opportunities to help in the conservation and preservation of the culture, the folklore, the dance, drama, the music, the opportunities of employment, both temporary and permanent, spin off business opportunities, that can grow as silos to promote quality tourism...the list actually becomes endless... but then where does it start is the question.

CONCLUSION

What more can one say about a festival that has been so long organized by the people of Bangalore/ Bengaluru, for the world to pay witness to historical event has been going on seamlessly. Year after year, the Thigalas have put all their might to ensure that traditions are upheld come what may. In the day and age of technology, where everything is questioned by the people, the young adults, some of the mores, however, esoteric they are, have been followed, with all sincerity and all efforts are made to evoke the best blessings of the presiding deity and the *Adi Shakti* who lends herself to her people on these eleven days, at the Karaga festival.

What needs to be done now: A few steps that if taken quickly will ensure the festival of Karaga will be noticed, not only at the city and state level, but also receive a National and International status.

i) The State and Central Government Tourism systems should ensure adequate marketing of the festival during various road shows that are organized in India and overseas. This will surely help usher in a comparison and an interest in the minds of the prospective visitors. The government stakeholders need to look at the festival as a product and make efforts to get the necessary branding and inclusion in various International Heritage oriented bodies as well.
ii) Educating the tourist officers across the country and overseas will help in ushering in a comparison of the festival, which will ensure an interest in the mind of the visitors. The State, National and local guides, should be giving more information and training, which will ensure capacity building to talk about the festival and various other destinations involved. Building of infrastructure for parking of tourist vehicles during regular visits will also help.
iii) The Tourism and Travel ecosystem should encourage visits to the temple site of the festival all through the year, which will ensure a discussion about the festival when it arrives. It is the intrigue that makes the visitors to come and get to experience the destination.
iv) Encourage social media participants and influencers with quality information that will help them to create artefacts which in turn will engage the prospective visitors.
v) By billing the festival as a 'peoples festival,' will enthuse the local community members to participate in larger number and get to know the vibe, the culture, the ethos that the festival communicates.

As rightly mentioned by Prof. Mastan, "…it will be difficult just to erase a socio-cultural construct that has been in existence for more than four hundred years; which is being carried on the people, hence the Karaga will stay on and ensure it continues its juggernaut for posterity and for all the future generations to participate, see and appreciate." As a final word one can say that world over, festivals may be losing sheen on account of the apathy of the generations to be; but then those who still witness them from a perspective of religiosity and but a philosophy to the festival, will make all efforts to ensure the continuation of the same. However, time will be a witness, as the Priest in the temple premises stated to the researchers, "….time is the essence, and time will tell us, whether such events will continue or not..."

REFERENCES

Balaji (2022) *Personal Communication* [WhatsApp]. Whatsapp.

Flowers, A. (2021) Lalbagh Botanical Garden Bangalore, History and Attraction. *Arena Flowers*. https://www.arenaflowers.co.in/blogs/news/lalbagh-botanical-garden-bangalore

Govindarajan, N. (2014). The Tigalas of Bendakaluru. *Deccan Herald*. https://www.deccanherald.com/content/378918/tigalas-bendakaluru.html

Hangal, R. (2022). *Personal Communication*. [Discussion].

Jayaram, S. (2021). *Bangalore's Lalbagh: A Chronicle of the Garden and the City*. Visual Arts Collective.

Kamath, S. U. (1983). *Karnataka State Gazetteer, Part II*. Government of Karnataka Publication.

Khan, S. & M., F. M. (2022). Indian Tourism and Hospitality, the FTAs, Domestic Tourism, GDP and Tourism Promotion. In Chowdhary, N., (Ed.), *Indian Tourism, Diaspora Perspectives* (pp. 3-19). Emerald Publishing Limited. doi:10.1108/978-1-80262-937-820221001

KTVG. (2014) *Karnataka Tourism Vision Group Recommendations to the Government of Karnataka: January 2014*. Kreto. https://www.kreto.in/images/Vision_Group.pdf

Kumar, S. (2022). Personal Communication [Discussion]

Mastan (2022). Personal Communication [Discussion].

News Nine. (2022). Bengaluru Karaga Festival: procession to visit Mastan Saab Dargah amid objections by Hindutva outfits. *News Nine*. https://www.news9live.com/state/karnataka/bengaluru-karaga-festival-procession-to-visit-mastan-saab-dargah-amid-objection-by-hindutva-outfits-163427?infinitescroll=1

Prasad, A. (2014). Will Nurseries become just a memory. *The New Indian Express, Bangalore Edition*.

Rakshitha, R. (2019). This year, Karaga may follow a different route. *Deccan Herald*. https://www.deccanherald.com/city/top-bengaluru-stories/this-year-karaga-may-follow-a-different-route-728440.html

Shivagupta, K. (2022). Personal Communication [WhatsApp].

Srinivas (2022). Personal Communication [Discussion].

Tourism, K. (n.d.) *Bengaluru Karaga*. Karnataka Tourism. https://www.karnatakatourism.org/destinations/bengaluru-karaga/

Venue, S. (2021). *Karaga festival and its rituals*. Select Venue. https://selectvenue.in/blog/karaga-festival-and-its-rituals/

Yamini, C. S. (2022). Karaga festival of Bengaluru: All you need to know. *Hindustan Times*. https://www.hindustantimes.com/cities/bengaluru-news/karaga-festival-of-bengaluru-all-you-need-to-know-101648034899002.html

KEY TERMS AND DEFINITIONS

Adi-Shakti: Literally, the primal, first power.

Dekho Apna Desh: This means 'See your own Country' (this was to primarily promote Domestic Tourism in India).

Habba: This is a festival.

Hakki: This is a bird (Bird Festival/*Hakki Habba*).

Jatre: This is a procession.

Katti: Sword. That is carried by the Veerakumaras.

Kere: This is a pond.

Kote: This is a fort.

Mahime: 'The Glory,' Here it is the glory of the place, the glory of Dharmaraya Temple dedicated the Pandavas of the Mahabharat.

Naadu: This is country. Here the reference is to the State of Karnataka.

Naari-Shakti: This means 'Power of the Woman.'

Pete (pet): Town/downtown. A place in the chapter: Thigalarapete, Ulsoorpet, Nagarthpet, Cubbonpet, Ganigarapet, Doddapet, Aralepet, Balepet, Kumbarpet and Gollarpet (see Illustration 4 to understand the *pete* concept).

Pooja (Puje/Puja/Poojah): Hinduism, ceremonial worship, ranging from brief daily rites in the home to elaborate temple rituals. The word puja is derived from the Dravidian *pu* ("flower"). In its simplest form, puja usually consists of making an offering of flowers or fruit to an image of a god.

Punya Bhoomi: This means Holy Land.

Swadesh Darshan: Sight-seeing of the home country, herein case, India.

Thigalas (Tigala): A Community of warriors and gardeners who are considered as the Veerakumars protecting Draupadi/Draupadamma.

Tota: This is a garden.

Veerakumaras: A community of warrirors who protect Draupadi/Draupadamma. The Thigalas considered themselves as Veerakumaras.

Chapter 11
Social Media Networks as an Effective Way to Promote Events

Betül Kodaş
Mardin Artuklu University, Turkey

Ela Nazlı Köz
İstanbul Gedik Üniversitesi, Turkey

ABSTRACT

Today, we have witnessed many events and festivals organized by different countries throughout the years. Some of these are organized locally, while the others are international. Social media networks are one of the essential marketing strategies which included advertising and promotion for event organizers. In this chapter, social media platforms, which can be used in promoting events and festivals, will be discussed conceptually by making a comprehensive literature review based on relevant literature. After reviewing the literature, this chapter will report how social media networks can be used as a promotional strategy for the event industry. So, this chapter aims to analyze in depth the relationship between social media platforms and promotions of events, and assessed how social media applications can be used to promote tourism events and festivals. At the end of the chapter, some practical implications will be presented to events marketers about social media tools used effectively to build effective promotional strategies for events and festivals.

INTRODUCTION

The need for the socialization of people is shaped in line with the opportunities offered by the world. Man, who is a social being, chooses within these possibilities to spend his leisure time. Leisure activities of people have moved away from the classical holiday perception and turned to the point of trying the difference. New destinations and innovations that will create an attractive element in people's leisure time have become the reason for the preference for people who have tried the classic holiday concept which is comprised of the sea, sand and sun. In this context, people sometimes prefer to participate in various activities to spend their limited time and in variety-seeking behavior (Quan and Wang, 2004).

DOI: 10.4018/978-1-6684-6356-7.ch011

The change in consumer expectations has enabled providers to take action to meet these expectations. At this point, events and festivals are one of the most important touristic attractions of a destination (Getz and Page, 2016). Moreover, destinations come to the forefront with the events they organized, creating a successful brand and having some important competitive advantage between rival destinations.

With the technological advancement, people have an idea instantly about a destination away from where they are today and get valuable information about tourist destinations or tourism business. When it is considered that this technological advancement has enormous impacts on destination and event promotion, social media networks have played an important role in tourist decision about visiting a tourist destination or attending an event held in a particular location. Instant information sharing and interacting with other users are realized with social media tools. In recent years, the interest in social media networks, which is an important branch of Web 2.0 technologies has been increasing.

People from all age groups create profiles on various social media networks so as to interact with others. These profiles have been frequently posted comments towards retrospective or with instant sharing. Some high-follower profiles play a promotional role, like a publication with high circulation, with interactions, likes and comments. It is seen as an efficacious element in many aspects that people carry it to the virtual world while socializing in the real world. In a way, this situation can also be considered as the elimination of distances. With the same perspective, social media has a great influence on the promotion of special events. Social media, which provides direct access to consumers, also forms the basis for creating an effective marketing strategy based on user comments. In this context, this chapter focuses on obtaining an in-depth understanding of the relationship between social media applications and event promotions. Moreover, this chapter aims to reveal how social media applications can be used to promote tourism events. In the chapter prepared in this line, social media platforms that can be used in the promotion of events and festivals will be discussed conceptually based on the existing literature. Subsequently, it will be discussed how social media networks can be used as a promotional strategy tool for the event industry. Finally, some practical implications will be presented for event marketers and organizers about using social media tools for building strategies for event promotions.

THEORETICAL BACKGROUND

Events Concept

There is no accepted definition for the concept of events, which has been a part of people's lives for centuries. Because when it is considered that the events organized due to various special events may consist of different types of events, it is very difficult to provide a terminological clarification (Jago, 1997). But there are some attempts to define the concepts of the event. For example, Silvers (2004, p. 2) described an 'event' as a detailedly designed experience to make an impact on people through participation. In another definition of events, Shone and Parry (2004, p. 4) defined events as 'non-routine events other than normal events that people have in their daily lives.

It is possible to state that the event concept is flexible and multifaceted, as the researchers who approached the concept of the event, investigate the subject from different perspectives and different scopes (Oklobdžija, 2015). The word of the event includes the following elements (Soteriades & Dimou, 2011):

- It consists of activities aimed at the desire of people to come together,
- Occurring in a specific place and time,
- Competitive environment,
- Search for Adventure
- Evaluation of opportunities.

An event with the above-mentioned features is carried out for sports, culture, entertainment, business or political purposes (Soteriades & Dimou, 2011). The types of planned activities carried out for a specific purpose are included; Cultural celebrations, political and state, arts and entertainment, business and trade, educational and scientific, sport competition recreational, and private events like weddings, and parties (Getz, 2005). It is considered that people come together through these events centuries ago. It is known that exhibitions and commercial events were held in ancient times. The support of some events such as the ancient Olympic Games and Festivals, which are seen as a prestigious social event, by the church shows that people pay serious attention to such organizations each different period of history (Oklobdžija, 2015). Because events can provide an important economic income for the host destination and local residents; can increase social interaction between different societies; can promote the arts; can create leisure opportunities and can be seen as an awesome communication tool (Getz, 1998). Events and festivals which are special events are a significant component of the tourism industry (Getz, 2008; Quinn, 2009; Getz & Page, 2016; Kodaş, 2021a). Tourism is seen as an important sector in many regions of the world due to its economic benefits to destinations and communities. Factors such as the increase in local employment opportunities and visitor expenditures and local economic development have made societies give importance to the tourism sector. In this context, various strategies are used to attract visitors to destinations. While some focus only on the natural and cultural values they possess, others have chosen to host various events such as concerts, sports events, congresses, or commercial activities (Hodur & Leistritz, 2006). Such that some popular destinations are well known as branded destinations worldwide with events such as Cannes Film Festival, Venice Film Festival and Berlin Film Festival (Nikolaeva, Bogoliubova, Fokin, Baryshnikov, Klevtsov & Eltc, 2017).

The events have positive impacts on the regional economy, destination marketing, destination image, social/cultural identity, urban development, and tourism. Many businesses and destination marketers focus on planning, developing, and marketing festivals and special events in order to increase tourist attractiveness and create a destination image (Getz, 1998). Each of the events is unique because of its temporal and spatial characteristics, and it is necessary to participate in those events to experience this attractive experience for event-goers (Getz, 2008). Therefore, events play a key role in the promotion to attract tourists to destinations, given the increasing global competition factor (Getz & Page, 2016). In terms of seasonality, which is one of the most important problems of the tourism sector, event organizations are one of the most effective solutions in solving seasonality problems and provide businesses with the opportunity to gain a competitive advantage with a differentiated product strategy (Connell et al., 2015). On the other hand, the use and commercialization of special events in the promotion of destinations as attraction centres may cause the risk of deviation from the real purpose of the events. This means that cultural elements are assimilated and damaged for the new type of tourist seeking authenticity (Getz, 1998). Furthermore, events have some negative impacts such as noise pollution, traffic and overcrowding (Soteriades & Dimou, 2011). Herein, it is important that event organizations are evaluated from a professional perspective and that they are carried out by professional practitioners and brought together with the consumer with appropriate marketing strategies.

Social Media Networks

The rapid development of Information Communication and Technologies (ICT) has brought considerable changes in the twenty-first century, as well as affected the demands of modern societies and supply sides like tourism business and destinations (Buhalis & Row, 2008; Buabeng-Andoh, 2012). Therefore, developments in technologies attract tourists and tourism practitioners to use the innovations brought by technology. Social media applications, which are seen as an extension of Web 2.0, are among the most used technological developments (Kodaş, 2021). Each technological development takes the e previous technology further. For example, Web 1.0 is a technologically based network that includes one-way information sharing sites prepared by experts and allows users to access passively, while Web 2.0, unlike Web 1.0, is a product of technology and dynamic network that allows for mutual information flow, where social interaction becomes prominent and users actively access. It is one of the user-oriented innovations brought by Web 2.0 technology, that it offers easy participation to many users and that a large number of contents is open and accessible in the virtual world (Sever & Ataizi, 2017). It is thought that Web 3.0 technology, which is the continuation of Web 2.0, will understand user requests and provide information that will meet the needs of the user in line with these requests (Choudhury, 2014; Kodaş, 2021). While Web 3.0 technology is still in the development stage in human life, Web 2.0 technology is at the focal point of daily life. Web 2.0 applications have two interrelated promotional roles in the marketplace. At first social media platforms allow businesses to communicate with their customers, on the other hand, they allow customers to communicate with other customers (Davidson, 2002: p.120). Social media has emerged as a new way for people to connect by integrating words, photos, videos and sounds in a virtual environment (Zeng & Gerritsen, 2014). This new way of communication includes blogs, podcasts and social networking sites created by consumers to share ideas and opinions (Drury, 2008). There are many different types of social media tools such as Facebook, Youtube, Flickr, Wikipedia, LinkedIn, Twitter and iStockphoto with millions of users (Mangold & Faulds, 2009).

Applications such as Wikipedia, which are described as blogs and collaboration projects, are generally seen as weak in terms of content richness, as they are limited to text-based content. In addition to the content of blogs written by people, social networks that allow pictures, videos and similar sharing are richer compared to blog and content pages. On the other hand, virtual social worlds and virtual game worlds are in the high group in terms of media richness as they offer a high level of face-to-face communication experience. However, blogs and social networks offer users the opportunity to become popular. Because these platforms could provide people with the creation of content on certain topics daily (Kaplan & Haenlein, 2010). Users can share their thoughts instantly and in a way that allows feedback easily via blogs and social media. The fact that Facebook has approximately 1 billion monthly active users and Twitter has 650 million active users shows that social media tools are an integral part of daily life today (Ohajionu & Mathews, 2016). The ongoing developments in the field of social media and the diversification of usage types make it essential to update the definition of the concept. While social media is considered a "tool" that allows people to broadcast widely, reach large audiences and influence them, social networks refer to the use of "social media tools" (Zeng & Gerritsen, 2014). These Web-based platforms also provide the dissemination and storage of information (Valls et al., 2012). Information sharing in social media tools provides many benefits to both businesses and consumers. For instance, it is important to know how many people like the online posts, the number of reposts and the comments about the sharing. Thus, consumer trust in business will be enhanced and affects positively the success of businesses through word-of-mouth communication.

Alkaya and Şahin (2018) examined the relationship between information sharing and Facebook social networks in the context of the Technology Acceptance Model. Findings of the study indicate that the trust factor increased social interaction. Moreover, it has been concluded that the more user-friendly a social network is designed, the more the attitude towards use and information sharing will increase. This new formation brought about by technology provides an opportunity for marketers to promote their products to large populations and to build close relationships with consumers (Kotler, Bowen, & Makens, 2010). Social media is seen as an excellent strategy for tourism marketing, especially in information seeking, purchasing decision behavior, promotion, and interaction with consumers (Zeng & Gerritsen, 2014). Although the greatest impact of social networks in marketing is in the pre-purchase stage (Valls et al., 2012), It is possible to discourse that social media has positive impacts during the experience phase (Jacobsen & Munar, 2012). This new virtual relationship between businesses and consumers is shaped in the light of the following elements (Valls et al., 2012):

- Establishing a corporate profile that ensures the reliability of the business, improves the brand image, and facilitates constant updates,
- Providing accurate information about what the customer wants in the target market,
- Engagement of users (who in reality engage themselves) so as to widen their knowledge of business products and services and gain greater confidence. Monitoring recommendations and reputation-enhancing consumption experiences related to the various communities.
- Enabling consumers to reach their goals with a lower cost and effort than traditional media channels,
- Realization of commercial activities supported by offline channels.

From the perspective of the supplier, social media provides an opportunity to understand consumer demands and monitor current developments. Thus, businesses can provide quality service by adapting to changing consumer demands (Nezakati et al., 2015). Social media provides the opportunity to follow consumers wherever they come together (Kotler et al., 2010). Contrary to traditional media, which is used only as a one-way source of information, users in social media actively participate in the communication process by sharing information and content. The fact that the followers give feedback to the instant sharing of the users led this form of communication to turn into a dialogue (Chan et al., 2013).

Given all the benefits it provides, an event organizer should use social media to organize and promote successful events; should encourage many existing and potential attendees to participate in the event (Kodaş, 2021). Crofton and Parker (2012) state that the use of social media as a marketing tool will increase the number of tourists and contribute positively to the tourism sector. From a different perspective, Fotis et al. (2010), draws attention to the fact that the active presence of social media in consumer life today causes an increase in distrust towards traditional marketing strategies and a decrease in the influence of traditional mass media. However, in order to efficiently use social media tools, which are among the promotional tools in today's technology era, it is required to perform promotional activities in a planned manner and to consider the growing importance of social media in promotion (Atçeken et al., 2018; Yavuz & Enes, 2021).

MAIN FOCUS OF CHAPTER AND METHOD

The purpose of this chapter is to reveal a linkage between social media and tourism event and festivals by analyzing literature that focused on using of social media in promoting the tourism events. For this objective, existing literature related social media platform was carefully reviewed by conducting comprehensive literature review. The quantitative and qualitative studies were selected to understand the link between social media networks and tourism event and festival such as culinary, cultural etc... The results of these studies interpreted to examine how social media tools play role in promoting events and festivals which are related to tourism.

Social Media as a Tool for Promoting Tourism Events and Festivals

A touristic product can only be experienced in the geography where the tourism supply is located. Due to the lack of possibility to try beforehand, the decisions of potential tourists are shaped by the tourism experiences of people who have been to the same destination or attended the same event before (Dalgın & Oruç, 2015). Simms (2012) found that many tourists who will visit a destination for the first time and who prefer an international destination investigate comments posted by tourists in social media platforms. Today it is increasingly popular for tourists to offer their feedback regarding their holidays and travels on social media. For this reason, many researchers point out the role of social media in the travel planning process (Varkaris and Neuhofer; 2017; Nezakati et al., 2015; Law et al., 2014; Leung, Law, Van Hoof & Buhalis, 2013; Jacobsen & Munar, 2012).

The increasing use of social media tools is pushing tourism businesses and destinations to create social media strategies for advertising and promotional programs. Parallel developments in the field of technology have led academicians studying in the field of tourism and events to conduct research in this field (Paris, Lee and Seery, 2010; Hsu, 2012; Simms, 2012; Moise & Crucero, 2014; Eşitti, Işık, 2015: Hudson, Roth, Madden & Hudson, 2015; Harb, Fowler, Chang, Blum & Alakaleek, 2019; Yavuz & Enes, 2021; Unurlu, 2021; Kodaş; 2021; Cobanoğlu, Doğan & Güngör, 2021). Studies show that information and communication technologies can be unique marketing tools for tourism businesses and destinations. Especially thanks to the social media platforms used as promotional tools, businesses and destinations can easily reach a large number of tourists and market their products. Because social media networks are virtual platforms that bring consumers who have similar interests together (Coursaris Osch & Balogh, 2016). These virtual platforms are tools that serve the four functions of promotion: informing, persuading, supporting and reminding (Köksal & Özdemir, 2013). It has been concluded that the persuasive effect of consumers is 22 times more effective than marketers using classical marketing strategies (Goh et al., 2012). In this context, it is ensured that the functions of promotion are fulfilled effectively through the content that consumers share on social media networks.

Most consumers have used online resources to obtain information on business trips. However, for holiday travel, it has been determined that the advice of friends and colleagues is a priority and is followed by websites and search engines (Verma et al., 2012). Research on the relationship between tourism and technology shows that the use of technology plays a critical role in tourists' choice of destinations or hotel choices. For instance, in the study of Leung, Law, Van Hoof and Buhalis (2013), a situational analysis was carried out on the use of social media from the perspective of the supplier and consumer. In this research, which examined 44 scientific research articles, it was determined that social media is a strategic tool for promotion, distribution, communication, management and market research and that consumers use social media during the phase of the travel planning process (Leung Law, Van Hoof and Buhalis, 2013). In a study which investigates the relationship between social media and purchasing intentions, it was concluded that the informative and fun dimensions of social media marketing have a positive effect on the consumer's intention to buy (Demirci Orel & Arik, 2020). Varkaris and Neuhofer (2017) examine the relationship of social media with different variables and states that social media influences hotel research and the way they create reservations, shaping the decision-making stage of the consumer. For this reason, it is necessary that hospitality business managers, which are of great importance for event organizations, provide the essential technological infrastructure and encourage users to make "real-time" posts. Research in this direction shows the significant impacts of technology, especially the use of social media, in attracting tourists to destinations or hotel businesses since it is considered a valuable source to get information all about hotels and tourist destinations. Studies regarding technology in the hospitality and tourism literature have provided a basis to perform studies related to the event and festival. Thus, despite limited empirical studies, social media platforms are considered as one of the best ways to promote events. Because social media is good at encouraging a large audience to become interested in the events. (Harb, et al., 2019). To identify a special event, Jago and Shaw (2000) used the number of participants, interest, destination image and exciting experience components. In this context, it is worth remembering that the effective use of social media networks has an impact on directing consumers' attention to the event destination and improving the image of the destination.

Social media platforms which event marketer and organizer use for planning, promoting and building events are remarkable marketing tool although it is considered new marketing tool (Harb, et al., 2019). Tourism destinations that organize tourism events and festivals can frequently use social media applications. For example, destinations or event organizers can announce their events or festivals on social media platforms months in advance and provide all necessary information to the participants. Thanks to online platforms, you can present all the news about the event to the participants and determine their requirements of the participants. To illustrate this, organizers of the international Edinburgh Festival use social networking platforms effectively. 101 thousand people follow the Facebook profile created for the event, 64.6 thousand people follow the Twitter profile of the festival, and 32.5 thousand people follow the Instagram profile of the organization. The YouTube profile, which has 2,626.091 views, is followed by 7,64 thousand subscribers. Thus, the organizers of this festival provide hundreds of thousands of people with instant access to the Edinburgh International Festival which is one of the largest festivals worldwide (Edinburgh International Festival, 2022).

Aydın (2020) states that the shares carried out for promotional purposes on social media networks in the context of studies carried out in digital marketing literature have a positive effect on consumers' interaction with content. With his research, he emphasizes that the announcement of events on the Facebook social network can result in relatively lower costs and higher interactions by using them in digital marketing. Many private and public organizations use social networks for promotional purposes to attract consumers' attention to events and destinations. A review of the Facebook page of a religious event called the Fòcara Festival in Puglia, Italy, indicates that Facebook can be an effective promotional tool (Pioni et. al., 2019).

A different study indicates that sharing Facebook event pages created by professional event organizers by followers will increase the consumer's sense of involvement, which may affect their intention to participate in events. For this reason, businesses need to encourage consumers who are social media users to create content (Lee, Xiong & Hu, 2012). For example, an account called @GamesTravel2014 has been set up on the Twitter social network to promote Games Travel 2014 events and to inform them about the event. Cottrill et. al. (2017) state that a Twitter user's tweeting encourages other users to participate in events, and with their research, practitioners use that Twitter account for this purpose. In a study conducted in Turkey, how the participants of the Bozcaada Local Tastes Festival were aware of the festival was investigated. It was observed that the advice of relatives/friends as information sources was 32.5% and internet/social media was 25.5% (Beyaz Kömürcü, 2013). For this reason, the researcher emphasizes that social media should be used especially in the promotion of festivals and destinations (White Coalman, 2013). Paris Lee and Serry (2010) emphasize that it would be the right approach to build trust in consumers using businesses' Facebook pages. In their study, the authors revealed that users'" trust and expected relationship through Facebook had a significant effect on users" acceptance of Facebook and Facebook posts about events may affect users' intentions to attend an event. A study of Facebook, Twitter and Instagram users show that event marketers presenting content on social media networks in a fun and user-friendly way will have a positive impact on perceived enjoyment, while perceived enjoyment will create positive attitudes towards the social media pages of events. This means consumer feedback and also information sharing between each other (Harb et al., 2019). Finally, based on existing literature, the factors influenced by social media in the context of event and festival tourism are as follows (Ngernyuang & Wu, 2020: p,25):

- Strengthens the image of the destination.
- An accessible information search tool used to obtain information and communicate with others.
- It is an effective promotional tool as one of the most powerful advertising methods.
- It is a reliable source with millions of users reached today.
- It ensures the consumer loyalty with a strong destination image.
- It is a decision-making tool that has a strong impact on people's decision-making.
- It offers a multifaceted social interaction between users and business managers.
- It stimulates emotions that encourage participation in events and festivals.

DISCUSSION AND CONCLUSION

This chapter theoretically aims to discuss the role of social media networks in promoting the tourism events. To do this a comprehensive literature review conducted to reveal the close relationship social media and events and identify that how social media applications can be used to promote tourism events and festivals. It is known that leisure events, which are an important component of destination marketing, have existed in people's lives for centuries and will continue to exist due to the need to socialize. Changes in living conditions also cause variations in people's types of socialization, making it necessary to adapt to new formations. Eventgoers are aware of events via social media platforms. Event organizers make successful marketing efforts to reach their audiences. Destinations and event planners are using frequently social media platforms such as Facebook, Twitter, Instagram, Linkedin etc… to give information about their events or festivals for their attendees to participate in the events. In today's world, where digital transformation manifests itself in all areas of life, the integration of business and social life with the virtual environment has become inevitable. Considering generation Y, which constitutes a large part of today's consumers, it is seen that these individuals with purchasing power frequently use social media and other virtual environments for information seeking, information sharing and socialization

(Levickaite, 2010). The point of view of the Y generation and the next generation on the virtual world is that it is an integral part of the real world.

Technological developments that have led to a revolutionary effect and transformation in human life have had a guiding effect on the marketing activities of businesses with web 2.0 infrastructure. Just as traditional media evolves into dynamic, interactive and user-oriented new media, traditional marketing approaches have taken into account the customer's behaviour of participating in the production. Social media networks, which are covered in the context of new media, are intriguing to others by offering users the opportunity to share their knowledge and experiences instantly. However, social media networks are one of the important components that allow people to interact with each other according to their interests and directly influence the decision-making phase. This convenience experienced by consumers in accessing information is also seen as beneficial in terms of reducing the risk of uncertainty during the decision-making process. This innovative form of word-of-mouth communication benefits businesses and event planners in many ways. The presence of suppliers on social media networks; analyzing the interests and requirements of the target market is seen as a necessity in promoting and marketing the events planned in this direction. It is possible to inspire confidence in consumers who are part of the virtual world at the point where quality service delivery is maintained with the right marketing strategies. Quality service and trust factors bring customer loyalty. By taking advantage of the instant feedback made on social networks, there is a chance to intervene in the disruptions that may occur in an event organization without losing time. In addition, through instant comments, producers create an environment of continuous improvement, avoiding repetitions and adapting to innovations. Therefore, social media networks are seen as one of the tools that will enable businesses and event planners to develop brand image at this point.

The tourism industry provides a high level of knowledge-based service. Therefore, all these developments in the field of information and communication have a direct impact on the tourism sector. Accessibility to information has a guiding effect in the decision-making process for touristic activities that have not been possible to try before. Potential consumers buy a touristic product or experience touristic facilities according to reviews posted by a tourist who has experienced before regarding that tourism product. This shows that each development in information and communication technologies will affect tourism demand. It is essential that a destination not only offers opportunities that will lead tourists to prefer that region but also raises awareness by highlighting these opportunities. For example, special events and festivals are one of the important components that have a direct positive effect on the image of the destination. For this reason, special events and festivals for various purposes have become an important part of destination promotion policies and have become important tools that contribute to destination branding studies today. As a result of the transformation of social media in human life and business, event and festival organizers need to adopt new marketing strategies, and constantly renew themselves on issues such as social networking and information management. Thus, it is possible to create demand for undiscovered touristic destinations. Social media networks which are considered as both geographically crossing borders with visual information sharing and eliminating social communication boundaries are seen as an effective tool for an interactive marketing strategy, and finally, event organizers should know attendee's requirements and demands to satisfy them and reach large number attendees if they want to promote their events/festivals via social media platforms since these platforms are a valuable source to get tourist information. (Moise & Crucero, 2014).

REFERENCES

Alkaya, A., & Şahin, F. (2018). Bilgi paylaşımının teknoloji kabul modeli üzerinden incelenmesi: Bir sosyal ağ uygulaması. *Uşak Üniversitesi Sosyal Bilimler Dergisi*, *11*(1), 11–21.

Aydın, G. (2020). Social media engagement and organic post effectiveness: A roadmap for increasing the effectiveness of social media use in hospitality industry. *Journal of Hospitality Marketing & Management*, *29*(1), 1–21. doi:10.1080/19368623.2019.1588824

Beyaz Kömürcü, G. (2013). *Etkinlik turizmi çeşidi olarak festivaller: Bozcaada Yerel Tatlar Festivali örneği, Çanakkale Onsekiz Mart Üniversitesi, Turizm İşletmeciliği anabilim dalı*. Basılmamış Yüksek Lisans Tezi.

Buabeng-Andoh, C. (2012). Factors influencing teachers' adoption and integration of information and communication technology into teaching: A review of the literature. *International Journal of Education and Development using ICT, 8*(1).

Buhalis, D., & Law, R. (2008). Progress in information technology and tourism management: 20 years on and 10 years after the Internet—The state of eTourism research. *Tourism Management*, *29*(4), 609–623. doi:10.1016/j.tourman.2008.01.005

Chan-Olmsted, S. M., Cho, M., & Lee, S. (2013). User perceptions of social media: A comparative study of perceived characteristics and user profiles by social media. *Online Journal of Communication and Media Technologies*, *3*(4), 149–178. doi:10.29333/ojcmt/2451

Choudhury, N. (2014). World wide web and its journey from web 1.0 to web 4.0. *International Journal of Computer Science and Information Technologies*, *5*(6), 8096–8100.

Cobanoglu, C., Doğan, S., & Güngör, M. Y. (2021). Emerging Technologies at the Events. In *Impact of ICTs on Event Management and Marketing* (pp. 53–68). IGI Global. doi:10.4018/978-1-7998-4954-4.ch004

Connell, J., Page, S. J., & Meyer, D. (2015). Visitor attractions and events: Responding to seasonality. *Tourism Management*, *46*, 283–298. doi:10.1016/j.tourman.2014.06.013

Cottrill, C., Gault, P., Yeboah, G., Nelson, J. D., Anable, J., & Budd, T. (2017). Tweeting Transit: An examination of social media strategies for transport information management during a large event. *Transportation Research Part C, Emerging Technologies*, *77*, 421–432. doi:10.1016/j.trc.2017.02.008

Coursaris, C. K., Osch, W. V., & Balogh, B. A. (2016 January). Do Facebook likes lead to shares or sales? Exploring the empirical links between social media content, brand equity, purchase intention, and engagement. *Proceedings of the 49th Hawaii International Conference on System Sciences (HICSS)*,(pp. 3546-3555). IEEE. 10.1109/HICSS.2016.444

Crofton, S. O., & Parker, R. D. (2012). Do Twitter and Facebook matter? Examining the economic impact of social media marketing in tourism websites of Atlantic Canada. *Journal of Tourism Research & Hospitality*, *1*(4). doi:10.4172/2324-8807.1000107

Dalgın, T., & Oruç, M. C. (2015). *Turistik tüketicilerin tercihlerinde sosyal medyanın etkisi: Amasya ilindeki 4 yıldızlı bir konaklama işletmesinde uygulama. Proceedings of the 16th National Tourism Congress*, Çanakkale.

Davidson, R. (2011). Web 2.0 as a marketing tool for conference centres. *International Journal of Event and Festival Management*, *2*(2), 117–138. doi:10.1108/17582951111136559

Demirci Orel, F., & Arık, A. (2020). Sosyal medya pazarlamasının tüketici satın alma niyetine etkisinin teknoloji kabul modeli aracılığıyla incelenmesi. *Erciyes Üniversitesi İktisadi ve İdari Bilimler Fakültesi Dergisi*, *57*(57), 205–232. doi:10.18070/erciyesiibd.701115

Drury, G. (2008). Social media: Should marketers engage and how can it be done effectively? *Journal of Direct, Data and Digital Marketing Practice*, *9*(3), 274–277. doi:10.1057/palgrave.dddmp.4350096

Edinburgh International Festival. (2022). *About.* EIF. https://www.eif.co.uk/about/media .

Eşitti, Ş., & Işık, M. (2015). Sosyal medyanın yabancı turistlerin Türkiye'yi tatil destinasyonu olarak tercih etmelerine etkisi. *Karadeniz Uluslararası Bilimsel Dergi*, *1*(27), 11–33.

Fotis, J., Buhalis, D., & Rossides, N. (2012). Social media use and impact during the holiday travel planning process. Springer-Verlag, 13-24.

Getz, D. (1998). Event tourism and the authenticity dilemma, Ed. William F. Theobald Global Tourism, Butterworth-Heinemann; 2nd edition.

Getz, D. (2005). *Event management and event tourism* (2nd ed.). Cognizant.

Getz, D. (2008). Event tourism: Definition, evolution, and research. *Tourism Management*, *29*(3), 403–428. doi:10.1016/j.tourman.2007.07.017

Getz, D., & Page, S. (2016). Progress and prospects for event tourism research. *Tourism Management*, *52*, 593–631. doi:10.1016/j.tourman.2015.03.007

Goh, K. Y., Heng, C. S., & Lin, Z. (2013). Social media brand community and consumer behavior: quantifying the relative ımpact of user- and marketer- generated content, In *2016 49th Hawaii international conference on system sciences (HICSS),* IEEE.

Harb, A. A., Fowler, D., Chang, H. J. J., Blum, S. C., & Alakaleek, W. (2019). Social media as a marketing tool for events. *Journal of Hospitality and Tourism Technology*, *10*(1), 28–44. doi:10.1108/JHTT-03-2017-0027

Hodur, N. M., & Leistritz, F. L. (2006). Estimating the economic impact of event tourism: A review of issues and methods. *Journal of Convention & Event Tourism*, *8*(4), 63–79. doi:10.1300/J452v08n04_05

Hsu, Y. L. (2012). Facebook as international eMarketing strategy of Taiwan hotels. *International Journal of Hospitality Management*, *31*(3), 972–980. doi:10.1016/j.ijhm.2011.11.005

Hudson, S., Roth, M. S., Madden, T. J., & Hudson, R. (2015). The effects of social media on emotions, brand relationship quality, and word of mouth: An empirical study of music festival attendees. *Tourism Management*, *47*, 68–76. doi:10.1016/j.tourman.2014.09.001

Jacobsen, J., Kr, S., & Munar, A. M. (2012). Tourist information search and destination choice in a digital age. *Tourism Management Perspectives*, *1*, 39–47. doi:10.1016/j.tmp.2011.12.005

Jago, K. L. (1997). Special events and tourism behaviour: A conceptualisation and an empirical analysis from a values perspective [Unpublished doctoral dissertation]. Victoria University.

Jago, K. L., & Shaw, R. N. (2000). Consumer Perceptions of Special Events: A Multi-Stimulus Validation. *Journal of Travel & Tourism Marketing*, *8*(4), 1–24. doi:10.1300/J073v08n04_01

Kaplan, A. M., & Haenlein, M. (2010). Users of the world, unite! The challenges and opportunities of Social Media. *Business Horizons*, *53*(1), 59–68. doi:10.1016/j.bushor.2009.09.003

Kodaş, B. (2021). Web Revolution and Events: Development and Progress, Birdir, K, Birdir, S, Dalgıç, A, and Toksöz, D. (Eds), In Impact of ICTs on Event Management and Marketing (pp. 18-32). IGI Global

Kodaş, D. (2021a). Festival Turizmi ve Festival Turizmi Motivasyonları. Tören, E. (Ed) Kültür Turizmi Motivasyonları. (pp. 119-147). Gazi Kitabevi: Ankara.

Köksal, Y., & Özdemir, Ş. (2013). Bir İletişim Aracı Olarak Sosyal Medya'nın Tutundurma Karması İçerisindeki Yeri Üzerine Bir İnceleme. *Süleyman Demirel Üniversitesi İktisadi ve İdari Bilimler Fakültesi Dergisi*, *18*(1), 323–337.

Kotler, P., & Bowen, J. T. ve Makens, J.C. (2010). Marketing for Hospitality and Tourism (5th edition). Pearson.

Law, R., Buhalis, D., & Cobanoglu, C. (2014). Progress on information and communication technologies in hospitality and tourism. *International Journal of Contemporary Hospitality Management*, *26*(5), 727–750. doi:10.1108/IJCHM-08-2013-0367

Lee, W., Xiong, L., & Hu, C. (2012). The effect of Facebook users' arousal and valence on intention to go to the festival: Applying an extension of the technology acceptance model. *International Journal of Hospitality Management*, *31*(3), 819–827. doi:10.1016/j.ijhm.2011.09.018

Leung, D., Law, R., Van Hoof, H., & Buhalis, D. (2013). Social media in tourism and hospitality: A literature review. *Journal of Travel & Tourism Marketing*, *30*(1-2), 3–22. doi:10.1080/10548408.2013.750919

Levickaite, R. (2010). Generations X, Y, Z: How social networks form the concept of the world without borders (the case of Lithuania). *LIMES: Cultural Regionalistics*, *3*(2), 170–183.

Mangold, W. G., & Faulds, D. J. (2009). Social media: The new hybrid element of the promotion mix. *Business Horizons*, *52*(4), 357–365. doi:10.1016/j.bushor.2009.03.002

Moise, D., & Cruceru, A. F. (2014). An empirical study of promoting different kinds of events through various social media networks websites. *Procedia: Social and Behavioral Sciences*, *109*, 98–102. doi:10.1016/j.sbspro.2013.12.426

Nezakati, H., Amidi, A., Jusoh, Y. Y., Moghadas, S., Aziz, Y. A., & Sohrabinezhadtalemi, R. (2015). Review of social media potential on knowledge sharing and collaboration in tourism industry. *Procedia: Social and Behavioral Sciences*, *172*, 120–125. doi:10.1016/j.sbspro.2015.01.344

Ngernyuang, K., & Wu, P. Y. (2020). Using social media as a tool for promoting festival tourism. *AIRCC's. International Journal of Computer Science and Information Technologies, 12*(3), 17–32. doi:10.5121/ijcsit.2020.12302

Nikolaeva, J., Bogoliubova, N., Fokin, V., Baryshnikov, V., Klevtsov, P., & Eltc, E. (2017). Cultural mega-events as an international, cultural and political tool. *Codrul Cosminului, 23*(2), 291–304.

Ohajionu, U. C., & Mathews, S. (2016). Social media and advertising: An exploratory study on the lifestyle of generation Y consumers in Malaysia. *TIJ's Research Journal of Social Science & Management, 6*(1), 31–39.

Oklobdžija, S. (2015). The role of events in tourism development. *Bizinfo (Blace), 6*(2), 83–97. doi:10.5937/BIZINFO1502083O

Paris, C. M., Lee, W., & Seery, P. (2010, February 10-12). The role of social media in promoting special events: acceptance of Facebook 'events'. [Conference presentation]. ENTER2010 eTourism Conference, Lugano, Switzerland.

Quan, S., & Wang, N. (2004). Towards a structural model of the tourist experience: An illustration from food experiences in tourism. *Tourism Management, 25*(3), 297–305. doi:10.1016/S0261-5177(03)00130-4

Quinn, B. (2009). Festivals, events and tourism. Festivals, events and tourism. In T. Jamal & M. Robinson (Eds.), *The SAGE Handbook of Tourism Studies* (pp. 483–503). Sage. doi:10.4135/9780857021076.n27

Sever, N. S., & Ataizi, M. (2017). Yeni medya ortam ve araçlarının bireylerin yeni yerler öğrenmeleri ve deneyimlemelerine katkıları üzerine öneriler. *Kurgu, 25*(1), 51–64.

Shone, A., & Parry, B. (2004). *Successful event management: A practical handbook.* Thomson Learning.

Silvers, J. R. (2004). *Professional event coordination.* John Wiley and Sons.

Simms, A. (2012). Online user-generated content for travel planning - different for different kinds of trips?. *e-Review of Tourism Research,* 10(3), 76-85.

Soteriades, M. D., & Dimou, I. (2011). Special events: A framework for efficient management. *Journal of Hospitality Marketing & Management, 20*(3-4), 329–346. doi:10.1080/19368623.2011.562418

Unurlu, C. (2021). The Integration of social media Into Event Tourism. In *Impact of ICTs on Event Management and Marketing* (pp. 69–85). IGI Global. doi:10.4018/978-1-7998-4954-4.ch005

Valls, J. F., Ouro, A., Freund, D., & Andrade Suárez, M. J. (2012). Analysis of social media platforms and their potential value for the tourism industry. 2nd Advances in Hospitality & Tourism Marketing & Management Conference (AHTMM) 2012, Corfu (Greece).

Varkaris, E., & Neuhofer, B. (2017). The influence of social media on the consumers' hotel decision journey. *Journal of Hospitality and Tourism Technology, 8*(1), 101–118. doi:10.1108/JHTT-09-2016-0058

Verma, R., Stock, D., & McCarthy, L. (2012). Customer preferences for online, social media, and mobile innovations in the hospitality industry [Electronic version]. *Cornell Hospitality Quarterly, 53*(3), 183–186. doi:10.1177/1938965512445161

Yavuz, G., & Enes, K. (2021). Social Media and Events: Before, During, and After. In Impact of ICTs on Event Management and Marketing (pp. 139-155). IGI Global.

Zeng, B., & Gerritsen, R. (2014). What do we know about social media in tourism? *A review. Tourism Management*, *10*, 27–36.

ADDITIONAL READING

Hossain, S. F. A., Ahsan, F. T., Nadi, A. H., Ahmed, M., & Neyamah, H. (2022). Exploring the Role of Technology Application in Tourism Events, Festivals and Fairs in the United Arab Emirates: Strategies in the Post Pandemic Period. In *Technology Application in Tourism Fairs, Festivals and Events in Asia* (pp. 313–330). Springer Singapore. doi:10.1007/978-981-16-8070-0_19

Karacaer, S. S. (2022). The Role of Social Media in Event Marketing: Outcomes for Practitioners and Participants. In *Research Anthology on Social Media Advertising and Building Consumer Relationships* (pp. 967–986). IGI Global. doi:10.4018/978-1-6684-6287-4.ch053

Rahman, M. K., Hossain, M. M., Akter, S., & Hassan, A. (2022). Technology innovation and social media as a global platform for tourism events. In *Technology application in tourism fairs, festivals and events in Asia* (pp. 121–132). Springer Singapore. doi:10.1007/978-981-16-8070-0_7

Turna, G. B. (2022). Impact of social media on tourism, hospitality and events. In *Handbook on Tourism and Social Media* (pp. 475–488). Edward Elgar Publishing.

Yavuz, G., & Enes, K. (2022). Social media and events: Before, during, and after. In *Research Anthology on Social Media Advertising and Building Consumer Relationships* (pp. 106–122). IGI Global. doi:10.4018/978-1-6684-6287-4.ch008

Chapter 12
Conceptual Framework to Position Devika River (Gupt Ganga) as a Religious Festival Tourist Destination

Biswabhusan Pradhan
Central University of Jammu, India

Ayush
Central University of Jammu, India

Krishma Rana
Central University of Jammu, India

Amit Gangotia
Central University of Jammu, India

ABSTRACT

In Hinduism, rivers are considered holy and sacred as they have various historical, religious, and social importance. The aim of this research is to explore the historical, religious, and socio-economic importance of the Devika River (Gupt Ganga) and the surrounding area and its potential for being developed as a religious festival tourism destination. This study employed qualitative method to explore the potential of research sites by using both primary data and secondary data. The results of this study revealed that the conservation of rivers is very important to save this culture and history. Also, the importance of festivals for the regional development through tourism was highlighted. Furthermore, it has been found that Devika River and the surrounding area has enormous potential to be developed as a religious festival tourist destination, but lack of planning and support from the government and minimal participation from other tourism stakeholders has always deprived this destination from exploring its true potential.

DOI: 10.4018/978-1-6684-6356-7.ch012

INTRODUCTION

India is known as the land of rivers, as the nation itself is named after the river 'Indus,' the Indus valley has an historical significance as Alexander the Great had camped on the banks of the river Beas, among the five important rivers of Punjab in northwest India (Agoramoorthy, 2015). Other significant river of the country are Ganges in the north part, Cauvery and the Godavari in the south, and Narmada, which creates a natural barrier between north and south, the Ganges, or Ganga, is India's national river and the most respected of all rivers (Keay, 2000). Understanding the cultural, religious, and societal value of rivers has therefore become vital and only then can essential limits be put in place to decelerate river exploitation that exceeds nature's threshold, human wisdom has long recognized the necessity for restrictions to steer human activities(Quinn, 2010), over millennia, the ancient Hindu texts have emphasized the need of such constraints (Agoramoorthy, 2015). However, Jammu and Kashmir region of India is famed for its natural beauty and as a heaven for its god given natural treasures. River Devika is a small holy river that flows through Udhampur district and has immense religious and cultural significance in the region; it is regarded as the elder sister of the holy Ganga, the river continues to flow through the village Mandal in the district of Samba, Jammu and Kashmir (Pundit Anil Goswami, personal communication, Dec 05, 2021). As many researchers highlighted the relevance of rivers in many contexts (Agoramoorthy, 2015; Gupta & Gulla, 2010; Sharpley & Sundaram, 2005; Singh, 2001), it led to the exploration of the river's untapped latent potential for sustainable development. Current study aims to highlight the untapped and undervalued hidden potential of Devika River.

LITERATURE REVIEW

Rivers have played a significant role in human history since the beginning of time and have been enlarged by civilization (Wang & He, 2022). Rivers have been essential to human survival, modernization, and, more recently, economic prosperity (Anim & Titus, 2012; Do Thi & Dombroski, 2022; Lee et al., 2022). Rivers supported long- and short-distance transit, trade, and hunting in ancient times and also rivers enabled great civilizations of the past to thrive (Akpan & Obang, 2012; Spoolman, 2022). Cities and cultural landscapes in the contemporary world can be categorized according to the fabric of river life to generate unique urban settings (Davis, 2020; Dowling, 2006). Rivers are among the most significant natural resources for tourism and pleasure. They, like most other waterways, have become popular destinations for water sports and other activities, that attracts a huge group of individuals to travel long distances in order to get a sense of it (Timothy, 2009). As Prideaux et al. (2009) illustrated how the River system long-term stability that assist tourism is critical but has been a major neglected subject of tourism and hospitality study until now. Rivers have a variety of important direct and indirect functions in tourism activities. Direct roles consists of:- place for activities and tourism attractions, including natural and built environments; transportation, including river cruising and cargo barges; recreational and pleasure activities including fishing and water sports and supply of potable drinking water, while indirect contribution consists of: food source, either directly through the supply of fish and other food sources or indirectly through the agriculture that the river supports. As the holy river Ganga is regard as Ganga Mata in India, the Ganges and its tributaries are intrinsically tied with Hinduism, India's primary religion (Agoramoorthy, 2015). The ancient Vedic scriptures, which constitute the heart of Hindu religion, were penned on the banks of this river, as people began to migrate from the Indus Valley to the Ganges plains towards the end of the Indus Valley civilization era(Cooper, 2009; Darian, 1978; Sen, 2022), The river Ganga is vital to the clean Hindu way of life since it is thought that the soul of Hindu people is not liberated from earth after death until the deceased's ashes are immersed in the Ganges' holy waters (Alley, 2002; Cooper, 2009; Flood, 2003; Reiss et al., 2001).

METHODOLOGY

This study used descriptive methods to investigate the tourism resources in the area surrounding the Devika River. The data utilized to support the study are based on primary data and secondary data. To acquire primary data, data collection procedures such as field observation, unstructured interviews were used. Secondary data is gathered from books, scientific publications, and articles on the Internet.

FINDINGS

Devika River (Gupt Ganga)

River Devika is a small holy river flowing mainly through Udhampur district which has got very great religious and cultural importance to the region; it is considered as the elder sister of holy Ganga. The river originates from Sudh Mahadev then flows through Chenani- Udhampur -Jandra-Purmandal-Uttar Vehini- Vijapur-Basantar and finally contributes to the river Ravi and then flows to Pakistan (Mr. S.M Sahni, Convener, INATCH, Jammu Chapter explained during the personal interaction with him on Dec 05, 2021).

The river is not visible on ground and mostly flows below the ground. "The river originates from a spot where the wife of the ruler of the mountain (Himalaya) Menaka gave birth to a kid known as Mainak; soon after its emergence, Lord Mahadeva dispatched her to the sea," according to the Kalika Purana's mythology. Deeg, a right bank tributary of Ravi, has been recognised as this river. Some scholars, however, tried to identify it with Deva or Devika, a tributary of the southern bank of river Sarayu, the river is known by various names which are: Devak, Devika, Gupt Ganga, Deva (Patowary, 2015). Devika has got great religious importance too. Besides cremating the dead, Hindus from neighboring villages immerse the ashes of the deceased in the holy water of this river, besides performing the last rituals many people also perform rites related to Kriya (yearly death anniversary) of their relatives, along with various Hindu religious and cultural festival takes place throughout the year near the bank of the river and area surrounded by it. (Pundit Anil Goswami, personal communication, Dec 05, 2021).

Religious and Historical Significance

Religious tourism is practiced in order to get ibrah or teachings and learn more about God's creation or the history of human civilization, as well as to open hearts in order to raise awareness that life in this world is not forever (Kim et al., 2020). In India, travel, particularly religious tourism, has a long history ; "if travel and tourism are ways of enlightening the spirit . . . they have been practiced for centuries in India, whose peoples have long expressed a common sense of community in organized religion" (Singh, 2001, p.138). However, tirthayatra, or pilgrimage, has been a part of Indian social life from ancient times (Sharpley & Sundaram, 2005). According to ancient saying Rishi Kashyap, after whom Kashmir is named, prayed to Lord Shiva for twelve years but didn't pose anything for him but plead Lord Shiva for a place where people can cleanse their soul just by taking a bath. Lord Shiva, along with Goddess Parvati, said that Ganga, Yamuna and Saraswati have already settled at their places, hence leaving no place for them. Then Rishi Kashyap requested the Goddess to come in the form of a river. Goddess Parvati agreed although with a condition for Lord Shiva she spoke, "I can go and will go down to earth dear lord, but, I will stay hidden from plain sight (Gupt roop) and wherever come out, there should be a temple of yours." ("The 2600-Year-Old History of Purmandal Temple," n.d.). However, Vaishno Devi Shrine is one of the most prominent religious places in Northern India, with millions of pilgrims visiting each year, these pilgrims come from all over India, driven by their religious beliefs reasons. People considered that Vaishno Devi as the second holiest destination after Chaar Dhaam Yatra (Gupta & Gulla, 2010). The distance between Purmandal and Vaishno Devi temple is about 92KM, despite great potential of religious tourism and natural landscape this destination still has not come to the notice of visitors so, intervention of concerned stakeholders can develop it as a religious tourism destination (Mr. S M Sahani (JKAS Retd.) Ex. Director Tourism, Jammu, personal communication, Dec 05, 2021). During interview Mr. S M Sahani (JKAS Retd.) Ex. Director Tourism, Jammu answer to the question- "can Purmandal and the surrounding areas along the Devika River be developed as a religious tourist attraction after Vaishno Devi, Katra" he said:

...yes, The initial development of the Vaishno Devi temple and the surrounding area was in bad shape, but after a few years of development, that area became a very popular tourist destination in India. Similarly, Purmandal and the area surrounded by the Devika River (Gupt Ganga) can be a very appealing religious tourism destination because the area is full of various types of tourism resources other than the temple and the holy river, so if concerned stakeholders' involvement occurs, the area Purmandal and the area surrounded can be developed as an indigenous or spiritual tourism circuit.

Purmandal Temple (Jammu)

Located thirty kilometers to the east of Jammu on the bank of Devika River, Purmandal Temples are pious and historic temples. This holy site is also referred as Uttar - Ki – Kashi/Chhota Kashi. Thousands of devotees visit this revered religious attraction in Jammu & Kashmir all year round. The presiding divinity of the temple is Goddess Parvati. 2500 years old Shiva- Kund is also prominent in the temple. The temple is extremely praised by the visitors for its outstanding stone architecture. To celebrate the holy union of Lord Shiva and Devi Parvati, Purmandal Mela (Shivratri) is celebrated every year during February/March (Purmandal Temple Jammu | Its History, Mythology & Purmandal Fair | Jammu & Kashmir, n.d.; Mr. S.M Sahni, personal communication, Dec 05, 2021). Also some other prominent fairs and festivals are observed throughout the year, these include – Chaitra Chudashi (In the mounth March), Makar Chaudashi (In the month of December), Samapati Amavasya (twice or thrice a year) and a monthly fair is also observed every Amavasya. (Pundit Anil Goswami, personal communication, Dec 05, 2021).

History of Purmandal Temple

Purmandal Temple is a historic stone temple. The temple is built by Raja Avanti Verman, also known as "Venidutt", who was the ruler of Avantipur princely state, presently in Pulwama district of Jammu and Kashmir. It is also famous, as is believed to be visited by Guru Nanak Dev, Maharaja Ranjit Singh, Ustad Bismillah Khan and Saint Kabir. The temple however has recent history. Raja Avanti Verman used to hunt for days. After many unprolific days, the king reached 'Pur' near Mandal (Purmandal) and shot an arrow at a female Jackal. The female Jackal however did not die instantly and hid from the hunters. It went to the "Swayambu-Kund" (today the Shiv Ling – Naag Dev Kund) died there after drinking water. For a long time, the monarch had no offspring until a girl kid was born into his family. The child, on the other hand, used to cry all of the time. When the boy grew older, it was found that he suffered from chronic headaches. The priests were called after seeing no physical problem. The Priests then prophesied that the child cries because you have killed an animal that has died at a spot dedicated to Lord Shiva. In order to solve this issue, you will have to find the animal, remove the arrow and perform its rites. After this the soldiers retrieved the skeleton and the girl cured as soon as the arrow was pulled out ("The 2600-Year-Old History of Purmandal Temple," n.d.). Another prominent temple in the premises is the "Ekadashi Rudra Pindi Temple" which is built by Raja Maan Singh of Jaipur, a total of 121 temples were constructed by Maharaja Gulab Singh hence attaining the name of Uttar-Ki-Kashi, in the pre-independence era the Vaishnodevi Yatra used to commence from Purmandal only (Mr. S.M Sahni, personal communication, Dec 05, 2021; Pundit Anil Goswami, personal communication, Dec 05, 2021).

Significance of Festival for Tourism Development

Festivals are important in the tourism industry as they attract visitors, they are key tourism assets in a specific location, as well as tourist products that meet the demands of tourists (Cudny, 2013). Quinn (2010) furthermore says that tourists are drawn to festivals, which leads to the development of festival tourism. Many factors entice visitors, the most prominent of which appears to be the intriguing, often exotic culture displayed during events. Tourists also want to attend uncommon events, meet individuals with similar interests, and learn more about the globe. As the importance of festival for the tourism development emphasized in previous research, Purmandal Mela has a great potential to contribute for tourism development in various ways. Because "Purmandal Mela is associated with a fascinating legend, according to which, Lord Shiva was a homeless person who went to the Himalayas, begging from door to door. Goddess Parvati, on the other hand, was raised amidst prosperity and surrounded by riches of every kind. King Himalaya had no objection in getting his daughter married to a beggar but was unaware of the fact that Lord Shiva had renounced all worldly desires and materialistic comforts. Goddess Parvati, however, remained insistent, and ultimately persuaded her father in agreeing to the marriage. So in order to celebrate the holy union, Purmandal Mela is celebrated every year" (Purmandal Temple Jammu | Its History, Mythology & Purmandal Fair | Jammu & Kashmir, n.d.). However, Purmandal Mela has always been deprived of adequate attention from concerned stakeholders, hence there is a need for stakeholders to provide proper attention to position and promote on primary basis for regional development through tourism.

Socioeconomic Importance

Tourism is seen as a driver for poverty reduction and provides economic benefits to local communities and other stakeholders, tourism generates major economic advantages for local people in alpine locations. In order to help economic development, most governments encourage the rise of tourism in their own nations(Dada et al., 2022; Ji et al., 2022; Lagos & Wang, 2022). Binns & Nel (2002) investigated the effects of tourism in South Africa and discovered impressive growth and favorable implications on local communities' economic prosperity. Tourism is viewed as beneficial to poor nations, regions, towns, and cities as a means of accelerating development (Mbaiwa, 2003). As the Devika River has a footfall of the domestic tourist because of its mythological importance, local residents operate small shops on the river bed for the devotees The river flows mostly underground so there is no or minimal usage of river water of the holy river however the water is used only for the Holy Bathes and last ritualistic ceremonies.

Figure 1. Religious Tourism Model of Purmandal

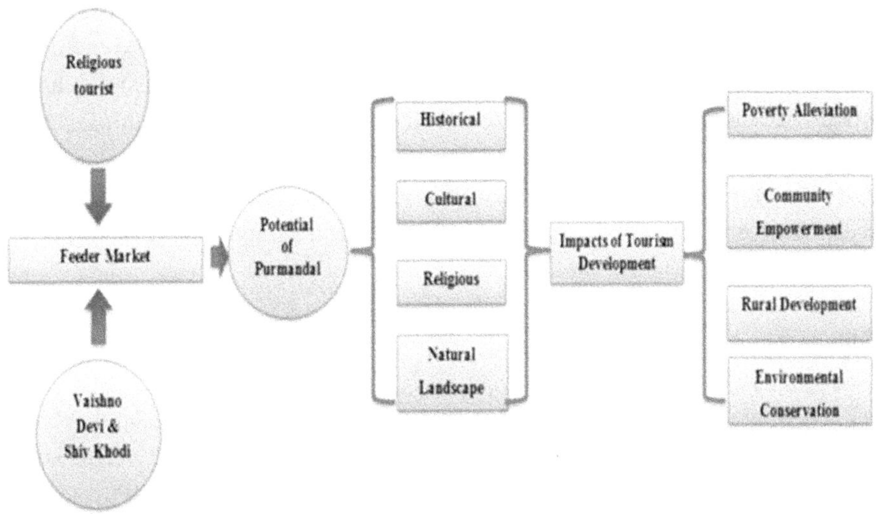

DISCUSSION

Various researchers have emphasized the importance of rivers in various dimensions such as location, potential, benefit, historical, religious, socioeconomic, and so on (Agoramoorthy, 2015; Dowling, 2006; Prideaux et al., 2009; Timothy, 2009). This study has attempted to highlight the potential of Devika River and area surrounded by it in terms of historical, religious, socio-economic importance to develop as a religious tourism destination. However, Tourism scholars have argued that tourism improves the social standards of any given community; induces change in both the environment and the lives of members of that community if the right approach is employed (Jude et al., 2018). Vukonić (1998) said pilgrims have made significant contributions to the idea of religious unity; they have made significant contributions to the prosperity of pilgrimage sites and larger area surrounding them, particularly along the routes leading to them, Pilgrimages and religious tourism have an impact on the demography of the pilgrimage site and its surroundings. This influence is manifested first and foremost in employment opportunities, which lead to community development. Karar (2010) highlighted that religious tourism has the great potential of generating revenue compare to what other form of tourism does, also explained that religious tourism has a significant advantage over other types of tourism owing to the attraction of large crowds in the form of visitors. Subedi (2015) explained tourism as a vital instrument for reducing rural poverty, raising living standards, and promoting conservation and national development, but only if it is understood deeply and broadly, tourism is essential to a country's economic prosperity, tourism is India's second greatest earner of foreign currency, tourism business employs a vast number of people, both skilled and unskilled it fosters national cohesion and worldwide fraternity. Tourism is a key catalyst in modern socioeconomic growth, helping in a variety of ways and strengthening the interconnected processes. While frequently promoted as a remedy for various issues such as underdevelopment, unemployment, poverty eradication, social injustice, and so on, the socioeconomic influence of religious tourism may be observed in its contribution to a vast population of the nation, with the biggest contributions to religious tourism found in the various religious locations in India (Srivastava, 2016). The Devika River and the area surrounding it, specifically Purmandal, have huge feeder market of religious tourists who are coming to Mata Vaishno Devi Shrine and Shiv Khodi in big number, if these religious tourists are provider with an alternative & valuable religious tourist site then it has great potential to generate multifold benefits to the destination. The hidden potential of Purmandal such as historical, cultural, social, and natural land scape can be tapped with the proper infrastructure development, marketing, planning and its execution. The destination holds a tremendous potential as a tourism resource and can be converted into valuable tourism product. This can result in poverty reduction, community development, job creation, and environmental benefits. More broadly, it will contribute to the region's long-term growth.

CONCLUSION

Tourism is a vital instrument for reducing rural poverty, raising living standards, and promoting conservation and national development, but only if it is understood deeply and broadly. This study attempted to explore the potential of Devika River and area surrounded by it particularly Purmandal to develop as a tourist destinations. This study employed qualitative method to explore the potential of research sites by using both primary and secondary data. This study used descriptive methods to investigate the tourism resources in the area surrounding the Devika River. Various tourism destinations in India have been organizing festivals to attract tourists, turning festivals into important tourist attractions However, existing research highlights the relevance of rivers in many contexts, it led to the exploration of the river's untapped latent potential for sustainable development. However, based on qualitative research, we determined the four potential dimensions of Purmandal in terms of historical, cultural, religious and natural landscape. Thus, the impact of tourism development result in poverty alleviation, community empowerment, rural development and environmental conservation. Moreover, the overall findings of this study highlight the necessity of uncovering and bringing to light the river's latent potential, which may benefit society and the environment in a multitude of ways as rivers are among the most significant natural resources for tourism activity and pleasure. They, like most other waterways, have become popular destinations for water sports and other activities. That attracts a huge group of individuals to travel long distances in order to get a sense illustrating how the sustainability of the river systems that assist tourism is critical but has been a major neglected subject of tourism study until now. Furthermore, the findings of this study also highlight the various benefits of tourism development in study area includes poverty reduction, community development, job creation, and environmental benefits. More broadly, it will contribute to the region's long-term growth. The tourism business employs a vast number of people, both skilled and unskilled. To avail all this benefit conservation of river is very important to save our culture and history as it has been found that Devika River and area surrounded by the river has enormous potential to develop as religious festival tourism destination.

IMPLICATION

Policymakers can begin the process of involving the local community and the stakeholders in the process of sensitization, keeping in mind the Triple Bottom Line Measure of carrying out ethical and sustainable practices of tourism. This will allow the policymakers to sensitize the local community and the stakeholders regarding the cultural and natural environment of the area that is surrounded by the Devika River. There is ample of scope of festival tourism in the region that is encircled by the Devika River. The promotion of this region as religious festival tourist destination will certainly generate multiple benefit for local community. In terms of economic and cultural empowerment. The policy maker needs to ensure the involvement of local community in all decision making process in order to develop the destination holistic manner. Though there is a dire need of sensitizing and capacity building of all related stakeholders which can be done with the help of proper planning.

REFERENCES

Agoramoorthy, G. (2015). Sacred rivers: Their spiritual significance in Hindu religion. *Journal of Religion and Health*, *54*(3), 1080–1090. doi:10.100710943-014-9934-z PMID:25183514

Akpan, E. I., & Obang, C. E. (2012). Tourism: A strategy for sustainable economic development in Cross River State, Nigeria. *International Journal of Business and Social Science*, *3*(5).

Alley, K. D. (2002). *On the banks of the Gaṅgā: When wastewater meets a sacred river*. University of Michigan Press. doi:10.3998/mpub.12072

Anim, O., & Titus, E. (2012). The relevance of tourism on the economic development of Cross River state, Nigeria. *Journal of Geography and Regional Planning*, *5*(1), 14–20. doi:10.5897/JGRP11.122

Binns, T., & Nel, E. (2002). Supporting local economic development in post-apartheid South Africa. *Local Economy*, *17*(1), 8–24. doi:10.1080/02690940110073800

Cooper, M. (2009). In B. Prideaux & M. Cooper (Eds.), *River tourism in the south Asian subcontinent. River Tourism* (pp. 23–40).

Cudny, W. (2013). Festival tourism–the concept, key functions and dysfunctions in the context of tourism geography studies. *The Geographical Journal*, *65*(2), 105–118.

Dada, Z. A., Najar, A. H., & Gupta, S. K. (2022). Pro-Poor Tourism as an Antecedent of Poverty Alleviation: An Assessment of the Local Community Perception. *International Journal of Hospitality and Tourism Systems*, *15*(1).

Darian, S. G. (1978). The Ganges in myth and history. University of Hawaii Press. .

Davis, W. (2020). Magdalena: River of dreams. Random House.

Do Thi, H., & Dombroski, K. (2022). Diverse more-than-human approaches to climate change adaptation in Thai Binh, Vietnam. *Asia Pacific Viewpoint*, *63*(1), 25–39. doi:10.1111/apv.12329

Dowling, R. K. (2006). *Cruise ship tourism*. Cabi. doi:10.1079/9781845930486.0000

Flood, G. (2003). The Saiva Traditions. *The Blackwell Companion to Hinduism*, 200–228.

Gupta, K., & Gulla, A. (2010). Internet deployment in the spiritual tourism industry: The case of Vaishno Devi Shrine. *Worldwide Hospitality and Tourism Themes*, *2*(5), 507–519. doi:10.1108/17554211011090120

Ji, Z., Xu, H., & Cui, Q. (2022). Tourism and poverty alleviation in Tibet, China: The role of government in enhancing local linkages. *Asia Pacific Journal of Tourism Research*, *27*(2), 173–191. doi:10.1080/10941665.2021.2020312

Jude, O. C., Uchenna, O., & Ngozi, E. (2018). Impact of religious tourism in host communities: The case of Awhum Monastery. *American Journal of Social Sciences*, *6*(3), 39–47.

Karar, A. (2010). Impact of pilgrim tourism at Haridwar. *The Anthropologist*, *12*(2), 99–105. doi:10.1080/09720073.2010.11891138

Kim, B., Kim, S., & King, B. (2020). Religious tourism studies: Evolution, progress, and future prospects. *Tourism Recreation Research*, *45*(2), 185–203. doi:10.1080/02508281.2019.1664084

Krakhmal, K. (2023). To The Study Of The Origins Of The Origin Of Local Civilizations In The Length Of The Syrdarya River. *Galaxy International Interdisciplinary Research Journal*, *11*(2), 208–218.

Lagos, K., & Wang, Y. (2022). International tourism and poverty alleviation: Cross-country evidence using panel quantile fixed effects approach. *Journal of Travel Research*, 00472875221119978. doi:10.1177/00472875221119978

Lee, C.-C., Zeng, M., & Luo, K. (2022). The Impact of Green Development on Modernization in China: Evidence from 108 Cities in the Yangtze River Economic Belt. *Emerging Markets Finance & Trade*, *58*(13), 3664–3688. doi:10.1080/1540496X.2022.2067475

Mbaiwa, J. E. (2003). The socio-economic and environmental impacts of tourism development on the Okavango Delta, north-western Botswana. *Journal of Arid Environments*, *54*(2), 447–467. doi:10.1006/jare.2002.1101

Patowary, D. K. N. (n.d.). *The Kalika Purana and the Yogini Tantra: A Comparative Study in Reference to Great Indian Rivers*. 4.

Prideaux, B., Timothy, D. J., & Cooper, M. (2009). Introducing river tourism: Physical, ecological and human aspects. *River Tourism, 1*.

Tour My India. (n.d.). Purmandal Temple Jammu: Its History, Mythology & Purmandal Fair. *Tour My India*. https://www.tourmyindia.com/states/jammu-kashmir/purmandal.html

Quinn, B. (2010). Arts festivals, urban tourism and cultural policy. *Journal of Policy Research in Tourism, Leisure & Events*, *2*(3), 264–279. doi:10.1080/19407963.2010.512207

Reiss, J., Abbott, C., & Gold, S. F. (2001). SACRED WATERS: A Pilgrimage Up the Ganges River to the Source of Hindu Culture (Book Review). *Publishers Weekly*, *248*(42), 68–68.

Sen, C. T. (2022). *Ashoka and the Maurya Dynasty: The History and Legacy of Ancient India's Greatest Empire*. Reaktion Books.

Sharpley, R., & Sundaram, P. (2005). Tourism: A sacred journey? The case of ashram tourism, India. *International Journal of Tourism Research*, *7*(3), 161–171. doi:10.1002/jtr.522

Singh, S. (2001). Indian tourism: Policy, performance and pitfalls. *Tourism and the Less Developed World: Issues and Case Studies*, 137–149.

Spoolman, S. (2022). Wisconsin Waters: The Ancient History of Lakes, Rivers, and Waterfalls. Wisconsin Historical Society.

Srivastava, S. (2016). Socio economic impact of religious tourism-a case study of mahakumbh mela 2013 at Allahabad. *International Journal of Management and Social Sciences*, *4*(1), 205–212.

Steinbach, J. (1995). River related tourism in Europe—An overview. *GeoJournal*, *35*(4), 443–458. doi:10.1007/BF00824354

Subedi, Y. (2015). Prospects and Problems of Religious Tourism in Nepal A Case Study of RamnadiDham (Ramdi), MalungaTunibotVDC, Syangja District, Nepal [PhD Thesis, Central Department of Rural Development Tribhuvan University, Kathmandu].

The 2600-year-old history of Purmandal Temple. (n.d.). U4U-Voice. http://u4uvoice.com/the-2600-year-old-history-of-purmandal-temple/

Timothy, D. J. (2009). River-based Tourism in the USA: Tourism and Recreation on the Colorado and Mississippi Rivers. *River Tourism*, 41–54.

Vukonić, B. (1998). Religious tourism: Economic value or an empty box? *Zagreb International Review of Economics & Business*, *1*(1), 83–94.

Wang, H., & He, G. (2022). Rivers: Linking nature, life, and civilization. *River*, *1*(1), 25–36. doi:10.1002/rvr2.7

Yu, S.-Y., Li, W.-J., Zhou, L., Yu, X., Zhang, Q., & Shen, Z. (2023). Human disturbances dominated the unprecedentedly high frequency of Yellow River flood over the last millennium. *Science Advances*, *9*(8), eadf8576. doi:10.1126ciadv.adf8576 PMID:36812330

Chapter 13
Destination Marketing and Branding of Nashik:
A Case Study of Sula Wine Festival

Gayatri Gawde
Amity University, Mumbai, India

Sachin Kumar Behera
Amity University, Mumbai, India

Sharad Kumar Kulshreshtha
Department of Tourism and Hotel Management, North-Eastern Hill University, Shillong, India

ABSTRACT

This chapter presents a case study, highlighting the importance of wine tourism with its connection with the Sula Wine Festival that happens in Nashik, Maharashtra. This case study presents an exploratory study of the Sula Wine Festival, with a focus on educating readers on its prominent features, USP, and attractions for the people interested in wine festivals. The study further explains some of the distinct advantages of Nashik such as its soil, location, and wide range of attractions nearby. The key objective of the case study is to make the readers aware of the power of wine festivals in tourism and encourage the blending of tourism and wine consumption for marketing and branding Nashik.

INTRODUCTION

UNWTO asserts that wine tourism, also called eno-tourism, is tourism that involves going to vineyards and wineries to taste, drink, and/or buy wine, most often at it or near the source.

Many formerly underserved regions have started to realise the full potential of wine tourism as a source of income for their local economies as more and more cities draw wine visitors. As a result, both the regional economy and the destination's brand are strengthened. The key elements of wine tourism

DOI: 10.4018/978-1-6684-6356-7.ch013

offerings are wine, gastronomy, art, and festivals, and they provide the kind of lifestyle that wine tourists typically aspire to.

The case study in this chapter, which relates to the Sula Wine Festival in Nashik, Maharashtra, emphasises the significance of wine tourism. The Sula Wine Festival is the subject of this case study, which aims to inform readers about its key characteristics, USP, and draw for wine festival enthusiasts. The study goes on to discuss some of Nashik's unique advantages, including as its soil, location, and plenty of local attractions. The case study's main goal is to raise readers' awareness of the importance of wine festivals in the tourist industry and to promote the fusion of tourism and wine consumption for the marketing and branding of Nashik.

ENOTOURISM

Considered as an exciting form of Special Interest Tourism, Wine Tourism also known as Eno Tourism is emerging globally (Bibicioiu & Cretu, 2013). Wine Tourism has a vital contribution for initiating regional development resulting into higher employment and economical advancement of the area. This can be achieved through effecting Destination Management techniques.

As a result, the wine industry has seen a significant increase in the number of Wine Festivals and Events. To celebrate their love of wine, as well as other leisure activities, Wine Tourists may visit this location. Attracting as many tourists as possible to wine tourism destinations is essential to the development of the destination's image, and wine festivals play an important part in that process. One of the fastest growing kinds of leisure and wine tourism has been events like these, which are also a big factor in the region's tourist growth. Aside from the fact that wine festivals help promote the Wine Tourism Destination, they also help increase the sales of wine brands and generate loyal clients for the participating wineries. Since there are so many festivals and events to choose from, the destination's image is enhanced greatly. These vineyards have recently begun catering to destination weddings as a result of rising demand. This also boosts tourist traffic and enhances the value of culturally focused items.

The services provided by wine tourism, such as wine festivals and tours, lead to close bonds between winemakers and patrons, which later translate into direct or indirect sales via word-of-mouth marketing. Increased economic benefits emerge from these direct or indirect sales. As a result, more tourists visit the wineries, which benefits from the added by-products. Additionally, the increased tourist traffic raises brand awareness and associations in customers' thoughts. As a result, these customers have a tendency to be brand loyal, leading to the development of a relationship between winemakers and customers for the purchase of these by-products. Along with these advantages, the proliferation of wineries and investment into the industry have made wine tourism a significant aspect of the tourism sector.

WINE TOURISM IN INDIA

With globalization and increasing economy of India, Indian wine industry is growing at a CAGR of 20-25 percent and estimated to US$150 million. Uniformity in wine laws, new regulatory policies, tax incentive and increasing spending power of middle class widely spotting wine is a lifestyle drink and increased acceptability in India (Kumar, 2021).

While wine consumption in India is currently very low (an estimated annual per capita consumption of less than 10 millilitres), the global annual per capita consumption of wine is around four litres compared to this, but the picture is rapidly changing. The Indian wine industry reported a double-digit annual compound growth rate of over 14 percent between 2010 and 2015 (Economic Times, Mar. 15, 2017).

Indian Wine Tourism market has seen 18% growth over few years, according to PwC, executive director of Corporate Finance and Investment Banking, Anand Dikshit. The Indian wine market currently has a value of nearly 17 million litres of Rs. 1400 crore. In line with other specialists and Total wine sales in India may increase to 8.3 million cases (one case equal to 9 litres) by 2020, according to other experts and industry players. According to Rajeev Samant, Sula Vineyard's founder and CEO, India's leading wine company, and Kapil Grover, Grover Vineyard's second-largest wine producer, the Indian wine market is expected to grow to 25 percent YoY over the next 25 years.

For the last several years, India has experienced an increase in wine tourism as prominent companies from all around the nation recognize its potential. As a whole, tourists nowadays are taking a deep interest in how best to drink or taste a wine, which wine matches best with what meal, and so on. For Indians, this has now become a major fashion statement. There are picture booths, excellent local and continental food, and large open areas at the major wine sites. As a result, all wineries strive to engage their customers with a variety of events, including wine tours, food pairings, grape stomping, and, of course, wine tastings. These offerings serve as an excellent introduction to the winery's goods, thus enhancing the brand's reputation.

There are a variety of causes driving the growth of Wine Tourism in India, including a growing need for new travel experiences and a willingness to pay to meet those needs, as well as a sense of honour among Indian tourists to visit these kinds of locations.

Wine Tourism results into regional development and the upliftment of society through the Indian wine brands and its awareness amongst the consumers. Hence it is one of the latest lifestyle trends in India. Furthermore, one of the mainstream factors of influence being international travel of certain segments such as upper middle class, elites, etc., led to diversification of new taste palates hence also resulting into introducing new and updated cuisines or fusion foods into the daily consumptions. Finally, this boosted wine tourism and often associate it with relaxation, communicating with loved ones, and exploring the wines in a delightful environment. Hence, Wine Tourism is also considered in one of the Experiential sectors of Tourism Industry. Nevertheless, wine tourism was recently introduced, India as a longstanding tradition of wine consumption by the Kings and other authorities and later dissipated during colonial period.

As discussed, the socio-economic benefits from these wineries for the local communities are getting benefited at both social and economic level. Wine Tourism has been expanding resulting into fast Gross Domestic Product development, flourishing metropolitan population, all resulting into changing lifestyles. These Wineries use various push – pull strategies for the brand establishment. The aesthetics of these wineries, attributes towards the experiential part of marketing resulting into more tourist footfalls, and Brand awareness. Furthermore, they foster a relationship with the consumers with regards to wine tourism and its by-products resulting into a new experience of the destination. Concluding, Destination marketing and brand establishment are crucial factors promoting wine tourism growth. Furthermore, wineries use these strategies to build their brands image and loyalty.

Considering the growth of wine tourism in India, one of the crucial components can be considered the health benefits when differentiated with several other alcoholic beverages with higher percentages of alcohol. It is now scientifically proven that wine reduces the risks associated with heart diseases, boosts immunity, etc resulting into human longevity. This benefits all the stakeholders of wine Industry. In this day and age, wine producers combine the tangible and intangible services and offer them as the wholesome package such as a bottle of wine with wine festivals, aesthetically pleasing vineyards, etc. resulting into higher consumer indulgence. Hence, these packages are a combination of leisure and luxury conveying a new domain for the satisfaction of wine lovers.

A lot of effort has been put into comprehending the new market of wine tourism by destination marketers and wine producers. They are aiming at ways for understanding the expectations of the market for optimum value creation. Hence these wineries brand themselves as leisure service providers and replicate international holiday standards to capture the market. This unique destination marketing approach is boosting Indian wine market when differentiated with other forms of tourism.

The states of Maharashtra and Karnataka are considered as one of the most significant destinations who are building their Wine Tourism Brands aggressively over a past decade. Nevertheless, with increasing competition, these brands are adding various services along with traditional wine tasting offerings. These services have transformed the traditional wineries into destination for Parties, weddings, events, etc apart from places for rejuvenation hence surpassing the traditional experience of wine tasting and wine drinking into novel experiences. In a consequence, relaxation and rejuvenation merged with novel experiences helps in building a lifestyle and higher level of consumer satisfaction can be defined as modern day Wine Tourism.

Statistically, Annual alcohol consumption in India is 33 litres by males and 11 litres by women. The hard liquor market in India is 93% of whisky or vodka with only 7% of beer consumption in India leaving only 1% population with wine consumption. Nonetheless, 60% of Indian men and 90% of women refrain from alcohol consumption in their lives, a higher shift has been witnessed over a few years in Indian drinking habits. Consequently, this demonstrates the growth potential of Indian Wine market. Even though India is amongst the lowest wine markets, a drastic shift has been witnessed due to the evolving updated lifestyle patterns of the Indians with more than 25% hike over past few years. (WHO, 2017).

A dramatic change has been witnessed in Indian Wine Industry since Liberalization. With automatic Foreign Direct Investments (FDI) for wineries, there is no requirement for a central licence with this automation easing the state level licensing process. Furthermore, the governments along with destination marketers are promoting Wine Tourism. Due to changing social values, a drastic increase in concerns regarding environmental issues have been observed resulting into Green Tourist Activities. Consequently, Argo and Eco tourism gaining impetus. The three important wine producing regions being Maharashtra, Karnataka and Himachal Pradesh, Maharashtra is the largest producer of Wine. With wine consumptions growing in urban areas of India, Mumbai accounts for 32% followed by Delhi NCR and Gurugram at 25%, Bangalore for 20% and Pune, Hyderabad at 5% and 3% each (Ms. Hiral Sheth, 2014).

ROLE OF WINE TOURISM IN DESTINATION MARKETING AND BRANDING

A festival can simply be termed as an event with certain traditional symbol that has been followed by the people. These festivals have a positive impact on the destination with regards to higher global awareness, and hence is also one of the crucial destination promotion techniques which works even during off and shoulder seasons of the destination. Hence these festivals are also considered of impetus for branding the destination at a national as well as international level.

Hence the tourist footfall in these destinations as well as festivals, can be further defined by the characteristics such as heterogenicity, degree of enjoyment, etc. Hence these festival planners, plan the festivals for a shorter duration and market strategically. This results into increased tourist footfalls and a potential tourism growth and awareness of the destination by effectively enhancing the brand image of the destination. Conversely, destination marketers also have an undifferentiated marketing strategy to attract tourists for these festivals. Hence, they always have aggressive marketing campaigns days and months before the festivals and events.

Wine Festivals are the buzz these days as the have a unique synergy between wine, events, travelling which ultimately add value and build the brand of the tourism at that particular destination. Hence these festivals are actually opportunities for social and cultural exchanges. Wine festivals and events is also an exchange between wineries and the destinations to market and boost product awareness. Hence, post these events, the tourists have a positive image not only for the Wineries but also for the destination and leads to revisitation and increased tourists' footfalls. Wine Festivals, hence, have a huge potential and are itself used as promotional strategies itself by destination markets itself for brand building.

Wine Festivals is actually considered as one of the components of wine tourism. Wine Tourism operates with the collaboration of Wine producers, Tour Operators and agencies and with the tourists. Hence, wine tourism is a collaboration of service providers and destination marketers via which destinations market the wine related products. Furthermore, wineries and their regions aggressively market their Wine brands and wine festivals play a role of promoting and boosting the awareness of these products at national and international levels. This automatically results into selling of wine products, brands, increases the aesthetically pleasing wineries, and build customers.

Branding is a result of aggressive advertising and is always prioritized by destination marketers. They focus on creation and establishment of brand and sustain it in the market for a long run. Hence it is evident that without the awareness of the brand, advertising is not effective. Even though the ultimate goal of branding is to promote the sales of any particular product, it is important to target creating brand awareness in the long run (Baker, 1986) (Obermiller, 1985). As a result, well known brands have a higher chance of selection when compared to the lesser-known brands as it's an important trait of consumer behaviour. Thus, a strong brand awareness results into higher sales of products. Hence branding can be stated as awareness or the knowledge of the potential consumer. This is also a similar definition of destination marketers i.e., fundamental level of brand knowledge (JANG, 2013).

Concluding this, if the consumer is not aware of the brands of wines in a wine destination, through effective destination marketing techniques such as hosting events, wine festivals, etc., their perception can be changed, and more awareness can be generated with respect to the Brand and Destination. Furthermore, this concept is can be used to test the efficiency of the advertising campaigns with optimum focus on Brand building. Hence, the destination marketers, wineries and regions get an opportunity in the form of Wine Festivals to promote and market the captivating destination and generate Brand awareness.

Concerning Wine Festivals, one of the components for ensuring the consumer satisfaction is the relationship between consumers and the distinctive services offered by the Wineries. These services include proper crowd management, minimum waiting times, impeccable selection of cuisines, all at a reasonable price. Other than these services, mainstream event services must also be of utmost importance including music, arts, hospitality, amenities, etc. Wine domain is represented by various wine brands and wineries. Hence, based on these factors, the wine festival management has standard procedures to enhance the experience of the tourists visiting for the festival and indirectly gaining their loyalty towards the Brand.

A venue for direct trial of Wines and other by-products is generated through these Wine Festivals. This decreases the chances of buying wines from local small and lesser-known wineries and increases brand awareness among the consumers and their chances of revisitation.

Wine tourism represents a marketing opportunity that increases the value of the wine destinations (Charters, 2002). Hence, Wineries and Tourism Organisations collaboratively develop co-marketing strategies. These two industries make a sustainable and a positive contribution for the destination. Building brand awareness and the image of a winery/wine region may be the most important success factor in the wine industry, and wine tourism is very much about achieving these goals. The wine consumer needs information about the origins of the wine and the geographical imagery of a wine region. Wine festivals would provide a convenient and effective vehicle to spread the message to target markets (Bruwer, 2002). A wine and food festival are an opportunity for Marketing and creating brand awareness and begin new consumer relationships.

To promote destinations culture, wine festivals are targeted to build the destination as a brand by destination marketers. These wine festivals play a trivial role in development of these tourism destinations. Hence, nowadays, cities have started emphasizing on opportunities for social and cultural interactions. Furthermore, to make the wine festival stand out, it needs distinctive planning with regards to the highlights, money invested, festival timeline, and these must be different in comparison with other events and festivals. The impacts of planned events within tourism are of increasing importance for destination competitiveness and in this sense those events are being included in the various tourist products (Getz, 2006). Hence while of these festivals, tourists constitute a prospective market for calculated events and the tourism industry has become an important player in their success and desirability (Camilleri, 2018) (Mazanek, 2010). Hence these festival planning should focus on tourists and visitor satisfaction concerning their motivations and behaviours.

Cultural destinations both directly and indirectly influences the impacts of the products on the potential visitors and also their intention of travel and hence leads to building the brand image in a certain way. Hence, destination marketers and organizations should always focus on creation of brand and destination image effectively. This is a crucial factor that influences the decision-making process, psychological satisfaction, revisitation and behavioural intentions (Chaulagain, 2019).

Wine Festival is a significantly recognised form of tourist attraction globally. Earlier studies have identified wine, food and cultural festivals not only play a vital role in economic development but also in regional tourism development with addition to already existing products (Quan S, 2004). These wine festivals promote local commodities and differentiates themselves from other festivals. Nevertheless, these festivals promote regional destination branding and generate a competitive benefit at state level further reflecting on a national level. Hence, it can be said that festivals are the one of the vital building pillars of destination branding. Destination wine image is a unique reflection of the regions cultural and wine tourism aspect. This image reflects the perception of visitors and is usually differentiated from general image of the destination as it involves visitors and tourists whose soul focus is on wine festivals and tourism. Hence it is vital for these festivals to be of shorter time frame.

A new way of marketing for destinations which are gaining popularity these days with the buzz of social media marketing is Integrated Marketing Communications referred to as IMC. Integration of marketing communications tools through a single message for consistency to build a brand is also applied to destinations. The destination marketers highlight arts and culture by these festivals and disseminate appropriate message through IMC keeping in mind the destination branding concept. In this regard, the harmony between the brand's focus as a destination and the festival theme should exist in order to create a favourable effect in the end. (Türkmenoğlu, 2017). Today, a tourist attraction has gained a broader meaning and thus events and festivals too are considered as attractions. (Gartner, 1996)

(Yuan, 2008) emphasises how prior experiences and the level of enjoyment with these experiences have an impact on the choice of locations, visits to attractions, and attendance at festivals. Festivals should therefore be interesting in topic and setting, well-organized, and promoted in order to draw crowds as short-term attractions. In reality, promoting festivals has shown to be a successful strategy for increasing tourism at the area and boosting its reputation. Promotion for festivals should be more targeted to a certain audience with the same desire to attend one, in contrast to other attractions or major events. Therefore, with the exception of recurring events, festivals are highly publicised beginning months before the actual event. Fans of periodic festivals closely monitor the media in order to fit it into their schedules and look forward to experiencing it once more.

The city's branding may have an impact on festival attendees' choices made with the goal of gratifying a specific social or cultural demand. In actuality, festivals that strengthen the branding concept are a part of the community, add vitality and vibrancy to the city, and serve as a key to differentiation. The consequences for destination marketing would be more significant if DMOs were to ascertain what kind of festivals would be effective to facilitate destination branding and improve competitive advantage. The theme of a festival is chosen by the festival organisers, either public or private, in accordance with the community's culture in order to improve the destination's image. It is clear that, although though destination branding is a constant process, festivals are only planned and advertised for a short while, but they nonetheless have a significant impact on destination branding. Therefore, long-running festivals should never lose sight of their association with the destination brand and should always use IMC. IMC of festivals that convey a consistent message to various demographics enhances the value of the destination brand.

We can thus draw the conclusion that the yearly wine festival serves as a tactical instrument for many wineries and wine regions to promote cellar door visits. Wine festivals give attendees the chance to interact socially—possibly with friends and family—while discovering and savouring an agricultural, natural environment and product. The participating wineries receive income and notoriety, the region's resources are better known, and the general public and outside service providers find new sources of clients.

REGIONAL LEVEL DESTINATION BRANDING

Maharashtra has long been thought of as a commercial state because it is home to the important semi-urban centres, historical sites, and the commercial metropolis of Mumbai. However, one needs to diversify tourism products and provide items of an international grade in order to truly advertise it as a tourism haven on a global scale. With its increasing global appeal, wine tourism would be the ideal solution, especially in terms of satisfying demand on a global scale. At this point, the Maharashtra Tourism Development Corporation (MTDC) is on a very high drive to do that, and the state government is quite supportive of this effort.

The grape is regarded as one of the significant commercial fruit crops in Maharashtra, one of the major grape-growing states in India. According to the Agricultural and Processed Food Products Export Development Authority (APEDA), Maharashtra's grape growing area is thought to be over 35,000 ha, and the state produces about 9 lakh metric tonnes of grapes annually. 70% of all Indian vineyard land is cultivated in Maharashtra, primarily in the districts of Nasik, Solapur, Sangli, Pune, and Ahmednagar.

Thomson Seedless, Tas-A-Ganesh, and Sonoka are the three main cultivars farmed in Maharashtra. Manik Chaman, Kishmish, Chorni, Sharad seedless, Perlette, Gulabi (Muscat Hamburg), and Flame seedless are more types that are becoming more popular (Madhav B. Thakare, 2015).

They split 80% of the state's overall production. Following the government of Maharashtra's initiative, the state of Maharashtra's tourism policy was presented in 2006 and was most recently acknowledged in 2016. Government officials have recognised the value of wine tourism in India. Wine tourism first motivated farmers to boost wine production and expand their vines close to wine industry. Because of the introduction of new cultivation techniques and the farmer's progressive attitude, the grape coverage is constantly expanding. Second generation farmers are drawn to the entrepreneurial side of growing grapes and processing them into value-added goods for sale to European markets, such as raisins, juice, and wine.

About 100,000 cases of wine are consumed annually in India, with Sula Wineries dominating the market with a 35% share. 5 India is currently the 11th-largest wine-consuming nation in the Asia-Pacific region, and by 2017 wine consumption in India is expected to reach 2.1 million cases. In the upcoming years, red wine is anticipated to expand by 71.6%. Additionally, the proportion of imported wine is anticipated to rise from 280,000 cases in 2013 to 480,000 cases in 2017.

After 15 years, Sula was the first winery to receive a licence in 1999; nevertheless, 75 more licences have since been granted. So, the current scenario is entirely liberalised. The main barrier in India is the regulatory burden placed on the wine industry. Alcohol abuse is a national problem. All of the retail establishments in Tamil Nadu are state-owned and only allowed to sell locally made or bottled goods. Tamil Nadu does not produce wine, so there is no inventory outside of the five-star hotels, leaving the market barren. Similar to Nashik, Karnataka is a wine-producing region that places significant state taxes on Maharashtra wine and vice versa. Lowering the taxes and creating free trade zones for wine within India are necessary to expand the Indian market. Other states, like as Andhra Pradesh, impose a 70% sales tax. Such problems make it challenging to market wine widely across the nation.

In the recent years, the cost of grapes has decreased as well, from USD 118 to USD 73 per metric tonne. Many wineries are still holding considerable quantities of wines from their last harvest in their tanks, which makes them reluctant to invest in new tankage. This is primarily due to cash flow difficulties. By purchasing some grapes from the promoters—Dindori vineyards—and some from contract producers, Sula achieves financial balance.

The wine sector has recently seen a shakeout, with many businesses unable to sell their wine. This restructuring was inevitable. Some of these new entrants are grape producers who saw this as an opportunity to integrate vertically, as many people have rushed into this industry in the last five years with very little grasp of what it takes to make good wine and market it effectively in India. But an entirely other set of skills is needed. Some of these players have suffered severe losses, especially those that over-crushed grapes in the previous two vintages in anticipation of a continuous massive boom. Currently, there is a divergence trend in the sector, with Sula and one or two other competitors growing, while many other players are filing for bankruptcy.

The fact that the wineries help relatively younger groups of residents in addition to hiring locals shows that they have the ability to have a good impact on the neighbourhood. It has been discovered that wineries actively support local development. It was also noted that the Wineries Association in Nashik produced a geographic indicator in order to imprint the name and image of a Nashik wine product—which aspires to build a foothold in the market—in the minds of consumers. Although the wine business in India's Nashik District has only recently begun to take shape, all of the aforementioned factors are attractive and help to support the local community's way of life. The creation of jobs and the diversification of rural economies are likely to be the advantages of the expansion of the wine business. Wine production combined with the incorporation of cutting-edge business ideas can help to reenergize rural economies by generating employment. Appropriate infrastructure is probably going to be very important in attracting tourists to wine-producing regions, which can help local economies. The potential for combined effects between vineyards, local communities, and the economy has been suggested by studies into the emerging Indian wine sector in the Nashik district.

ZONAL BRAND DEVELOPMENT

About Nashik

The Indian state of Maharashtra is home to Nashik or Nasik. It is referred to as the "Wine capital of India" and is the third-largest and third-most industrialised city in Maharashtra. At 565 metres above mean sea level, the Nashik district is situated between 18.33- and 20.53-degrees North latitude and 73.16- and 75.16-degrees East longitude in the northwest region of Maharashtra state. The Indian Census of 2015 indicated that Nashik has a population of 61,07,187.

The Nashik district's climate is typically mild and ideal for growing grapes. Winters are generally nice with a maximum and minimum temperature of 32°C and 8°C, respectively, while summer temperatures are above 42°C on an annual average. The relative humidity varies from 43% to 62% (IMD weather report, 2011), despite the district's average annual rainfall of 1056.5 mm. Peninsular Maharashtra is not the best region for viniculture in general. Wine production is hampered by the diversity of the soil, climate, and population. Due to its higher altitudes along slopes and hillsides, milder breezes, and ideal temperature for picking grapes, ageing wine, and storing wine, Nashik district is an exception to this rule.

A particular policy has been established by the Maharashtra Tourism Development Corporation (MTDC) for the Maharashtra wine circuit. The proximity to Mumbai, the capital of Maharashtra, has given Nashik winemakers access to the wine market. Due to the semi-urban regions that surround the Nasik district, many tourists from all over India and the world visit there to stroll around wineries on the weekends and during extended vacations. Every year, the period from harvest to wine packaging is between December and March. It's said that now is the busiest time of year to tour vineyards. The owners of the vineyards organise unique wine tasting routes, wine festivals, and vineyard tours. For the entire wine sector as well as wine tourism, this season is the most lucrative.

Government initiatives and programmes for grape research and development: The Indian government provides the following kind of assistance to the nation's grape industry:

1. Motivate and assist farmers in developing vines. Give money and subsidies for the construction of storage facilities and drip irrigation systems.
2. Offer research assistance, a quick response strategy, and mitigation to maintain grape productivity in the face of challenging conditions like hailstorms (in 2013 and 2014), droughts (in 2002 and 2011), etc.
3. Encourage and support the export of high-quality, fresh grapes by providing growers, producers, and suppliers with training.

4. Under the All India Coordinated Research Project on Grapes, the Indian Council for Agricultural Research (ICAR), Institutes, and State Agricultural Universities conduct research on grapes at various locations. The National Research Centre for Grapes (NRCG), which is based in Pune, Maharashtra, serves as the focal point for organising and conducting grape research across the nation (Madhav B. Thakare, 2015).

BUILDING SULA WINERY AND SULA WINE FESTIVAL AS A BRAND

More than 30 nations import Sula's wine, which is produced on 3000 acres of land in Nashik, the country's biggest winery. With a resort and vineyard, this winery was the first in India to popularize Wine Tourism. Visitors to Sula may participate in a program called Wine Education, which allows them to learn more about wine tasting and the winemaking process. This is an interaction that makes one feel like they've made a lasting impression on the place they're visiting.

More than seventy-five percent of visitors to the Nashik area come to visit Sula Vineyards, making it the most popular wine tourism destination in India. Sula was the first winery to receive a licence in 1999 with the vision of making the product ideal in terms of quality/taste and price/value and continued to use it. In India, advertising for alcoholic beverages is subject to numerous prohibitions. It cannot be used to endorse branded items or be a part of sponsorship for important sporting events. Thus, used entertainment events and wine tastings as part of its main promotional strategy.

Wine tours and "experimental tasting" at Sula Vineyards' winery helped to introduce people to wine. All visitors are welcome to participate in the wine tours, which give them a look at the vineyards, the wine-making process, wine tastings, courses, and special events. People were able to connect with wine as a result of letting go of their wine-related inhibitions. It greatly aided in placing Brand Sula with Indian consumers as it was the first of its type in India.

Sula offers roughly 1600 wine-tasting classes a year to teach people about the finer qualities of enjoying a glass of wine, especially with Indian cuisine, in an effort to make the Indian palate more receptive to wine. It also underlines that Indian wines are more compatible with the Indian palate and complement spicy Indian food than wines from other countries.

Focusing on Indian women as their target demographic is their most audacious strategy for capturing the Indian market. With expanding horizons and an open mind to accepting change, Indian women have become more economically and socially independent, and they have recognised the huge potential in this market. Sula produced a wine that was lower in alcohol, a little more effervescent, and sweeter than other varietals specifically for ladies. As ladies began to accept this wine, it became a big success.

Wine sales in India were also boosted by increased exposure to Western culture, international travel, and experiences in other nations where drinking wine is a way of life. About 5% of Sula's production, which came in red, white, rose, sparkling, and dessert kinds, was exported. Japan was the company's biggest market, followed by the United States and the United Kingdom, but it also exported to other regions of Europe and the Middle East, and it has hopes to grow in Africa, China, and Russia.

The first grape resort in India, Beyond, which boasted of a roomy 2000 sq. ft. (again, the first of its type) wine-tasting lounge with a view of the Gangapur lake and gorgeous, rolling vineyards. Sula Vineyards partnered with five-star hotels and upscale restaurants in addition to wine shops in order to give wine an exclusive product association in India. Sula wines are given a special place in the restaurants by being elevated to a high-class social joy. Additionally, it established links between diverse cuisines and various wine types.

The wines are served in restaurants decorated in a European style to create a wow factor and to reinforce brand recognition long after customers return. During their visit to Sula Vineyards, wines are discounted in an effort to persuade non-buyers to become purchasers. Sula Vineyards has given international wine brands a significant amount of exposure in competition when combined with food and wine festivals in significant Indian metropolises.

SULA WINE FEST

Every year, the Sula Vineyards host Sula Fest, a celebration commemorating all kinds of Sula. India's Largest Gourmet World Music Festival with Wine and other Alcoholic Beverages included in Gourmet. Sula Fest is not just a music festival, but it's also a wine festival, with Beluga Vodka imported by Sula, Jägermeister, Cointreau and Budweiser Beer making up the rest of the drinks on offer at the event.

In terms of its entertainment event, a Wine Weekend with Sula in the form of Sula Fest—a music festival that is hosted at their winery every February—the company Sula has made a significant effort to build an exclusive identity. Giving brand Sula a hip and interesting image is the main goal. To foster an upbeat atmosphere, the festival offers a wide range of activities like grape stomping, wine sampling, foot massages, and tattooing.

Sula Fest, India's largest vineyard celebration, takes place over the course of two days. In addition to music from across the globe, tasty cuisine and wine, and branded souvenirs, the festival has an array of other attractions. Celebrities, actors, musicians, writers, journalists, bloggers, wine specialists, journalists, and people from all walks of life from India and even abroad attend the Fest. Tourists can participate in activities such as grape picking, wine tasting, and knee-deep grape stomping as part of a masterclass, as well as enjoy performances by top national and international bands and artists on the main stage of the festival. In addition to camping with nature, tourists may also stay at the adventurous campground. A tree is planted at Sula Vineyards for each and every one of the festival's guests, earning it the title of India's first sustainable music festival. Since its inception in 2008, the Sula Fest has grown into a major event, attracting anywhere from 16,000 to 18,000 attendees annually, in addition to the 350,000 people that visit Sula Vineyards each year for wine tourism.

The event is expected to bring in more than Rs 7 crore in income (about Rs 6,000 per person), which includes ticket sales, sales of food, wine, and mementos, transportation, and lodging. Even though Sula receives only a little portion of that (they likely end up just breaking even), the positive publicity that is brought to their brands is immeasurable.

The venue comprises of two performing stages with an amphitheatres main stage where musical performances take place; it may accommodate 3,000–4,000 spectators, a DJ's arena with another stage, food, water, and beverage stalls. The beverage partners included Asahi, Cointreau, Jägermeister, Stolichnaya, Jack Daniel's, etc. including a recently expanded wine-tasting terrace and sampling area, both days saw full occupancy of its 500-person capacity (Alok, 2016)

It could possibly be said that there aren't many people who enjoy both wine and music. But that completely misses the point: People between the ages of 25 and 45 who are prepared to pay more than Rs 6,000 to attend something similar on a voluntary basis Each of them will connect on social media with an additional 300 friends, some of whom will share the postings with their own networks, giving you a potential audience of more than five million buyers and decision-makers. In addition to a sophisticated and upscale audience, Sula Fest also draws its share of celebrities, who might potentially sell to millions of potential customers.

Towns that host festivals can also make money off-site from overnight guests. For instance, Sula Fest, which draws 10,000 tourists on average each year, helps Nashik book around 80–90% of its hotel rooms. Local businesses benefit from free marketing and advertising at festivals as well since attendee's remark about their enjoyable experiences when they return home. It would be ideal if visitors shared their experiences in the form of comments and images on Facebook or other social media. Successful festivals have a positive economic impact on the entire community's economy, benefiting both tourism- and non-tourism-related businesses (Walimbe, 2019).

Wine Festival and Wine Tourism is also great for real estate development, tax revenues, and employment growth in any community, in addition to its entertainment value. Studies now demonstrate that travel and tourism generate jobs, both directly within the travel and tourist business and indirectly in industries like retail and transportation. The "multiplier effect," which results from tourists spending their money on goods and services, contributes to the creation of additional jobs. Nashik's Sula fest generates a sizable number of job possibilities for residents of the city and other adjacent locations. Vendors, art installations, musical workers, fans, show directors, and medical support are just a few of the fantastic job opportunities that these festivals in cities and small villages generate. A sizable contingent of trainees is required on all occasions for these festivals to be successful. The trainees must work throughout the festival concert as well as in the pre- and post-festival periods. They help with setup, vending, catering, staffing the information booth, and other tasks.

CONCLUSION

A region's or society's economic development is correlated with rising revenues and the corresponding rises in savings, investment, and consumption. Nashik may have higher levels of revenue, higher levels of consumption, better levels of employment, and increased investment plus saving if such festivals are successful and produce positive revenues. A considerable decrease in tourists was seen in the area as a result of the cancellation of Sula Fest in 2021 and 2022, which was caused by the Covid - 19 pandemic. This serves as more evidence of the festival's significance for the region's wine industry. If the benefits, and indeed the chances of survival, of small and medium wineries and small tourism firms that cater to wine tourists are to be maximized, one's grasp of wine tourism from both the product and destination perspectives must be greatly expanded.

The wine industry's production orientation and the tourist industry's service quality can be combined to deepen our understanding of the possibilities and avoid difficulties of wine tourism as a whole.

REFERENCES

Alok, C. (2016, February 20). A toast to SulaFest. *Business Standard*. https://www.business-standard.com/article/beyond-business/a-toast-to-sulafest-116021901007_1.html

Baker, W. J. (1986). Brand Familiarity and Advertising: Effects on the Evoked Set and Brand Preferences. *NA: Advances in Consumer Research, 13*, 637–642.

Bibicioiu, S., & Cretu, R. (2013). Enotourism: A niche tendency within the tourism market. *Scientific Papers. Series Management, Economic, Engineering in Agriculture and Rural Development, 13*(2), 31–40.

Bruwer, J. (2002). Wine and Food Events: A Golden Opportunity to Learn More about Wine Consumers. *NA: The Australian and New Zealand Wine Industry Journal, 17*(3), 92–99.

Camilleri, M. (2018). Travel marketing, tourism economics and the airline product: An introduction to theory and practice. Springer Nature. doi:10.1007/978-3-319-49849-2

Charters, S. K. (2002). Who is the Wine Tourist? *NA: Tourism Management, 23*, 311–319.

Chaulagain, S. W., Wiitala, J., & Fu, X. (2019). The impact of country image and destination image on US tourists ' travel intention. *NA: Journal of Destination Marketing and Management, 12*, 1–11. doi:10.1016/j.jdmm.2019.01.005

Gartner, W. C. (1996). Tourism Development: Principles processed and policies. Van Nostrand Reinhold.

Getz, D., & Brown, G. (2006). Critical Success Factors for Wine Tourism Regions: A Demand Analysis. *NA: Tourism Management, 27*(1), 146–158. doi:10.1016/j.tourman.2004.08.002

Ms. Hiral Sheth, M. A. (2014). Wine Festival and Wine Consumerism - A Perspective Study on Influence and awareness in the Growth and Acceptance of Bangalore. *International Journal of Engineering Technology, Management and Applied Sciences, 2.*

Jang, J. (2013). The Effects of Quality and Satisfaction on Awareness and Behavioral Intentions:Exploring the Role of a Wine Festival. *Kentucky: Journal Of Travel Research.*

Kumar, D. A. (2021). *Socio-Economic Benefits from Indian Winery*. International Journal of Engineering and Management Research. doi:10.31033/ijemr.11.4.3

Madhav, B., & Thakare, M. L. (2015). *A perspective study of wine tourism in Nashik district*. Journal of contemporary research.

Mazanek, J. (2010). Managing the heterogeneity of city tourists. In J. Manzanek & K. Wöber (Eds.), *Analysing International City Tourism*. Springler Verlag. doi:10.1007/978-3-211-09416-7_6

Obermiller, C. (1985). Varieties of Mere Exposure: The Effects of Processing Style on Affective Response. *NA: Journal of Consumer Research, 12*(June), 17–30. doi:10.1086/209032

Quan, S. W. N., & Wang, N. (2004). Towards a structural model of the tourist experience: An illustration from food experiences in tourism. *NA: Tourism Management, 25*(3), 297–305. doi:10.1016/S0261-5177(03)00130-4

Times, E. (2017, March 15). Wine Tourism. *Economic Times*, p. 1.

Türkmenoğlu, G. Ö. (2017). *IMC Strategies of Festivals in Destination.* NA: Strategic Innovative Marketing, Springer Proceedings.

Walimbe, O. (2019, January 10). How events like SulaFest are helping boost Nashik's growth. *Just News*. https://justnashik.com/2019/01/10/how-events-like-sulafest-are-helping-boost-nashiks-growth/

WHO. (2017, October 10). WHO. *Times of India*. https://timesofindia.indiatimes.com/india/Average-Indian-male-consumes33-litres-of-alcohol-/year-WHO/articleshow/35139360.cms

Yuan, J. M., Morrison, A. M., Cai, L. A., & Linton, S. (2008). *A model of wine tourist behaviour: A festival approach*. NA. *International Journal of Tourism Research*, *10*(3), 10. doi:10.1002/jtr.651

Chapter 14
Destination Branding Through Tourism Festivals of Khasi Hills and Garo Hills Region Meghalaya in India:
Case of Cherry Blossom Festivals and Me.gong Festival

Sharad Kumar Kulshreshtha
Department of Tourism and Hotel Management, North-Eastern Hill University, Shillong, India

Iadonlang Tynsong
Martin Luther Christian University, India

Ashok Kumar
Institute of Hotel Management, Catering Technology, and Applied Nutrition, India

ABSTRACT

Meghalaya is the most beautiful and scenic state among the northeastern regions. In Sanskrit, 'Meghalaya' means 'Abode of Clouds.' This state has blessed with natural and cultural attractions. Meghalaya tourism has conducted two more yearly tourism festival associated with their nature and culture to attract more tourists Khasi and Garo hills region, a sincere initiative for destination branding through events. This chapter is an attempt to showcase the potential these two tourism festivals, i.e., Cherry Blossom Festivals and Me'gong Festivals, which are associated with festival destination marketing of these two regions of Meghalaya. The purpose of this chapter is to investigate the emergence of such tourism festivals with mix of traditions, culture, heritage, indigenous characterizes of local destination, in order to transform it into a festival and attract a large number of tourists to build-up iconic destination branding.

DOI: 10.4018/978-1-6684-6356-7.ch014

INTRODUCTION

Festival is the most fascinating aspect to showcase the culture, rituals, customs and traditions, all kinds of seasonal variations i.e., summer, winter, spring, autumn, and monsoon which also have directly connected with economic and social activities of our life associated with agriculture and horticulture-based activities like seeding, sowing, reaping, and harvesting, etc. The festivals always start with ceremonial synergy, happiness, enthusiasm, and rituals, and connect with a particular destination and local culture as well as pro-active participation of festival stakeholders. Hosting destinations and festival attractiveness are inextricably linked (Li., Dai., Tang., & Chen, 2020). In the early studies 1990s on cultural festivals, festival management simultaneously was a vital part of event management research. Festival and event tourism has been one of the fastest-growing area of research and practice worldwide (Getz, 1991; Nicholson & Pearce, 2001),

The concept of destination branding came into being because of the tourist industry's increasing rivalry. Destination branding, often known as destination marketing, is a package of tourism offerings, experiences, and events delivered through an innovative tourist-friendly strategy. The development of destination branding is always a difficult process. It creates a favourable impression of the location in the eyes of tourists. Moreover, travellers form their expectations based on previous experiences (Rahman., Masud., Akhtar., & Hossain, 2021). The goal of destination branding is to advertise a certain place to visitors from other nations (Rahman et al., 2021) by providing a lucrative package of travel-related goods and services in the most upscale way possible. To create a favourable impression of that site in the minds of potential consumers and support them in remembering it, destination branding portrays and distinguishes a specific location amid a multitude of competing destinations (Hankinson, 2005). When it comes to establishing a marketing plan, branding is critical. Just as products, services, and organisations are all branded. Cities can also be promoted in this manner. Cities have a variety of elements, such as culture, art, legacy, and natural resources. translates to complex identities that exist in the minds of customers. Meanwhile, these characteristics work together to create a distinct picture in the eyes of customers. Because a brand represents what exists in the minds of customers rather than what the marketer seeks to communicate through the brand (de Chernatony and McDonald, 1992). In the present context, most destinations are using events and festivals as tools to promote the destination. Since, festivals and touristic events boost the visibility of host communities, drawing tourists and driving up spending, which helps local economies. Local governments provide the majority of the funding for these events because they want to demonstrate the value of their expenditures by evaluating the events' economic effects. Numerous studies have been conducted on the effects of tourism and tourist events, festivals in particular. However, there is still disagreement over the most accurate manner to assess their contributions to regional economies.

The Cherry blossom festival is a kind of event where we welcome the blooming of cherry tree flowers. Years ago, this festival was usually celebrated in Japan, South Korea, Australia and America. Now Meghalaya is also celebrating it on a very large scale just like other countries of the world. In the first three days, people and also the organiser were sad since there is no cherry blossom flowers in the trees but their hope and expectation didn't come to waste since the flower started to bloom right on time before the closing of the festival. Now almost all the trees are blooming and bring beauty and picturesque that cover the city in pink and white colour (Demetriou, 2019). There are the following seven features of festival tourism: *1.) short period of time visit; 2.) specific festival destination; 3.) having interest in such festivals-oriented holidays;4.) rich history of festivals and associated culture; 5.) very differs purpose rather than cognitive tourism; 6.) linkages with other types of tourism; 7.) system of classifications of such festival tourism* (Stavska. 2021).

Events and cultural festivals are being used more often by territories to improve their image and advance tourism. As a deliberate move to draw tourists and strengthen their image, several places throughout the world have developed event portfolios. A destination brand is the result of a dynamic interplay between the primary assets of the place and how potential tourists view them (Piva., Cerutti., Prats., Raj, 2017). In the Context of the Meghalaya, Cherry Blossom Festival and Me'gong Festival had able to attract tourists from across the globe to witness this event. Cherry Blossom Festival and Me'gong Festival combine, able to bring a number of high artistic standards delivered by an accomplished musician with many musical genres, from rock to modern music, in this enthralling setting. The festival offers fresh talents the chance to gain popularity, and also to gain experience. These festivals not only attract Tourists in terms of Music events but food and craft are the major components that draw tourists and participants to be part of the events.

The purpose of this research is to emphasise the significance of events and festivals by branding two cities in promoting the tourist industry in Meghalaya. With this research, the local community and stakeholders will be able to grasp the significance of festivals and how to leverage them to attract visitors to the state. The present study highlights the real gaps in comparison between the two festivals. As per the Observation and also the data from the field both the festival highpoint the local culture of their own tribes such as the food, the dance form and handicraft. However, there is study had been done on destination branding through such festivals in the study area. Local Non-Governmental Organizations (NGOs) are always outraged that the government spends so much public money on festivals that provide no economic benefit to unemployed youth. This perspective clearly demonstrates the state's lack of study in this sector of the event.

REVIEW OF LITERATURE

Festivalization

'Festivalization' is a buzzword which a very emerging event management practice worldwide. The contemporary phenomena in festival cities, such as the rise of 'festivalization' as a term to describe the tangible and intangible effects of festivals on the places that host them (Richards, Greg & Londoño, Maria del Pilar Leal, 2022). The *millennials* and *Generation Z*, as youths are also very passionate about the festival and which is the most vibrant and emerging sector globally. Especially, in metropolitan and Tier-II cities as urban and commercial centre organises such festivals for trade and business promotion as for branding destination. Nowadays it is considered a most powerful tool to promote destination and product experiences which associate with the brand. The Festivalization of Culture explores the links between various local and global cultures, communities, identities and lifestyles.

Festival Tourism

The word "festival" derives from the French word festival, which means "holiday". The concepts of "festival tourism" or "festival tour" are increasingly appearing in the list of tourist services offered and are gaining a wider audience (Stavska. 2021). Festivals are a rapidly developing phenomenon, which plays a significant role in the development of tourism (Cudny, 2013). Festivals embrace catalytic roles in any destination which brings societal benefits with positive outcomes economic and social objectives

Destination Branding and Marketing Through Festivals

Events are chronological phenomena and every event is unique in nature because of place-people interaction. According to Getz 1997, Cultural events are held in different forms but most of them are appearing in the form of festivals, carnivals, exhibitions, concerts, and performances. Festivals are largely about preserving the originality and individuality of a place or region, and they play important roles in the city or region and are characterised by attractiveness, image, animation, and development. They help to lengthen the tourist season, the primary season, or to create a new season in the life of the region or resort (Yeoman et al. 2004). Festivals with an appealing event schedule may position the destination in the market and attract more tourists in the future, resulting in longer stays in the destination and exploration of tourist attractions (Getz 1997). Presently, there is fierce rivalry among destinations, and we can observe the efforts they have done to develop their country as a tourism destination. Tourists establish the picture of the place in general based on their experience acquired from attending the event. Festivals may be effective in anchoring and even transforming a destination's image. The festival has firmly established the city's cultural credentials as a distinguishing feature of the city's brand. The festival's publicity can help to improve the destination's image. Festivals can alter potential visitors' perceptions of the host community and its physical setting (Yeoman et al. 2004).

Festivals Vis-A-Vis Destination Marketing

A festival is an event that is conducted on the same day and in the same month every year and has been designated by locals as a traditional icon of a place. (Revilla Hernández, 2016). Festivals are acknowledged to have a significant impact on human life. As it reflects the community lifestyle, the attendee. On the one hand, they offer a variety of social activities to participants while also serving as a source of revenue for the host cities. The most well-known sort of festival is the artistic festival, and these festivals can play an important role in bringing together several fields of art in one location. As a result, the Marketing of events or festivals has expanded significantly, in order to achieve their marketing Objectives (celik and Cetinkaya, 2013).

Festivals often provide socio-economic advantages for the destinations where they are organized. Since events and festivals provide mass promotional opportunities about the place because of the different exposure they receive in numerous sources of social media. (Pirnar, Ige and Kurtural, Sinem and Tutuncuoglu, Melih, 2019). In each and every Festival or event, there are always Positive and negative impacts and these impacts can either help in proper marketing of the place through social attachment and the emotion of belonging to a particular group or community, Increasing the connections between people and improving creativity, innovativeness and image of the destinations (George, 2017). Through events, people in the community can able to support their livelihood and improve the standard of living in the community and the residents and their environment are connected via a sense of joy and longingness among each other. In the recent past, Cultural celebrations centred on the local community evolved so as to reflect the aspirations, passions, and objectives of the community (Derrett, 2003). Many authors pointed out that they are different types of festivals as highlighted in table 1. However, when we talk about festivals, we normally focus on three things: the participants, the locale, and the type of audience.

Table 1. Festivals and related destination

Types of Events/Festivals	Attributes and examples
Seasonal festival	*The National Cherry Blossom Festival, Washington, D.C* *Cherry Blossom Festival Shillong India* *Cherry Blossom Festival, South Korea* *Cherry Blossom Festival Japan*
Activity Oriented Festivals	*Boryeong Mud Festival, South Korea, Songkran Water Festival, Thailand, Holi Festival, Wangala (Festival of 100 Drums) Meghalaya, India.*
Destination based festivals	*Jaisalmer Desert Festival, India,* *Rio Carnival of Brazil, Venice Carnival, Italy, and Cannes film festival, France Edinburgh Fringe Festival, Scotland, Sydney New Year's Eve, Australia, Albuquerque International Balloon Festival, USA Lazarim Carnaval Portugal, Quebec Winter Carnival, Canada*
Iconic Festival	*La Tomatina Festival Spain. Gilroy Garlic Festival, California* *Bordeaux Wine Festival, France* *FIFA World Cup, ICC Cricket World Cup and Olympic Games are examples.*
Film Festival	*International Film Festival of India, Mumbai International Film Festival, Jaipur International Film Festival* *Toronto International Film Festival* *Berlin International Film Festival, Venice Film Festival* *Chicago International Film Festival, Melbourne International Film Festival*
Tourism Festival	*International Kite Festival, Taj Mahotsav, Ran of Kutch Festival, India*

Source: Authors

Research Questions

1. What are the potential and prospects of *tourism festivals of Khasi Hills and Garo Hills Region of Meghalaya in India?*
2. What are the *Unique selling propositions (USPs) of tourism festivals of Cherry Blossom Festivals and Me'gong Festival in destination marketing and branding of Meghalaya?*

OBJECTIVES

1. To study the potential and prospects of tourism festivals of Khasi Hills and Garo Hills Region Meghalaya in India.
2. To study the Unique selling propositions (USPs)of tourism festivals of Cherry Blossom Festivals and Me'gong Festival in destination marketing and branding of Meghalaya.
3. To discuss the strategic role of Meghalaya Tourism, Destination Management Organisations (DMOs) and Non-Government Organisation (NGOs) and the local community for managing these festivals.

RESEARCH METHODOLOGY

This study investigates the potential of local festivals that act as a pull factor for tourist attractions and encourage tourists to visit a particular tourist destination of Khasi Hills and Garo Hills of Meghalaya. In-depth interviews with festival stakeholders from both festivals were the primary method for gathering qualitative information. Observation techniques were applied as visitors to these two festivals. This chapter is based on a qualitative research method to study the most emerging festivals organised by the Department of Tourism, Government of Meghalaya promote an intangible and indigenous culture of Khasi Hills and Garo Hills Region of Meghalaya. The extensive literature review and in-depth focus interview of stakeholders have been done to assess the perception of visitors, event organisers, and local communities in both the respective destinations. Henceforth, the investigators have used case studies. The case study-based methodology has been used for the comparative analysis of the two festivals to emphasise the issues and challenges associated with these two festivals by finding the impact of destination marketing and destination branding. Given that the study will evaluate the two types of events' existence in Meghalaya and their effects on destination branding, a causal-comparative research methodology would be the most appropriate. As a consequence, the fundamental structure of the study goals was determined via causal-comparative research.

DESTINATION BRANDING THROUGH EVENTS

Although the concept of branding has been widely applied to goods and services, it has only just begun to expand into travel destinations and emerged as a subject of research in the late 1990s (Tasci & Kozak, 2006). Destination branding, as defined by Morrison & Anderson (2002, p. 17), is "the technique utilised to build a unique character and personality that is different from all competitor destinations." Another definition of destination branding is "a set of marketing activities that support the development of a name, symbol, logo, wordmark, or other graphics that readily identifies and differentiates a destination; that consistently conveys the expectation of an unforgettable travel experience that is exclusively associated with the destination; that serves to consolidate and reinforce the emotional connection between the visitor and the destination; and that reduce (Blain, Levy & Ritchie, 2005, p. 337)

Events involving established brands may be leveraged to strengthen the destination's brand identity. Co-branding, as defined by Chalip & Costa (2005), is the process of combining the brand of the event and the brand of the place. Consequently, it is hoped that elements of the event's brand will translate into the destination's brand, and vice versa (Xing & Chalip, 2006). The key to an effective destination event co-branding strategy, according to Jayaswal (2009, p. 23), is to locate the network of links between competing places before holding an occasion that might strengthen, alter, or introduce positive associations. The *Cherry Blossom Festivals* had objective of effective marketing and branding of "Destination Meghalaya" as most potential state for its unique bio-resources, nature, eco-tourism.

CASE STUDY 1: CHERRY BLOSSOM FESTIVAL

Genesis of Cherry Blossom Festival

As a gesture of friendship between the people of Japan and the people of the United States, Cherry trees were given by the people of Japan in the year 1912 as a gift of friendship. Cherry trees are revered in Japan as an exalted flowering plant known as Sakura, or "flowering cherry trees". The cherry blossom represents the evanescence of human life and epitomizes the transformation of Japanese culture throughout history. The first Cherry Blossom Festival was started in the year 1927 by school children in America who took the initiative in planting cherry trees and others activities. In 1935 civic bodies

Figure 1. Model of Destination Branding Through Tourism Festivals

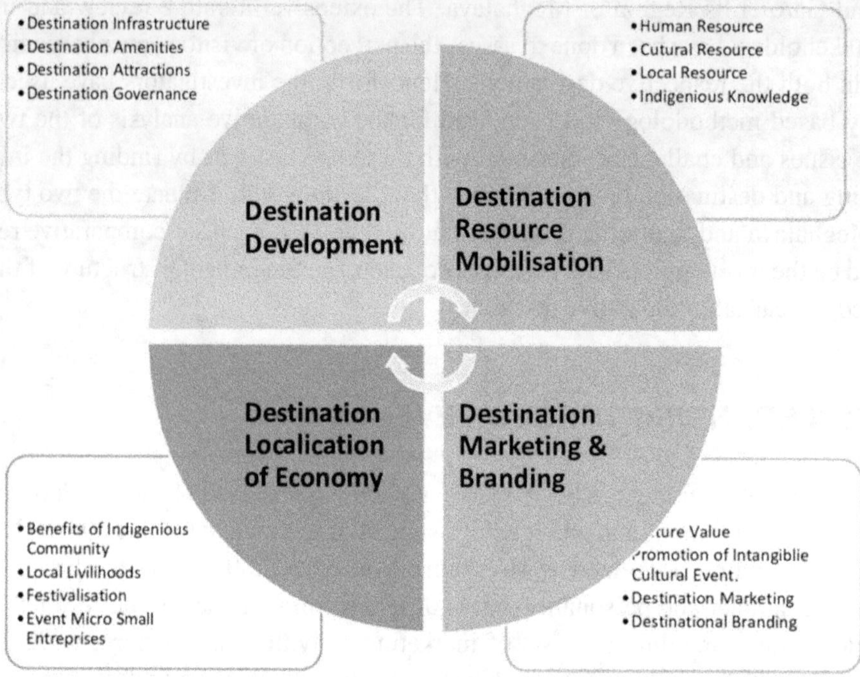

helped in expanding the festival by sponsoring the events and became an annual event in subsequent years. And in 1940 Cherry Blossom Pageant was introduced. Since its inception in 1994, the National Cherry Blossom Festival has grown from modest beginnings to be one of the biggest springtime celebrations in the world. It was originally scheduled to take place over a week, but today the festival has been extended to two weeks. In today's Festival, more than 1.5 million people visit the festival spread out over four weekends to take in the diverse programming and to see the magnificent trees on display. Over the years, millions have participated in the annual event that heralds spring in the capital of America. (National Park Service, 2021).

Meghalaya's first Cherry Blossom Festival was initiated by the State Government and the Institute of Bio-resources and Sustainable Development (IBSD) and the Indian Council for Cultural Relations (ICCR) in 2016. The event was developed to encourage tourism in the area. During the festival's debut season in 2016, a huge number of tourists attended the event. This year, organisations such as IBSD and ICCR, in collaboration with the government of Meghalaya, planned a larger-scale event with more effect than in 2016. International and national delegates, as well as an open public appeal to national and local visitors, were invited to this festival.

Importance of the Cherry Blossom Festival

This was the First International Cherry Blossom Festival which is celebrated in Autumn during November, *but* in other countries usually blooms in spring during March-April (Forest and Environment

Department, Government of Meghalaya). Cherry Blossom Festival is organised based on two aspects firstly it promotes tourism activities in the state by highlighting the vibrant culture and tradition of the state. Apart from that Cherry Blossom Festival can help give a platform to various artists, first-generation start-ups and local entrepreneurs from across the state to showcase their talents, their products and so on. Not only that Cherry Blossom Festival is providing opportunities to come together to learn and exchange ideas (Choudhury, 2021).

The investigators had the opportunity to interact with different vendors and visitors during the Cherry Blossom Festival from 2016 to 2020. Through the interaction, they have been able to gather different views and suggestions that they would like to put forward. One of the dissatisfactions that the people who visited this event on the first two days of the festival was because they could not find the blooming cherry blossom in the venue, especially in wards lake (Nan Palo). Even the foreign visitors mentioned this when we got to interact with them. Singh (2017) in an article written in Hindu Newspaper states that there is "No carpet of colours at Shillong festival, as cherry blossoms fail to appear" and he did also mention that all the necessary arrangements from the organization were in place such as scores of stalls centred around the theme of Cherry Blossoms at the inauguration venue, dancers from the three major tribes of Meghalaya, Garo, Khasi and Jaintia and others from North Eastern State are ready to take the stage. As many as 15 different cultural events, and many other musical bands and other items were planned. Among several events on opening day, a storytelling session was organized at picturesque Ward's Lake in the evening. The lake, in the heart of town, had another cluster of about 100 cherry trees gently lit up, and hundreds of people gathered at the venue.

The organiser received a very positive response from the stall vendors since they are delighted to be part of the event; some of them are first-time participants in such a festival, while others have been a part of it for many years. Among the other things we enjoy at the festival are many artisanal products from the state. Among the items on display are handicrafts, local foods, Khasi costumes, tattoos, bike and automobile exhibitions, as well as fruits and wine products from the local area. An initiative like this could offer economic opportunities for vendors of local cuisine, according to the vendor. Additionally, she stated that the first two days of the festival are normally dry, while the following two days are always crowded, especially at night.

Impact of Cherry Blossom Festival

Like every major event, the Cherry Blossom Festival has its pros and cons. It would be naïve to assume that an event of such scale would inevitably have a positive impact on society and vice versa. Some of the positive impacts it had on the state are as follows:

This festival has brought a lot of resources to the state. With the large influx of people coming from various parts of the world, the local community is allowed to grow their business. This festival generates employment opportunities for local people. This helps contribute to and generates the economy of the state. Several opportunities have been created for hoteliers, restaurants, homestays, souvenir shopkeepers, tourist guides, taxi operators, drivers, and rural tourist spots as a result of this event.

1. It allowed local musicians to showcase their talents and gain exposure.
2. It was an excellent opportunity for local vendors to sell their merchandise.
3. The local cuisines and handicrafts of the state were given proper exposure.

4. It creates a strong relationship between the two governments of Meghalaya and the Government of Japan.
5. With the coming of this festival, there have been changes in the lifestyle of the people.

People from various parts of the world come to Shillong to celebrate this festival. Along with them comes a deep understanding of the vastly different cultures of the world. Through cultural exchange, there is a noticeable change in the day-to-day lives of the locals.

The Cherry Blossom Festival serves as a catalyst for promoting several tourist destinations in the state. Meghalaya attracts people from all over the world. In between their stays, they often do a lot of sightseeing which helps boost the economy of the state further. This also helps spread awareness of Meghalaya as a tourist destination all over the world through word of mouth as well as various media platforms. Some of the more negative points include:

1. Possibility for abuse of drugs and alcohol among the youth demographic.
2. The festival was another cause of traffic jams and parking issues, cleanliness and waste management and some environment concerns,
3. Inconvenience for local residents near the venue by laud sound and crowded locality.
4. Post event excitement among youths sometime create conflicts therefore, safety and security issues and maintain law and order condition challenging.

Table 2. Basic facts of cherry blossom festival

Name	Shillong Cherry Blossom Festival
Type of event	Entertainment Events
Organizer	Department of Tourism Government of Meghalaya In collaboration with the Institute of Bio-resources and Sustainable Development (IBSD) and the Indian Council for Cultural Relation (ICCR) and Manage by Rock ski EMG.
Event History	The cherry blossoms festival first was started in the year 2016. To showcase the beauty of the Himalayan cherry blossom bloom and several other cultural activities, Shillong began to hold the Cherry Blossom Festival. These unspoiled sights are revered as both a gift from the Himalayas and a lucky charm.
Event Held	November (4 days of Events)
Event Management Company	Roclski Events and Media Group, Shillong North East India
Event venue	Polo 2 Ground, JN Complex, Wards Lake Golf Course and Tara Ghar
Event Content	The four-day festival features a variety of programming, from individual competitions like a fashion show, singing, cosplay and many more, to the display of state products through concerts and other forms of entertainment.
Visitors Statistics	Over 15,000 visitors attended all the venues sites prepared for the event per day
Awards and certificates:	Gold Awards: Festival of the Year-Art/Culture/Lifestyle and Gold Awards: - Best Tourism Event categories Silver Awards: Concert of the Year.

Source: Compiled by Authors

CASE STUDY 2: ME'GONG FESTIVAL

In Meghalaya, the festival of Me'gong is named after one of the flowers (Bauhinia Purpurea), which bloom in abundance in the Garo Hills region of Meghalaya at the onset of autumn and last until about the end of February. The plant grows in abundance in the Garo Hills region of Meghalaya. As part of the local cuisine, young shoots that emerge during the jhum season, when the first rains fall, are collected and eaten by the people as a form of nutrition. Through games, shows, and performances, the event highlights the traditions and cultures of the people. The three-day festival showcases indigenous brews and beverages from all over the area and aims to promote the culture, diversity and lifestyle of the Garo Tribes of Meghalaya.

The first edition of Me'gong festival started in the year 2021 and it was inaugurated by the Hon'ble Chief Minister of State of Meghalaya, Mr. Conrad Sangma at Babadamgre on the outskirt of Tura. The Chief Minister said, *"The festival is part of the various celebrations to commemorate Meghalaya's 50th year of statehood and India's 75th year of Independence The three-day event includes include several varieties of brew coming from all over the region".* Apart from Meghalaya, participants from Arunachal Pradesh, Sikkim, and Assam are displaying their traditional beverages, which are generally composed of rice and fruits, during this year's festival. At the occasion, the Chief Minister launched the logo for *'Meghalaya at 50,'* in the presence of Honourable Member of Legislative Assembly of Meghalaya State Mr. Thomas A Sangma and Mr. Marcuise Marak, as well as Deputy Head of Mission, *Embassy of Sweden* in India Gautam Bhattacharya. The Tribal Research Institute (TRI) was declared by the Chief Minister, who also stated that Me'gong will be held at the TRI campus the next year as an annual event. Together with brew competitions and cooking demonstrations, this year's festival will feature rock performances by local artists and an exhibition of various North-Eastern dances.

The Second Edition of Me'gong Festival sponsored by the Government of Meghalaya was organised from December 1 to 3 at Jengjal Airport, located along NH-51, about 32 kilometres from Tura. To make

Table 3. Profile of Me'gong Festival

Name of Festival	*Me'gong Festival*
Type of event	*Entertainment Events*
Organizer	*Department of Tourism (DOT), Government of Meghalaya.*
Event Management Company	*Springboard Surprises* Event Production, Management, Promotion & Consultancy Firm in North East India
Event History	*The Me'gong festival was started in the year 2021. The event showcases the traditions and cultures of the people through games, shows and performances. The three-day festival showcases indigenous brews and beverages from all over the area and aims to promote the culture, diversity and lifestyle of the Garos.*
Event Held	*November- December (3 days of Events)*
Event Venue	*Babadamgre and Jengjal Airport, Tura, Garo Hills, Meghalaya*
Event Content	*The three-day festival features a variety of programs, such as Brew competitions, cultural performances, fashion shows, singing, traditional and modern music performance and many more.*
Visitors statistics	*Over 5,000 visitors.*
Awards and certificates:	*Awards are given to the best brews out of 50 brewers who participated in the event.*

Source: Authors

Table 4. A comparative analysis of cherry blossom festival and Me'gong Festival

Basis of Study	*Cherry Blossom Festival*	*Me'gong Festival*
Type of Event	Entertainment Event	Entertainment Event
Organizing Body	Department of Tourism Government of Meghalaya in collaboration with the Institute of Bio-resources and Sustainable Development (IBSD) and the Indian Council for Cultural Relation (ICCR) and Manage by Rock ski EMG.	Department of Tourism Government of Meghalaya.
Nomenclature & Theme	Cherry Blossom Flower (*PrunusSerrulata*)	Me'gong Flower(*Bauhinia Purpurea)*
Event Content	Musical showcase by local musicians. Indigenous produce by local vendors. Local Cuisines and Beverages. Traditional handicrafts.	The event showcases the traditions and cultures of the people through games, shows and performances. The three-day festival showcases indigenous brews and beverages from all over the area and aims to promote the culture, diversity and lifestyle of the Garos.
Impact on Tourism Development	Cherry Blossom Festival act as a catalyst to promote various tourist destination in the state. Since people from all over the world come to Meghalaya. In between their stay, they often do a lot of sightseeing which helps boost the economy of the state further. This also helps spread awareness of Meghalaya as a tourist destination all over the world through word of mouth as well as various media platforms.	Me'gong festival acts as a catalyst to promote the various tourist destinations in the Garo Hills Region of Meghalaya.

Sources: Authors

Figure 2. Six Cs of Tourism Festivals
Source: Authors

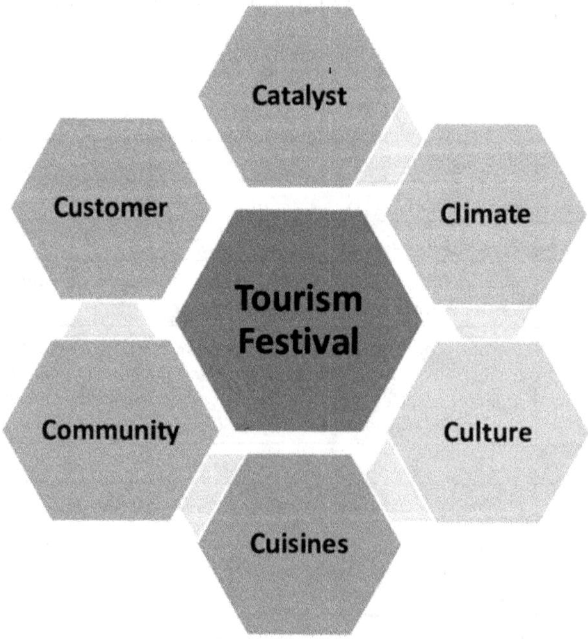

the festival even grander this year, the famous international soft rock/pop band, Michael Learns to Rock (MLTR), has been invited to enthral the crowd.

Activities conducted during Me'gong Festival: (i) Brew Competition, Strongman Competition, (ii) International Circus Act, (iii) Cultural Performances from across the Khasi, Jaintia & Garo Hills, as well as Sikkim, (iv) Fashion Show featuring the best established & up and coming designers from the State, (v) Traditional & Modern Music Performances, (vi) Laser & Light Show, (vii) Pyrotechnic Show, (viii) World class VIP Campsite, (ix) Public Campsite set up.

ROLE OF EVENT MANAGEMENT ORGANISER FOR THE PROMOTION OF TOURISM FESTIVALS

Here in both the festivals' the role of event management organisers is very crucial and strategic to create a unique ambience and festival design with effective all sorts of management functions and operations with coordinate by the most experienced and professional team who had paid the most attention to making visitor friendly environment.

- *Planning, designing, and organising tourism festivals*
- *Applying creativity, unbox ideas for the tourism festivals*
- *Strategic in event tourism planning and execution*
- *Support to develop destination's image by promoting, positioning, and branding it across the globe.*
- *Finding event most suited event locations and managing logistics*
- *Arranging light and sound, LEDs, innovative event technologies, drone shooting coverage of events, drone shows, music, artists, singers, catering, fireworks etc.*
- *Making strong festival marketing promotional strategies, and selecting various social media platforms.*
- *Post-festival feedback and communication*
- *Ensuring overall festival performance and progress review.*

Role of Meghalaya Tourism in the Promotion of Tourism Festivals

The Meghalaya Tourism organises and promotes various tourism festivals in three respective regions of Meghalaya i.e. Khasi region, Jaintia region, and Garo region from time to time as per the festivities or tourism seasons. The state has blessed lots of tourism potential due to its hilly terrains and most favourable climatic condition monsoon season, summer season, spring, autumn, and winter season. Meghalaya Tourism planned such tourism-centric festivals in a well-planned manner involving the proactive participation of event planners, tourism and hospitality stakeholders and the local community of the restive region to organise or celebrate such festivals.

Most recently Meghalaya tourism has organised various festivals in the state these are the following:

- *Monolith Festivals*
- *Winter Tales Festival,*
- *The Meghalayan Age Festival*

- *Shillong Autumn Festival*
- *The Hills Festivals*
- *Cherry Blossom Festival*
- *Me'gong Festival*
- *Meghalaya Grassroot Music Project*

Figure 3. Role of Meghalaya Tourism for the Promotion of Tourism Festivals
Source: Authors

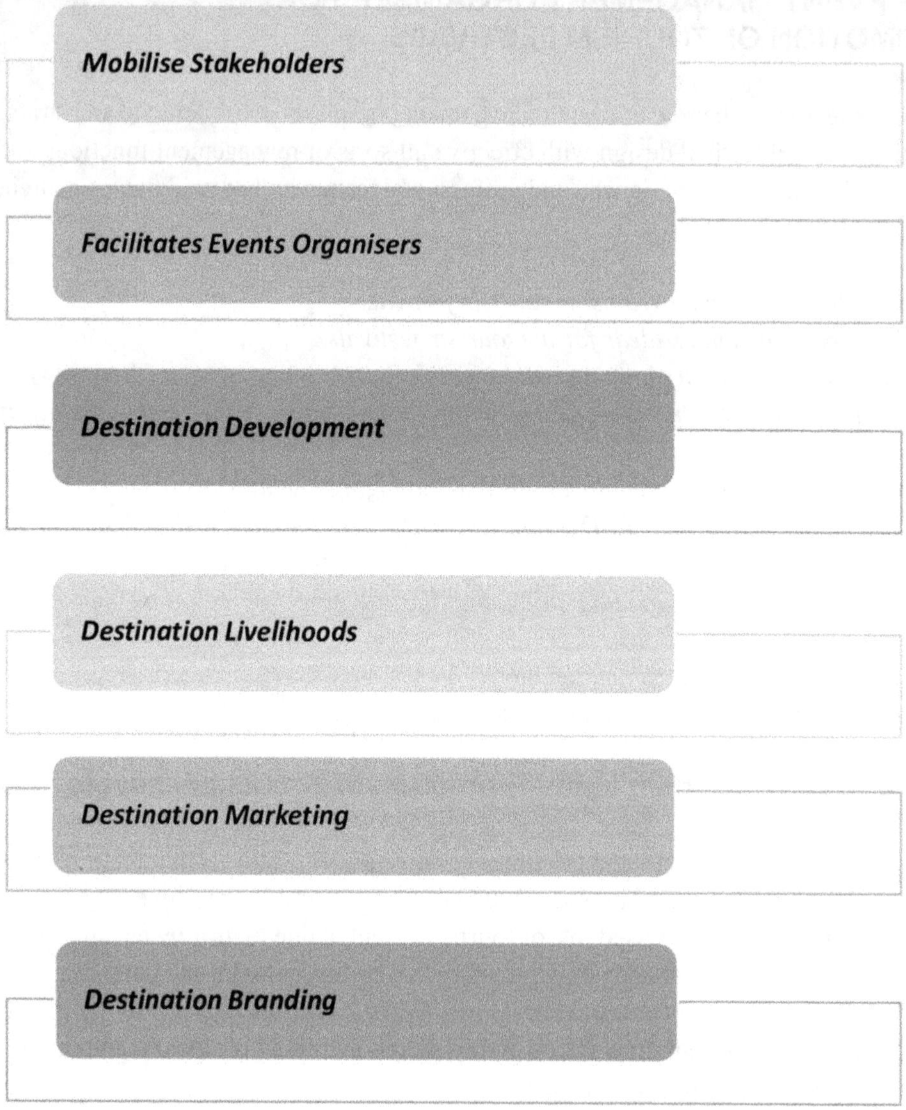

STUDY FINDINGS AND OUTCOMES OF CHERRY BLOSSOM AND ME'GONG FESTIVAL

The guests, and visitors who visited and participated in the 1st edition of the Me'gong festival organised by the Department of Tourism, Government of Meghalaya included Special Interest Tourists (SIT) such as Gastronomic Tourists who are interested in Indigenous Food and Beverages, Tourists and visitors interested in local brews, Eating competition, Traditional Games, Traditional and modern Music, Indigenous art and crafts, Traditional attire fashion shows, Wangala Dance (Hundred drums) etc. The host communities included local brewers from the whole North-eastern states of India, Fruit Vintners, musical performers etc. Overall, the response from guests and hosts was overwhelming. The 1st edition of the festival witnessed an impressive crowd and the feedback was satisfactory. Even though the organisers took care of all the aspects of successful event planning, some constraints needed to be dealt with. The guests and visitors suggested having more stalls and outlets for indigenous cuisines of North East India. They also looked satisfied with the accommodation facilities available in the form of campsites, however, they were not sufficient for all visitors and tourists. The visitors were satisfied with the level of hygiene and sanitation, and the availability of volunteers in case of any assistance.

Sustainable and Green Practices in Tourism Festivals of Meghalaya

The Tourism festivals of Meghalaya promote sustainable and green practices. The venue of the Me'gong festival was Babadamgre which is surrounded by lush green cashew nuts orchards. The stalls and the outlets showcased the traditional setup using bamboo, canes and shrubs. The traditional and rudimentary tools and equipment used in indigenous cuisines, brewing, and serving the local brews were well showcased during the event. The traditional leaf plates, bowls, spoons, bamboo containers and mats, earthen pots and potteries were abundantly used to showcase the sustainable and green practices adopted by the host community of Meghalaya. The portrayal of tree houses, *"Borang"* shows the natural touch and sensitiveness towards the nature of the Garo Tribes. The local delicacies were served in leaf plates and bowls, whereas the brews were poured and tested using a natural dispenser called "*pong*". The drinking water facilities were provided with glasses made out of bamboo. The sitting arrangements are made so that the visitors can sit and enjoy the food, fashion shows and live musical performances, except for the VVIP visitors and delegates.

Digital Implication in Tourism Festivals

Digital media and social media platforms play a major role in creating awareness and publicity of tourism festivals. The social media handlers of the Department of Tourism, Government of Meghalaya, and event management teams play an active role in facilitating and mobilising the needs and requests of the hosts, tourists and visitors. The accommodation facilities offered by the organisers were booked Online and the Digital payment gateways were also accepted. However, the stalls were not offering and accepting any digital payments. This is one area organisers can consider for increased visitor satisfaction. When the pandemic impacted Meghalaya's economy and the nation as a whole, many events in Meghalaya have been encouraged to be hybrid events, where everyone may attend the event both physically and digitally. Today's festivals and events are also focused on minimising their negative environmental effects by opting for paperless transactions and utilising the functions of event apps. Such an example is an interactive computerised schedule and map: Guests can easily find what they're looking for in-app rather than sifting through several email attachments or utilising paper maps to choose where to go next. Digital flyers, QR code scanning, WhatsApp and swap cards etc are some other digital tools that have been employed in the festival and events. Instead of exchanging business cards that are easily lost and end up in the garbage, attendees and sponsors may connect through the event app.

CONCLUSION

The Government of Meghalaya envisages establishing Meghalaya as one of the top 10 states of the country and tourism plays a pivotal role in achieving this goal. The policies and programs implemented by the Government work towards trying to make Meghalaya the most sought-after tourist destination. Meghalaya, also known as the Abode of Clouds, is known for its abundant biodiversity areas, famed monsoons, and dramatic terrain. However, Meghalaya also has a rich cultural and natural heritage, and many festivals are celebrated all around the year to commemorate the same. This makes Meghalaya an exemplary destination for avid travellers around the globe. They can immerse themselves in the vivid ethnic heritage of the state These festivals form an integral part of showcasing the cultural wealth of the State and unlocking its potential through tourism. The most prominent festivals taken into consideration for the present study are the Cherry Blossom festival and the Me'gong Festival. One of the most vibrant and beautiful celebrations held each year in Meghalaya is the Cherry Blossom Festival, which takes place in Shillong. The Cherry Blossom Festival is celebrated in Shillong in November to mark the blossoming of pink cherry flowers. It is a 10-day celebration comprising a series of events including music and literature festivals. The Me'gong festival is celebrated during the first week of December in the Garo Hills region of Meghalaya. This celebration is given the name Me.going after a flower that grows plentifully in the Garo Hills. Through games, shows, and performances, the event showcases the traditions and cultures of the people. The three-day festival also showcases beverages from all over the area. The overall analysis shows that regional festivals can play a major role in regional destination branding and in improving the image of local food and culture. It suggests, however, that government support and sponsorship, and festival management skills are essential to enhance the role of festivals in regional destination branding.

REFERENCES

Blain, C., Levy, S. E., & Ritchie, J. R. B. (2005). Destination Branding: Insights and Practices from Destination Management Organizations. *Journal of Travel Research*, *43*(4), 328–338. doi:10.1177/0047287505274646

Celik, S., & Cetinkaya, M. Y. (2013). Festivals In Event Tourism: The Case of International Izmir Art Festival. *International Journal of Contemporary Economics & Administrative Sciences*, *3*(1), 1–21.

Chalip, L., & Costa, C. A. (2005). Sport Event Tourism and the Destination Brand: Towards a General Theory ". *Sport in Society*, *8*(2), 218–237. doi:10.1080/17430430500108579

Choudhury, T. (2021). Experience the Magical Cherry Blossom Festival 2021 In Shillong. [Online News]. *East Mojo*. https://www.eastmojo.com/meghalaya/2021/11/26/experience-the-magical-cherry-blossom-festival-2021-in-shillong/

Cudny, W. (2013). Festival tourism – the concept, key functions and dysfunctions in the context of tourism geography studies. *The Geographical Journal*, *65*(2), 105–118.

De Chernatony, L., & McDonald, M. (1992). *Creating powerful brands*. Butterworth-Heinemann.

Demetriou, D. (February 12, 2019). Everything you need to know about seeing cherry blossoms in Japan. [online newspaper]. *Telegraph*. https://www.telegraph.co.uk/travel/destinations/asia/japan/articles/japan-cherry-blossom-sakura-guide

Derrett, R. (2003). Festivals & regional destinations: How festivals demonstrate a sense of community & place. *Rural Society*, *13*(1), 35–53. doi:10.5172/rsj.351.13.1.35

George, B. P. (2017). The evolution of destination branding: A review of branding literature in tourism. *Journal of Tourism. Heritage & Services Marketing*, *3*(1), 9–17.

Getz, D. (1991). *Festivals, special events, and tourism*. Van Nostrand Reinhold.

Getz, D. (1997). *Event management and Event Tourism*. Cognizant Communication Corporation.

Hankinson, G. (2005). Destination brand images: A business tourism perspective. *Journal of Services Marketing*, *19*(1), 24–32. doi:10.1108/08876040510579361

Ian, Y., Robertson, M., Ali-Knight, J., Drummond, S., & McMahon-Beattie, U. (2004). *Festival and Events Management*. Elsevier.

Jayaswal, T. (2009). *Events-induced tourism: A protocol analysis*. [Master's Thesis, Auckland University of Technology].

Li, J., Dai, G., Tang, J., & Chen, Y. (2020). Conceptualizing Festival Attractiveness and Its Impact on Festival Hosting Destination Loyalty: A Mixed Method Approach. *Sustainability, 12*(8), 3082. doi:10.3390/su12083082

Morrison, A. M., & Anderson, D. J. (2002). Destination branding. Paper presented at the *Missouri Association of Convention & Visitor Bureaus Annual Meeting*. MAC.

National Park Service. (2021). *History of Cherry trees*. NPS. https://www.nps.gov/subjects/cherryblossom/history-of-the-cherry-trees.htm

Nicholson, R., & Pearce, D. G. (2001). Why do people attend events: A comparative analysis of visitor motivations at four South Island events. *Journal of Travel Research*, *39*(4), 449–460. doi:10.1177/004728750103900412

Pirnar, I., Kurtural, S., & Tutuncuoglu, M. (2019). Festivals and destination marketing: An application from Izmir City. *Journal of Tourism. Heritage & Services Marketing*, *5*(1), 9–14.

Piva E., Cerutti S., Prats L., & Raj R. (2017). *Enhancing Brand Image through Events and Cultural Festivals: The Perspective of the Stresa Festival's Visitors.*

Rahman, M. K., Masud, M. M., Akhtar, R., & Hossain, M. M. (2021). Impact of community participation on sustainable development of marine protected areas: Assessment of ecotour -ism development. *International Journal of Tourism Research.*

Revilla Hernández, M., Santana Talavera, A., & Parra López, E. (2016). Effects of co-creation in a tourism destination brand image through Twitter. *Journal of Tourism. Heritage & Services Marketing*, *2*(2), 3–10. doi:10.5281/zenodo.376341

Richards, G., & Leal Londoño, M. P. (2022). Festival cities and tourism: Challenges and prospects. *Journal of Policy Research in Tourism, Leisure & Events*, *14*(3), 219–228. doi:10.1080/19407963.2022.2087664

Singh. S. S. (2017). No carpet of colours at Shillong festival, as cherry blossoms fail to appear [online newspaper]. *The Hindu.* https://www.thehindu.com/news/national/other-states/meghalaya-no-carpet-of-colours-at-shillong-international-festival-as-cherry-blossoms/

Stavska, Y. (2021). The Features of Festival Tourism as a Component of Event Tourism. *The Scientific Heritage*, (66), 23–32. doi:10.24412/9215-0365-2021-66-3-23-32

Tasci, A. D. A., & Kozak, M. (2006). Destination brands vs. destination image – do we know what we mean? *Journal of Vacation Marketing*, *12*(4), 299–317. doi:10.1177/1356766706067603

Xing, X., & Chalip, L. (2006). Effects of Hosting a Sports Event on Destination Brand: A Test of Co-branding and Match-up Models. *Sport Management Review*, *9*(1), 49–78. doi:10.1016/S1441-3523(06)70019-5

Chapter 15
Value Creation of Destinations by Destination Management Organisations Through Strategic Planning and Promotion of Fairs and Festivals of NE India

Angshumi Goswami
North-Eastern Hill University, India

ABSTRACT

This chapter aims to add value to available existing literature in term of highlighting the catalytic roles and responsibilities of destination management organisations (DMOs). Moreover, it aims to study the involvement of DMOs in strategic planning and promotion of local fairs and festivals. Destination management organisations (DMOs) act as torchbearers of every global and local tourism destination and are indispensable when promoting a destination through events and festivals. In this context, the sustainable development of the destination is imperative and a crucial factor for prospective visitors and guests. This chapter shall comprehensively discuss DMOs and their underlying roles and responsibilities. Moreover, it shall highlight the operational activities undertaken by city-level DMOs in relation to the strategic planning and promotion of events being hosted. This study shall be purely based on reviews done on existing literature supported by cases of major fairs and festivals specifically with reference to North Eastern India.

INTRODUCTION

In the past few years, the tourism industry has managed to be among the top three export industries contributing approximately US$ 5.81 trillion to the worldwide GDP in 2022. This figure attributed to

DOI: 10.4018/978-1-6684-6356-7.ch015

almost 6.1% of the total GDP. According to the World Travel and Tourism Council, in the pre-pandemic scenario i.e., in 2019, almost 10.3% jobs were created by tourism and international visitor spendings amounted to US$ 1.8 trillion. However, with the surge of covid-19, these numbers dropped sharply in 2020. Tourism statistics depict the international tourist arrivals at 917 million in 2022 in contrast to the pre-pandemic times (WTTC, 2022). Tourism being one of the major industries in the world plays and important role in the economic development of many countries, especially the developing ones. It supports nations in overcoming pressing challenges of unemployment and monetary instabilities thereby fostering economic growth. Understanding the vast potential of tourism, destinations are now intensely competing with each other to be recognised as the most attractive ones and to be at the forefront. Due to the ever-increased competition it is crucial that destinations retain their competitive advantage and achieve its underlying objectives. To this end, many destinations are led by destination management organisations (DMOs) to oversee the management of tourism of that region (Bornhorst et al., 2010).

Tourism destinations are complex structures which cannot easily be managed. In such times, a balanced mix of human expertise, organisational collaboration and innovation deems necessary to ensure critical success of a destination. Destination management and marketing functions are attributed as pillars of destination success in this globalised and competitive tourist market (Pike & Page, 2014). As per Bornhorst et al. (2010), a DMO's role can be broadly described as working towards the well-being of destination residents on one hand and thriving to offer visitor satisfaction which is highly satisfactory and memorable on the other. DMOs are aimed at generating tourist visitation to a region and create a positive image in the minds of prospective visitors – both mental and visual. The global circulation of images of destinations around the world is now a powerful tool to showcase the touristic resources (natural, cultural, religious and so on) inherent to a destination. The current chapter aims to assess how exactly DMOs operating in a region work in collaboration with tourism stakeholders to bring about tourism growth in a destination. The context of study however is limited to fairs and festivals celebrated by communities and cities throughout the year which holds high values for tourists visiting a new destination (here, the NER). These fairs and festivals are representations of a region's traditional and cultural elements which make it unique from the rest. Customs, social practices, believes, cuisines, dance forms, art forms, rituals – are all subsets of cultural tourism. The typology of events that can be celebrated is significantly diverse. These can be cultural celebrations, political, arts and entertainment, scientific and educational, business and trade and recreational. Although in recent times, Assam Tourism Development Corporation in alliance with other bodies, are organising various types of events, the prominence of cultural events is undoubtedly more.

The chapter starts with a background study on DMOs and their underlying roles and responsibilities followed by a brief on the stature of DMOs in India specifically North East India. Further, it focuses on the innovative strategies undertaken by DMOs to successfully plan and implement fairs and festivals. Later in the chapter, the challenges faced by DMOs while managing and implementing events shall be discussed and it shall end with the practical/ managerial implications for the DMOs. The region under study is the North Eastern Region (NER) of India with two case studies on Ambubachi Mela of Kamakhya Temple, Assam and Hornbill Festival of Khonoma, Nagaland. The main limitation was observed as unavailability of adequate literatures on overall DMOs operating in the NER let alone subjective literatures on DMOs and their involvement with any fairs and festivals conducted in the region. Also, only two cases as examples have been mentioned in this chapter. Although case studies were found related to the area of work however, the mention of DMOs as leaders were not made. This was a major gap that has been addressed and suggestions made through this chapter can be put into action to overcome challenges of management and sustainability.

Methods

This segment highlights the research methodology chosen for achieving the objectives of the study. The objectives of this study are: *a)* to determine the roles and responsibilities of DMOs at a national and regional level; *b)* to determine the prominence of DMOs in NER; *c)* to study the involvement of DMOs in strategic planning and promotion of fairs and festivals in NE; *d)* to analyse the operational activities undertaken by DMOs while hosting events, and finally, *e)* to highlight practical implications of this study for DMO reference. An attempt was made to showcase the underlying roles and functions of DMOs across the country in comparison to countries abroad. Through this, the relevance of such roles in the North-eastern region is understood. Strategies for planning and promoting an event and the related impacts on destination growth and sustainability is analysed. Finally, a few challenges facing DMOs and recommendations for practical application is reported. The chapter is purely descriptive in nature and literature-based. Materials for the study have been collected through journals, internet sources, published research papers and articles, newspaper, etc. No such interviews were conducted with DMOs instead meaningful data were extracted from recent literatures. One major limitation of the methodology used was the lack of very recent updated literatures on fairs and festival in North East Region.

LITERATURE REVIEW

Destination Management Organizations and Destination Development

Tourism is one of the fastest growing sectors globally (Chaturvedi, 2010; Chaudhary, 2010; Sharpley, 2006; Sheehan et al., 2007) comprising a complex system of interrelated but independent parts. According to the United Nations World Tourism Organization (UNWTO), with the increase in the number of destinations investing in tourism development, modern tourism has become a key driver for socio-economic progress through the creation of jobs and enterprises, infrastructure development and revenues through exports (2012). In addition, every tourism destination is an amalgamation of several elements and stakeholders (hotels and government primarily) (Sheehan et al., 2007) and including the local indigenous communities or the service providers. As such, it is vital that the elements are coordinated and managed in the best possible way so as to provide positive visitor experiences and increased destination brand value. On the above backdrop, the tourism sector has evolved tremendously through the changes in consumer tastes and preferences, technological advancements, increased demand and supply of services and the tourist's involvement in purchase decision-making. The achievement of successful tourism development depends on tourism planning as the starting point and this is to be incorporated with any other local planning (e.g., Municipality) (Department of Tourism; South Africa, 2018).

Tourism development refers to the growth and maintenance of the tourism industry in a given locality (Stainton, 2018). Being termed as the 'destination developers' (Sheehan et al., 2016), DMOs have become the key organization which is responsible for leading, stimulating, co-ordinating, as well as monitoring tourism development and marketing of any destination. DMOs provide leadership and direction (through planning and strategies) for tourism development (Nergusa & Coros, 2016). Hence, proper destination management through collaboration of stakeholders, benefit to local communities, destination branding and positive visitor experience not only enhances destination competitiveness but also paves the way to planned tourism development of a region (Varghese, 2016). Destination management organizations (DMOs) act as an essential marketing and management tool for tourism destinations responding to increases in demand, global competition and fragmentation within the industry. DMOs are often entitled with the task of managing a tourist destination with promotion and marketing of cities and states to consumers and other stakeholders being a regular feature (Morrison, 1998). It has been defined as the actor that is in charge of strategic planning, managing and organizing destination resources (Pike, 2004; Presenza et al., 2005). Owing to the paradigm shift in tourism due to globalisation, a destination's sustainability and competitiveness primarily depends on the ability of its DMOs to manage destinations (Pike & Page, 2014; Ritchie & Crouch, 2003; Volgger & Pechlaner, 2014). A DMO's presence is crucial for its ability to strategically handle tourism destination marketing and development as well as facilitate partnerships and collaborations with industry players (WTO, 2007). This management should be able to satisfy the economic, social, and aesthetic needs. At the same time, it should maintain cultural integrity, essential ecological processes, biological diversity and life supporting ecosystems (Aleksandrov, 2014).

Evolution of DMOs can be pointed out to the year 1864 when the first Regional Tourism Organisation (RTO) was set up at St. Mortiz in Switzerland. The first convention and visitor's bureau (CVB) were set up in the USA at Detroit in the year 1896 and the world's first National Tourism Organisation (NTO) was reported in New Zealand in 1901. From the research conducted by Pike over the years, as taken from Kumar et al. (2020), the first mention of destination marketing was done back in the year 1973. Ever since, these organisations became catalysts of tourism development by mid of 20th century. Years later, the Australian National Tourist Authority (1929), Canadian Bureau of Tourism (1934), French NTO (1910), Italian NTO (1919) and a few others were established. According to the author, most DMOs came into existence during 1970's and 1980's. In Asian countries, tourism associations in Hong Kong, Japan, Thailand, Singapore, and India were formed as DMOs. Indian Tourism Development

Table 1. Highlights the evolution of the concept of DMO and formation of entities over the years in India (adapted from Kumar et al., 2020)

Sl. No.	Committee/Program	Year	Purpose
1	Sir John Sargeant Committee	1945	Promotion of tourism in India
2	Department of Tourism	1958	Promotion of tourism in India and abroad
3	L. K. Jha Committee	1963	Investigation of decline in tourist arrival
4	India Tourism Development Corporation (ITDC)	1966	Management of hotels, transport, entertainment and promotion of tourism
5	Ministry of Tourism	1967	Formulation of policies and programs for government agencies, private players for development and promotion of tourism.
6	Estimates Committee	1969	Effective and efficient central organisation, Selection of person for DoT, India
7	Report of Indian Institute of Public Administration (IIPA)	1970	Type of organisation and staffing pattern in DoT, India
8	First Tourism policy	1982	Priority of tourism in India
9	National Committee on Tourism	1986	National Tourism Board and IITTM
10	National Action Plan	1992	Increase foreign exchange and increase employment
11	Second tourism policy	2002	The growth of tourism in India

Tourism (ITDC) was established in 1966 and was the first NTO of the nation operating in collaboration with the Ministry of Tourism. Noteworthy, the organisational structure of Indian DMOs is unique when compared to other countries. Ministry of Tourism is the apex body controlling the functions of ITDC (NTO) and other State/ Union Territory Tourism Departments. Since State level DMOs promote their geographical areas, they are called RTOs. The need for development of an NTO in India was laid down by the inclusion of tourism in the Five Year Plans after which the Sargeant Committee was formed for tourism promotion in 1945 by the Indian Government. This committee recommended the setting up of representative tourism organisations and its regional branches and therefore, the Department of Tourism came into effect. Table 1 illustrates the Evolution of DMOs in India by Kumar et al. (2020).

Overall Roles and Responsibilities of DMOs

The DMO focuses on marketing as the principal management function, but they also play a role in product development and operations (Dore & Crouch, 2003). Presenza et al. (2005) classified the functions of DMOs on basis of two significant roles namely, external destination marketing and internal destination development. In general, the primary role of a DMO consists of marketing, promotion and sales, coordinated destination planning and long-term management (Bieger et al., 2009; Pearce, 1992; van Harssel, 2005). However, the roles of a DMO in terms of sustainability are far more complex than mere promotion and marketing. It involves conserving a destination's richness by minimizing all the probable negative impacts of tourism as well as preserving its natural resources (Bonardi et al., 2006; Buhalis, 2000; Fennel, 2008; Honey, 1999), developing a destination's image and reputation as sustainable and ecofriendly (Crouch, 2007; Osmankovic et al., 2010; Wight, 1998; Wray, et al., 2010), contributing the maximum share of economic benefits to the local community (Crouch & Ritchie, 1999; Ritchie & Crouch, 2003; Simpson, 2008; Tosun & Timothy, 2003), satisfying visitors' needs and shifting their focus to sustainable products requiring optimal consumption of resources (Budeanu, 2007; Thidell & Leire, 2004; Lohmann, 2004; Miller, 2003; Paul et al., 2002; Sharpley, 2006) and laying importance on local population and public-private stakeholders (Aas et al., 2005; Bornhorst, et al., 2010).

Tourism destination management functions broadly involves the following (Bornhorst et al., 2010; Morrison, 2013):

- Leading and coordinating the tourism efforts of all stakeholders/ service providers within the destination.
- Marketing the destination for image building. Promoting the destination, information facilitation, managing bookings and maintaining customer relations.
- Making local communities aware of the importance of visitor industry through proper leadership and tourism advocacy.
- Assisting the development of tourism facilities, events and program that helps in destination positioning.
- Ensuring seamless customer experience by providing best visitor services.
- Working in collaboration with key stakeholders to enhance visitors.

Figure 1 illustrates the DMO roles given by UNWTO as understood by Morrison (2013).

Figure 1. A flowchart indicating the management roles of a DMO as defined by the United Nations World Tourism Organization (UNWTO)

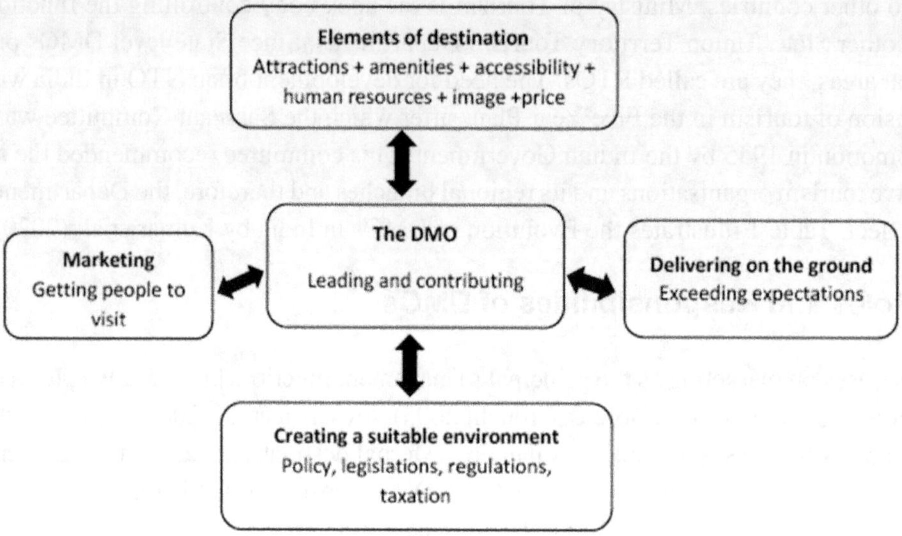

Successful destination management simply means establishing equilibrium between the tourism region, its stakeholders and the local residents. The major aim is to have a common vision that addresses the needs of a diverse set of stakeholders. Moreover, striking a balance between the three major pillar of sustainability namely, social/cultural (especially the local community), environmental (use and preserve natural resources) and economic (focussed on tourism-based organisations) is crucial. Morrison (2013) argues that NTOs are the most powerful organisations in the nation operating the tourism sector. Its roles range from introducing regulations and legislations, policymaking, planning and developing strategies, tourism research, capacity building through trainings, and so on. Regional organisations (RTOs) also perform similar roles focussing more on domestic/ state tourism.

DMOs in Strategic Planning and Promotion of Fairs and Festivals of North East India

The increased global competition in tourism in both the domestic and overseas markets has forced tourism destinations to apply fundamental marketing practices to destinations, notably the application of destination branding (Hanna & Rowley, 2008). According to Zhang and Murphy (2009) successful marketing of a destination requires an effective and healthy management mechanism. Marketing helps in adding value to the destination by developing comprehensive promotional strategies. These strategies are formulated and implemented in isolation or in unison with other destination stakeholders. As such, the specific activities of DMOs vis-à-vis marketing include identification of key tourist markets that shows keen interest in visiting the destination; devising branding techniques to compel such target markets and image creation through branding. Buhalis (2001) points out three strategies for enhancing destination marketing efficiency namely, enhanced tourist satisfaction and visitor happiness; building up profitability and competitiveness of the local tourism industry, local small enterprises and medium sized tourism enterprises and developing destination sustainability and ensuring prosperity of the host population.

India being a diversified nation, is an assimilation of social and cultural ethnicity and traditional beliefs. It inspires awe amongst tourists far and wide. The traditions and customs prevalent across the country urges the local community to live in harmony in the society. Distinct with an assortment of dialects, attires, religious practices and scrumptious delicacies, the Indian culture though fluctuates, shares similar characteristics with one another. It represents different religious thoughts of different communities. The socio-cultural diversities are found in the folklore, religion, customs and traditions, conventional practices, indigenous celebrations, and fairs. Fairs and festivals happen to be the backbone of all communities. The cultural aspect of communities is supported and strengthened by such regional events. In India, local fairs and festivals are being held quite frequently. Such celebrations are embraced by people with open hearts as it encourages social bonding. North East India is a gateway to the South Asian region and has the potentiality to be developed as the country's powerhouse. The region falls in the Eastern Himalayas mega biodiversity hotspot and has abundant natural and mineral resources. It is a conglomerate of eight states and is home to various tribes with variations in language, cuisines, ideologies, social organisations, cultural traditions and economic life. The variety in crops, medicinal herbs, agricultural produce, flora and fauna makes it a desired destination. The might river Brahmaputra and its tributaries also makes tourism viable in the region. This easternmost part of the country has a lot to offer to anyone visiting the region. The huge cultural and economic potential that is offered by the fairs and festivals is noteworthy (Thomas, 2020). The NER could strategically exploit its geographical significance to reap benefits for the region as well as the country on whole.

North East India has unique cultural and historic elements which forms a major part of the tourism industry of the region. It is believed that tourists spend more on cultural tourism at any point in time. Cultural tourism is about travelling towards experiencing a new art, heritage, and a special character of a place while being able to learn the history, heritage and lifestyle of that community and also contributing to the economic well-being of the hosts. NER celebrates its tribes through fairs and festivals showcasing the lifestyle, traditions and rituals of the indigenous cultural communities which gives a good opportunity to the tourists. Bihu, Jonbeel Mela, Shivratri, Raas in Majuli are some of the noteworthy events observed in Assam. Given that Assam also has a mix of tribes, some other tribal festivals are also celebrated (Das & Acharjee, 2013). Ambubachi Mela is an annual event that celebrates the bleeding Goddess Kamakhya, at Kamakhya Temple of Assam. The temple constructed atop Nilachal hill is one of the 52 Shakti Peeths (shrines) and as per Hindu mythology, has been built on the 'genitalia' of Goddess Sati which fell on this spot. Ambubachi Mela is held during the Ahaar month of the Assamese calendar for a period of four days. Thousands of tourists and devotees throng the temple on these days. Ambubachi celebrates the annual cycle of menstruation of the goddess and can also be related to cult worship. In practice, all kinds of agricultural activities like ploughing and sowing are ceased. The temple is closed for devotees for the first three days and finally reopens on the fourth day. The 'mela' is so called because on these four days, a 'mela' or 'fair' takes place in the temple premise. Kamakhya temple is considered a prominent religious centre for Indian as well as inbound tourists. A study was conducted by Sharma (2019) to assess the current scenario of tourism in Kamakhya temple during Ambubachi mela. An attempt was also made to assess the socio-economic impacts of the fair on the local residents alongwith the roles played by the local DMOs for its promotion. The author studied a mixture of eight positive and negative variables to study impact which was ranked after necessary analysis. Results showed that the positive variables from highest to the lowest were ranked as 'increase in resident income', 'creation of employment opportunities', 'provision of market for local produce' and 'improved standard of living' respectively. The negative impacts from highest to lowest ranking were 'problems of pollution', 'shortage of drinking water', 'excessive dependence of local communities on tourism' and 'frequency of crimes' respectively. From the marketing perspective, significant measures have been taken by Assam Tourism Development Corporation (ATDC) in collaboration with India Tourism as well as in PPP model, however, a lot is yet to be done. An independent Board of Trustees is also appointed for managing the temple and all events related with it. The ATDC has promoted the mela through its official website over the years. Another website kamakhyatourism.com was opened and Facebook pages were opened and operated with the same objective of better promotion. Many innovative strategies were formulated by the Govt. like conducting road shows across major states like West Bengal, Delhi, Maharashtra and others to promote Ambubachi Mela. Hoardings in over 300 religious destinations of India were erected. Advertisements were made in cinema halls and FM radio for better audience reach. Different parts of the city such as the temple, airport, parks and bus stands were illuminated. A cultural event named 'Katha Vashak' was organised with the participation of renowned artistes and was aired in various television channels (Sharma, 2019). After a hiatus of two years due to Covid, in 2022, Ambubachi Mela was celebrated again following strict guidelines issued by the Kamrup Metro District Administration. The authority restricted lodging facilities around the temple premise. Unauthorised distribution of packaged food and water other than the authorised organisations was also prohibited. Every year, ATDC with other administrative bodies like Guwahati Municipal Corporation (GMC), PWD, and concerned departments work full swing to ensure things are in place for smooth functioning of the fair.

Hornbill Festival of Nagaland has become a brand name amongst tourists worldwide. It was first celebrated in the year 2000 under the aegis of Government of Nagaland who promoted it as a cultural event to give audiences a pan-Naga experience. Kohima was initially selected as the festival ground which was later shifted to its permanent location in Kisama. The festival is held in the month of December every year for ten days straight and performers enact the traditional and modern lives of Nagas through performances. The state is home to sixteen different tribes and a fixed slot is given to each and every tribe to perform their songs, dances and narrate stories to the audience. In addition, car rallies, rock shows, fashion shows, horticultural produce, book festivals and children's carnival. Hornbill is a platform which seeks to preserve Naga culture by showcasing the Naga traditional identity to the world. So far, this festival has been a commercial success among the local public as well as the tourists. Every December, the otherwise quiet state hustles and bustles with tourists and taxis all over the place. The city is decorated with numerous billboards of the festival which gives a colourful edge to the place (Longkumer, 2016). The author in his studies highlight how organisers and locals are aware of revenues generated through the Hornbill Festival. Different businesses flourish at the end of every year as inflow of tourists is directly proportional to their interest in buying local products. It fosters the entrepreneurial spirit among local traders to sell their goods like pickles, jam, locally brewed beer, handloom and handicrafts. The festival becomes an educational tool for the younger generation of Nagas who are not as well acquainted with their own traditions as their elders. It is a way of familiarising them to the older ways of life thereby ensuring continuity of their cultural distinctiveness. Emphasis on how the local tourism authority has managed and organised the event was made. Amidst all the positive aspects of Hornbill, scholars over the years as Longkumer (2016) says, have had rather conflicting views about this event. It was viewed as 'encouraging antisocial activities and sexual immorality'. The open trade and consumption of alcohol and local brew was debated upon by a segment of the public. Littering and garbage following the ten days long festival was also seen as a serious issue. However, keeping tourism as a pivot, and with certain regulations in place, the event has successfully been carried on all these years.

Value Creation of Destinations by DMOs Through Events

The concept of value has several approaches. Service value can be the customer's experience during the delivery or any behavioural changes due to innovating services. Lately, product-and-sales oriented strategies are replaced by customer-oriented ones. The sustainability of a service provider's competitive advantage in totality depends on the value of services delivered by them. Therefore, in order to achieve a higher level of value through events, it is necessary for event planners and business managers to comprehend the process of creating and capturing value. Lepak et al. (2007) states that "value creation depends on the relative amount of value subjectively realized by a buyer or user. This subjective value realization translates into the user's willingness to exchange a monetary amount for the value received." As per Bowman and Ambrosini (2000), use value (V) can be defined as the "subjective valuation of consumption benefits by a consumer (i.e., the maximum willingness to pay for a firm's products or services)" whereas exchange value (P) is the "monetary amount the consumer actually pays in a market exchange (i.e., the purchase price)". Value creation, however, happens in several levels: individual (the tourist), organisation (the event organisers), and societal or regional. This chapter primarily studies value created at the organisation level, i.e., in DMOs.

As is already understood, destination value is created through a sequence of primary and support activities which are fundamental for the performance of the tourism sector. These activities are collectively named as 'value-chain' and the concept is integral for value addition in destinations. Tourism value-chain activities are an integration of related activities which produces values for both the tourists and the service providers. It is a two-way process and is deemed successful only when it generates positive experiences for tourists consuming the service. Furthermore, since tourism is experiential in nature and services are produced and consumed simultaneously, the customer's role is quite significant. Activities such as product development, sales and destination operations and services are the key primary activities of the tourism value chain and has high impact on value creation of a destination. Value chain activities, hence, are the building blocks of competitive advantage of any tourism destination (Mai, 2021).

DMOs are one amongst the many organisations engaged in planning, organising and implementing events in a tourism destination. Firms like such create value through innovating new methods of production and other technological advancements in order to establish consumer's value towards the service delivered. To innovate successfully, firms need to replace new capabilities for old ones, engage in strategic networking and work in collaboration with other destination stakeholders (Lee et al., 2020). While planning and managing events in a destination, DMOs need to ascertain variables such as motivations, preferences, work and leisure contexts, and cultural as well as community influences accordingly. Other activities for value creation as stated by Lee et al. (2020) includes "adequate resource acquisition and allocation, partnerships and collaboration, image building of organised events and media coverage, co-branding of events and destinations and identifying primary cultural variables that shall affect the value and attractiveness of tourism events." In concert to this, government involvement in event tourism is crucial. Coherence between government's tourism-related activities with those of other stakeholders can go a long way into creating destination value. Higher customer value perceptions can be achieved through knowledge and resource sharing among stakeholders of the tourism industry thereby collectively developing and administering event-based policies (Lee et al., 2020). Organizations should also recognize that the customer experience is often more important than the event itself. In other words, customers may be willing to pay a premium for an event that offers a great experience, even if the event itself is not particularly unique or special. Thus, it is important for organizations to focus on creating a positive customer experience from the very beginning.

FINDINGS AND CHALLENGES

Destination Growth and Sustainability Through Fairs and Festivals

Festivals and celebrations help to strengthen the social and cultural roots and give communities the authority to preserve and uphold their traditions, and bring financial values. Events are perceived as a tool for economic and social progress. Each of these occasions is associated with the fulfilment of financial values, such as agricultural bliss, prosperous trade, innovative entrepreneurial growth, and social and community advancement. Local and regional fairs and festivals are increasingly being used to advance the tourism industry and regional economic development. The celebrations and festivals can serve as a major impetus for the advancement and improvement of cottage industries, safeguarding of rural culture, traditions and beliefs, ethnicities and more. The events and celebrations give opportunities to the community members to exhibit their rare skills in creating unique ethnic commodities and furthermore help for its promotion (Thomas, 2020). "Rural religious celebrations and fairs provide opportunities to enhance social, financial and cultural trade" (Hu, 2007). They have transformational impacts on societies by ensuring group cohesion and knowledge transmission. Community-based events and festivals provide an opportunity to celebrate the rural distinctiveness of the local community while also strengthening the region's tourism industry. The events and celebrations can act as a catalyst to supplement the region's infrastructure needs. Events and celebrations provide opportunities for rural communities to showcase their talent, craftsmanship, craft skill, dance, cuisines, and various types of societal landscape, as well as promote noteworthy livelihood through enterprising and entrepreneurial activities (Edensor, 2016). During these festival times, a large number of business people who rely on temporary trade explore new markets.

With the advent of Covid 19 pandemic in 2020, economies all across the world were brought to a staggering halt. Amongst the businesses profoundly affected by covid, tourism and hospitality were the most hit. As a measure of keeping the rising cases in check, Indian government issued strict guidelines and restrictions entailing nationwide lockdown, curfew, work-from-home, social distancing, masking and quarantines. Not just this, rail and air travel as well as inter-state travel were curbed to and from the country. India being a popular and favoured destination for tourists worldwide, faced a major blow. Huge losses were incurred with the loss of tourism demand. Consequently, tourism jobs were lost and thousands became unemployed. Tourism and hospitality businesses, especially startups were forced to shut down their operations permanently. The demand-side and supply-side of the tourism industry slumped significantly with economic slowdown and limited tourism activity. This translated into reduction of tourism expenditure: domestic, outbound and inbound. After the first and second wave of covid, there was gradual reopening of international borders. Successful vaccination drives and tapering down of positive cases and mortalities revived the confidence to travel with which tourism activity started gaining normalcy. Outbound tourists switched to exotic domestic destinations while following covid-appropriate behaviour. To this end the tourism industry is reviving, but at a slower pace, and is likely to reach pre-pandemic levels in due course of time. Covid brought about a 'new normal' in the contemporary world which have changed the way people live and travel. New tourism trends have replaced the past ones. Travel preferences and tourist behaviour have shifted to new paradigms. Studies show that travellers today are most likely to choose sustainable destinations secluded in the lap of nature to satisfy their accessibility and cleanliness, ecotravel, money value, security and safety, privacy and aesthetic needs (Narmadha and Anuradha, 2022). There is a strong demand for immersive tours with short itineraries which considers the financial, health and safety and other such aspects of tourists. Enhanced usage of technology for bookings, host-guest interactions, service delivery and payments has become a necessity after 'physical distancing' was accepted as a norm post-covid. A digitally driven industry can address the travel risk perceptions of tourists and ensure safe travel. Today, QR codes for scanning menus, app-based room bookings and payments, contactless check in and check out and app-based food orders are frequently used by travellers. Therefore, moving forward, through proper communication and introduction of digital tools, a business can manage to secure the trust of their incoming guests and eventually reach the maximum levels of success.

Community-based events and celebrations improve the social and traditional perceptions of destinations while also supporting additional infrastructure development in neighbouring regions. Events and festivals can be used to strengthen existing firms and organisations as well as to launch new ventures. Furthermore, such celebrations provide opportunities to demonstrate a variety of social and cultural artistic skills, including drawing Rangoli, home decorations, music, dance, and dramatization, all of which have a religious connotation. As a result of tourism growth, life of locals has changed in many ways through better access to education, employment, healthcare, and other services. Festivals act as mood elevators and to bring happiness and enthusiasm in one's life every person in the community eagerly waits to celebrate it. "The pleasure of festivals equally touches the heart of all people without having any discrimination on the basis of gender, caste, creed, religion, and geographical distribution. Moreover, celebrations and festivals replicate social and cultural activities of various communities and thereby, renew many internal matters by collective observations and mutual considerations" (Thomas, 2021). A DMO, whether local or regional, is not devoid of problems and challenges. To be a successful destination marketer and manager, DMOs constantly work on the challenges they face. Some challenges faced by DMOs in planning and implementation of fairs and festivals can be stated as under:

- Lack of adequate finances to support DMO activities is a major drawback. Although huge funds are allocated by the Central Ministry towards the development of North Eastern region, these funds are not appropriately allocated. Money raised by local DMOs through various other ways stands inadequate to organise a large-scale event. For instance, under the scheme "Development of Kamakhya Temple and Pilgrimage Destination in and around Guwahati under PRASAD" funded by the Ministry of Tourism, Govt. of India, Rs. 2990.00 lakh has been received by the State Government through ATDC Ltd (Economic Survey of Assam, 2020-21). But this is specific to Ambubachi only. Other events are somehow sidelined.
- Coordinated partnerships among destination stakeholders appear at minimum levels. Organisations, whether private or public, seems to be moving in different directions when it comes to achieving their objectives. Therefore, there is an absence of collaboration and communication among parties. DMOs tend to work independently in marketing the destination which is not ideal.
- Destination brand image is not well built. Despite the fact that certain fairs and festivals like the ones mentioned in this chapter are quite prominent, several other festivals are not as famous. Moreover, visiting the destination during an event in NER does not necessarily motivate tourists to travel across the entire Northeast region thereafter.

- No standard procedure of organising the whole event. Planning happens to be the most crucial phase of an event. It decides the ultimate success of the event. From setting up objectives, collaborating with stakeholders, gathering necessary funds, targeting definite markets, promotion and branding, up to receiving tourists and delivering seamless customer experience – every activity needs standardisation and testing. This is lacking in every major fair and festival celebrated in NER.
- Inadequate industry knowledge of DMO personnels. Decisions on planning and implementation is often taken by officials without proper knowledge about the current industry scenario in the region as well as the nation. An event should usually be an extension or a better version of the previous one. However, it is often observed that lack of innovation in marketing and promotion strategies fail to yield desired benefits.
- The NER mostly suffers from socio-political unrest. Although it is not at all frequent, tourist decision-making is immensely affected. Domestic as well as inbound tourists find it difficult to invest their time and money into a destination characterised by political insurgencies.

CONCLUSION

Practical Implications

The North-eastern region of India is a storehouse of religions, tribes and cultures. More than 200 ethnic tribes with distinct socio-cultural identity live in harmony all across the eight north eastern states. Respectively, various fairs and festivals representing their traditions and beliefs are observed all throughout the year. Moreover, since the region falls under the Indo-Burma biodiversity hotspot, it has a rich floristic diversity. Some endemic flora as well as fauna are also celebrated to mark its value in the region. This constitutes a major part of all fairs and festivals that are organised by communities of which a significant chunk has received tremendous touristic response which subsequently has motivated the governmental as well as non-governmental tourism bodies to invest in such events. Some noteworthy festivals other than the ones already mentioned are Ziro Music Festival (celebrates the music of Arunachal Pradesh), Shirui Festival (celebrates the 'siroi lily' flower), Bihu (the biggest festival of Assam), Saga Dawa of Sikkim (celebrated by the Buddhists to commemorate the birth, enlightenment and death of Budhha), Kharchi Puja of Tripura and many more. The Brahmaputra Beach Festival, Kite Festival, Kaziranga Golf Competition etc. are organized in collaboration with the Tourism Department and are important endeavour to attract tourism. Under the Swadesh Darshan scheme funded by the Central ministry places like Tezpur, Sibsagar and Majuli in Assam has been allocated Rs. 9097.68 through ATDC to the State to be developed as a Heritage Circuit. Another Wildlife Circuit funded by the Centre covers Manas National Park, Pobitora NP, Nameri NP, Kaziranga NP and Dibru Saikhowa NP and is allocated Rs. 9467.94 lakhs for development (Economic Survey of Assam, 2020-21). As such, many fairs and festivals are gaining popularity in these circuits.

It should not be forgotten that while an event might be a reason to draw tourists to a region, it might also be the reason why a certain category of tourists avoid the region. From the study highlighted in this chapter, a few recommendations can be made to address any underlying challenge that DMOs are currently facing. Local DMOs of the NER can put these recommendations into practice.

- **Research Cell and Standardisation:** A research and development cell is deemed necessary in many cases. This specific cell would be responsible to plan out the entire agenda of conducting an event. Right from the decision to organise an event to roping in suppliers and finances to market the event and conduct it, shall be done by the research cell. Tourism strategies implemented at a national level is often adapted by local DMOs. However, it is crucial to mould these strategies to suit their product offering, tourist type and industry players predominant in that region. The national strategies should be created through stakeholder involvement and local DMOs should be supported in regulating it at their local levels.
- **Supplier:** DMOs can earmark definite suppliers as their permanent partners to deliver their products and services without fail during the course of an event. Use of proper suppliers mean standardised delivery each year. Suppliers can include specific hotels, homestays or lodges to accommodate incoming pilgrims. It also includes the entities facilitating drinking water and hygienic food, cleaning the event arena, disposing garbage and emergency services. Supplier authorisation is key to ensure systematic delivery of services to the consumers.
- **Well-Planned Process:** Devising a destination-adequate process by DMOs for organising a large-scale event is mandatory. Every destination is identified by a unique set of characteristics. The type of stakeholders, service providers, locational features and resources attract a unique set of visitors to that destination. Thus, standard operating procedures of one destination cannot be used on another. It is therefore necessary to plan a process which is appropriate to that region. DMOs shall also fix the budget and determine how and where money will be spent. Areas in which the DMOs will collaborate in marketing are also determined.
- **Promotion of Potential Tourism Products:** While promoting a particular fair or festival to its prospective tourists, DMOs should also market related tourism products of that region. These products can be nearby famous locations, local traditional handlooms, handcrafted items, ethnic cuisines and drinks and folk culture of the region. DMOs along with travel agents and tour operators can prepare short itineraries which can be covered in two or three days by the tourists. A one-day city tour can be planned including visits to some museums or art galleries. Intrastate or interstate tour can also be provided. This shall indirectly promote other tourism products of the destination as well generate extra tourism revenue.
- **Connectivity and Physical Infrastructure:** The most important element of a destination is accessibility. Certainly, the NER has commendable infrastructure which has been made better in the last few years. However, to improve tourist inflow, convenience is a priority. DMOs can link up with Indian Railways to facilitate special trains at concessional rates to the region during an event. Tie-ups with more and more hotels and homestays with clean and hygienic facilities to put up tourists during the event. Facilities of food stalls and drinking water should be provided at specific points around the venue to avoid litter inside the venue. Sanitation facilities should be built in event sites in addition to the existing ones. The bigger the crowd, higher are the problems of improper sanitation.
- **Full Utilization of DMO Websites as Well as Other Organisations:** In this age of digitisation, state websites should be fully utilized for promotion. Not just this, websites should facilitate real-time interactions among host and guest to handle queries. Sharing or posting about the true nature of a destination is mandatory. Tourists should be provided with information that moulds their purchase intention. Therefore, information needed to plan a trip and information needed during the trip should be posted in the websites. Post covid, tourists can be motivated only by way of

transparency in operations. Website handlers can get creative about the structure of the website and what information to focus upon. DMOs should not forget the after-travel services as this can keep tourists coming back to the region.

- **Familiarisation Tours:** DMOs should approach travel agencies and tour operators of this region to collaborate with travel organisations across India as well as abroad so as to set up North East at an international platform. Special itineraries related to any event can be curated and members of these organisations can be brought into the region and given a familiarisation tour. As such, the same shall be highlighted in international itineraries which might be capable of bringing inbound tourists to the North East.

Destination Management Organisations of North East India is responsible to manage any event being organised across the eight states. Although measures have been taken by these organisations this far, a lot is yet to be done for promoting the fairs and festivals at a global scale. It is observed that while on one hand a few festivals have managed to generate quite a crowd, on the other hand, few more have been restricted to a particular locality. This imbalance needs to be checked to ensure equality. Moreover, DMOs should keep a tab on the number of tourists and visitors every year as these numbers will reflect the rise and fall of any event over the years.

REFERENCES

Aas, C., Ladkin, A., & Fletcher, J. (2005). Stakeholder collaboration and heritage management. *Annals of Tourism Research*, *32*(1), 28–48. doi:10.1016/j.annals.2004.04.005

Aleksandrov, K. (2014). The role of DMO for sustainable development of a tourist destination–Bulgaria case study. *Journal of Tourism Research*, *9*(2), 198–209.

Bieger, T., Beritelli, P., & Laesser, C. (2009). Size matters! Increasing DMO effectiveness and extending tourist destination boundaries. *Tourism: An International Interdisciplinary Journal*, *57*(3), 309–327.

Bonardi, L., Ludovici, A. A., & Furlani, R. (2006). *Alps and tourism: Finding the balance point.* Academic Press.

Bornhorst, T., Ritchie, J. B., & Sheehan, L. (2010). Determinants of tourism success for DMOs & destinations: An empirical examination of stakeholders' perspectives. *Tourism Management*, *31*(5), 572–589. doi:10.1016/j.tourman.2009.06.008

Bowman, C., & Ambrosini, V. (2000). Value creation versus value capture: Towards a coherent definition of value in strategy. *British Journal of Management*, *11*(1), 1–15. doi:10.1111/1467-8551.00147

Budeanu, A. (2007). Sustainable tourist behaviour–a discussion of opportunities for change. *International Journal of Consumer Studies*, *31*(5), 499–508. doi:10.1111/j.1470-6431.2007.00606.x

Buhalis, D. (2000). Marketing the competitive destination of the future. *Tourism Management*, *21*(1), 97–116. doi:10.1016/S0261-5177(99)00095-3

Buhalis, D. (2001). Tourism in Greece: Strategic analysis and challenges. *Current Issues in Tourism*, *4*(5), 440–480. doi:10.1080/13683500108667898

Chaturvedi, D. (2010). Tourism in India: Ensuring buoyancy and sustainability. *Yojana*, *54*, 5–7.

Chaudhary, M. (2010). Training Men to Manage Tourism. *Yojana*, *54*, 8–10.

Crouch, G. I. (2007). *Modelling destination competitiveness. A survey and analysis of the impact of competitiveness attributes*. Academic Press.

Crouch, G. I., & Ritchie, J. B. (1999). Tourism, competitiveness, and societal prosperity. *Journal of Business Research*, *44*(3), 137–152. doi:10.1016/S0148-2963(97)00196-3

Das, B., & Acharjee, S. (2013). Problems and prospects of cultural tourism: A case study of Assam, India. *International Journal of Physical and Social Sciences, 3*, 455-462.

Department of Tourism Annual Report. (2018/2019). https://www.gov.za/sites/default/files/gcis_document/202002/department-tourism-annual-report-2018-2019.pdf

Director of Economics and Statistics. (2022). Economic Survey of Assam, 2021-22. Assam.

Dore, L., & Crouch, G. I. (2003). Promoting destinations: An exploratory study of publicity programmes used by national tourism organisations. *Journal of Vacation Marketing*, *9*(2), 137–151. doi:10.1177/135676670300900203

Edensor, T. (2016). *National Identity, Popular Culture and Everyday Life*. Bloomsbury Publishing.

Fennel, D. (2008). *Ecotourism* (3rd ed.). Routledge.

Hanna, S., & Rowley, J. (2015). Towards a model of the place brand web. *Tourism Management*, *48*, 100–112. doi:10.1016/j.tourman.2014.10.012

Honey, M. (1999). *Ecotourism and sustainable development: Who owns paradise?* Island Press.

Hu, B. (2007). *Informal Institutions and Rural Development in China*. Routledge. doi:10.4324/9780203947449

Kumar, P., Mishra, J. M., & Kumar, G. S., & R, R. (2020). Destination marketing & management organizations in India. *Solid State Technology*, *63*(6).

Lee, C. S., Martin, D., Hsieh, P. F., & Yu, W. C. (2020). Principles of value creation in event tourism: Enhancing the competitiveness of regional clusters. *Journal of Global Scholars of Marketing Science*, *30*(4), 437–453. doi:10.1080/21639159.2020.1784771

Lepak, D. P., Smith, K. G., & Taylor, M. S. (2007). Value creation and value capture: A multilevel perspective. *Academy of Management Review, 32*(1), 180–194. doi:.2007.23464011 doi:10.5465/amr

Lohmann, M. (2004). New demand factors in tourism. *European Tourism Forum Budapest/Hungary*, 10.

Longkumer, A. (2016). *Visualising national life, the Hornbill festival as culture and politics*. doi:10.23912/978-1-910158-55-5-3012

Mai, N. (2021). Increase tourism value through value chain concept. *Destination: Business and Management Review*. https://destination-review.com/en/increase-tourism-value-through-value-chain-concept/

Miller, G. A. (2003). Consumerism in sustainable tourism: A survey of UK consumers. *Journal of Sustainable Tourism, 11*(1), 17–39. doi:10.1080/09669580308667191

Morrison, A. (1998). Small firm statistics: A hotel sector focus. *Service Industries Journal, 18*(1), 132–142. doi:10.1080/02642069800000008

Morrison, A. (2013). *Marketing and managing tourism destinations*. Routledge. doi:10.4324/9780203081976

Narmadha, V., & Anuradha, A. (2022). Paradigm change in tourist preferences towards evolving new normal tourism and travel trends. In Handbook of Research on Changing Dynamics in Responsible and Sustainable Business in the Post-COVID-19 Era. IGI Global. doi:10.4018/978-1-6684-2523-7.ch019

Nergusa, A. L., & Coros, M. M. (Eds.). (2016). Destination management organizations (DMO's) roles and performance – literature review. In *Proceedings of the 2nd International Scientific Conference SAMRO 2016 – News, Challenges and Trends in Management of Knowledge-based Organizations.* Editura Economica.

Osmankovic, J., Kenjic, V., & Zrnic, R. (2010). Destinantion managment: Consensus for competitiveness. *Proceedings of Tourism and Hospitality Management Conference*, 513-525.

Paul, K. I., Polglase, P. J., Nyakuengama, J. G., & Khanna, P. K. (2002). Change in soil carbon following afforestation. *Forest Ecology and Management, 168*(1-3), 241–257. doi:10.1016/S0378-1127(01)00740-X

Pearce, D. G. (1992). *Tourist organizations*. Longman Group UK Ltd.

Pike, S. (2004). *Destination marketing organisations*. Routledge. doi:10.4324/9780080494463

Pike, S., & Page, S. J. (2014). Destination marketing organizations and destination marketing: A narrative analysis of the literature. *Tourism Management, 41*, 202–227. doi:10.1016/j.tourman.2013.09.009

Presenza, A., Sheehan, L., & Ritchie, J. B. (2005). Towards a model of the roles and activities of destination management organizations. *Journal of Hospitality, Tourism, and Leisure Science*, *3*(1), 1–16.

Ritchie, J. R. B., & Crouch, G. I. (2003). *The competitive destination: A sustainable tourism perspective*. CABI. doi:10.1079/9780851996646.0000

Sharpley, R. (2006). Tourism - A vehicle for development? In R. Sharpley & D. J. Telfer (Eds.), *Tourism and development - Concepts and issues* (p. 214). Viva Books Private Limited. doi:10.21832/9781845414740-003

Sheehan, L., Vargas-Sánchez, A., Presenza, A., & Abbate, T. (2016). The use of intelligence in tourism destination management: An emerging role for DMOs. *International Journal of Tourism Research, 18*(6), 549–557. doi:10.1002/jtr.2072

Sheehan, L. R., Ritchie, J. R. B., & Hudson, S. (2007). The destination promotion triad: Understanding the asymmetric stakeholder interdependencies between the city, the hotels and the DMO. *Journal of Travel Research, 46*(1), 64–74. doi:10.1177/0047287507302383

Simpson, M. C. (2008). Community benefit tourism initiatives—A conceptual oxymoron? *Tourism Management, 29*(1), 1–18. doi:10.1016/j.tourman.2007.06.005

Stainton, H. (2018). *TEFL tourism in South East Asia: A sustainable solution?* Tourism and Sustainable Development.

Thidell, Å., & Leire, C. (2004). *Consumer perceptions, understanding and use of product related environmental information-A literature review of the Nordic knowledge base*. Academic Press.

Thomas, A. (Ed.). (2020). *Participation and role of women in events and festivals: A study on the contributions of women in revitalizing and sustaining the folk potential of fairs and festivals in India. In Event Tourism in Asian Countries: Challenges and Prospects*. Apple Academic Press Incorporated.

Tosun, C., & Timothy, D. J. (2003). Arguments for community participation in the tourism development process. *Journal of Tourism Studies*, *14*(2), 2–14.

UNWTO-United Nations World Tourism Organization. (2012). UNWTO Tourism Highlights. UNWTO.

Van Harssel, J. (2005). *Glossary-Destination Management Organization. In Fundamentals of Destination Management and Marketing*. Educational Institute American Hotel & Lodging Association.

Varghese, B. (2016). A strategic evaluation on competency of Karnataka destinations through Destination Management Organizations. *American Journal of Industrial and Business Management*, *6*(02), 102–108. doi:10.4236/ajibm.2016.62010

Volgger, M., & Pechlaner, H. (2014). Requirements for destination management organizations in destination governance: Understanding DMO success. *Tourism Management*, *41*, 64–75. doi:10.1016/j.tourman.2013.09.001

Wight, P. (1998). *Tools for sustainability analysis in planning and managing tourism and recreation in the destination*. Academic Press.

World Tourism and Travel Council. (2022). *Economic Impact Reports*. Author.

Wray, M., Dredge, D., Cox, C., Buultjens, J., Hollick, M., Lee, D., ... Lacroix, C. (2010). *Sustainable regional tourism destinations: Best practice for management, development and marketing*. CRC for Sustainable Tourism Pty Ltd.

Zhang, Y., & Murphy, P. (2009). Supply-chain considerations in marketing underdeveloped regional destinations: A case study of Chinese tourism to the Goldfields region of Victoria. *Tourism Management*, *30*(2), 278–287. doi:10.1016/j.tourman.2008.07.004

Chapter 16
Long-Term Forecasting of Tourism Festivals Post COVID-19

Rachel Chanel Clarke
Trigo Networks, USA

ABSTRACT

Forecasting festivals and destination tourism does not involve a secret recipe held by industry gatekeepers but is the applying tools, techniques, and systems. The significance that the COVID-19 pandemic contributes to the greater landscape of destination festivals is multifaceted as time progresses. Sales determinants are affected by a change in the labor force, logistic delays, health and safety concerns, and many factors still to be monitored and readapted through forecasting. Changes in social interactions, such as norms of personal space between family, friends, and strangers, have shifted, causing festival planners to be aware of directional signals that affect festival gatherings (McClure et al., 2020). Long-term forecasting tools assess the cultural shifts of consumer segmentation, revealing the adverse effects that COVID-19 has on demographic trends. Consequently, this chapter provides context while identifying forecasting methodologies and a framework for festivals and destination tourism post-COVID-19.

INTRODUCTION

The COVID-19 pandemic shifted consumer tastes, geopolitical risks, industry trends, policy, socio-economic challenges, and the overall global economy, inevitably impacting the tourism industry and festivals. Several studies (Johnson, 2023; Kakaei et al., 2022; Sharoff, 2021; World Health Organization, 2022) recognize the profound impacts of COVID-19 on society and business, including global travel restrictions, an unknown death toll estimated between 7 million and 14.9 million (excess death toll) from January 2020 to December 2021, global shortages of food and essential medicines, high inflation, a fluctuation in global currencies, strained relations between the two largest economies (the US and China), worldwide protests, an increase in global conflict and sanctions, transitions to remote work, and shift in values. Historically, disruptions to industries and businesses, such as recessions, regulations,

DOI: 10.4018/978-1-6684-6356-7.ch016

new developments, or change in norms, are anticipated. "The sectors that have been the most affected [by COVID-19]—like personal services, leisure, entertainment, and travel—are the ones that in the past have been considered recession-proof," as stated by the University of Chicago Professor Austan Goolsbee (Sharoff, 2021, Divergent responses to COVID-19 section, para. 2). Consequently, the ability of leaders within the tourism festival space to implement best practices, prepare for such disruptive events through effective forecasting, and pivot to meet the needs of policymakers and preferences of consumers will be crucial to their survival and determine whether they flourish or flounder.

Forecasting helps industry leaders assess marketplace conditions, evaluate past and current performance, and leverage strengths and resources to support to predict future performance. Forecasting is more than a single act; but is a consistent act using tools to track, observe, monitor, learn, and report in preparation for potential outcomes. One of the most important forecasting tools is asking questions in a skilled and strategic manner, allowing festival organizers to gain a strategic competitive advantage and assist their teams in preparing for long-term outcomes effectively. In this chapter, as we discuss different topics that make up the forecasting objective, we will pose questions that can be used in a framework for destination tourism and festival planning. The chapter begins by exploring the forecasting of sales determinants and long-term stability past the COVID-19 pandemic while examining different social and cultural aspects of festival marketing and planning. The chapter concludes by identifying a forecasting framework that festival marketers, organizers, practitioners, and other stakeholders can utilize as the baseline for festivals and destination tourism as the world recovers from the Coronavirus pandemic.

FORECASTING SALES DETERMINANTS

Examining the surrounding factors that affect the overall objective can aid in the long-term forecasting of future sales. Ancillary products and services, or the components that complement and support the execution and success of the main attraction, are also selling points to entice consumer sales. One of the vital ancillary components relative to festivals and tourism is lodging. Whereas traditionally, customers would stay in hotels or motels, forecasters now agree that the trend, or the transition in one direction, currently favors marketplaces that allow people to rent their properties short-term to consumers. However, the COVID-19 pandemic interrupted the trend and altered forecasters' analyses:

Airbnb is a major player in short term-letting, listing more rooms than "Marriott International, Hilton Worldwide, Intercontinental Hotel Group, Wyndham Worldwide and Accor Hotel Group" (The Spaces, 2020). Airbnb has disrupted the hotel sector by facilitating trading of space between ordinary citizens (Zachet al., 2020). The question we ask is: has COVID-19 now disrupted the disruptor? We argue that this is indeed the case. (Dolnicar & Zare, 2020, p. 1)

As with Airbnb, the shift and complete interruption of trends from pre- to post-COVID-19 are replicated in many areas that affect destination tourism and festival planning, including marketing, branding, social interaction, and sales. Organizers experienced an alteration to "business as usual," even down to the minute details such as how many attendees are allowed in an entrance queue at a time to the increased specificity in contractual agreements formed in the planning and preparation stages, protecting festival organizers against "unforeseeable" happenings or acts of God (De Ly & Fontaine, 2006, p. 2).

Yet, the COVID-19-induced shock is different in three critical ways. First, the economic shock and the consequent travel decline are global; "international tourist arrivals dropped by 20%-30%" (UNWTO, 2020, p. 2). Second, the economic shock is more dramatic, with reductions to economic growth twice as big as those caused by regular shocks. Third, the shock has the potential to trigger structural changes in certain sectors of the industry. (Dolnicar & Zare, 2020, p. 1)

The Airbnb example illustrates the pitfall of having a product without applying forecasting tools to prepare for the future of the industry.

According to Airbnb, the company now has over 7 million accommodations worldwide and over 50,000 handcrafted activities, across over 220 countries and regions, amounting to over 750 million Airbnb guest arrivals, which is more than the top five hotel's three brands combined. Since the onset of the outbreak, the number of Airbnb listings on offer has remained relatively steady, but the number of bookings has plummeted due to travel bans and health concerns, according to the vacation-rental site AirDNA. The reduction in bookings is most significant in COVID-19 epicenters. (Hu & Lee, 2020, p. 2)

Airbnb hosts failed to monitor the cues relating to their business and the shift in the global landscape of tourism as it pertained to the pandemic. Following the cues revealed by utilizing predictive methodology helps business leaders foresee the current situation based on a historical and analytical context and to apply the results in each area of their business. For the music and culture festival Afrochella in Accra, Ghana, the ancillary services provided through neighboring businesses were the driving force for producing an alternative block party version of the destination festival in December 2020 (Ghana Tourism Authority, 2020). In an interview with Afrochella's co-founder, Abdul Karim Abdulla, in Black Enterprise magazine, he explains how the workforce depended on tourism fostered by the festival (Givens, 2020). Through a partnership with the Ministry of Tourism, they hosted a block party version of the event to help the local businesses and keep attendees safe (Givens, 2020). Comparatively, local laws caused cancellations or postponements of festivals around the world. In many locations, holding gatherings and public events was temporarily against the law. In 2018, France ranked number one in the world for top destinations by international tourist arrivals, with 89 million tourists, according to the World Tourism Organization (World Tourism Organization, 2019). In 2019, 90 million international tourist visited France as tourism makes up 8% of their GDP (Ministère de l'Europe et des Affaires étrangères, 2021). Coronavirus immediately halted France's major economic enterprise as President Macron made a public address on March 16, 2020, and took a "drastic measure, which was put into effect by a decree of the same date, is mandatory home confinement" (Boring, 2020, para. 1). Such governmental movements result in institutionalized social change, or the deliberate planning and organization from large operating bodies, such as governments, universities, or corporations, to shift the norms in society. Herein, the institutionalized social change plan is to prevent a global health crisis.

Assessing the correlation between laws that governed the pandemic and sales, or the correlation tool, highlights the changes needed in the business. For Festival de Cannes in France, the organizers refused to immediately cancel or postpone the festival date of May 12-23, 2020, even though the French government banned public gatherings (Lattanzio, 2020). Once the government extended the mandate, Festival de Cannes further postponed its' event date, rejecting the festival taking on any other shape or form, such as a virtual festival, as did its peers like Sundance Film Festival or the Toronto International Film Festival, which primarily took place on an online streaming platform (Glasner & Weaver, 2020; Ward, 2021). As seen with the Government of France taking solid measures to shut down all movement throughout the country with the onset of the COVID-19 outbreak, a swath of governments reacted to implement post-COVID-19 plans to counteract the economic downturn actualized by the pandemic. On January 9, 2023, Washington DC Mayor Muriel Bowser and the Office of the Deputy Mayor for Planning and Economic Development (DMPED) introduced DC's Comeback Plan (Executive Office of the Mayor, 2023). The plan "create[s] 35,000 new jobs in high-growth target sectors… [including] Hospitality, Tourism, and Entertainment" (WUSA 9, 2023). Observing governmental behavior regarding festivals and tourism can help planners determine the future steps for growth, expansion plans, or even budgetary concerns.

Long-term forecasting, or utilizing tools, techniques, and systems to predict outcomes in the distant future, five to ten years ahead, points to the possibility that COVID-19 will have a lasting influence on tourism festivals, particularly as determined by the government. The United States is developing an "inclusive disability policy" for those diagnosed with Long COVID, the "debilitating, lasting symptoms that can persist long after the acute COVID-19 infection has resolved" (The United States Government, 2022, paras. 1 & 12). Business owners in the United States will have to provide accommodations to meet the needs of this demographic of people with disabilities. However, understanding this new demographic illuminates the need for festival producers to provide safe, fun, and consciously curated experiences for those with Long COVID, ensuring that their attendees have an equally enjoyable time as other attendees and retaining or increasing their consumer base.

An essential component of the sales model for festivals includes a sales and marketing team to attract customers and partners, then convert those interactions into sales of tickets, sponsorships, and other assets. The effectiveness of the sales staff relies upon multiple factors, including training and resources. The organization must identify the leader that will take on the responsibility of educating themselves about the sales determinants with a forecasting lens and identifying other leaders that can steer the tourism festival's objectives in the anticipated direction years ahead. Sales and marketing leaders must develop an efficient training program that allows each employee the ability to perfect their designated duties and responsibilities. "Individual abilities strongly predict individual job performance" (Ostroff & Judge, 2007, p. 319). Forecasting is powerful even in developing staff training. By developing the capabilities of staff members by focusing on their current strengths and weakness, one can guarantee that once the staff meets their benchmark of proficiency, then they will meet performance objectives. This idea is similar to the human capital theory in that investments in people, such as the staff, make them more productive and that the choice by individuals and employers to invest depends on their expected outcome (Ostroff & Judge, 2007).

In response to COVID-19, marketers may find themselves delivering marketing plans that seem antithetical. Keukenhof in the Netherlands, one of the world's largest and most prominent flower parks, had to close its doors and cancel its festival in 2020 due to COVID-19, with a brief opening in 2021 (Keukenhof, 2020). Marketers had to keep communications accessible to their customer base and potential customers during the shutdown. "Over the past year, millions of people have enjoyed videos of our blooming park. But admiring the flowers live is something no video can compete with as you will miss the smells and the atmosphere in the park," stated the director, Bart Siemerink (Töre, 2022, para. 7). The festival of flowers in an experience that incorporates parts of the five senses that are not possible via the internet. In just visiting the park by May of 2022, "More than 1.1 million visitors entered the Keukenhof botanical garden... Our expectations have been exceeded," a spokesperson said (ANP, 2022, para. 1). Approximately 75% of attendees are from other countries, with Chinese nationals decreasing sharply due to consistent COVID-19 travel restrictions maintained in China (ANP, 2022). Staying true to the general trend and sign of the times, or the zeitgeist, which for the pandemic period is maintaining relevance through the marketing and promotion of visuals online, proved beneficial in sales for Keukenhof.

Many other festivals, such as food festivals, faced the same obstacle, looking to experienced staff to retain customers' interest during shutdowns and create marketing tactics for longevity after the pandemic. Fresh Fest Digi Fest in Pittsburg, Pennsylvania, a craft beer festival founded in 2018 that celebrates the diversity of craft beer aficionados, recreated its initial marketing strategy once the COVID-19 pandemic struck. Through a partnership with their event sponsor, Work Hard Digital, they were able to:

Put together the Fresh Fest app, which provides users with a festival schedule, a marketplace for both tickets and swag, instructions for getting Black-owned beer shipped to your door through Tavour, workshops, and so many other cool resources. Once someone buys a ticket, which is just $10, they will get a private link to six streaming channels on YouTube, which we recorded ahead of time and goes live the first day of the fest. (Bracey, 2020, para. 7)

It is not just returning to normal; the perspective of attending and experiencing festivals has changed.

Forecasting the festival landscape after COVID-19, practitioners must consider that when the past is mentioned, concerning normalcy, some destination festivals, as in life, were already experiencing tumultuous situations exacerbated by the pandemic. The Gilroy Garlic Festival in California closed indefinitely after four decades of attracting foodies and garlic fans worldwide, as festival organizers say the event is no longer feasible because of increased insurance costs and "lingering uncertainties" from COVID-19 (Boniello & Byrne, 2022, para. 2). Some festival producers may find themselves not aligned or comfortable with changes that are necessary to maintain or increase sales in the future. Using forecasting tools help organizations effectively pivot and adapt. Others may have experienced setbacks or even disasters before the pandemic, in which the COVID-19 pandemic became the tipping point for some festivals. The Gilroy Garlic Festival was the victim of a mass shooting in 2019, killing three people, injuring over a dozen attendees, devastating a vibrant community, and causing the City of Gilroy to intensify the insurance requirements of the festival (Stone et al., 2019).

Destination festivals experienced many blockades that halted businesses and threatened the existence of live events. Travel restrictions, large gathering bans, and other mandates hit every aspect of many destination festivals. For destination festival marketers, knowing the customer holistically, understanding their buying patterns, commitment to lifestyle choices and brands, experiences during the pandemic, and other sentiments about COVID-19 in the public space will contribute to developing the best possible long-term marketing plan. Essentially, the plan must attract consumers, solidify sales, and turn consumers into repeat buyers.

LONG-TERM STABILITY AND PIVOTS PAST COVID-19

Some governmental bodies have organized to provide financial assistance to live entertainment companies by keeping many businesses open and creating new opportunities for others. Economists' assessment of this type of single influx of capital without a long-term plan for the individual businesses creates a micro-macro paradox, or a "short-term efficacious intervention [having] few discernible, sustainable long-term benefits…which can unintentionally undermine whatever fragile chance for sustainable development may already be in play" (Moyo, 2009, p. 44). Tourism forecasters can assess many likely long-term outcomes, including an initial vast amount of competition post-COVID-19 but a sharp decline due to a lack of sustainability in the long term. Considering the micro-macro paradox, selling festivals to a larger owner or joining similar festivals in the short term to avoid obsolescence may be prominent in the future.

Homing in on the new barriers or opportunities offered by fast-changing laws pertaining to COVID-19 answers the questions such as: "Is the timing for entry into the industry right?" Or "Is this a product or service that customers want and can use?" The correlation tool leads to innovation, as the business leader can see its limits and shift. The trend showed that live music festivals began to adapt to new laws and shifted to releasing digital content. In mid-2020, "virtual events [were] up 1,000% since the Coronavirus" (Koetsier, 2021, para. 1), which was one indicator of festivals shifting to a virtual platform; however, the one determinant does not reveal a timeframe, so it is difficult to know if this shift is to be definite or finite. Another aspect to consider is the attendees' reception to virtual events versus in-person festivals. In Canada, despite many digital alternatives being available, most say that digital content will not replace the feeling of seeing live music, including 89% who attend live music events regularly (Coletto, 2020).

Additionally, Least Developed Countries or low-income countries with little to no major exports that are recognized and determined by the United Nations, and emerging nations, countries that do not concretely fit the definition of a developing nation but have close similarities or have advanced slightly past the definition, have limited Internet connectivity (United Nations Department of Economic and Social Affairs, n.d.).

To capitalise on [virtual festivals], there is a need to address the digital skills shortages within the sector and improve digital access beyond large metropolitan areas, with the additional consideration that digital access does not replace a live cultural experience or all the jobs that go with it. (Travkina & Sacco, 2020, p. 3)

If producing festivals in areas with international guests, consideration of a forecasting analysis of guests' expectations enhances the creation of a hybrid destination festival experience. Festival producers can gain innovation by observing past adaptations that industry leaders in diverse ancillary markets had to employ to circumvent this obstacle. For example, in Ethiopia, the company Hybrid Designs has developed an integrated technology called RIDE to provide a means for citizens to request rides at any time through SMS text messaging, eliminating Wi-Fi connectivity issues or the need for a data plan, and avoiding inflated Wi-Fi prices, as common in some low-income countries. Applications that have gained global success, such as multinational companies Uber and Lyft, are the blueprint of the innovative technology that adapted to the needs of those in Ethiopia. With a new engine of income in the country through RIDE, the company can produce outcomes that will impact society more greatly, like employing 90% of women in the technology firm (Aidoo, 2019).

The type of festival will help the forecaster accurately determine the reception and success of a virtual experience. Although music festivals did not do well through live streaming, festival types such as sports or gaming, historically viewed via television, show promising results. Action sports event ESPN X Live experienced double-digit viewership increases on television and triple-digit growth on @Xgames social and digital media platforms in 2021 (Wilson, 2021). The live event decided to adapt, as many other event producers were forced to do, and they continued the event and canceled live audiences. "X Games production staff were able to leverage their ESPN colleagues' experience mitigating COVID-19 risk with other productions, such as college football games" (Bruton, 2021, para. 2). To an audience accustomed to experiencing sports via digital delivery, viewing a virtual, live sports event was not a far stretch. However, keeping the festival-feel of the event remained a challenge. Vice President of X Games, Tim Reed, stated that he "thinks some of the innovations that emerged during [2021]'s event—including X Fest, a virtual media hub and using Zoom for athlete podium interviews—will become fixtures even when we move beyond COVID-19 mitigations" (Wilson, 2021, para. 1). If successful in the adaptation to producing events past the pandemic onset, many festival producers may find permanence that contributes to the sustainability of the event.

Massive digitalization coupled with emerging technologies, such as virtual and augmented realities, can create new forms of cultural experience, dissemination, and new business models with market potential. (Travkina & Sacco, 2020, p. 3)

Forecasters understand the value that technology has played in the marketplace: giving consumers what they want instantaneously; the ability to interact with people from any walk of life–be that a celebrity, a corporate executive, a fellow fan, or a reporter; and the ability to relive an experience again and again. Translating these ideas to provide festivals digitally and enhance the festival experience, the post-COVID-19 lifestyle calls for innovation in this area. Like X Games, many festivals aim to provide a virtual experience indefinitely. The Sundance Film Festival halted its in-person offerings in 2022 to strictly online; however, it will offer a hybrid presence in its return-to-normal plan, allowing film fans unable to visit Park City, Utah, to participate in the event (Lang, 2022). Technology is undoubtedly the future, with or without COVID-19. The pandemic only accelerated the need for diversification among products and services in the festival scape, which relied almost solely on in-person experiences.

Expanding festivals into cyberspace poses challenges to producers despite many benefits. Content and experiences online increase the number of patent, trademark, and copyright infringements to a global scale, cyber-attack vulnerabilities, privacy and data breaches, misinformation, and impersonation. The popular festival Tomorrowland in Belgium reported a data breach in 2018 with its electronic ticketing system through a company called Paylogic, in which "data [was] taken [including] name, email, gender, age, and post code" (Boyd, 2018, "Sensitive" versus "not sensitive" section, para. 4). The expanding reliance on technological systems increases the number of possible occurrences throughout the planning and execution of a festival. Forecasters look at past examples to accurately predict and safeguard their entities as they advance toward their expected goals. Beer and Rosini (2022) report on Documentary.org:

With the innovations in technology and distribution, [film] festival audience members are no longer restricted to the confines of a screening calendar or a single theater. Instead, they may have the option to view films online via on-demand, pay-per-view transactions, or through streaming content over a festival-branded digital network. Industry insiders foresee a future where, with so many ways to present films, festivals morph into the potentially lucrative realm of content distribution. (para. 5)

The lucrative nature of streaming films also makes it necessary to increase legal precautions. "The grant in a standard festival license may now need to include digital distribution rights, errors, and omissions insurance, and expanded representations and warranties that cover broader commercial screenings" (Beer & Rosini, 2022, para. 5).

SOCIAL INTERACTION

As the trek along the unprecedented post-COVID-19 journey continues, festival organizations must examine many critical components of the strategic planning process with a forecasting perspective. Instead of shooting in the dark by trying a new idea and if it does not work within a couple of years, trying something different, the festival organizers should conduct surveys from consumers, collect general data on public attitudes, and keep a database of inquiries from potential customers gives festival stakeholders a clear view of the social movements and temperatures of their consumers. The collected observations contribute to creating aligned products and services for the future.

Artis-Naples in Southwest Florida experienced a protest of about 40 concertgoers that tried to convince other attendees to demand their money back due to a COVID-19 policy that required a professionally administered test within a set amount of time for admission to its halls and museum (Heithaus, 2021). Temperaments in Southwest Florida showed signs that most citizens refused compliance with COVID-19 prevention policies. The Supreme Court mandate in January 2022 forced healthcare professionals in Florida to follow COVID-19 laws, which some staff persistently refused to comply with after the ruling and quit their jobs (Persing, 2022). One can assess public attitudes by the actions taken by elected officials of a given region in democratic nations, such as Florida's Governor, Ron DeSantis, who publicly abhorred some COVID-19 prevention policies and took action to stop "mask mandates at schools and to shield businesses from fines for allowing unvaccinated and unmasked patrons" (Amrhein et al., 2021, para. 11). The resistance to COVID-19 prevention protocols that led to a protest of the concertgoers was predictable.

Before COVID-19, social media already dominated public discourse. Being unable to socialize and gather in public places and being advised not to meet in person with distant family, friends, and colleagues led to a boom in online interaction. A new-to-market social media outlet founded in 2016, TikTok, saw a sharp increase of users of 85.3% in 2020 as people looked for new and exciting ways to connect in lieu of the worldwide COVID-19 shutdowns and public gathering bans (Dixon, 2022). Of note, the increased use of TikTok did not erase the desire for other social media platforms. "In Italy, the Instagram live viewership has doubled in two weeks" from the onset of the pandemic in March 2020 (Garcia, 2020, para. 21). The timeframe in which people had to adapt to this method of communication shows that it is not used only out of necessity, but also enjoyment. DJ D-Nice, an active musician since the 1980s, increased in popularity as he founded Club Quarantine, an online gathering of over 100,000 attendees consistently tuning in to his DJ set and chatting with friends, followers, and celebrities (Gibbs, 2020). The title of a New York Times article focused on his event stated that "The Hottest Parties in Town Are Now Online" (Garcia, 2020). For festival shareholders, this means that consumers are increasingly accepting new ways to communicate, interact, participate, and fellowship online. Club Quarantine offered the harmony of music enjoyment plus fellowship through the Instagram Live chat feature. Instead of attempting to recreate a live music event, the experience added a benefit to social media participation. Socially, this was the closest thousands of people could get to being in an event with a celebrity or influencer, free of charge, an offering many events cannot produce physically. Observations of these ancillary components identify the balance that festival planners must obtain in moving past COVID-19 to identify areas of ingenuity with social capabilities that festivalgoers experience and accept in addition to their expectations instead of as a substitution for their expectations.

Another social consideration is the act of people bringing or convincing others to attend the festival. People attend festivals because of the camaraderie of others who are similar to them or because those whom they want to affiliate with are attending. Additionally, people do not attend festivals when other like-minded individuals, or individuals they want to be like, shy away from the event. As mentioned in the Southwest Florida protest incident, people avoided the entertainment event but preferred rallying outside with like-minded individuals. This example is the concept of socialization, when an individual enters a group and transitions from being an outsider to an affiliate (Jex & Britt, 2014). Essentially, festival marketers want to position their brands and products in ways that create an atmosphere that sells the festival as an inclusive and inviting experience for the target consumer.

It is important to recognize the festivalgoers' interdependence in interacting with each other. A group, by definition, is a "collective of people [that] must, in some way, be interdependent. [Meaning] the outcomes each member of a collective receives depend, to some degree, on the other members of the collective" (Jex & Britt, 2014, p. 341). Evaluation of the group is essential considering that most groups are not homogenous, nor do they have similarities in personal attributes across each individual in the environment, and some members of one group may also identify with members of another group of people that the festival serves (Ostroff & Judge, 2007). Therefore, the interaction between all attendees will determine the level of enjoyment, safety, prevention or spread of illness, and even the retention of the festival's future attendance. In the post-COVID-19 landscape, tourism festival organizers must consider the audiences traveling from different countries. One may evaluate procedures that keep all attendees safe by enacting COVID-19-related protocols for all attendees, as advised for those from the most high-risk countries. Although countries with an elevated number of COVID-19 cases might need more vetting, the organizers must ensure the equality of all attendees. The organization may consider how this will affect their consumers in the long term, observing that the pandemic may only subside in certain areas of the world while other regions manage the spread of the virus. Developing a plan with a legal team or legal representative is crucial, considering that the area where the tourism festival operates governs the protocols implemented but must accommodate international customers.

Festival planners' determination of the appropriate amount of time that a festival should wait before increasing the size and scale, the implementation of preventive measures, and the level of other restrictions all depend on the demographics of the consumer and laws governing the festival location. As customers differ in emotions and attitudes towards the pandemic and what that means for public gatherings, closely assessing the consumer base is key.

CULTURAL DYNAMICS

In the planning of festivals, just like in real estate, 'location, location, location' determines market demand, client, investor, and stakeholder profiles, as well as the trajectory of a wide array of quantitative measures. Moreover, local demographics often align with the cultural preferences and consumer tastes of festivalgoers and residents, including music, food, climate, and ambiance. However, those demographics can shift. During the COVID-19 pandemic, there were shifts in the population composition in cities like New Orleans, reinforcing and accelerating changes and influxes of wealth that had already been underway since Hurricane Katrina in 2005. Katrina was the previous disruptive event that transformed the economic and cultural landscape.

The [displaced] population has returned and—in some cases exceeded—its pre-Katrina level, but only for some racial groups. The one racial group where this is unquestionably, and unfortunately, not the case is Black New Orleanians. In 2019, there were nearly 100,000 fewer Black New Orleanians in the city than in 2000, accounting for a nearly six percent drop in the share of the city's population. This confirmed a hypothesis that New Orleans would appear "whiter and wealthier as a result of selective return migration" following Hurricane Katrina. (Babb, 2021, Introduction and Motivation Section, para. 5)

Festival planners in New Orleans may ask questions regarding the shifting culture of the city and how it affects the products and services related to the event. Additionally, in what ways could organizers continue to communicate with traditional attendees who may have left the city of New Orleans to retain them and grow upon the existing customer base? Technology, via e-newsletters or social media, serves as a means to continue interacting with clients who moved to another location, as well as a caveat to allow festivals to stay true to the culture it represents and attract a larger audience.

Like many large-scale festivals in 2020, Mardi Gras was deemed as a "super-spreader event," taking place during the onset of the COVID-19 pandemic, on February 25th, and "[drawing] more than a million visitors, and a recent study found that Mardi Gras resulted in nearly 50,000 coronavirus infections" (Juhasz, 2021, para. 2). At the time of Mardi Gras 2020, there were no COVID-19 mandates in place that prevented large gatherings. The cultural connection to festivals impacts the audience's behaviors when a disruptor like COVID-19 interrupts the festivities. The New Orleans Police Department had to monitor the streets where people lived and frequented in case Mardi Gras loyalists gathered to celebrate. "The contribution of culture can also be important in other areas of local development, from rebuilding public trust to contributing to post-pandemic urban design adapted to social distancing rules" (Travkina & Sacco, 2020, p. 4). However, another festival that draws large, dedicated audiences from around the world, Coachella, a music and arts festival that draws affluent guests, "celebrities, floods social media feeds, and sparks trends in fashion" had no incidents of avid fans showing up to the venue looking for others who wanted to party despite the shutdown (Canal, 2016, para. 1). Proximity to large communities associated with the origin of the celebration connects guests that are traveling from other cities to the heart of destination festivals, such as Mardi Gras in New Orleans, Louisiana. The Mardi Gras celebration is on a Tuesday, also called "Fat Tuesday, before Ash Wednesday and the start of Lent," a 40-day season of fasting and prayer (History.com Editors, 2010, When Is Mardi Gras? section). Traditionally, Mardi Gras celebrates the confluence of African, Latin, and European cultures and illustrates the influence of the Roman Catholic religion, with which 35.9% of New Orleanians identify (World Population Review, 2022). "Mardi Gras has an economic impact of more than $1 billion for the city of New Orleans" (Kiernan, 2021, as cited in Chavez, 2022).

DESTINATION TOURISM AND FESTIVAL FORECASTING FRAMEWORK

Moving past COVID-19 has many opportunities, possibilities, and avenues forward. A framework for how to move forward for businesses and entities will minimize risks and failures and create a clear path that will guide the stakeholders toward more significant successes.

Assess and create a plan of action by asking and answering questions. The framework below (see Figure 1) is a guidepost to every aspect of a festival's long-term planning process. Asking questions about each element of the festival gives a thorough analysis of the past, present, and future of those components. As the iconic poet Maya Angelou famously said in a phone interview with The Arizona Republic, "You can't really know where you are going until you know where you have been" (Cordova,

2014). Event organizers and marketers must gain a clear understanding of the practices in the past and the current reception of those practices by society today to make predictions of the best practices in the long term. The team working through this framework would consider what transpired in the past and how the team previously handled each part before the COVID-19 pandemic. Incorporate a mix of closed-ended questions that lead to a list of answers, usually "yes" or "no," and open-ended questions, which give room for explanation. For this exercise, the 'present' represents only how the components were affected during the transitionary phases of the COVID-19 pandemic in order to evaluate the impact of the Coronavirus. However, use the forecasting framework at any given point in time.

This exercise evaluates the festival's long-term future when the transitionary period is complete. As developed by this exercise, the future has the organization's end goal or vision in mind. It shines a light on the connectors that bridge the past to where the organization wants to be in 10 years or more with accuracy. By objectively looking at the evaluated data collected about the past and present, the observed future in which the organization is heading can be altered or enhanced. When developing answers or responses, ensure a near-even mix of opinions and assumptions to facts and figures. Forecasting is essentially part intuition and part math (Allspaw, 2008). There may be more facts but not more opinions, as it is necessary to base future predictions upon concrete data. Facts are comprised of qualitative data, or "' types,' and may be represented by a name, symbol, or a number code," and quantitative data, defined as "measures of values or counts and are expressed as numbers" (Australian Bureau of Statistics, n.d., Quantitative and Qualitative Data Section).

The Who

"Who" questions should focus on diverse identifying demographics of stakeholders, customers, employees, community, and anyone affiliated with the overall topic. If evaluating the topic of "Consumers and COVID-19 protocols," as in Figure 1, then the "who" is predominantly asking questions about the consumer. Additionally, ask "who" questions about those stakeholders that affect the consumer, such

Figure 1. Rachel Chanel Clarke forecasting framework

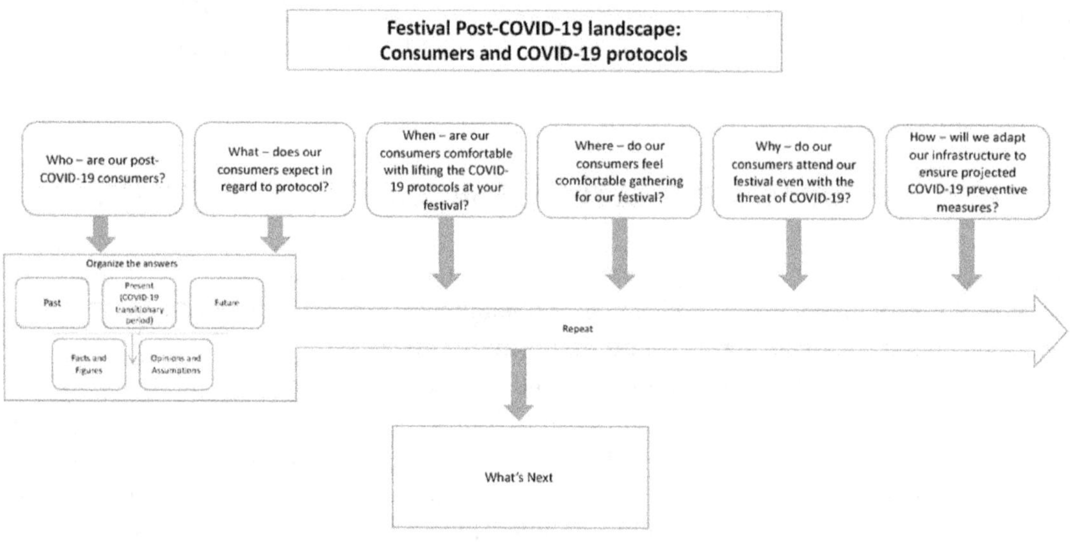

as, "Who would align in sponsoring an event with this audience?" or "Who is best suited to perform for this audience?" As previously discussed in this chapter, historical background information about the people that the festival serves can highlight the reception of the festival post-COVID-19 to the targeted clientele, including how people historically vote or the predominant religion of the attendees—being thoughtful and factual about the "who" allows the team to initiate and develop programs, products, and services to serve customers better while integrating the necessary steps that the organization must take in the long term.

The What

As defined by Oxford Languages (2022), the use of "what" is to ask for information specifying something. Figure 1. asks, "What do our consumers expect regarding protocol?" In answering that question, festival planners will explore different possibilities related to clarification or specificity of the COVID-19 protocols. When thinking about the past, planners can assess what the consumers did at the pandemic's start. Suppose the festival's demographic is predominantly college-aged attendees. In that case, one can factually derive that most attendees expected to isolate at home in the past, as the pandemic completely halted their weekly schedules. "At least 1.5 billion students were out of school at the peak of the pandemic" (Blake & Wadhwa, 2020, Closing Classrooms section). Conversely, if the festival attendees had many unemployed attendees, they might not have been acclimated with the early stay-at-home orders.

The When

"When" is pertinent in this exercise as it is the basis of the entire framework. The assessment of "when" the situation will take shape and unfold concerning different variables affecting the organization lends to the overall forecasting outcome or the predicted result. Identifying the past "when" versus the "when" in transition during the COVID-19 pandemic helps festival managers better understand how the "when" will shape up in the future. For example, in Figure 1, "when" can give the organization a timeline of execution and offer guidance on diversification of products and services and the receptiveness among the people involved in the festival. If through this exercise it is noticed that the attendees had resistance to the COVID-19 protocols in the beginning stages or are historically against vaccinations and medications, as are some religious groups, then one can ascertain that implementing protocols related to vaccinations might threaten the future of the festival. Stakeholders must make complicated decisions on how to proceed, shift, or remain.

The Where

"Where" is in reference to the location. The location can be the venue, the city of the event, an indoor or outdoor setting, in-person, online, or hybrid, and any other spatial determinant. In Figure 1, the question is posed, "Where do our consumers feel comfortable gathering for our festival?" Crowded places make it difficult to social distance and have an abundance of frequently touched surfaces, and intense movement, such as dancing in confined spaces, which increase the risk of COVID-19 infection (McClure et al., 2020). Practitioners must objectively examine what spaces are safe and what environment their participants would deem suitable during the festival. Some "where" questions may relate to logistics, such as "Where can a surplus of merchandise be stored?" Being honest and realistic in the questioning will create vast opportunities for festival marketers. Suppose that COVID-19 caused a severe surplus in merchandise, costing the organization money to store off-site. In that case, the marketer might meet with the community engagement department to incorporate that merchandise with outreach efforts and increase the festival's branding.

The Why

Asking "why" "can shift your perspective, create opportunities, and take people and businesses to a level that was once thought to be impossible. Questions starting with why are typically thought provoking and lead to more in-depth conversations and more follow up questions" (Prossack, 2021, The Power of Why Section). In Figure 1, the question, "Why do our consumers attend our festival even with the threat of COVID-19?" poses a plethora of answers pertaining to different parts of the festival's benefits. One might say that in the past, consumers attended festivals because they did not understand how the infection was transmitted. Another answer could be that presently attendees are exhausted from quarantining and isolation. Practitioners can predict that their audience will not fear the risk of infection because they have contracted the virus with little to no symptoms.

Exploring the question of "why" also opens up more questions that can be filtered into other categories. For instance, in understanding "why" a trend of desensitization among people to the idea of Coronavirus infection is because they had the virus with little to no symptoms, one may want to specify what percentage of their festival's target audience has been infected with COVID-19 to determine if their audience follows the same trajectory. The answer to this question is likely a numeric, closed-ended response. Once it is clear that there is about the same number of facts as there are assumptions or more facts than opinions, assess all answers by asking, "what's next."

The How

The how usually incorporates the other "w" questions. For example, "what adaptations to infrastructure, such as equipment, training, or parking, need to be established or maintained, and how to make those adaptations work?" Here is where the festival organizer will pose the question to develop answers on "how" that will take place. Figure 1. asks, "How will we adapt our infrastructure to ensure projected COVID-19 preventive measures?". What does the team predict? Will there be more technological advances in which the festival will plan to have all the latest equipment that prevents the spread of COVID-19? Will the festival seek sponsorship now from those who are on the cutting edge of vaccinations and cures in order to be able to offer medicines to attendees free-of-charge prior to attending the event? The "how" is the means by which the organization will utilize the forecasting outcome and spark innovation. "How" is a transformative question, as it leads to realistic, concrete, actionable steps based upon innovation and forethought for the long-term future and ties in directly with the list compiled in the "What's Next" section.

WHAT'S NEXT?

After collecting multiple perspectives and answers to the questions posed, the data collected will be compiled and analyzed by asking, "what's next"? Organizers pick the evaluated information from the previous questions to put their plan into action. The organization goes from asking the questions to determining how does the festival shift, pivot, or stay the same. This question-asking process develops actionable steps, predominantly from the "how" questions, clarifies the actionable plan, and strengthens the organization and festival as a whole.

In the example above, mentioned in the section titled "The How", sponsorship was identified as a reasonable means to protect concerned consumers against COVID-19. The festival team predicted that ("who") their post-COVID-19 customers with varying demographics are those concerned about the virus and are comfortable with vaccinations and other preventive measures. Their consumers expect ("what") to seek out activities in which they feel protected in public spaces. The consumers are comfortable with attending festivals in the post-COVID-19 period ("when"). Their customers want to gather in safe spaces, particularly well-ventilated areas or outdoors ("where"). Their post-COVID-19 customer wants to continue to enjoy the activities that they did in the past ("why"). Back to the "how" question, "How will we adapt our infrastructure to ensure projected COVID-19 preventive measures?". One of the identified conclusions led to a follow-up question, "Will the festival seek sponsorship now from those who are on the cutting edge of vaccinations and cures in order to be able to offer medicines to attendees free of charge prior to attending the event?". The exercise exposes a base of information in which festival planners can build upon a timeline to determine that this consumer would likely attend an event with a sponsor that develops or sells a vaccine or cure for COVID-19 as an added safety benefit. With the new foresight found through the forecasting exercise, members of the festival planning team can identify their roles, responsibilities, and tasks that will materialize the forecasted future.

As the "who," "what," "when," "where," "why," and "how" are posed, the festival managers will notice that the exploration of the questions and answers can overlap and enhance one another, which leads to innovation. The series of questions illuminate the path to accurate forecasting. Warren Buffet said innovation is seeing opportunities that others do not (Taylor, 2014). Unless one utilizes forecasting tools, the future is not front and center for others to see. The composite of all answers and outcomes is transferable into the What's next, or the forecast. Periodically, the organization should revisit its plan with regard to evaluating variables such as current events, finances, access to resources, and other factors (Allspaw, 2008).

CONCLUSION

One thing most people will agree with is that gathering in festival-style events will not go away. As what we consider a festival will shift, change, and evolve, stakeholders must look to forecasting to fully develop a path forward. Forecasting is an ongoing process (Allspaw, 2008). With forecasting, destination festivals must have a realistic plan to move past the many obstacles created by COVID-19, including travel restrictions, large gathering bans, and other mandates. The forecasting tools used in this chapter give insight into why forecasting is critical for destination festivals moving past the pitfalls of the Coronavirus. Long-term forecasting also elucidates a spectrum of possible outcomes that festival marketers analyze to make educated choices on the future and stability of the festival and then adapt accordingly. Through forecasting, marketers anticipate and understand the trends affecting their customers and their affinities, understand their sentiments about COVID-19 in a public space and can develop an effective plan to attract and retain consumers and solidify sales.

REFERENCES

Aidoo, T. (2019, November 28). *Samrawit Fikru, the tech genius behind Ethiopia's version of Uber whose staff is 90 percent female.* Face 2 Face Africa. www.face2faceafrica.com/article/samrawit-fikru-the-tech-genius-behind-ethiopias-version-of-uber-whose-staff-is-90-percent-female

Allspaw, J. (2008). *The Art of Capacity Planning*. O'Reilly Media, Inc.

Amrhein, S., Nirappil, F., Leone, J., & Dupree, J. (2021, September 6). In Florida, a summer of death and resistance as the coronavirus rampaged. *The Washington Post*. www.washingtonpost.com/health/2021/09/05/florida-coronavirus-delta-surge

ANP. (2022, May 16). More than 1.1 million people visited the Keukenhof in 2022. *NL Times*. www.nltimes.nl/2022/05/16/11-million-people-visited-keukenhof-2022

Associated Press. (2022, June 29). Essence fest is back in New Orleans after two-year hiatus. *U.S. News & World Report*. www.usnews.com/news/entertainment/articles/2022-06-29/essence-fest-is-back-in-new-orleans-after-two-year-hiatus

Australian Bureau of Statistics. (n.d.) *Statistical language - Quantitative and qualitative data*. Australian Bureau of Statistics. www.abs.gov.au/ websitedbs/D3310114.nsf/Home/Statistical+Language+-+quantitative+and+qualitative+data

Babb, N. (2021, May 5). "Baby won't you please come home": Studying ethnoracial segregation trends in New Orleans pre and Post Hurricane Katrina. *Journal of Public and International Affairs*. jpia.princeton.edu/news/baby-wont-you-please-come-home-studying-ethnoracial-segregation-trends-new-orleans-pre-and-post

Beer, S., & Rosini, N. J. (2022, January 14). *Licenses and rights in the post-COVID festival landscape*. Documentary.org. www.documentary. org/column/licenses-and-rights-post-covid-festival-landscape

Boniello, K., & Byrne, K. J. (2022, April 24). Beloved Gilroy Garlic Festival shutting down after four decades. *New York Post*. www.nypost.com/ 2022/04/23/gilroy-garlic-festival-shutting-down-indefinitely/

Boring, N. (2020, March 18). *France: Government imposes confinement measures to stem spread of coronavirus*. The Library of Congress. www.loc.gov/item/global-legal-monitor/2020-03-18/france-government-imposes-confinement-measures-to-stem-spread-of-coronavirus/

Boyd, C. (2018, October 31). *Tomorrowland festival goers affected by Data Breach: Malwarebytes labs*. Malwarebytes. www.malwarebytes.com/ blog/news/2018/10/tomorrowland-data-breach

Bracey, D. (2020, August 23). *From pandemic to protests: How food businesses are responding*. Bon Appétit. www.bonappetit.com/story/food-businesses-covid-19

Bruton, M. (2021, February 9). X Games' Pandemic Pivot, Embrace Of New Social Media Platforms Pay Off With Tv And Digital Viewership Spike. *Forbes*. www.forbes.com/sites/michellebruton/2021/02/09/x-games-pandemic-pivot-embrace-of-new-social-media-platforms-pay-off-with-tv-and-digital-viewership-spike/

Centers for Disease Control and Prevention. (2022, August 16). *CDC Museum Covid-19 Timeline*. Centers for Disease Control and Prevention. www.cdc.gov/museum/timeline/covid19.html

Chavez, R. (2022, February 16). *For New Orleans, the return of Mardi Gras is critical 'for our pocketbooks and our souls'.* PBS. www.pbs.org/newshour/nation/for-new-orleans-the-return-of-mardi-gras-is-critical-for-our-pocketbooks-and-our-souls

Coletto, D. (2020). As the pandemic goes on, the threat to live music grows. *Abacus*. abacusdata.ca/live-music-threat-pandemic-music-canada/

Cordova, R. (2014, May 28). *Maya Angelou's 2011 'Arizona Republic' interview*. AZCentral.com. www.azcentral.com/story/entertainment/books/2014/05/28/maya-angelou-arizona-republic-interview/9682587/

De Ly, F., & Fontaine, M. (2006). *Drafting international Contracts: An Analysis of Contract Clauses*. Academic Press.

Dixon, S. (2022, February 8). *Topic: Social media use during coronavirus (COVID-19) worldwide*. Statista. www.statista.com/topics/7863/social-media-use-during-coronavirus-covid-19-worldwide/#topicHeader__wrapper

Dolnicar, S., & Zare, S. (2020). COVID19 and Airbnb – disrupting the disruptor. *Annals of Tourism Research*, *83*, 102961. doi:10.1016/j.annals.2020.102961 PMID:32834218

Executive Office of the Mayor. (2023, January 9). *Mayor Bowser unveils DC's comeback plan*. Government of the District of Columbia Muriel Bowser, Mayor. mayor.dc.gov/release/mayor-bowser-unveils-dcs-comeback-plan

Garcia, S. E. (2020, March 21). The hottest parties in town are now online. *The New York Times*. www.nytimes.com/2020/03/21/arts/d-nice-instagram.html

Ghana Tourism Authority. (2020, December 21). *Afrochella Block Party 2020*. Visit Ghana. www.visitghana.com/events/afrochella-block-party/

Givens, D. (2020, December 30). *Afrochella co-founder talks the future of live events amid covid-19*. Black Enterprise. www.blackenterprise.com/afrochella-co-founder-abdul-karim-abdulla-talks-the-future-of-live-events-amid-covid-19/

Glasner, E., & Weaver, J. (2020, September 19). *What worked and what wowed the mostly virtual TIFF 2020 film fest*. CBCnews. www.cbc.ca/news/entertainment/tiff-2020-wrap-1.5730261

Gonzalez, C. (2022, April 16). *Coachella 2022: Festival's return leaving local businesses dancing*. FOX 11 Los Angeles. www.foxla.com/news/coachella-2022-festivals-return-leaving-local-businesses-dancing

Heithaus, H. H. (2021, November 13). Artis-naples protest decries COVID-19 mandates, but for small audience. *Naples Daily News*. www.naplesnews.com/story/news/local/2021/11/13/rally-artis-naples-protests-covid-policy-and-evokes-nazi-germany/85950560
02/

History.com Editors. (2010, January 25). *Mardi Gras 2022*. www.history.com/topics/holidays/mardi-gras

Hu, M. R., & Lee, A. D. (2020). Airbnb, COVID-19 risk and lockdowns: Global evidence. SSRN *Electronic Journal, 2*. www.doi.org/10.2139/ssrn.3589141 doi:10.2139/ssrn.3589141

Jex, S. M., & Britt, T. W. (2014). *Organizational psychology: A scientist-practitioner approach*. John Wiley & Sons.

Johnson, C. K. (2023, March 10). Pandemic 3 years later: Has the COVID-19 virus won? *The Washington Post*. www.washingtonpost.com/health/2023/03/10/covid-pandemic-anniversary/f3252752-bf55-11ed-9350-7c5fccd598ad_story.html

Juhasz, A. (2021, February 12). *What Mardi Gras looks like during the COVID-19 pandemic*. NPR. www.npr.org/2021/02/12/967458788/what-mardi-gras-looks-like-during-the-covid-19-pandemic

Kakaei, H., Nourmoradi, H., Bakhtiyari, S., Jalilian, M., & Mirzaei, A. (2022). Effect of covid-19 on food security, hunger, and food crisis. *COVID-19 and the Sustainable Development Goals*, 3–29. doi:10.1016/B978-0-323-91307-2.00005-5

Keukenhof. (2020, May 12). *Keukenhof closes successful virtual season*. Keukenhof. www.keukenhof.nl/en/news/keukenhof-sluit-succesvol-virtueel-seizoen/

Koetsier, J. (2021, June 30). Virtual events up 1000% since COVID-19, with 52,000 on just one platform. *Forbes*. www.forbes.com/sites/johnkoetsier/2020/05/27/virtual-events-up-1000-with-52000-on-just-one-platform/?sh=29792fa7a233

Lang, B. (2022, August 30). Sundance will return in-person in 2023 with later Digital Component, unveils Covid Safety Measures. *Variety*. www.variety.com/2022/film/news/sundance-dates-2023-digital-covid-safety-measures-1235354472/

Lattanzio, R. (2020, March 14). *Cannes won't decide whether to cancel the festival until April 15 as France shutters theaters*. IndieWire. www.indiewire.com/2020/03/cannes-coronavirus-cancellation-decision-1202217815/

Mardi Gras for families: New Orleans. (n.d.). www.neworleans.com/events/holidays-seasonal/mardi-gras/mardi
-gras-for-families/

McClure, J., Wu, T., & Scarr, S. (2020, July 1). New normal: How far is safe enough? *Reuters*. www.graphics.reuters.com/HEALTH-CORONAVIRUS/SOCIALDISTANCING/qzjvqenmyvx/

Ministère de l'Europe et des Affaires étrangères. (2021, November). *Tourism*. France Diplomacy - Ministry for Europe and Foreign Affairs. www.diplomatie.gouv.fr/en/french-foreign-policy/tourism/

Moyo, D. (2009). *Dead aid: Why aid is not working and how there is a better way for Africa*. Macmillan.

New Orleans & Company. (n.d.). *Late night itinerary: New Orleans*. www.neworleans.com/plan/itineraries/late-night/

Ostroff, C., & Judge, T. A. (2007). *Perspectives on organizational fit*. Psychology Press. doi:10.4324/9780203810026

Persing, S. (2022, January 14). *Hospitals in southwest Florida complying with vaccine mandate after Supreme Court decision*. Wink News. www.winknews.com/2022/01/14/hospitals-in-southwest-florida-complying-with-vaccine-mandate-after-supreme-court-decision/

Prossack, A. (2021, November 6). The power of asking the right questions. *Forbes*. www.forbes.com/sites/ashiraprossack1/2021/11/05/the-power-of-asking-the-right-questions/?sh=73d5e308217e

Sharoff, R. (2021, April 2). *How covid-19 and trade tensions have affected the U.S. and China*. University of Chicago News. news.uchicago.edu/story/how-covid-19-and-trade-tensions-have-affected-us-and-china

Stone, A., Winsor, M., Margolin, J., & Doom, J. (2019, July 29). *Gilroy Garlic Festival shooting: Suspect Santino William Legan, 19, identified as gunman who allegedly killed 3 people*. ABC News. www.abcnews.go.com/US/active-shooter-incident-garlic-festival-california/story?id=64624542

Taylor, B. (2014, July 23). Wisdom of Warren Buffett: On innovators, imitators, and idiots. *Harvard Business Review*. www.hbr.org/2008/10/wisdom-of-warren-buffet-on-imi

The Mayor's Office of Cultural Economy. (2017, May). *2016 New Orleans Cultural Economy Snapshot*. nola.gov/cultural-economy/documents/2016-ce-snapshot-electronic/

The New Orleans Jazz & Heritage Festival and Foundation, Inc. (2021, October 14). *Our history*. The New Orleans Jazz & Heritage Foundation. www.jazzandheritage.org/about-us/our-history/

The United States Government. (2022, April 5). *Fact sheet: The Biden administration accelerates whole-of-government effort to prevent, detect, and treat long covid*. The White House. www.whitehouse.gov/briefing-room/statements-releases/2022/04/05/fact-sheet-the-biden-administration-accelerates-whole-of-government-effort-to-prevent-detect-and-treat-long-covid/

Töre, Ö. (2022, March 30). *World's biggest flower park in Netherlands opens*. Focus on Travel News. www.ftnnews.com/tours/43688-world-s-biggest-flower-park-in-netherlands-opens

Travkina, E., & Sacco, P. (2020, September 7). *Culture shock: Covid-19 and the cultural and creative sectors*. www.oecd.org/coronavirus/policy-responses/culture-shock-covid-19-and-the-cultural-and-creative-sectors-08da9e0e/

United Nations Department of Economic and Social Affairs. (n.d.). *Least developed countries (ldcs) | Department of Economic and Social Affairs*. United Nations. www.un.org/development/desa/dpad/least-developed-country-category.html

Ward, B. (2021, December 10). *2021 Sundance Film Festival Announces Additional Films*. Sundance.org. www.sundance.org/blogs/news/2021-sundance-film-festival-announces-additional-films/

Wilson, O. (2021, April 5). *Driven by young talent, X Games Aspen 2021 TV viewership, social and digital content experience significant growth*. ESPN Press Room U.S. www.espnpressroom.com/us/press-releases/2021/02/driven-by-young-talent-x-games-aspen-2021-tv-viewership-social-and-digital-content-experience-significant-growth/

World Health Organization. (2022, May). *Global excess deaths associated with covid-19, January 2020 - December 2021*. World Health Organization. www.who.int/data/stories/global-excess-deaths-associated-with-covid-19-january-2020-december-2021

World Population Review. (n.d.). *New Orleans, Louisiana population 2022*. New Orleans, Louisiana Population 2022 (Demographics, Maps, Graphs). www.worldpopulationreview.com/us-cities/new-orleans-la-population

World Tourism Organization. (2019). *International Tourism Highlights* (2019 Edition). UNWTO. doi:10.18111/9789284421152

WUSA 9. (2023, January 9). *Mayor Bowser reveals DC comeback plan*. www.wusa9.com/article/news/local/dc/mayor-muriel-bowser-reveals-dc-comeback-plan/65-05ecee64-8758-47d2-ac21-de8b9e00135d

Zaczkiewicz, A. (2017, April 26). *Researchers: Use coachella, burning man to test new concepts*. WWD. www.wwd.com/business-news/marketing-promotion/coachella-trendalytics-report-10875558

Chapter 17
Investigation of Local People's Perceptions of Festival Tourism With Cultural Interaction Dimensions

Sibel Sü Eröz
Kırklareli University, Turkey

Kaplan Uğurlu
Kırklareli University, Turkey

Aydemir Ay
Trakya University, Turkey

ABSTRACT

Festivals are events that enable the transmission of past ties to the future without severing them. Festivals serve as a conduit for cultural exchange between the residents of a region and the local and international visitors who attend the festivals. The primary objective of this study is to investigate the perceptions of Edirne, Turkey locals regarding festival tourism with cultural exchange and cultural balance interaction dimensions, as well as to establish whether the dimensions of cultural interaction differ based on demographic variables. A total of 390 individuals were surveyed to obtain data for the study, and SPSS 22.0 was used to analyze the acquired data. This study reveals that festivals significantly contribute to regional tourism and the local population, as well as to the development of intercultural communication, and that the sub-dimensions of cultural exchange and cultural balance differ depending on the type of events.

INTRODUCTION

The Turkish Nation, whose history dates back thousands of years, has established the most influential, powerful, and long-lasting states in the history of the world on the lands where it resides, and has wiped

DOI: 10.4018/978-1-6684-6356-7.ch017

out numerous cultures and civilizations, possesses a vast cultural accumulation that has been shaped by the historical accumulation. Festivals serve a crucial part in the transmission of this cultural accumulation from the past to the present. Festivals can be evaluated as events that reinforce the concept of continuity between the past, present, and future and function as a bridge. One of the primary functions of festivals, according to Zhang et al. (2019), is to "permit communal representation, collective celebration, and, in most cases, the collective expression of jointly expressed forms of socio-cultural identity". In a sense, festivals provide a setting for the formation of the cultural memory of the local populace and for social contact based on the impression of a shared past that they generate. This social interaction setting is conducive to cultural exchange. Ürkmez and Akbulut (2020) note that festivals, which play a significant role in the production of cultural heritage and its transmission to future generations, are areas of intercultural and intercommunal contact with the participation of both locals and international tourists. Gibson et al. (2011) assert that festivals serve as both a venue of enjoyment and identity performance. According to the Festival Experience Prism Model established by Morgan (2008) to investigate how festival participants evaluate their experiences, social connection, and cultural dialogue are crucial to the success of a festival. Social contact can be defined as the exchange of information, emotions, and behaviour between people, individuals-groups, and groups. According to Pamir (1997), the encounter between tourists and natives creates a social and cultural change on both sides.

Due to the fact that culture is a versatile actor for tourism, the literature contains research that explicate the dynamics between tourism and culture from various perspectives. The dimension that distinguishes this study from others is its investigation of the presentation of perceptions associated to cultural interaction with the festival tourism dimension from the perspective of locals, rather than the general effects of all sorts of tourism. The study consists of three sections in this dimension. In the first section, festival tourism and cultural interaction are explored, and in the second, festival tourism supply sources in the province of Edirne are examined. In the final component, an application study was conducted to determine the function that festival tourism plays in cultural interaction. In the concluding section of the study, the outcomes of the performed analyses are discussed.

Festival Tourism and Cultural Interaction

Festivals share a common characteristic. Gibson et al. (2011) note that festivals are filled with entertainment, deceptions, and rituals of memory, as well as their capacity to unite people. By the usage of festivals as a tourist attraction, individuals of vastly different cultures participate in cultural exchange. Tylor (1871) provided the first definition of the concept of culture. Culture is "the comprehensive whole that encompasses knowledge, faith, art, law, morals, conventions, and all other abilities and habits acquired by a member of society". Currently, the increasing mobility of tourism influences the culture in its various dimensions by increasing the frequency of locals' interactions with tourists. Emir and Toker (2020) and Bozyer (2012) define the cultural effects of tourism as "all the changes made or maintained in cultures' value systems, individual behaviour patterns, family relationships, shared lifestyle and habits, inventiveness, traditional rites, and social structure".

It is conceivable to say that intercultural encounter is the catalyst for cultural transformation among locals in the tourism industry. It is expected that there are positive and negative effects of this change that can be evaluated in both directions. In addition, as cultural interaction between locals and tourists contributes to the development of cultures, it is believed that the element of cultural balance should be addressed in a respectable manner. In this context, the research focuses on the cultural balance and cultural exchange components of cultural interaction.

Cultural Exchange

The phenomenon of culture is a dynamic concept that changes through time in response to numerous variables (globalization, education, technology, etc.). Çakmakcı and Öztürk (2019) also define cultural influences as the causes of long-term, incremental changes in a society's values, beliefs, and cultural practices. Tourism also creates the stage for cultural transformation by increasing the number of interactions between locals and tourists. Tourism, according to Uçar (2010), causes the fusion of societies with different cultural and social structures by allowing the local community and the tourist to interact, so that these communities with different knowledge, tendencies, cultures, customs, and traditions influence one another and their social structures change over time. According to Zhuang et al. (2019), tourism influences the social habits, customs, values, beliefs, and way of life of persons who reside in a tourist location. Civelek (2014) also asserts that the way of life and behaviour of tourists on vacation, the structure of the clusters they form, their relationships with the people in the places they visit, and the officials who serve them all result in substantial changes to traditions, beliefs, value judgments, family structure, political structure, and economic structure. In his study, Pala (2001) categorizes the effects of tourism on culture as changes in religious beliefs and behaviours, value judgments, lifestyle, and language. He argues that this is likewise the result of a direct connection (encouragement, economic change, change of art). The great inclination of locals to copy tourists may also contribute to cultural transformation.

Although the literature emphasizes the negative consequences of tourism on culture, it is important to remember that its positive effects are also contested. Understanding, preserving, and conserving local culture, arts, and customs, as well as empowering host communities and strengthening cultural values, are a few of the positive benefits listed by Shahzalal (2016). Linderová et al. (2021) describe tourism's positive cultural effects as renewing local culture and fostering a sense of local pride, self-respect, and identity. According to Civelek (2014), the social relations established by the indigenous people with tourists through cultural interaction cause the indigenous people to abandon their old conservative attitudes and develop models of behaviour that are open to the outside. At the same time, these relationships liberate the community's value judgments from their old patterns and allow them to take on a more rational nature.

Cultural Balance

In the course of history, cultures have merged in a manner appropriate to the economic, social, and cultural levels of the people living in those civilizations, resulting in the formation of a distinct culture over time. Also, tourism affords locals the chance to interact with individuals of other lifestyles. There is little doubt that the cultural exchange process that occurs through tourism with visitors, who see the world differently and carry a distinct culture, will also serve to enrich the culture of the local population. The cultural interaction that occurs as a result of the festivals will also contribute to the development of intercultural empathy, where similarities and contrasts across cultures can be noted. The contact between persons from different cultures will play a significant part in achieving cultural harmony by reducing cultural incompatibilities over time and laying the groundwork for finding a common ground of agreement and transforming negative views towards immigrants. According to Civelek (2014), the fact that people from various societies share information and get more familiar with one another's cultural structures results in the establishment of new friendships. As a result of these friendships and the establishment of new ones, the potential to contribute to world peace will develop. Research and language acquisition will be motivated by the urge to speak with strangers. Through interacting with people of other cultures, they will enrich their own. Bakan and Canöz (2017) and Yağbasan and Demirbağ (2017) emphasize that it will be functional for groups to come together in equal settings and mutually appreciate each other's cultures in order to lessen the negative effects of stereotyped attitudes and biases. According to Reisinger (2009), the interaction between locals and tourists can result in mutual appreciation, understanding, respect, tolerance, liking, the development of positive attitudes, and the decline of ethnic biases, stereotypes, and racial tension, as well as contribute to cultural enrichment and learning about others. Keleş (2021) asserts that the local population has gained cultural wealth as a result of encountering a number of multifaceted interactions with individuals from different cultures, and that a more tolerant attitude toward different cultures has been fostered in society, as well as a broader cultural perspective. It is conceivable to state that cultural interaction prepares the road for obtaining cultural balance by fostering a deeper understanding of individuals with various cultural patterns.

Festivals in Edirne Destination

It may be argued that festivals, together with other appeal dimensions, contribute to Edirne's capacity to reach a prominent position in the tourism industry. There is little doubt that the festivals have significant economic effects by enhancing the province of Edirne's tourism appeal and giving it an edge over its competitors. Local government entities have an active role in the organization of festivals where cultural transmission occurs actively, and they are empowered by law to provide such cultural and artistic services (Gündüz Alptürker & Alptürker, 2021). In terms of festivals, the province of Edirne offers a wide range. It has been observed that the number of festivals conducted by the Edirne Governorate, Edirne Municipality, district municipalities, non-governmental organizations, and educational institutions in the Edirne destination has increased during the past three years.

Table 1. Festivals organized by the Edirne City Municipality in Edirne

Name of the Festival	Place	Responsible
Kirkpinar Oil Wrestling Festival	Edirne city center	Edirne Municipality
Kakava and Hidirellez Festivals	Edirne city center	Edirne Municipality
International Edirne Band and Liver Festival	Edirne city center	Edirne Municipality
Seed Exchange Festival	Edirne city center	Edirne Municipality
9/8 Music Festival	Edirne city center	Edirne Municipality
Edirne 1st International Folk Dance Festival	Edirne city center	Edirne Municipality
Thrace Producing Women Festival	Edirne city center	Edirne Municipality
International Edirne 23 April Children and Science Festival	Edirne city center	Edirne Municipality
Edirne Souvenir Festival	Edirne city center	Edirne Municipality
Enez 1st National Cycling Festival	Enez	Edirne Municipality, Edirne Saros Tourism Infrastructure Services Association

Source: The table has been created according to the information obtained from the Edirne Municipality web page.

In addition to supporting the image of the city of Edirne through the organization of numerous types of festivals, the central municipality functions as a catalyst for festival tourism in Edirne, which holds a total of ten such events. Moreover, the Kirkpinar Oil Wrestling Festival, the Kakava and Hidirellez Festival, and the Marching Band and Liver Festival stand out among these events due to their international prominence. Table 1 displays the events held by the central municipality in the destination of Edirne.

Kirkpinar Oil Wrestling Festival

Since 1950, Edirne Municipality has hosted the seven-day Kirkpinar Oil Wrestling Festival, which has been held annually since 1361. The festival is held annually between the final week of June and the first week of July. The event begins on Friday with prayers for the wrestlers, and the matches are staged on the field at Sarayiçi, one of Edirne's most prominent promenade areas. The champion wrestler earns the title of Chief Wrestler and the famous golden belt for a duration of one year. A wrestler who has been the champion for three years in a row receives the golden belt permanently. In Turkey, wrestling, which has attracted attention throughout history and is historically practiced, is classified as an equestrian sport (Turkish Ministry of Culture and Tourism General Directorate of Research and Development, 2022). Although an exact figure cannot be confirmed, between 300 and 400 thousand people attend the festival annually.

There are many legends about the emergence of this festival, which has been held since 1361. The best known of these legends is as follows: In 1357, during the expeditions organized by the Ottoman Sultan Orhan Gazi to conquer Rumeli (Thrace Region), his son Süleyman Pasha embarked on a reconnaissance trip surrounding Edirne after having temporarily captured the city. When returning to Greece, the pioneer soldiers take a vacation on the island of Samona, which remains on the territory of Greece today. Here, forty courageous men, who have come to find love during their time off, begin to wrestle. Ali and Selim, who are brothers, are unable to win after several hours of wrestling. Later, at Ahiköy, two brothers resumed their fighting. But they are unable to overcome once more, and they expire with no remaining breath (Bilar, 2022). Their companions then bury them behind a fig tree in the same location. Years later, when they return to the same location, they find a verdant spring where the tombs of the two Pehlivans (Wrestlers) formerly stood. In remembrance of those who were buried there, the region was afterwards given the name Kirkpinar (Edirne Municipality, 2022). Sultan Murad, who conquered Edirne in 1361, organizes a wrestling in remembrance of forty courageous raiders in the summer of the same year with the order he gave. This organized wrestling will go down in history as "Kirkpinar Wrestling" (Bilar, 2022). After the Balkan War and World War I, the place where wrestling was held shifted and moved to Virantekke. with the foundation of the Republic of Turkey in 1923, wrestlers were taken to Sarayiçi (Ay, 2010). Medieval Kirkpinar Oil Wrestling got the "European Excellent Destinations Award" in 2008, with the theme of "Tourism and Intangible Cultural Heritage" (Turkish Ministry of Culture and Tourism, 2022). It was included to the List of Intangible Cultural Treasures in 2010. (UNESCO Turkish National Commission, 2022). In 2021, Guinness World Records (2021) registered it as the oldest continuously organized sports competition in the world, and a distinguishing feature of Turkish wrestling is that the wrestlers are soaked in oil prior to the competition, making it difficult to hold each other and necessitating a high level of sport mastery. In addition, the fact that the Kirkpinar Oil Wrestling Festival has been held for 662 years, is based on a legend, and incorporates several religious and cultural rituals sets it apart from other sports and festivals. Kirkpinar Oil Wrestling Festival, one of the most prominent sports organizations in Turkey, is followed by local and foreign sports enthusiasts and continues for a week with tournaments, folklore shows, exhibits, fair organizations, and wrestling matches.

Kakava-Hidirellez Festival

Hidirellez, which dates back to pre-Islamic periods among Turks, and the Kakava Festival of the Romans, which dates back to ancient times and where comparable rituals were conducted, have been held by Edirne Municipality since 1950 and last two days between May 5 and May 6. The Hidirellez tradition, which is celebrated as a spring festival, is a festival, ceremony, and holiday comprised of rich cultural values nourished by Central Asian culture (Shamanism), ancient Anatolian culture (abundance-fertility ceremonies, immortality), Islamic culture (Hizir Ilyas motif), and common Balkan culture. Hidirellez is a spring celebration in which the entire Turkish population participates, celebrates, and observes certain traditions (Artun, 2007). With the admission of Islam, it is celebrated as the end of winter and the start of summer, since it is believed to be the day when Hizir and Ilyas meet. This day, which was once known as Ruz-i Hizir, is now known as Hizir-Ilyas day. In many regions of Anatolia, it is widely believed that on Hidirellez night, desires will come true, the sick will be healed, fortunes will be unlocked, and abundance will rise (Döğüş, 2015). Hidirellez festivals are held not just in Edirne, but also in numerous other Turkish cities. The celebration of the Spring Festival Hidirellez, which was added to the Intangible Cultural Heritage List in 2017 (UNESCO Turkey National Commission, 2022), combined with the Roman feast of Kakava, distinguishes Edirne. In his study, Paspati (1870) claims that when the Romans leave their winter quarters in the spring, they gather in the center of a green field near a water source and celebrate the only celebration fit for their race, the Kakava or Cauldron festival, away from the Greeks and Turks. Beginning with the lighting of the Kakava fire, the festival continues with many events and ceremonies. Prior to ten years ago, it was exclusively celebrated by the residents of Edirne. But, as a result of the media's coverage, a far larger number of people have become aware of the holiday. It exhibits both Turkish and Romani cultural influences.

International Marching Band and Liver Festival

Since 2011, the International Marching Band and Liver Festival has been organized by the Edirne Municipality, beginning on Friday and running for three days. In addition to national marching bands, foreign marching bands, particularly those from the Balkans, perform at the festival. At the tenth Marching Band and Liver Festival, held in 2022, the Istanbul Metropolitan Municipality City Orchestra, Izmir Seferihisar Municipality Marching Band, Bursa Metropolitan Municipality Marching Band, Rodna Band from Romania, Dobrich, Burgas and Tirgovishte Marching Bands from Bulgaria, Herceg Novi Band from Montenegro, Zaim Band from Moldova, Elbasan Band from Albania, Szekesfehervar and Duna from Hungary (Edirne Municipality, 2022). Edirne pan liver, one of the most significant gastronomic components of Edirne cuisine, was recorded in 2009 and granted a geographical indication (Turkish Patent Institute, 2022). The Ottoman palace cuisine of Edirne, which served as the capital of the Ottoman Empire for 92 years, is the origin of one of the world's quickest regional meals, Edirne pan liver. As a result of the impact of palace food on local cuisine, pan liver has become one of Edirne's defining characteristics. Lungs of calves with Edirne identity are cooked by smearing them with Edirne flour, which ranks second in wheat production in Turkey, and frying them in sunflower oil, which ranks third in wheat production. It is served as a side dish with Edirne/Karaağaç dried and fried peppers (Ürkmez and Akbulut, 2020). Edirne pan liver, which is manufactured and served at numerous food and beverage establishments in Edirne, is the most popular dish in terms of gastronomy tourism among visitors to Edirne (Kendir, 2020). In 2022, 1.5 tons of Edirne pan liver were given away for free to festival participants (Demirören News Agency, 2022). During the festival, additional events such as concerts and a hot pepper cooking competition are also held.

RESEARCH FRAMEWORK

In this section, the purpose of the research, the research model, the research universe and sample, data collection methods, and data produced as a result of the completed analyses are described.

Purpose of the Research

This research seeks to evaluate the nature and extent of cultural interaction among festival participants in Edirne, a city on the western border of Turkey. Additional purposes include assessing if the characteristics of cultural interaction vary according to demographic variables and investigating the perspectives of Edirne locals towards festival tourism. Saçlı (2020) investigated the relationship between cultural interaction and festival enjoyment using a sample of Silifke (Mersin, Turkey) International Music and Folklore Festival participants. This study also aims to contribute to Edirne Municipality, which organizes these events in Edirne, and to the literature by examining the perceptions of the local community

Figure 1. The research model

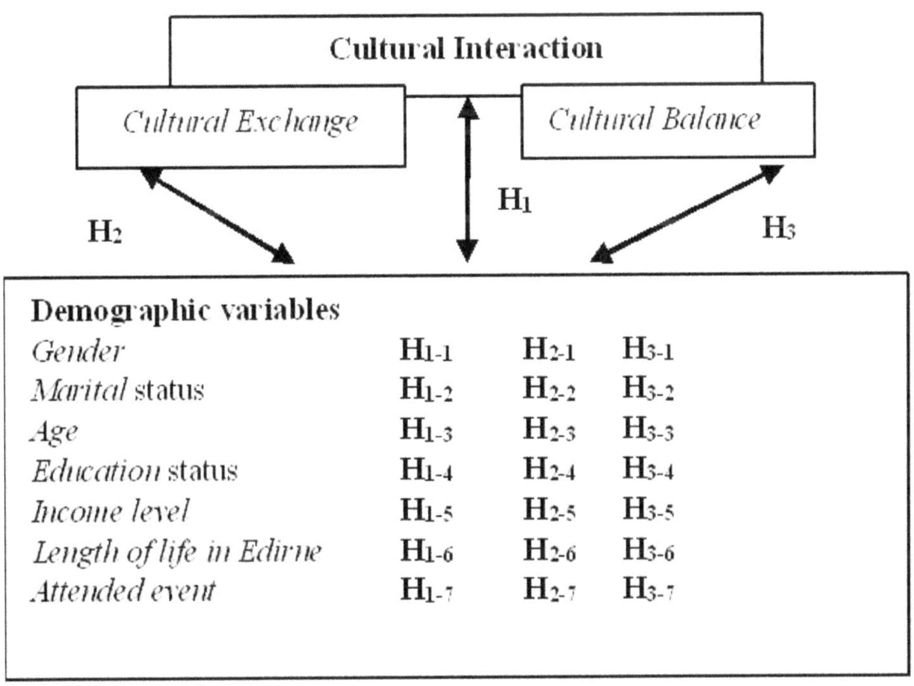

on cultural exchange in relation to the festivals. The following hypotheses will be examined as part of this study's scope.

H_1: There is a statistically significant relationship between the demographic variables and the local population's perceptions of cultural interaction in relation to festival tourism.
H_2: There is a statistically significant difference between the demographic variables and the local population's attitudes regarding the cultural exchange dimension of festival tourism.
H_3: There is a statistically significant difference between the demographic characteristics and the attitudes of the local community concerning the cultural balance dimension of festival tourism.

The model created in light of the theories is shown in Figure 1.

Research Method

The method of quantitative research was employed in this study, and the questionnaire was used to collect data. Those who participated in at least one of the most renowned Edirne festivals, including the Kirkpinar Oil Wrestling Festival, Kakava and Hidirellez Festival, and Marching Band and Liver Festival, and who named Edirne as their destination make up the universe of the study. The sample method was easily chosen during the investigation. The research data were gathered from 10.09.2022-10.11.2022. In total, 390 face-to-face and web-based questionnaire responses were evaluated. According to Sekaran (2003), a sample size of 384 individuals is sufficient for populations of one million or more.

The questionnaire consists of two parts. The first part consists of a *personal information form* containing the participants' gender, marital status, age, educational status, income level, demographic questions to determine the length of life in Edirne, and questions to determine which of the 2022 festivals they attended and the degree to which the festival met their expectations. The second part is made up of the *cultural interaction scale* created by Saçlı (2020) to measure the cultural interaction between tourists and locals. Saçlı calculated the entire Cronbach's Alpha coefficient of the scale to be 0.877. SPSS 22.0 was used to analyse the data acquired from the sample (Statistical Package for Social Sciences).

FINDINGS

Descriptive Statistics

As shown in Table 2, the demographic characteristics of the participants were as follows (according to the highest percentage ratios): 56.9% of the participants (n=222) were male, 57.9% (n=226) were single, 35.6% (n=139) were bachelor's degree holders, 30.3% (n=118) had an annual income between 5,001 and 10,000 TL, and 66.4% (n=259) were born in Edirne.

Table 3 contains information about the festivals that participants in the destination of Edirne attended in 2022. The festival with the biggest attendance, as shown in Table 3, is the Kakava-Hidirellez Festival, with 16.2% (n=63) of the total. This is followed by the Kirkpinar Oil Wrestling Festival with 13.6% (n=13.6) of the total population. On the other hand, a sizeable proportion of the participants, 37.4% (n=146), attended all the events.

Table 2. Demographic characteristics of the participants

Categories	Groups	n	%
Gender	Female	168	43.1
	Male	222	56.9
Marital status	Married	164	42.1
	Single	226	57.9
Age	18-27	136	34.9
	28-37	77	19.7
	38-47	101	25.9
	48-57	49	12.6
	58 and older	27	6.9
Educational status	Secondary school	45	11.5
	High school	35	9.0
	Vocational school	111	28.5
	Bachelor's degree	139	35.6
	Graduate degree	60	15.4
Income level	5,000 TL and less	69	17.7
	5,001-10,000 TL	118	30.3
	10,001-15,000 TL	107	27.4
	15,001-20,000 TL	29	7.4
	20,001 TL and more	67	17.2
Length of life in Edirne	I'm from here	259	66.4
	1-5 years	84	21.5
	6-11 years	19	4.9
	11 years and above	28	7.2

Table 3. Festivals attended by participants in Edirne destination in 2022

Festivals	n	%
Kakava-Hidirellez Festivals	63	16.2
Marching Band-Liver Festival	1	0.3
Kirkpinar Oil Wrestling Festival	53	13.6
Kakava-Hidirellez Festival and Marching Band-Liver Festival	8	2.1
Kakava-Hidirellez Festival and Kirkpinar Oil Wrestling Festival	119	30.5
Marching Band-Liver Festival and Kirkpinar Oil Wrestling Festival	-	-
All	146	37.4
Total	390	100

Table 4 shows the degree to which the festivals met the participants' perceptions. Examining Table 4 (according to the highest percentage ratios) reveals that satisfaction with perceptions for all three festivals is quite high. This ratio covers individuals who did not attend any of the three festivals, as well as those who attended more than one festival and rated their level of perception satisfaction for each festival.

Table 4. The degree of satisfaction with perceptions for the festivals

Festivals		n	%
Kakava-Hidirellez Festival	Non-Participant of the Festival	52	13.3
	Intermediate	17	4.4
	High	**215**	**55.1**
	Very high	106	27.2
	Total	***390***	***100***
Marching Band-Liver Festival	Non-Participant of the Festival	234	60.0
	Low	45	11.5
	Intermediate	15	3.8
	High	**95**	**24.4**
	Very high	1	0.3
	Total	***390***	***100***
Kirkpinar Oil Wrestling Festival	Non-Participant of the Festival	71	18.2
	Intermediate	14	3.6
	High	**181**	**46.4**
	Very high	124	31.8
	Total	***390***	***100***

The Results of Factor Analysis and Reliability Values of the Scale

In the study, exploratory factor analysis was used to examine the validity of the scale and identify the evaluation dimensions. Initially, a factor analysis was conducted by examining the Kaiser-Meyer-Olkin (KMO) value and the Bartlett sphericity test result to evaluate the suitability of the scale expressions for factor analysis and sample adequacy. The KMO coefficient value for the scale is 0.971>0.5; the Barlett sphericity test result [Approximate Chi-Square = 9959.885; df = 378; sig = 0.000 ($p<0.001$)] indicates that the scale is spherical. 75.976% of the variance is explained by this two-factor structure. As a consequence of the performed factor analyses, the statements contained in the same factor are extracted, and Table 5 shows the statements included in the research analysis, factor loads, and the findings of the reliability analysis of the factors. As shown in Table 5, the Cronbach's value for the cultural exchange component of the scale was 0.914 (n=4) and the cultural balance dimension was 0.860 (n=4). The total Cronbach's value of the scale was determined to be 0.923 (n=8) after factor analysis.

Table 5. The results of the factor analysis of the cultural interaction scale

Items	Item Load Values		% of Variance	Cronbach's α
	1	2		
Cultural Exchange				
It is effective in changing moral values and behaviour.	0.704		34.195	0.914
It is effective in the change of religious belief and behaviour.	0.856			
It is effective in changing eating and drinking habits.	0.703			
It is effective in changing the language of the local people.	0.734			
Cultural Balance				
It provides a balance between cultures.		0.822	41.781	0.860
It aims to eliminate cultural incompatibilities.		0.793		
It has a mission to create intercultural empathy.		0.841		
It enables individuals from different societies and cultures to exchange information with each other.		0.837		
Total Scale			75.976	0.923

The Normality Test Result of the Scale

Before determining whether the cultural interaction measure is affected by demographic variables, a normality test was conducted. Table 6 displays the results of the Kolmogorov-Smirnov and Shapiro-Wilk tests used to determine if the scale fits the normality assumption. As a result of the Kolmogorov Smirnov ($p<0.05$) and Shapiro-Wilk ($p<0.05$) tests, it was determined that the distribution was not normal; hence, non-parametric tests were utilized in the difference tests.

Table 6. The normality test result of the scale

Factors	Kolmogorov-Smirnov	Shapiro-Wilk
	p	p
Cultural Exchange	0.000	0.000
Cultural Balance	0.000	0.000
Total Scale	0.042	0.000

The Mean of the Scale

After evaluating the responses to the survey questions on a 5-point Likert scale, assuming that the gaps are equal, 0.80 was calculated as the score gap for arithmetic averages. (Score Difference = (Highest value - Lowest value) / 5 = (5 - 4) / 5 = 4 / 5 = 0.80) The evaluation gap of arithmetic averages was calculated as follows: the gap between 1.00 and 1.80 points was considered "very low," the gap between 1.81 and 2.60 points was considered "low," the gap between 2.61 and 3.40 points was considered "medium," the gap between 3.41 and 4.20 points was considered "high," and the gap between 4.21 and 5.00 points was considered "very high."

As shown in Table 7, the overall average of the scale was determined to be $\bar{x}$=3.0054 based on data collected from the participants. The averages for the cultural exchange ($\bar{x}$=2.7449) and cultural balance ($\bar{x}$=3.2660) dimensions are moderate. Yet, the average of the cultural balance component is greater than the average of the cultural exchange dimension.

Table 7. The mean of the scale

Scale	Minimum	Maximum	$\bar{x}$	Standard Deviation
Cultural Exchange	1	5	2.7449	1.08565
Cultural Balance	1	5	3.2660	1.03127
Total Scale	1	5	3.0054	0.98922

Hypothesis Testing Results

Table 8 displays the results of applying the Mann Whitney U test and the Kruskal Wallis test, both non-parametric tests, to determine the differences between the local population's perception of festival tourism in Edirne, the dimension of cultural exchange from the dimensions of cultural interaction, and demographic characteristics. As a result of the Kruskal-Wallis test, the Mann-Whitney U test was applied to the expressions that exhibited statistically significant differences based on demographic factors, and Table 8 displays the identified origins of the differences.

There was no statistically significant difference between the cultural exchange dimension and the gender (U=17304.000, $p>0.05$) or married status (U=18347.000, $p>0.05$) variables, as determined by the Mann-Whitney U test.

The Kruskal-Wallis test used in the analysis revealed a statistically significant difference between the magnitude of cultural change and income level (χ^2=13.761, $p<0.05$). In order to determine the source of the differences, the Mann-Whitney U test was applied to the expressions that displayed significant differences based on income level. The average rank of festival attendees earning less than 5,000 Turkish Lira was found to be lower than the average rank of festival participants earning between 5,001 and 10,000 Turkish Lira and 15,001 and 20,000 Turkish Lira. Similarly, the average rank of festival attendees with an income between 5,001 and 10,000 TL was lower than the average rank of festival attendees with an income of 20,001 TL or more.

According to the results of the Kruskal-Wallis test as shown in Table 8, there was also no statistically significant difference between the cultural exchange dimension and the variables of age (χ^2=2.347, $p>0.05$), educational status (χ^2=2.347, $p>0.05$), life expectancy in Edirne (χ^2=5.511, $p>0.05$), and participation activity (χ^2=9.842, $p>0.05$).

Table 9 shows the results of applying the Mann Whitney U test and the Kruskal Wallis test, both non-parametric tests, to determine the differences between the local population's perception of festival tourism in Edirne, cultural interaction dimensions, cultural balance dimensions, and demographic characteristics. As a result of the Kruskal-Wallis test, the Mann-Whitney U test was applied to the expressions

Table 8. The result of the discriminant analysis between the cultural exchange dimension and demographic characteristics

Categories	Groups	n	M. Rank	Sum of Ranks	U	Chi-Square	Sig	Sig[1]	Variation
Gender	Female	168	203.50	34188.00	17304.000	-	0.222	-	-
	Male	222	189.45	42057.00					
Marital status	Married[a]	164	194.37	31877.00	18347.000	-	0.866	-	-
	Single[b]	226	196.32	44368.00					
Age	18-27	136	204.18	-	-	2.347	0.672	-	-
	28-37	77	188.19						
	38-47	101	196.62						
	48-57	49	178.20						
	58 and older	27	199.80						
Education status	Secondary school	45	179.78	-	-	2.483	0.0648	-	-
	High school	35	193.09						
	Vocational school	111	198.74						
	Bachelor's degree	139	203.91						
	Graduate degree	60	183.22						
Income level	5,000 TL and less[a]	69	168.41	-	-	13.761	0.008	0.006 0.012	b>a d>a
	5,001-10,000 TL[b]	118	215.22					-	-
	10,001-15,000 TL[c]	107	194.87					-	-
	15,001-20,000 TL[d]	29	234.14					0.016	d>e
	20,001 TL and more[e]	67	172.96					0.011	e>b
Length of life in Edirne	I'm from here	259	194.55	-		5.511	0.138	-	
	1-5 years	84	181.23						
	6-11 years	19	241.92						
	11 years and above	28	215.64						
Participation activity	Marching Band-Liver Festival	1	195.50			9.842	0.080	-	
	Kirkpinar Oil Wrestling Festival	53	169.72						
	Kakava-Hidirellez Festival	63	182.73						
	Kakava-Hidirellez Festival and Marching Band-Liver Festival	8	118.06						
	Kakava-Hidirellez Festival and Marching Band-Liver Festival	119	203.37						
	All	146	208.20						

*p< 0.05

that exhibited statistically significant differences based on demographic factors, and Table 9 displays the identified origins of the differences.

There was no statistically significant difference between the cultural balance dimension and the gender (U=16927.500, p>0.05) and marital status (U=17325.000, p>0.05) variables, as determined by the Mann-Whitney U test.

According to the results of the Kruskal-Wallis test applied in the analyses, a statistically significant difference was found between the cultural balance dimension and the income level (χ^2=15.944, p<0,05). The Mann Whitney U test was applied to the expressions that showed significant differences according to the income level in order to determine the source of the differences. The average rank of festival participants whose income level is 5,000 TL and below was found to be lower than the average rank of festival participants whose income level is 5,001-10,000 TL, 10,001-15,000 TL and 15,001-20,000 TL. Similarly, the average rank of festival participants with an income level of 5,001-10,000 TL was found to be lower than the average rank of festival participants with an income level of 20,001 TL and more.

In the analyses, the Kruskal-Wallis test found a statistically significant difference between the cultural balance variable and the life expectancy variable in Edirne (χ^2= 8.02, p<0.05). The Mann-Whitney U test was applied to the expressions that differed significantly based on the length of life in Edirne in order to identify their source. The average rank of festival attendees with a life expectancy of 1-5 years in Edirne was found to be lower than the average rank of festival attendees with a life expectancy of 6-11 years and 11 years and above in Edirne.

The Kruskal-Wallis test performed in the analyses showed a statistically significant difference between the cultural balance dimension and the participation activity variable (χ^2= 17.151, p<0.05). In order to determine the source of the differences, the Mann-Whitney U test was applied to the expressions that showed significant differences based on the type of activity undertaken. It was discovered that the average rank of the participants of the Kirkpinar Oil Wrestling Festival and the Kakava-Hidirellez Festival was lower than the average rank of those who participated in both of these festivals and all of the festivals. Similarly, the average rank of Marching Band Liver Festival participants was found to be lower than the average rank of participants overall.

In addition, according to the results of the Kruskal-Wallis test as shown in Table 9, there is no statistically significant difference between the magnitude of cultural change and the average rank of the groups based on age (χ^2=3.339, p>0.05) and educational status (χ^2=0.585, p>0.05).

Table 10 displays the findings of two non-parametric tests, the Mann Whitney U test and the Kruskal Wallis test, which were used to examine the locals' perceptions of festival tourism in Edirne, as well as variations in overall cultural interaction and demographic variables. The Mann-Whitney U test was performed to the expressions that showed statistically significant differences based on demographic characteristics as established by the Kruskal-Wallis test, and Table 10 displays the differences' sources.

There was no statistically significant difference between the total cultural interaction and gender (U=17041.000, p>0.05) and marital status (U=17786.000, p>0.05), as determined by the Mann-Whitney U test.

The Kruskal-Wallis test performed in the analyses showed a statistically significant difference between the total cultural interaction variable and the income level variable (χ^2= 15.538, p<0.05). In order to determine the source of the differences, the Mann-Whitney U test was applied to the expressions that showed significant differences based on income level. The average rank of festival participants earning less than 5,000 Turkish Lira was found to be lower than the average rank of festival participants earning between 5,001 and 10,000 Turkish Lira and 15,001 and 20,000 Turkish Lira. Similarly, the average

Table 9. Results of discriminant analysis between cultural balance dimension and demographic characteristics

Categories	Groups	n	M. Rank	Sum of Ranks	U	Chi-Square	Sig	Sig[1]	Variation
Gender	Female	168	205.74	34564.50	16927.500	-	0.117	-	-
	Male	222	187.75	41680.50					
Marital status	Married	164	188.14	30855.00	17325.000	-	0.271	-	-
	Single[b]	226	200.84	45390.00					
Age	18-27	136	205.77	-	-	3.339	0.503	-	-
	28-37	77	189.43						
	38-47	101	187.73						
	48-57	49	181.99						
	58 and older	27	214.69						
Education status	Secondary school	45	195.62	-	-	0.585	0.965	-	-
	High school	35	192.39						
	Vocational school	111	200.55						
	Bachelor's degree	139	195.82						
	Graduate degree	60	187.14						
Income level	5,000 TL and less[a]	69	164.67	-	-	15.944	0.003	0.003 0.042 0.014	b>a c>a d>a
	5,001-10,000 TL[b]	118	216.64						
	10,001-15,000 TL[c]	107	200.70					0.024	d>e
	15,001-20,000 TL[d]	29	227.29						
	20,001 TL and more	67	167.96					0.004	e>b
Length of life in Edirne	I'm from here[a]	259	195.32	-	-	8.020	0.046		
	1-5 years[b]	84	175.08					0.035 0.026	c>b d>b
	6-11 yearsl[c]	19	235.26						
	11 years and above[d]	28	231.45						
Participation activity	Marching Band-Liver Festival [a]	1	323.50	-	-	17.151	0.004		
	Kirkpinar Oil Wrestling Festival[b]	53	163.24						
	Kakava-Hidirellez Festival[c]	63	176.13						
	Kakava-Hidirellez Festival and Marching Band-Liver Festival[d]	8	100.69						
	Kakava-Hidirellez Festival and Kirkpinar Oil Wrestling Festival[e]	119	206.23					0.014 0.004	e>b e>d
	All[f]	146	211.15					0.048 0.008 0.007	f>a f>b f>d

*p< 0.05

Table 10. Results of discriminant analysis between general cultural interaction dimension and demographic characteristics

Categories	Groups	n	M. Rank	Sum of Ranks	U	Chi-Square	Sig	Sig[1]	Variation
Gender	Female	168	205.07	34451.00	17041.000	-	0.145	-	-
	Male	222	188.26	41794.00					
Marital status	Married	164	190.95	31316.00	17786.000	-	0.497	-	-
	Single	226	198.80	44929.00					
Age	18-27	136	205.36	-	-	2.873	0.579	-	-
	28-37	77	187.33						
	38-47	101	193.05						
	48-57	49	179.08						
	58 and older	27	208.07						
Education status	Secondary school	45	187.24	-	-	1.333	0.856	-	-
	High school	35	192.56						
	Vocational school	111	200.16						
	Bachelor's degree	139	200.20						
	Graduate degree	60	183.91						
Income level	5,000 TL and less[a]	69	164.80	-	-	15.538	0.004	0.003 0.011	b>a d>a
	5,001-10,000 TL[b]	118	216.94						
	10,001-15,000 TL[c]	107	198.22						
	15,001-20,000 TL[d]	29	229.78					0.025	d>e
	20,001 TL and more	67	170.16					0.006	e>b
Length of life in Edirne	I'm from here[a]	259	194.40	-	-	6.406	0.093	-	-
	1-5 years[b]	84	179.44						
	6-11 years[c]	19	239.11						
	11 years and above[d]	28	224.27						
Participation activity	Marching Band-Liver Festival[a]	1	269.50	-	-	14.488	0.013		
	Kirkpinar Oil Wrestling Festival[b]	53	162.89						
	Kakava-Hidirellez Festival[c]	63	179.98						
	Kakava-Hidirellez Festival and Marching Band-Liver Festival[d]	8	106.13						
	Kakava-Hidirellez Festival and Kirkpinar Oil Wrestling Festival[e]	119	206.44					0.013 0.010	e>b e>d
	All[f]	146	209.51					0.010 0.014	f>b f>d

*p<0.05

rank of participants with an income between 5,001 and 10,000 TL was lower than the average rank of participants with an income of 20,001 TL or more.

The Kruskal-Wallis test applied to the analysis showed a statistically significant difference between the total cultural interaction variable and the participation activity variable (χ^2=14.488, p<0.05). In order to determine the source of the differences, the Mann-Whitney U test was applied to the expressions that showed significant differences based on the type of activity undertaken. It was discovered that the average rank of participants of the Kirkpinar Oil Wrestling Festival and the Kakava-Hidirellez Festival was lower than the average rank of individuals who participated both of these festivals and all of the festivals.

In addition, according to the results of the Kruskal-Wallis test as shown in Table 10, there is no statistically significant difference between the cultural exchange dimension and the average rank of the groups based on age (χ^2=2.873, p>0.05), educational status (χ^2=1.333, p>0.05), and participating activity (χ^2=14.488, p>0.05).

CONCLUSION AND RECOMMENDATIONS

Festivals can be evaluated as events that facilitate the transmission of historical, social, and cultural messages from the past to the present through cultural interaction. From the perspective of locals and tourists, festivals are a tourist activity for cultural exchange, education, and enjoyment. Application research was conducted with the objective of determining how the residents and tourists interact culturally during the festivals.

In 2022, Kakava-Hidirellez Festival and Kirkpinar Oil Wrestling Festival attracted the most locals in Edirne, according to data gathered on locals' attendance at festivals in the destination. In addition, a general evaluation reveals that a sizeable portion of the local populace attends all of the festivals. It should be mentioned that local participation in the Marching Band-Liver Festival, one of the most well-known festivals held by Edirne Municipality, falls short of expectations. It is believed that one of the most important reasons for this is that locals are more aware of the centuries-long history of the Kakava-Hidirellez Festival and the Kirkpinar Oil Wrestling Festival, both of which are on the UNESCO Intangible Cultural Heritage List, and view these two festivals as symbols of collective memory. Erbay, Demirkol, and Özmutlu (2020) assert that these events are a part of the region's living culture because they are viewed as cultural and social activities that are anticipated annually by the residents of the region. As a result, it is evident that Edirne Municipality needs research aimed at increasing locals' participation in the Marching Band-Liver Festival.

As a result of the performed analyses, it was discovered that the average attitudes of the local people about the cultural interaction dimensions, both cultural exchange and cultural balance, were at a moderate level. Yet, the cultural balance dimension is substantially greater than the average of the cultural exchange dimension. For a tourism destination to be successful and sustainable, it is crucial that the perceptions and attitudes of the local population towards tourists are positive. This study indicates that cultural exchange during festivals is not seen negatively. In Edirne, groups with diverse cultural backgrounds have created a synthesis culture over many years of living together. In addition, the inability of a single culture to live in a confined environment in Edirne, one of the opening border regions of Turkey, is viewed as a crucial element in the absence of a negative perspective of cultural exchange in cultural interactions with tourists. It also coincides with the findings that the indigenous people of Edirne create a cultural environment where tolerance and hospitality are at a high level by fusing their own culture with the cultures of the surrounding region. This result cannot, however, be interpreted as having no impact on the cultural interaction of festivals. In particular, the dynamic structure of culture necessitates a longer time frame for addressing its consequences.

The acceptable hypotheses resulting from the performed analyses are listed below.

H_{1-5}: There is a statistically significant difference between the cultural interaction perceptions of the local people regarding festival tourism and the income variable.
H_{2-5}: There is a statistically significant difference between the cultural exchange dimension perceptions of the local people regarding festival tourism and the income variable.
H_{3-5}: There is a statistically significant difference between the cultural balance dimension perceptions of the local people regarding festival tourism and the income variable.
H_{2-6}: There is a statistically significant difference between the perceptions of the cultural interaction dimension of the festival tourism of the local people and the event attended.
H_{3-6}: There is a statistically significant difference between the perceptions of the cultural balance dimension of the festival tourism of the local people and the event attended.

On the basis of the collected data, it was determined that the dimensions of cultural interaction, cultural exchange, and cultural balance vary with respect to the income level of the local population. The average perception of cultural interaction among locals, as well as the average perception of cultural exchange and cultural balance, declines as the economic level decreases. It is feasible to conclude, based on a broad evaluation, that the cultural interaction of festival participants with lower incomes is limited. This suggests that because the festivals are free of charge, those with lower incomes only participate as spectators. They are unable to take advantage of the paid facilities provided during the festival, which provide socialization opportunities that visitors enjoy (such as eating, sitting in a cafe, etc.), so their cultural interactions remain at a lower level. In other words, the low number of interaction points and reduced cultural and social interaction between two distinct cultures as a result of income disparities are effective in achieving the desired outcome.

A further outcome is the differentiation of cultural interaction and cultural balance dimensions from the variables evaluated in this interaction based on the locals' participation in an activity. The views of cultural interaction and the perceptions of the sub-dimension of cultural balance are greater among festival participants than among those who participate only one festival. There is no doubt that each festival that participates enhances the degree of interaction. In addition, participants of the Kakava-Hidirellez Festival and the Kirkpinar Oil Wrestling Festival, both of which are on the UNESCO Intangible Cultural Heritage List, have enhanced views of the subdimensions of cultural exchange and cultural balance. It is believed that these two events, which serve to preserve the shared historical memory, foster greater cultural interaction and cultural harmony by fostering a shared spirit and unifying individuals.

The study examines the perceptions of local people about cultural interaction at festivals organized by the municipality of Edirne. The study will cause limited comments to be made on the subject in this aspect. Because there is no one-sided cultural interaction during the festivals. Therefore, in future research, the examination of the perceptions of local and foreign visitors attending the festival and the cultural interaction between these visitors will help to provide a holistic perspective on the issue. Again, another proposed research topic is on the support of local people to festival events. Haviland (1999) says that since culture is a set of rules and standards that cannot be directly observed in studies related to culture, what anthropologists should do is to carefully monitor groups of people over a certain period of time, ask questions, listen, and make solutions. By taking this suggestion into consideration in tourism studies, the realization of qualitative research related to cultural interaction will add depth to the field. In the study, a research was conducted on the participants of the festival organized by the Edirne central municipality. Considering other festivals as a whole or each festival separately with the dimensions of cultural interaction will enrich the literature. Festival tourism has not only cultural effects on the perception of local people. Another proposed research topic is to examine the environmental, social, and economic dimensions of festival tourism in the Edirne destination.

REFERENCES

Artun, E. (2007). Türk Halk Kültüründe Hıdırellez. *Ç.Ü. Türkoloji, 1575*, 1-9. Retrieved from https://turkoloji.cu.edu.tr/HALKBILIM/erman_artun_turk_halk_kulturunde_hidrellez.pdf

Ay, A. (2010). Kirkpinar Oil Wrestling: Oil Wrestling, Edirne, Turkey. Ceren Kitapevi.

Bakan, Ö., & Canöz, K. (2017). Kültürlerarası İletişim Bağlamında Türk Üniversite Öğrencilerinin Japonlara Yönelik Kalıp Yargıları. *Selçuk Üniversitesi SBE Dergisi*, (37), 91-107. Retrieved from https://dergipark.org.tr/en/download/article-file/1724897

Bilar, E. (2022). Kırkpınar Yağlı Güreşleri Tarihi. *Edirne Kent ve Kültür Bilinci*. Retrieved from https://edirnekultur.com/detay/tarihi-863

Bozyer, Ü. (2012). Turizm Sosyolojisi. In S. G. S. T. Tartışmaları (Ed.), *Muammer Tuna*. Barkod.

Çakmakcı, E., & Öztürk, Ş. (2019). Toplumsal Değişimin Bir Faktörü Olarak Turizm. *Süleyman Demirel Üniversitesi, Fen-Edebiyat Fakültesi Sosyal Bilimler Dergisi*, *47*(2), 1-17. Retrieved from https://dergipark.org.tr/tr/download/article-file/841854 doi:10.35237/sufesosbil.544714

Civelek, A. (2014). Turizmin Sosyal Yapıya ve Sosyal Değişmeye Etkileri. *Selçuk Üniversitesi Sosyal Bilimler Meslek Yüksekokulu Dergisi, 13*(1-2), 331-350. Retrieved from https://dergipark.org.tr/en/pub/selcuksbmyd/issue/11301/135146

Demirören News Agency. (2022). *Edirne'de 1.5 ton bedava ciğer dağıtıldı, vatandaş birbiriyle yarıştı*. Retrieved from https://www.dha.com.tr/yerel-haberler/edirne/edirnede-1-5-ton-bedava-ciger-dagitildi-vatan-2069957

Döğüş, S. (2015). Anadolu'da Hızır-İlyas Kültür ve Hıdrellez Geleneği. *Türk Kültürü ve Hacı Bektaş Velî Araştırma Dergisi*, *74*, 77-100. Retrieved from https://hbvdergisi.hacibayram.edu.tr/index.php/TKHBVD/article/view/1721

Edirne Belediyesi. 2022). *10. Uluslararası Edirne bando ve ciğer festivali başlıyor*. Retrieved from https://www.edirne.bel.tr/icerik/10uluslararasi-edirne-bando-ve-ciger-festivali-basliyor

Edirne Municipality. (2022). *Kırkpınar Yağlı Güreşleri Festivali Tarihi*. Retrieved from https://edirnekirkpinar.com/tr/yazi/6-genel-bilgiler/36-tarihi

Emir, O., & Toker, A. (2020). Turizmin Toplumsal Etkileri I. *Açıköğretim Fakültesi Yayını, 2875*, 51-66. Retrieved from https://ets.anadolu.edu.tr/storage/nfs/SOS109U/ebook/SOS109U-20V1S1-8-0-0-SV1-ebook.pdf

Erbay, R., Demirkol, C., & Yıldırım Özmutlu, S. (2016). *Trakya Bölgesinde Festival, Fuar ve Panayır Organizasyonlarının Bölgenin Sosyo-Ekonomik Yapısına Etkileri*. Namik Kemal University Scientific Research Project. NKUBAP.00.13.AR.12.03. Retrieved from http://acikerisim.nku.edu.tr:8080/xmlui/bitstream/handle/20.500.11776/867/NKUBAP.00.13.AR.12.03.pdf?sequence=1

Gibson, C., Connell, J., Waitt, G., & Walmsley, J. (2011). The Extent and Significance of Rural Festivals. In *Festival Places: Revitalising Rural Australia (Tourism and Cultural Change)* (pp. 3–24). Channel View Publications. doi:10.21832/9781845411688-004

Guinness World Records. (2021). *Oldest Competition, Wrestling*. Retrieved from https://www.guinnessworldrecords.com/world-records/73717-oldest-competition-wrestling

Gündüz Alptürker, İ., & Alptürker, H. (2021). Gelenekten Geleceğe Yöresel Kültürün Taşınmasında Festivallerin Rolü. *MANAS Sosyal Araştırmalar Dergisi, 10*(2), 1409-1422. Retrieved from https://dergipark.org.tr/tr/download/article-file/1299408

Haviland, W. A. (1999). *Cultural Anthropology* (9th ed.). Harcourt Barace College Publishers.

Keleş, H. (2021). *Turizmin Yerel Halkın Yaşam Kalitesi, Memnuniyeti ve Turizme Desteği Üzerindeki Etkileri: Manavgat Örneği* [Doctoral Thesis]. Necmettin Erbakan Üniversitesi Sosyal Bilimler Enstitüsü Turizm İşletmeciliği Anabilim Dalı, Turizm İşletmeciliği Bilim Dalı, Konya. Tez no: 704246. Retrieved from https://tez.yok.gov.tr/UlusalTezMerkezi/tezSorguSonucYeni.jsp

Kendir, H. (2020). Gastronomik Lezzetlere Yönelik Çevrimiçi Turist Deneyimlerinin İncelenmesi: Edirne Tava Ciğeri Örneği. *Safran Kültür ve Turizm Araştırmaları Dergisi, 3*(3), 408-423. Retrieved from https://dergipark.org.tr/tr/pub/saktad/issue/59328/845512

Linderová, I., Scholz, P., & Almeida, N. (2021). Attitudes of Local Population Towards the Impacts of Tourism Development: Evidence From Czechia. *Frontiers in Psychology, 12*, 1-11. Retrieved from https://www.frontiersin.org/articles/10.3389/fpsyg.2021.684773/full doi:10.3389/fpsyg.2021.684773

Morgan, M. (2008). What Makes a Good Festival? Understanding the Event Experience. *Event Management, 12*(2), 81–93. doi:10.3727/152599509787992562

Pala, N. (2001). *Turizmin Sosyo-Kültürel Sonuçları Açısından Bir İnceleme (Ballıca Mağarası Örneği)* [Master Thesis]. Sakarya Üniversitesi Sosyal Bilimler Enstitüsü, Sosyoloji Ana Bilim Dalı, Kurumlar Sosyolojisi Bilim Dalı, Sakarya. Tez no: 111986. Retrieved from https://tez.yok.gov.tr/UlusalTezMerkezi/tezSorguSonucYeni.jsp

Pamir, H. Y. (1997). *Turizmin Sosyal Yapıya ve Sosyal Değişmeyi Etkileri* [Master Thesis]. İstanbul Üniversitesi Sosyal Bilimler Enstitüsü, Turizm Ana Bilim Dalı, İstanbul. Tez no 64564. Retrieved from https://tez.yok.gov.tr/UlusalTezMerkezi/tezSorguSonucYeni.jsp

Paspati, A. G. (1870). *Études sur les Tchingianés ou Bohémiens de l'Empire Ottoman*. Constantinopol.

Reisinger, Y. (2009). *International Tourism: Cultures and Behavior* (1st ed.). Elsevier. Retrieved from https://www.researchgate.net/publication/330919283_Sociocultural_Impacts_of_Tourism_on_Residents_of_World_Cultural_Heritage_Sites_in_China

Saçlı, Ç. (2020). Festival Memnuniyeti ile Kültürel Etkileşim Arasındaki İlişki Üzerine Bir Araştırma: Silifke Uluslararası Müzik ve Folklor Festivali Örneği. *Türk Turizm Araştırmaları Dergisi*, *4*(1), 505-521. Retrieved from https://www.tutad.org/index.php/tutad/article/view/189/188

Sekaran, U. (2003). Research Methods for Business: A Skill Bulding Approach (4th ed.). John Wiley & Sons Inc.

Shahzalal, M. (2016). Positive and Negative Impacts of Tourism on Culture: A Critical Review of Examples from the Contemporary Literature. *Journal of Tourism, Hospitality and Sports*, *20*, 30-34. Retrieved from https://core.ac.uk/download/pdf/234696865.pdf

Turkish Ministry of Culture and Tourism. (2022). *EDEN 2008-Turizm ve Somut Olmayan Kültürel Miras*. Retrieved from https://eden.ktb.gov.tr/TR-246372/eden-2008---turizm-ve-somut-olmayan-kulturel-miras.html

Turkish Ministry of Culture and Tourism General Directorate of Research and Development. (2022). *Kırkpınar Yağlı Güreş Festivali*. Retrieved from https://aregem.ktb.gov.tr/TR-268224/kirkpinar-yagli-gures-festivali.html

Turkish Patent Institute. (2022). *Coğrafi İşaret Tescil Belgesi*. Retrieved from https://ci.turkpatent.gov.tr/Files/GeographicalSigns/128.pdf

Tylor, E. B. (1871). *Primitive Culture: Researches into the Development of Mythology, Philosophy, Religion, Art and Custom*. John Murray. Retrieved from https://books.google.com.tr/books?hl=tr&lr=&id=RUMBAAAAQAAJ&oi=fnd&pg=PA1&dq=tylor+1871+culture+pdf&ots=X_u4fqQSiZ&sig=8LynCxMkaXp5CkcTzkJ9-cH7D-8&redir_esc=y#v=onepage&q=culture&f=false

Uçar, N. (2010). *Turizmin Kırsal Alana Etkisi: Gökçeada Örneği* [Master Thesis]. İstanbul Üniversitesi Sosyal Bilimler Enstitüsü Antropoloji Anabilim Dalı, İstanbul. Tez no: 262553. Retrieved from https://tez.yok.gov.tr/UlusalTezMerkezi/tezSorguSonucYeni.jsp

UNESCO Turkish National Commission. (2022). *UNESCO İnsanlığın Somut Olmayan Kültürel Mirası Temsilî Listesi.* Retrieved from https://www.unesco.org.tr/Pages/126/123/UNESCOİnsanlığın-Somut-Olmayan-Kültürel-Mirası-Temsilî-Listesi

Ürkmez, D., & Akbulut, D. (2020). Yerel Yönetimlerde Bir Halkla İlişkiler Aracı Olarak Festivallerin Kullanımı. *Karadeniz İletişim Araştırmaları Dergisi, 10*(2), 1-27. Retrieved from https://dergipark.org.tr/tr/download/article-file/1299100

Yağbasan, M., & Demirbağ, İ. (2017). Kültürlerarası İletişime Stratejik Bir Yaklaşım: Kültürlerarası Empati (Bir Kültürlerarası İletişim Projesinin Sonuçları). *Gümüşhane Üniversitesi İletişim Fakültesi Elektronik Dergisi, 5*(2), 619-640. Retrieved from https://dergipark.org.tr/tr/download/article-file/357717 doi:10.19145/e-gifder.320646

Zhang, C. X., Fong, L. H. N., Li, S., & Ly, T. P. (2019). National Identity and Cultural Festivals in Post-colonial Destinations. *Tourism Management, 73*, 94–104. doi:10.1016/j.tourman.2019.01.013

Zhuang, X., Yao, Y., & Li, J. (2019). Sociocultural Impacts of Tourism on Residents of World Cultural Heritage Sites in China. *Sustainability, 11*(840), 1–19. doi:10.3390u11030840

Chapter 18
Destination Branding and Mega Events:
A Case of FIFA World Cup 2022 and Sultanate of Oman

Mohit Kukreti
Independent Researcher, Oman

Amitabh Mishra
University of Technology and Applied Sciences, Oman

Aarti Dangwal
Chitkara University, India

Vishal Jain
University of Technology and Applied Sciences, Oman

Veena Tewari
University of Technology and Applied Sciences, Oman

ABSTRACT

Tourist attractions are often centered on mega sports events. Since the mega events have the potential to generate new revenue and enhance the appeal of the destination similarly, the FIFA World Cup 2022 in Qatar presented a unique opportunity for Oman to leverage the mega sporting event to promote its tourism industry globally. Oman used this mega event to brand its tourism industry by collaborating with FIFA, hosting fan zones, effectively using social and print media, highlighting accessibility and accommodation, and promoting local tourism attractions. The mega event resulted in increased international arrivals, revenue generation, and an enhanced image of Oman in the Middle East. The chapter is an attempt to explore various initiatives taken by the Sultanate of Oman to attract international visitors. Further details will review the methods adopted by public and private sectors in Oman in leveraging the recent mega event of FIFA World Cup 2022 to become a preferred global destination.

DOI: 10.4018/978-1-6684-6356-7.ch018

INTRODUCTION

Tourist attractions are often centered on mega sports events, and various destinations compete to either host or leverage the most prestigious ones. Hosting the mega-events can significantly spill over goals in terms of the economic benefits and the potential for improved infrastructure and international image. These events have the potential to generate new revenue and enhance the appeal of the destination.

Tourism destination branding attributable to mega events is essential to the tourism industry. Mega events of the Middle East such as FIFA Qatar World Cup 2022, Olympics, Asian Games, Dubai Expo 2020, Qatar International Food Festival, Dubai shopping festival, Abu Dhabi Grand Prix, Dubai World Cup, Sharjah Light Festival, Salalah Khreef Festival etc., provide destinations with a unique opportunity to enhance their brand identity and image by creating positive publicity and media exposure. Destination managers should carefully plan and manage mega events to align with the destination's branding strategy.

According to Pike and Page (2014), "Events have the potential to create, communicate, and deliver a destination's brand" (p. 133). According to a recent report, Oman is leveraging the upcoming FIFA 2022 Football Cup for destination branding purposes (Smith, 2022). The FIFA 2022 World Cup in Qatar allows Oman to leverage event tourism for destination branding. As a neighboring country, Oman can benefit from the influx of tourists attending the World Cup by promoting its unique tourist attractions and experiences. So, according to Al-Maamari and Al-Hinai (2021), a long time before, Oman had already begun to develop its tourism infrastructure in anticipation of the World Cup. The country has invested in new hotels and resorts and improved its transportation network to make it easier for visitors to travel between Oman and Qatar. This fact was further confirmed by the case study conducted by Al Mawali and Al Ghaithi in 2021, which also highlighted the upcoming hosting of the 2022 event's significant impact on the tourism industry in Oman.

Ghimire and Schewe (2017) noted that Qatar is expected to receive many tourists during the event, which could spill over to neighboring countries, including Oman. Additionally, the investment in infrastructure and facilities in Qatar could improve the overall tourism infrastructure in the region, which could benefit Oman (Ratten, 2017). As such, Oman did experience an influx of tourists who sought accommodation, food, and other tourist services and contributed towards tourism development.

The 2022 FIFA World Cup in Qatar had a positive impact on the tourism industry in Oman. The event increased the number of tourists visiting Oman, generating revenue for the country's tourism industry. The World Cup had also created employment opportunities in Oman. The Sultanate of Oman had successfully leveraged the World Cup to promote its tourism industry and attract more visitors to the country.

LITERATURE REVIEW

Mega Events

According to Getz and Page (2016), hosting mega sporting events such as the FIFA World Cup can drive tourism growth by creating a sense of urgency for tourists to visit the host country. Similarly, Kikuchi and Koo (2017) found that the FIFA World Cup in Brazil in 2014 significantly increased international tourism in the country. This event provided a unique opportunity for Brazil to showcase its tourism industry and attract more tourists to its shores. However, the impact of hosting mega sporting events on neighboring countries is less explored. Hosting major sporting events increases tourism, investment, and infrastructure development (Bialkowski, 2015). A study by Kasimati and Dawson (2002) found that hosting the Olympic Games positively impacted the tourism industry, with the number of tourists increasing by up to 9% in the years following the event. Similarly, a study by

Baade and Matheson (2016) found that the FIFA World Cup positively impacted the host country's economy, increasing GDP, employment, and investment. A study by Preuss (2014) examined the impact of the 2006 FIFA World Cup in Germany on neighboring countries such as the Netherlands, Belgium, and France. The study found that the event had a significant positive impact on the tourism industry. Similarly, a study by Preuss and Solberg (2017) examined the impact of the 2010 FIFA World Cup in South Africa on neighboring countries such as Botswana, Mozambique, Namibia, Swaziland, and Zimbabwe. The study suggests that hosting a mega-event like the FIFA World Cup can positively impact the tourism industry of neighboring countries.

On the other hand, some studies have shown that hosting mega-events such as the FIFA World Cup can negatively impact the tourism industry of neighboring countries. A study by Andriotis and Vaughan (2003) examined the impact of the 2004 Olympic Games in Athens on the tourism industry of neighboring countries such as Turkey and Cyprus. The study found that the event had a negative impact on the tourism industry of these countries, with a decrease in the number of visitors due to the fear of high prices and overcrowding. However, other potential challenges are associated with hosting major sporting events, such as high costs, infrastructure challenges, and environmental concerns (Guttentag, 2015). Additionally, there are concerns about the impact on the tourism industry of neighboring countries, which may face increased competition (Hall, 2011). Therefore, it is important to carefully consider the potential benefits and challenges before deciding to host such an event.

Strategies for Mega Event

Leveraging refers to the strategic use of something (such as an opportunity, resource, or asset) to achieve a desired outcome or advantage. In the context of the chapter, leveraging means utilizing this mega event to boost Oman's tourism industry by taking advantage of the increased attention and interest generated by the event to promote Oman as a tourist destination.

Specifically, in the context of Oman, several mega-tourism events have been held in recent years, including the Muscat Festival, now Muscat Nights, Salalah Tourism Festival. Al-Rawahi et al. (2018) conducted a study on the impacts of the Muscat Festival on Oman tourism. The study found that the festival had a significant positive impact on the local economy, generating revenue for local businesses and creating employment opportunities for the local population. However, the study also found that the festival had negative environmental impacts, such as increased waste generation and air pollution.

Similarly, Al-Hadhrami et al. (2019) conducted a study on the impacts of the Salalah Tourism Festival on Oman tourism. The study found that the festival had a positive impact on the local economy, particularly in terms of job creation and income generation. However, the study also found that the festival had negative impacts on the environment, such as increased waste generation and water consumption. In addition, several studies have also investigated the role of mega events in promoting Oman tourism. For example, Al-Lamki et al. (2017) conducted a study on the impacts of the Naseem A'Sabaa event on Oman tourism. According to Smith and Brent (2018) and Al Bulushi and Al Hinai (2019), the mega tourism events can have both positive and negative impacts on Oman's tourism industry. A study by Al Awa and Al Araimi (2020) examined the impacts of the Muscat Festival, a major tourism event in Oman. The authors found that the festival had a positive impact on the local economy, but also resulted in increased traffic congestion and environmental pollution. Another study by Al Hinai and Al Bulushi (2021) explored the impacts of the Salalah Tourism Festival on Oman's tourism industry. The authors found that the festival had a positive impact on the local economy and contributed to the development of the tourism industry in the region. Overall, the literature suggests that mega tourism events can have both positive and negative impacts on Oman's tourism industry. To maximize the benefits of such events and minimize their negative impacts, careful planning and management are essential.

Pike (2009, 2010) has noted that despite numerous academic contributions, destination branding from events is still a relatively new topic to investigate. Additionally, Freire (2012) has also highlighted this issue, focusing specifically on destination branding in the Middle East. Therefore, the objective of his chapter is to fill a void in the literature on destination branding by investigating the efficiency of Oman's brand within the context of mega event tourism. In doing so, this chapter aims to address the lack of research gap on the subject in the Middle Eastern region and the insufficient contribution to the literature on destination branding utilizing mega event tourism.

Yeoman et al. (2017) suggest that host countries should focus on promoting their unique cultural and natural attractions, as well as their hospitality and infrastructure, to capitalize on the World Cup's potential benefits. According to Balakrishnan and Michael (2020) branding is an essential aspect of successful tourism promotion, as it helps to differentiate a destination from its competitors and establish a unique identity by focusing on creating a strong brand image that reflects their unique culture, values, and offerings.

In the case of the Sultanate of Oman being one of Qatar's neighboring countries, the FIFA 2022 World Cup in Qatar presents a unique opportunity to promote its tourism industry. Oman is known for its natural beauty, rich history, and warm hospitality. By leveraging the World Cup to showcase its unique attractions and promote its tourism industry, Oman can attract a new wave of tourists and establish itself as a premier tourism destination. In terms of branding, Oman could focus on its unique selling points, such as its natural beauty, cultural heritage, and warm hospitality. Mishra and Kukreti (2021) highlighted Oman as a luxury tourism destination. So, Oman's tourism industry could create a campaign that showcases the country's stunning landscapes, rich history, and friendly locals, with a message that emphasizes the unique luxury tourism experiences that Oman offers. According to Li et al. (2019), a successful tourism brand is one that effectively communicates the unique attributes of a destination and creates a positive emotional connection with potential visitors. As mentioned above, in the context of the World Cup, Oman's tourism brand should highlight its proximity to Qatar, its cultural and natural attractions, and its unique hospitality and tourism offerings. Along with the Sultanate of Oman, it is equally important to assess the benefit of this mega sporting event in the GCC.

Sports Tourism

Sports tourism lately has come up as a significant contribution to global GDP. Sports tourism not only opened up the new stream of revenue, but also supported the destination branding by boosting their image and exposure. With increased tourism activity it also brings social and community development to the hosting and nearby geographical locations. Social exchange theory is one of the established frameworks which provides various advantages of sports tourism in terms of development, relationship to residents and contributions to the community (Giango et al., 2022).

Branding of destinations on the basis of sports events is a strategic way ahead for the governments of the host countries. Regional cooperation and understanding at the regulatory level may bring sustainable benefits to the society, as the mega events can't be conceived and implemented without the active support of the various stakeholders (Mohammad et al., 2012).

Sports tourism is the driver for both urban and rural development as it not only influences the economic but also socio -cultural hierarchy of any society. Future is promising for the sports industry tourism, as it provided foundation for the holistic development of by improving the quality of facilities, performing marketing activities both at national and international level, quality of hospitality providers and overall increased safety of people (Mohammad et al., 2012).

Destination branding can be termed as corporate branding. Ingle image or branding proposition is a complex task for the destinations. Destination branding shows the relationship between the image of the place and its people. It's about establishing different however positive image of the city (Irisi & Ariana, 2017).

The FIFA World Cup 2022 Benefiting the GCC Countries

The FIFA World Cup was one of the biggest sporting events in the world, and it had a significant impact on the host as well as neighboring destinations. Since all the available hotels in Qatar were occupied and priced high, therefore, several supporters chose to lodge in the adjacent countries, which provided them with transport to Doha via the newly introduced shuttle flights during match days.

The Gulf states, including Abu Dhabi, Dubai, and Qatar, have utilized sports events to promote tourism, development, and regeneration in recent years. These events range from Formula 1 races to major football tournaments and are held in state-of-the-art facilities constructed specifically for the events. Similarly, Qatar has hosted several significant sporting events, such as the 2022 FIFA World Cup, the first World Cup to be held in the Middle East. The country has invested heavily in infrastructure and stadium construction to prepare for the tournament, and it is expected to attract a considerable number of visitors to the region.

In addition to promoting tourism and development, sports events have also helped raise the profile of the Gulf states on the global stage. By hosting major sporting events, these countries have showcased their modern facilities and infrastructure, as well as their commitment to promoting sports and healthy living. Overall, sports events have become an essential tool for branding and promoting tourism development and regeneration in the Gulf states, including Abu Dhabi, Dubai, Oman, Saudi Arabia, and Qatar. Through investing in world-class facilities and hosting major sporting events, these countries have successfully attracted visitors from around the world and elevated their profile on the global stage.

Alpen Capital (2022) reported a rise in international tourists visiting the GCC between 2016 and 2019, with an annual growth rate of 2.8%. The UAE and Kuwait were the primary drivers of this growth, with compound annual growth rates (CAGRs) of 6.6% and 6.7%, respectively. The majority of the demand for tourism came from personal, leisure, and religious travel, followed by business and professional travel. However, Qatar experienced a decline in tourism with a CAGR of -10.1% over the same period due to the embargo imposed by neighboring GCC nations in 2017-18. In 2020, due to the COVID-19 pandemic, international tourist arrivals in the GCC declined by 75.0% year-on-year from 70.8 million in 2019 to 17.7 million.

Nonetheless, GCC countries acted quickly to combat the crisis by implementing strict measures to curb the spread of the virus. With the pandemic now reaching an endemic state, these countries have announced several measures to ease restrictions. Furthermore, the easing of cross-border travel restrictions by western regions has contributed to a recovery in the industry. The FIFA World Cup 2022 attracted more than 1.4 million fans from around the world to Qatar.

Figure 1. International tourist arrivals (mn) to GCC
Source: Department of Culture and Tourism – Abu Dhabi, Department of Tourism and Commerce Marketing (Dubai Tourism), Qatar Tourism Annual Report, Oman National Centre for Statistics and Information, Bahrain Tourism and Exhibition Authority, World Travel and Tourism Council (WTTC); extracted from Alpen Capital.

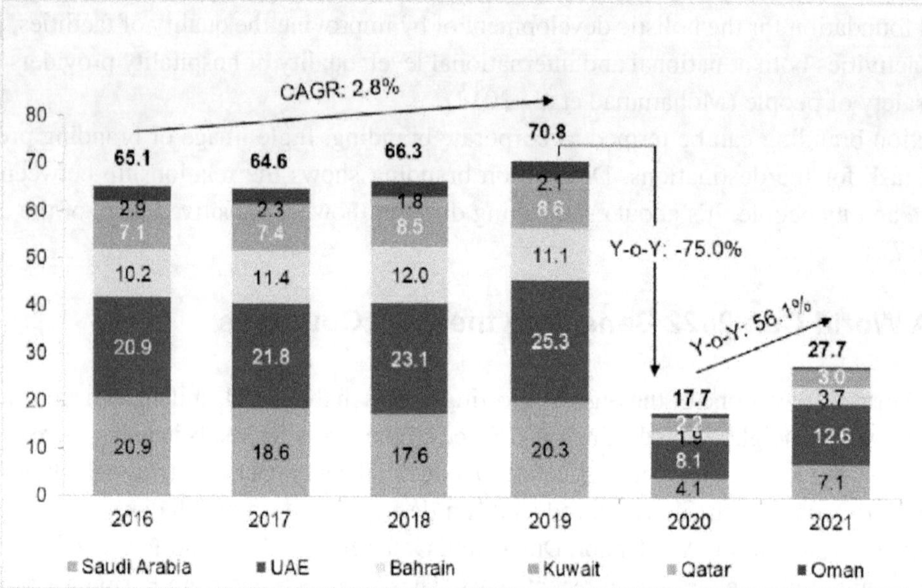

It is predicted that the hospitality sector in the GCC will bounce back in 2022 and achieve a year-on-year increase of 74.8% to reach US$ 26.3 billion by the end of the year, similar to pre-pandemic

Figure 2. GCC hospitality market forecast until 2026 (US$ BN)
Source: Alpen Capital.

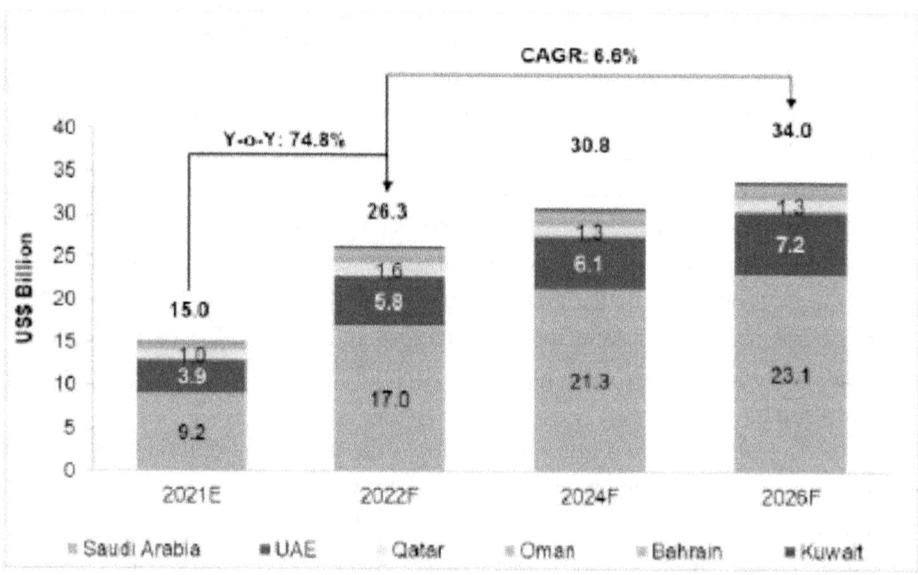

levels. The main reason for this substantial growth rate is due to the lower base of 2021 as the countries in the region recuperate from the pandemic. According to Figure 2, the industry is expected to continue to grow steadily at a compound annual growth rate (CAGR) of 6.6% between 2022 and 2026, reaching US$ 34.0 billion.

According to Ayeni (2023), during the World Cup, more than 2.5 million individuals traveled to the area. The hosts, which included Qatar, the United Arab Emirates, and Saudi Arabia, benefited significantly from increased tourism. Nonetheless, according to CSRGulf, Kuwait missed out on fully capitalizing on this tourism opportunity due to the delayed restructuring of its tourism industry and aviation fleet, as well as the delayed completion of expansion work on its international airport.

Qatar is the second-largest contributor to the nation's GDP through tourism, accounting for 10.3% of the total GDP. The UAE is ahead of Qatar in this regard. Qatar has greatly benefited from hosting the World Cup in terms of tourism infrastructure development and promotion of tourist destinations in the country, according to CSRGulf. The UAE leads the Gulf countries in terms of tourism revenues and contribution to the gross domestic product, with the sector contributing over 11.6% to the country's output. Bahrain ranks third with a 6.8% contribution to its GDP, while Saudi Arabia is an emerging destination for tourism and entertainment in the Middle East, with a 5.3% contribution to its GDP. Projections indicate a significant increase in this figure for Saudi Arabia. Kuwait and Oman have the lowest contribution from tourism, with 3.3% and 3% of their GDP, respectively (Ayeni, 2023).

It is worth mentioning that the Qatar FIFA 2022 World Cup has had a positive impact on the entire GCC region by contributing to increased tourism, improved infrastructure, and a boost to the local economy through the generation of significant revenue from ticket sales, sponsorships, and merchandise sales. Additionally, the event has enhanced the global image of GCC tourism and provided a platform for showcasing the region's culture and promoting cultural exchange.

Based on the previous review of existing literature and other data sources, this chapter tries to answer the following questions:

- What initiatives has the Sultanate of Oman taken to attract tourists to its territory?
- How the Sultanate of Oman leveraged the mega event of Qatar FIFA World Cup 2022 to boost its Tourism Industry's global brand image?
- What role do government policies and initiatives play in shaping Oman as a destination brand?

THE SULTANATE OF OMAN IN THE CONTEXT

Oman is a country located on the southeastern coast of the Arabian Peninsula, and its tourism industry has been growing rapidly in recent years. According to Al Hinai and Al Musalhi (2018), Oman's tourism industry has been expanding due to a number of factors, including government investments in infrastructure, marketing efforts, and the country's unique culture and natural beauty.

Tourism Arrivals in Oman

According to the National Centre for Statistics and Information (NCSI), Oman's tourism industry has been growing at a steady pace in recent years. In 2018, Oman received 3.2 million tourists, which increased to 3.5 million in 2019, an increase of 6.7% from the previous year. The majority of visitors came from neighboring countries such as the UAE, Saudi Arabia, and Qatar. However, the COVID-19 pandemic had a severe impact on the tourism industry worldwide, and Oman was no exception. The country's borders were closed, and international flights were suspended, leading to a sharp decline in tourist arrivals. In 2020, the number of tourists dropped to 1.6 million. In 2020, Oman recorded only 785,000 visitors, a significant decrease of 77.4% compared to the previous year. Despite the challenges posed by the pandemic, Oman's tourism sector has been showing signs of recovery in 2021. In January 2021, the country welcomed 25,000 visitors, a 37% increase compared to the same month in the previous year. The gradual easing of travel restrictions and the implementation of strict health and safety measures have helped Oman's tourism industry to regain momentum. As of September 2021, the number of tourists who visited Oman was 483,701, and in 2022 Oman welcomed 2.9 million international arrivals. This increase is attributed to the FIFA World Cup 2022 as well as the Ministry of Heritage and Tourism participation in several road shows and trade shows.

Oman has been steadily developing its tourism industry in recent years, with the government placing a strong emphasis on promoting the country's natural beauty, rich culture, and history to the world. With its natural beauty, infrastructure development, and diversification of tourism offerings by promoting adventure tourism, culture and heritage, ecotourism, and luxury tourism, focus on sustainability and responsible tourism, Oman has well positioned itself to become a leading tourism destination in the Middle East. The government has also introduced measures to promote sustainable tourism, such as the "Oman Green Awards" to encourage eco-friendly practices in the tourism industry.

Oman's hospitality industry has been a significant factor in the growth of tourism. The country has a range of accommodation options, from luxury resorts to budget-friendly hotels and guesthouses, to cater to the needs of different types of tourists.

Oman Tourism Strategy 2040

A long-term plan was developed by the Oman Ministry of Tourism to develop and promote sustainable tourism in the country called Tourism Strategy 2040. According to Oman Ministry of Tourism (2020), the strategy focuses on enhancing the tourism sector's competitiveness, diversifying tourism products, and promoting Oman as a top tourist destination in the region. The strategy also emphasizes the importance of sustainable tourism practices and aims to minimize the negative impact of tourism on the environment and local communities. This includes promoting eco-tourism and responsible tourism practices, developing sustainable tourism infrastructure, and supporting local communities through tourism development. Tourism Strategy 2040 is a comprehensive plan that outlines the direction for Oman's tourism sector's growth and development. The strategy's focus on sustainability and diversification will help Oman's tourism industry thrive in the coming years.

FIFA World Cup 2022 in Qatar presents a unique opportunity for Oman to leverage the event to promote its tourism industry. Here are some ways in which Oman leveraged this mega event to brand its tourism industry:

Collaboration With Qatar for FIFA 2022 for Branding Oman's Tourism

The Ministry of Heritage and Tourism, Oman Airports Authority, Oman Air, OMRAN, ROP, and several other public and private organizations in the Sultanate of Oman worked together with Qatar to encourage its tourism industry by presenting it as a complementary destination to Qatar. This included promoting tour packages that combine visits to Qatar and Oman, highlighting the unique attractions and experiences each country has to offer.

According to a report by the Oman Ministry of Tourism, "Oman's proximity to Qatar, coupled with its diverse range of tourism offerings, make it an ideal complement to the World Cup experience. Visitors can easily combine a visit to Oman with their World Cup journey, enjoying everything from beach resorts to desert adventures, mountain trekking to cultural immersion" (Ministry of Tourism, 2021). As noted by travel writer Kate Springer, "Oman has been working hard to position itself as a must-visit destination in the Middle East, and its efforts are paying off. With its stunning landscapes, rich culture, and modern amenities, Oman is a compelling complement to Qatar, offering visitors a chance to explore the region in a more diverse and immersive way" (Springer, 2022).

According to Oman News Agency (2022), the Sultanate of Oman showed its support for Qatar's hosting of the FIFA World Cup 2022 by cooperating in several areas. The cooperation focused on promoting Muscat as a major city for hosting World Cup fans and ensuring that they have an exceptional experience while visiting Oman. As part of this effort, GCC countries offered multi-entry visas for up to 90 days to Hayya Card holders, which allowed for easy travel between Muscat and Doha. Additionally, there were programs in place for airport and transportation facilities, hotels and tourist packages, and events and activities, all accompanied by promotional campaigns. Ultimately, these efforts were aimed at providing the best possible experience for World Cup fans visiting Oman.

Hosting World Cup 2022 Fan Zones for Branding Oman's Tourism

Oman hosted fan zones during the World Cup for fans who want to experience the tournament in a different setting. These fan zones showcased Omani culture and traditions, as well as promoted local businesses and attractions. The Fan zone was hosted at the Oman Convention and Exhibition Centre in Muscat, one of the few cities authorized to welcome tournament fans. The festival provided an opportunity for fans to enjoy the excitement of the tournament while also experiencing the culture and attractions of the Sultanate. The Oman Convention Centre's garden, which hosted the festival, spanned across a vast expanse of over 9,000 square meters, and the parking lot can accommodate up to 4,000 cars. The festival area was a collaborative effort between various partners, such as entrepreneurs and small-to-medium-sized businesses, and it seamlessly blended traditional Omani culture with modern technology, resulting in a remarkable experience for attendees.

The Undersecretary of the Ministry of Heritage and Tourism and Head of the Steering Committee overseeing the Oman World Cup Programme, HE Azzan Qasim Al Busaidi, expressed that the World Cup is a significant sporting event that will take place in the region in recent years. He mentioned that the 'Oman World Cup Programme' aims to take advantage of its position as a satellite city to support Qatar's World Cup efforts and attract fans who are looking for an ideal place to stay and explore while attending the tournament. The main attraction of the programme is the Football Fan Festival, which will bring the excitement and atmosphere of the World Cup to both local and visiting fans. Moreover, the programme intends to showcase Oman's unique history, heritage, beauty, and hospitality to the visitors (Ministry of Heritage and Tourism. 2022).

Additionally, the festival boasted a virtual reality zone where visitors showcased their football abilities, competed in the e-player contest for the Football Fan Fest title, and received exclusive souvenirs from the festival. Fan Zone broadcasted 64 matches on the mega screen upon the big stage of the world cup in relationship with the host broadcaster beIN Sports.

Oman's match readiness and enthusiasm can be easily understood by reviewing the following preparations of activities and entertainment.

- **Gaming and Entertainment:** Football Fan Festival welcomed well-known musical stars, comedians, dancers and performers, local artists. With Virtual Reality and PlayStation zones, Football Fan Festival showcased 'Fast Foot Challenge'; 'bend it like Beckham' challenge; Teqball; Foosball table, for entertainment.
- **International Cuisine and Tastes:** Football Fan Festival not only celebrated the best footballing nations from around the world but also international cuisine and tastes from all corners of the globe by a range of locally sourced suppliers.
- **Retail and Merchandis:** By the amazing artisans and magnificent makers at Football Fan Festival brought together the best possible mix of rich Omani Culture in the form of souvenirs and gifts, as a special memory of Oman for the visitors.
- **Technology:** Football Fan Festival integrated the perfect blending of history and culture with technology and innovation. This was reflected from the development of our Virtual Reality (VR) zones to test the football skills and connection via two-way portal to connect inside fans with those on the outside, allowing them to wave hello to friends in other locations. The FIFA World Cup 2022 made significant use of the technology innovation network and locally created solutions, such as immersive 360-degree virtual reality tours of the Khalifa International Stadium, teams introduction etc. and cultural sites.
- **Competition Activities and Incentives:** Football Fan Festival across 23 days of World Cup action and entertainment offered visitors a range of opportunities to get involved with a number of competitions and incentives. These included kids football competitions, a prize draw for everyone who attends, the chance for e-gamers to compete for the title of Football Fan Festival champion and a user-generated promotion which provides attendees the chance to win free goodies by posting about their Football Fan Festival experience on social media.

Usage of Allied Parallel Sporting Event for Branding

The Oman Observer (2022) reported that Oman participated in the Camel and World Cup 2022 events organized by the Arab Union for Camel Races (Mahari) in Al Sheehaniya as a component of the FIFA World Cup 2022 in Qatar. The events lasted until December 18, and the opening ceremony was attended by camel chariots from Qatar and Royal Camels from Oman. This event included displays of the Royal Camels, as well as matches of the first international championship of 'camel handball', and a camel race featuring a human rider. This type of sporting event, which was in alliance with the FIFA World Cup 2022, was very helpful in promoting and popularizing Oman's cultural sports and heritage tourism.

Usage of Promotion and Publicity

A decision was made to hold a major competition called #HalfTimeForOman, where a winner would be chosen at each match of the competition to receive a seven-night stay in a luxurious hotel in Oman. People in charge were encouraged to share their vacation ideas on social media. To boost the country's tourism sector, the Oman government has made it easier for World Cup spectators to enter the country by accepting their entry tickets as entry visas. The transport company Mowasalat 'Karwa' tagged its

741 buses manufactured in Oman as eco friendly and carbon neutral. These buses have emerged as a popular mode of transportation for fans commuting to the Qatar World Cup. This sustainability effort significantly contributed towards the promotion and brand building of Oman.

Social media is a powerful tool for promoting tourism, and Oman leveraged it during the World Cup by using hashtags related to the tournament and showcasing Omani attractions and experiences that would appeal to football fans. The fans followed '@FtblFanFestival' across social media in Oman to stay updated on the World Cup festivities. A unique partnership with print and social media and Oman's national football celebrities such as Ali Al Habsi as the official brand ambassador alongside favourite sports personalities Khalil Al Balushi, Fatma Al Nabhani and Mohammed Al Noufali highlighted a new approach for promoting the tournament and its events. The shopping malls in Oman announced the start of a promotional campaign named 'Football Unites' as a way to celebrate the FIFA world cup 2022. The malls initiated several promotional campaigns from October 6, 2022, to December 31, 2022, to create an environment of joy and enthusiasm for families, shoppers, and visitors of the malls.

According to Oman News Agency (2022), the collaborative initiative incorporated extraordinary advertising that coincided with the 2022 World Cup. This was achieved through the execution of two strategies aimed at promoting and marketing the distinctive tourism attractions that the Sultanate of Oman has to offer.

Branding Highlights via Accessibility and Accommodation

Oman's proximity to Qatar makes it an ideal destination for fans who want to experience the tournament but also explore other parts of the region. Oman can highlight its accessibility by promoting easy transportation options between the two countries.

Branding Efforts by Oman Air

Oman Air, as the national carrier of Oman, played a crucial role in branding Oman tourism during the Qatar World Cup 2022. Oman Air advertised and promoted Oman as a tourist destination during the Qatar World Cup with in-flight promotions, media campaigns, and offered special travel packages. The Oman Air showcased Omani hospitality by providing a warm welcome to visitors and promoting Omani customs and traditions. Eng. Abdulaziz Saud Al Raisi, CEO of Oman Air, stated that Oman Air has made available ticket packages that consist of lodging and discounted fares in over 120 reservation outlets globally (Oman News Agency, 2022).

According to Oman News Agency (2022), the cost of reserving flights for one day during the World Cup matches from November 20 to December 3, 2022, began at RO 99 for economy class and RO 305 for business class in Oman Air. Furthermore, to incentivize citizens, discounted prices have been introduced for them, with rates starting at RO 49 for economy class and RO 155 for business class. According to Nair (2022), the aviation sector in the Sultanate of Oman benefited from the FIFA Football World Cup, according to the latest figures released by the International Air Transport Organisation (IATA) the booking percentage for air ticket between Qatar and Oman increased from 5% to 33%.

Branding Efforts by the Muscat International Airport

According to Qatar News Agency (2022), Saud bin Nasser Al Habaishi, the Deputy CEO of Oman Airports, had outlined the plan for facilitating the movement of football fans through Muscat International

Airport during the World Cup. There were designated specific gates for planes carrying fans, providing special ports for verifying their identities, and installed dedicated customs inspection devices. Additionally, a customer service team was present to guide travelers through the dedicated routes. The airport also collaborated with partners to ensure a specific transportation plan for accommodating the flow of fans to and from the airport, with designated parking areas for public transport buses.

Branding Efforts by Accommodation Sector

Shabib Mohammed al Maamari, who served as the Managing Director of the National Travel Operator representing Omran Group, emphasized that the tourism offerings available in the program go beyond 20,000 hotel rooms and included over 200 tourism products provided by 35 local operators and 10 land transport companies. These companies have established partnerships with 52 airlines from different parts of the world that provided flights to the Sultanate of Oman (Oman News Agency, 2022).

Similar to its neighboring country UAE, Oman had experienced a significant rise in popularity due to the recent FIFA World Cup Qatar 2022. Although this may not have directly resulted in increased bookings, it has helped to put the Arab world on the global stage, which will have a long-term impact. The General Manager of Al Bustan Palace, Nabil Al Zadjali, believes that the international exposure generated by the FIFA World Cup will serve as a foundation for attracting a new group of travelers in 2023. Therefore, they plan to capitalize on this increased awareness to continue drawing visitors (Corder, 2023).

Promotion of Local Attractions

Oman has a wealth of natural and cultural attractions that could appeal to football fans. The country promoted these attractions during the World Cup by showcasing them through various channels, including online and offline media. The Ministry of Heritage and Tourism, Sultanate of Oman strongly believes that millions of fans arriving for the FIFA World Cup, which takes place every 4 years look to explore tourist attractions and learn about the cultures of the host country and the neighboring countries in the region.

The Sultanate of Oman's Programme for the FIFA World Cup Qatar 2022 offered lodging and tourism-related activities through accommodation options, tourism products, experiences, and guided tours. These services were provided by small and medium-sized tourism companies managed by experts in the field. A team comprising both government and private units, such as the Ministry of Culture, Sports and Youth, Royal Oman Police, Oman Tourism Development Company (Omran), Oman Air, and Oman Airports, is responsible for running the program. The team operated under the supervision of the Ministry of Heritage and Tourism (Oman News Agency, 2022). The Sultanate is famous for its diverse range of charming valleys, caves, shorelines, and mountains. It is also a favored destination for both local and foreign visitors who are interested in exploring its rich history and cultural heritage. As competition among Gulf Cooperation Council countries intensifies in the tourism and hospitality sector, Oman has the opportunity to stand out from its regional counterparts by focusing on developing its eco-tourism, wellness, adventure, and agritourism offerings.

Tourist spots in Oman such as Salalah, Jebel Akhdar, Jebel Shams, Al Hajjar Mountains, and Nizwa helped increase the number of visitors to the country. As a result of thrust from the FIFA World Cup tourist boost, Oman's Civil Aviation Authority granted permission to the Royal Balloon Company, a joint venture between a Turkish and Omani firm, to operate in the Sharqiyah Sands region, expanding its selection of adventure activities. Additionally, a cable car project spanning 3.5km in the Darbat area of Dhofar is set to open to the public this year (AGBI, 2023).

The FIFA World Cup took place during the winter season, at that time many cultural, heritage, and tourism events were held in different governorates of the Sultanate. In addition to the tournament, a Football Fan Festival was also hosted. The Dhofar Governorate hosted the Frankincense festival during the winter, which highlighted the historical significance of frankincense in Omani culture. The festival included an exhibition for selling and showcasing frankincense products, live demonstrations of frankincense distillation, traditional crafts, as well as cultural and entertainment events.

During the FIFA World Cup, the Oman Air Carnival was a must-see event, which took place in picturesque locations around Muscat, South Al Batinah, and A'Dakhiliyah governorate. The carnival featured paragliding parades at various tourist attractions. The Muscat Food Festival, which welcomed over 60,000 visitors and over 156 unique restaurants in its previous edition, was also held at the same time.

The Desert Adventure Festival was held in the North Sharqiyah Governorate, offering a range of activities for adventure enthusiasts, such as outdoor camping, off-roading, and the Oman Desert Marathon. Additionally, various events were held in different governorates throughout the winter season. Visiting Oman during the World Cup presented a unique opportunity to explore and appreciate the rare treasures that the country has to offer, creating delightful memories that will last a lifetime.

Branding via Meeting of Sustainability and UNSDGs and Oman Vision 2040 Goals From FIFA World Cup 2022 Promotion

UN Sustainable development goals (UNSDGs), which were outlined in the UN's report "Our Common Future" in 1987, as well as Oman vision 2040 referred to the kind of economic, environmental, and social development that can fulfill the needs of the present generation while also ensuring that the future generations can fulfill their own needs.

The FIFA World Cup 2022 presented a unique opportunity for branding via meeting sustainability and United Nations Sustainable Development Goals (UNSDGs) as well as Oman Vision 2040 goals. The Sultanate of Oman utilized some potential strategies in their promotional efforts for the FIFA World Cup 2020. Oman emphasized sustainable practices to align with sustainability goals and prioritized eco-friendly practices in all aspects of the event execution. The local and regional culture was highlighted and incorporated into branding efforts can help to build connections with the community, and promote a sense of pride in the event. To align with UNSDGs, Oman demonstrated a commitment to social and environmental responsibility, gender equality and diversity, education, healthcare, and enhancing employment via this event to reduce poverty. To further support Oman Vision 2040 goals, branding efforts aligning with Vision 2040's goals promoted sustainable tourism development in Oman by investing in local infrastructure and businesses, supporting job creation and economic growth, and promoting environmental conservation efforts. Oman's branding efforts that emphasized sustainability, culture, and social responsibility built a positive reputation for the FIFA World Cup 2022, and demonstrated Oman's commitment to promoting sustainable development in the region.

CONCLUSION

Overall, the FIFA World Cup 2022 presents a significant opportunity for Oman to promote its tourism industry and showcase its unique attractions and experiences to a global audience. By collaborating with FIFA and leveraging various marketing channels, Oman can effectively brand itself as a complementary destination to Qatar during the World Cup and beyond.

Oman's Ministry of Tourism stated that the country has a rich cultural heritage, diverse landscapes, and unique experiences to offer visitors, and sees Qatar as a complementary destination. They believe that working together can offer visitors a truly unforgettable experience.

Oman's growing tourism infrastructure and proximity to Qatar make it an attractive destination for visitors to extend their stay beyond the World Cup, according to a report by Colliers International. Oman has a variety of attractions and experiences to offer and has the potential to draw a significant number of visitors.

Oman's unique cultural heritage and diverse tourism offerings were highlighted by Brouder et al. in 2019. They also emphasized the importance of collaboration between Oman and Qatar to maximize the potential of their relationship to complement Qatar's World Cup offerings.

Qatar World cup offered several opportunities to Oman in the form of enhanced Economic impacts. Oman invited SMEs of more than 60 different categories to contribute to the construction, management and marketing of the Football Fan Festival, as a result generated income and created direct and indirect job opportunities to benefit its economy. Oman also got the advantage of enhanced national pride. Football Fan Festival marked Oman's role in relation to the World Cup and provided nationals with a cause and the means to celebrate this historic occasion.

Despite the challenges posed by the pandemic during 2020 and 2021, Oman's tourism industry performed exceptionally well by leveraging FIFA world cup event to brand its tourism destinations and increased inbound tourism to 2.9 million in 2022. In 2022, with 2.9 million international arrivals, Oman was lagging behind only 16% away from the pre COVID – 19 figure of 3.5 million international arrivals of 2019. There were two reasons that led to this growth, the first being the FIFA World Cup 2022, and the second being the involvement of Oman Ministry of Heritage and Tourism in various exhibitions and events. The camel race sporting event, which was in alliance with the FIFA World Cup 2022, also contributed towards promoting and popularizing Oman's cultural and heritage tourism.

An extraordinary advertising that coincided with the FIFA World Cup 2022 via social media and print media promoted and marketed the distinctive tourism attractions that the Sultanate of Oman has to offer. The Oman government has made the decision to relax the VISA rules. To boost the tourism industry in the country, a ticket to the World Cup served as an entry visa to Oman. In addition, Oman Air flights offered lower prices between Muscat and Doha. Oman's Programme for the FIFA World Cup Qatar 2022 offered lodging and tourism-related activities through accommodation options, tourism products, experiences, and guided tours.

The future is promising in terms of increased tourist arrivals in the Sultanate of Oman. The government needs to complement these upcoming opportunities with strategic steps for making The Sultanate as a preferred tourist destination globally.

The Sultanate of Oman utilized some potential strategies related to Oman vision 2040, UNSDGs in their promotional efforts for the FIFA World Cup 2020 to enhance the tourism brand image of Oman. However, as with any large-scale event, the FIFA World Cup 2022 also posed disadvantages or risks related to the depletion of limited natural resources, emissions, and the health, safety, and well-being of everyone involved, including workers, participants, attendees, and local communities. Through the process of stakeholder consultation, Oman made sure to identify the most important areas to prioritize in order to focus resources and efforts on enhancing the FIFA World Cup 2022 sustainability and the achievement of UNSDGs, Oman Vision 2040. This significantly reduced the disadvantages of the mega events and enhanced positive advantages. Future holds for public-private partnership at the regulatory and policy guideline level which will further facilitate the adoption of effective practices and strategic decisions towards making the Sultanate of Oman as a destination brand.

In summary, the FIFA World Cup 2022 offered the Sultanate of Oman a noteworthy chance to boost its tourism industry and display its exclusive attractions and experiences to the worldwide audience. By partnering with FIFA and utilizing different promotional platforms, Oman effectively established itself as a supplementary location to Qatar throughout and after the World Cup.

REFERENCES

AGBI. (2023, February 2). *Oman urged to differentiate itself to attract tourists*. Retrieved from https://www.agbi.com/analysis/oman-urged-to-look-to-adventure-tourism-to-boost-visitors/

Al Awa, A., & Al Araimi, F. (2020). Examining the Impacts of Muscat Festival: A Major Tourism Event in Oman. *Tourism Review International*, *24*(1), 45–55. doi:10.3727/154427220X15796808472023

Al Awa, S. A., & Al Araimi, M. H. (2020). Evaluating the economic impact of Muscat Festival on the tourism industry in Oman. *Journal of Tourism and Cultural Change*, *18*(3), 265–278.

Al Bulushi, A., & Al Hinai, Y. (2019). The impacts of the Salalah Tourism Festival on Oman's tourism industry. *Journal of Tourism and Cultural Change*, *17*(2), 197–209. doi:10.1080/14766825.2017.1369169

Al-Hadhrami, S. S., Al-Hinai, S. A., & Al-Busaidi, K. A. (2019). Impacts of Salalah Tourism Festival on Oman tourism. *Journal of Tourism and Cultural Change*, *17*(1), 1–17. doi:10.1080/14766825.2017.1382264

Al Hinai, M., & Al Musalhi, M. (2018). Factors contributing to the expansion of Oman's tourism industry. *Journal of Tourism and Hospitality Management*, *6*(1), 1–9.

Al-Maamari, A., & Al-Hinai, A. (2021). The role of tourism in promoting the Sultanate of Oman: An analytical study. *International Journal of Management, Technology, and Social Sciences*, *6*(2), 57–67.

Al Mawali, S. M., & Al Ghaithi, A. H. (2021). The Effect of Hosting the 2022 FIFA World Cup in Qatar on the Tourism Industry in Oman: A Comparative Study. *Journal of Tourism and Hospitality Management*, *9*(2), 55–67.

Al-Rawahi, N. A., Al-Badi, A. H., & Al-Shihi, H. (2018). The impact of the Muscat Festival on Oman tourism. *Journal of Tourism and Cultural Change*, *16*(4), 435–450.

Ayeni, O. (2023, January 5). World Cup a boon for tourism in Gulf countries: CSR Gulf report. *The Peninsula*. https://thepeninsulaqatar.com/article/05/01/2023/world-cup-a-boon-for-tourism-in-gulf-countries-csrgulf-report

Baade, R. A., & Matheson, V. A. (2016). Going for the Gold: The Economics of the Olympics. *The Journal of Economic Perspectives, 30*(2), 201–218. doi:10.1257/jep.30.2.201

Balakrishnan, M. S., & Michael, S. (2020). Branding Major Sporting Events as Tourism Products: The Case of FIFA World Cup. *Journal of Destination Marketing & Management, 15*, 100383. doi:10.1016/j.jdmm.2019.100383

Bialkowski, C. (2015). Economic impact of mega sports events on tourism in developing countries. *Procedia Economics and Finance, 24*, 343–348. doi:10.10162212-6571(15)00600-8

Brouder, P., Teixeira, R., Ioannides, D., & Ioannides, M. (2019). Complementary destinations and visitor experiences: The case of Oman and Qatar. *Journal of Destination Marketing & Management, 12*, 13–22.

Capital, A. (2022, August). *GCC Hospitality Industry Report.* Retrieved from https://alpencapital.com/research/2022/alpen-capital-gcc-hospitality-industry-report-aug-2022.php

Colliers International. (2020). *Oman: A complementary destination to Qatar.* Retrieved from https://www2.colliers.com/en-OM/Research/Oman-A-Complementary-Destination-to-Qatar

Corder, J. (2023, January 5). *Oman hotel industry outlook for 2023: A fresh start.* Hotelier Middle East. Retrieved from https://www.hoteliermiddleeast.com/business/oman-hotel-industry-outlook-for-2023

Ehsani, M., Sadeqi Sogdel, H., Heidary, A., & Jamshidi, A. (2012). The impact of hosting sporting events on sport tourism development: A case study. *International Journal of Research in Business and Social Science, 1*(1). Retrieved from http://www.ssbfnet.com

Freire, J. R. (2012). Destination branding in the Middle East: Politics, religion, and commerce. *Journal of Destination Marketing & Management, 1*(1-2), 74–82.

Getz, D., & Page, S. J. (2016). *Event tourism: Definition, evolution, and research.* Routledge.

Ghimire, K., & Schewe, C. D. (2017). Mega-events, tourism development and perceived community well-being. *Current Issues in Tourism, 20*(4), 401–417. doi:10.1080/13683500.2015.1126275

Giango, M. K., Hintapan, R., Suson, M., Batican, I., Quiño, L., Capuyan, L., Anoos, J. M., Batoon, J., Aro, J. L., Maturan, F., Yamagishi, K., Gonzales, G., Burdeos, A., & Ocampo, L. (2022). Local support on sports tourism development: An integration of emotional solidarity and social exchange theory. *Sustainability, 14*(19), 12898. doi:10.3390u141912898

Gulf News. (2021). *Oman sees potential in complementing Qatar's World Cup offerings.* Retrieved from https://gulfnews.com/oman/oman-sees-potential-in-complementing-qatars-world-cup-offerings-1.1622710583378

Guttentag, D. A. (2015). Airbnb: Disruptive innovation and the rise of an informal tourism accommodation sector. *Current Issues in Tourism, 18*(12), 1192–1217. doi:10.1080/13683500.2013.827159

Guttentag, D. A. (2015). Airbnb: disruptive innovation and the rise of an informal tourism accommodation sector. *Current Issues in Tourism, 18*(12), 1192-1217. . doi:10.1080/13683500.2013.827159

Kasapi, I., & Cela, A. (2017). Destination Branding: A Review of the City Branding Literature. *Mediterranean Journal of Social Sciences, 8*(4), 129–142. Advance online publication. doi:10.1515/mjss-2017-0012

Kasimati, E., & Dawson, P. (2002). Assessing the impact of the summer Olympics on Greece's tourism industry. *Journal of Travel Research, 41*(4), 385–390.

Kikuchi, T., & Koo, W. W. (2017). Mega sporting events and tourism: Brazil's image and the 2014 FIFA World Cup. *Journal of Travel Research, 56*(3), 292–303.

Li, X., Wang, D., & Li, Y. (2019). Exploring the effects of destination branding on tourist revisit intention: The role of emotional attachment. *Journal of Destination Marketing & Management, 12*, 29–39. doi:10.1016/j.jdmm.2019.01.002

Ministry of Heritage and Tourism. (2022, October 25). *The World Cup is Coming to Oman with the Football Fan Festival*. Retrieved from https://mht.gov.om/ar/news/82/THE-WORLD-CUP-IS-COMING-TO-OMAN-WITH-THE-FOOTBALL-FAN-FESTIVAL

Mishra, A., & Kukreti, M. (2021). *Challenges and prospects for Oman in the making of luxury tourism destination*. Routledge.

Nair, V. (2022, December 28). The aviation sector in the Sultanate of Oman benefited from the FIFA Football World Cup. *Oman Observer*. https://www.omanobserver.om/article/1130231/oman/oman-benefits-from-fifa-world-cup-travel-iata

National Centre for Statistics and Information. (2020). *Tourism in Figures 2019*. Retrieved from https://data.gov.om/OMN-2020/tourism-in-figures-2019

National Centre for Statistics and Information. (2021). *Tourism in Oman*. Retrieved from https://www.ncsi.gov.om/Elibrary/Tourism%20in%20Oman%202020%20.pdf

Observer, O. (2022, December 15). *Oman are participating at the Camel and World Cup 2022 activities*. Retrieved from https://www.omanobserver.om/article/1129800/sports/oman-taking-part-at-camel-and-world-cup-2022-event

Oman Ministry of Tourism. (2020). *Tourism Strategy 2040*. Retrieved from https://www.omantourism.gov.om/wps/portal/mot/tourism/home/about-us/our-strategy/tourism-strategy-2040/

Oman News Agency. (2022, September 7). *Oman unveils details of joint programme for FIFA World Cup Qatar 2022*. Retrieved from https://www.omannews.gov.om/NewsDescription/ArtMID/392/ArticleID/74803/Oman-unveils-details-of-joint-programme-for-FIFA-World-Cup-Qatar-2022

Pike, S. (2009). Destination brand positions of a competitive set of near-home destinations. *Journal of Travel Research, 48*(2), 139–151.

Pike, S. (2010). Destination brand positioning - A comparative analysis of the positioning of competitive sets. *Journal of Hospitality Marketing & Management*, *19*(2), 112–132.

Pike, S., & Page, S. J. (2014). Destination Marketing Organizations and destination branding: A narrative analysis of the literature. *Tourism Management*, *41*, 1–26. doi:10.1016/j.tourman.2013.09.009

Preuss, H., & Solberg, H. A. (2017). The impact of the 2010 FIFA World Cup in South Africa on neighboring countries. *Journal of African Business*, *18*(4), 483–501. doi:10.1080/15228916.2017.1365717

Qatar News Agency. (2022, September 7). *Oman Announces Accompanying Programs for FIFA World Cup Qatar 2022*. Qatar News Agency. https://www.qna.org.qa/en/News-Area/News/2022-09/07/0073-oman-announces-accompanying-programs-for-fifa-world-cup-qatar-2
022R

Ratten, V. (2017). Mega-events and entrepreneurship: A research agenda. *Journal of Entrepreneurship in Emerging Economies*, *9*(2), 246–259. doi:10.1108/JEEE-07-2016-0031

Smith, J. (2022). *Oman Leverages FIFA 2022 Football Cup for Destination Branding*. Oman Tourism News. Retrieved from https://www.omantourismnews.com/oman-leverages-fifa-2022-football-cup-for-destination-branding/

Times of Oman. (2021). *Oman sees 37% jump in tourist arrivals in January 2021*. Retrieved from https://timesofoman.com/article/oman-sees-37-jump-in-tourist-arrivals-in-january-2021

Yeoman, I., Robertson, M., Ali-Knight, J., Drummond, S., & McMahon-Beattie, U. (2017). *Festival and Events Management: An International Arts and Culture Perspective*. Routledge.

Compilation of References

Aaker, D. A. (1991). *Managing Brand Equity: Capitalizing on the Value of a Brand Name*. The Free Press.

Aaker, E., & Joachimsthaler, E. (2000). The brand relationship spectrum: The key to the brand architecture challenge. *California Management Review*, *42*(4), 8–23. doi:10.1177/000812560004200401

Aaker, J. L. (1999). The Malleable Self: The Role of Self-Expression in Persuasion. *JMR, Journal of Marketing Research*, *36*(1), 45–57. doi:10.1177/002224379903600104

Aakers, J. (1991). The negative attraction effect? A study of the attraction effect under judgment and choice. *Advances in Consumer Research. Association for Consumer Research (U. S.)*, *18*, 462–469.

Aas, C., Ladkin, A., & Fletcher, J. (2005). Stakeholder collaboration and heritage management. *Annals of Tourism Research*, *32*(1), 28–48. doi:10.1016/j.annals.2004.04.005

AGBI. (2023, February 2). *Oman urged to differentiate itself to attract tourists*. Retrieved from https://www.agbi.com/analysis/oman-urged-to-look-to-adventure-tourism-to-boost-visitors/

Agoramoorthy, G. (2015). Sacred rivers: Their spiritual significance in Hindu religion. *Journal of Religion and Health*, *54*(3), 1080–1090. doi:10.100710943-014-9934-z PMID:25183514

Aidoo, T. (2019, November 28). *Samrawit Fikru, the tech genius behind Ethiopia's version of Uber whose staff is 90 percent female*. Face 2 Face Africa. www.face2faceafrica.com/article/samrawit-fikru-the-tech-genius-behind-ethiopias-version-of-uber-whose-staff-is-90-percent-female

Ak, M. (2011). Marka yönetimi ve imaj. (1. Baskı), İstanbul: Kültür Sanat Basım Evi.

Akdoğan, Ç. (2017). Destinasyon markalama bağlamında yavaş şehir (cittaslow) hareketi üzerine bir değerlendirme. *Uluslararası Bilimsel Araştırmalar Dergisi*, 379-390.

Akhoondnejad, A. (2016). Tourist loyalty to a local cultural event: The case of Turkmen handicrafts festival. *Tourism Management*, *52*, 468–477. doi:10.1016/j.tourman.2015.06.027

Akpan, E. I., & Obang, C. E. (2012). Tourism: A strategy for sustainable economic development in Cross River State, Nigeria. *International Journal of Business and Social Science*, *3*(5).

Aksoy, Z., & Kodaş, B. (2021). Determining the food experiences of tourists visiting Istanbul through online reviews. In Cihan Cobanoglu, Ebru Gunlu Kucukaltan, Muharrem Tuna, Alaattin Basoda & Seden Dogan (Eds.), Advances in Managing Tourism Across Continents (pp. 623–636). USA: USF M3 Publishing

Akturan, O., & Oğuztimur, S. (2016). Kent markalaşması kavramının içeriği ve gelişimi: Farklı disiplinler farklı yaklaşımlar. *Planlama Dergisi*, *26*(2), 119.

Akturk, H., & Akbaba, A. (2021). Local food festivals within the scope of destination branding. In C. Cobanoglu, & V. Della Corte (Ed.), Advances in global services and retail management (pp. 1–9). USF M3 Publishing.

Al Awa, A., & Al Araimi, F. (2020). Examining the Impacts of Muscat Festival: A Major Tourism Event in Oman. *Tourism Review International*, *24*(1), 45–55. doi:10.3727/154427220X15796808472023

Al Awa, S. A., & Al Araimi, M. H. (2020). Evaluating the economic impact of Muscat Festival on the tourism industry in Oman. *Journal of Tourism and Cultural Change*, *18*(3), 265–278.

Al Bulushi, A., & Al Hinai, Y. (2019). The impacts of the Salalah Tourism Festival on Oman's tourism industry. *Journal of Tourism and Cultural Change*, *17*(2), 197–209. doi:10.1080/14766825.2017.1369169

Al Hinai, M., & Al Musalhi, M. (2018). Factors contributing to the expansion of Oman's tourism industry. *Journal of Tourism and Hospitality Management*, *6*(1), 1–9.

Al Mawali, S. M., & Al Ghaithi, A. H. (2021). The Effect of Hosting the 2022 FIFA World Cup in Qatar on the Tourism Industry in Oman: A Comparative Study. *Journal of Tourism and Hospitality Management*, *9*(2), 55–67.

Albayrak, T., & Caber, M. (2018). Examining the relationship between tourist motivation and satisfaction by two competing methods. *Tourism Management*, *69*, 201–213. doi:10.1016/j.tourman.2018.06.015

Alegre, J., & Cladera, M. (2010). Tourist expenditure and quality: Why repeat tourists can spend less than first-timers. *Tourism Economics*, *16*(3), 517–533. doi:10.5367/000000010792278419

Aleksandrov, K. (2014). The role of DMO for sustainable development of a tourist destination–Bulgaria case study. *Journal of Tourism Research*, *9*(2), 198–209.

Al-Hadhrami, S. S., Al-Hinai, S. A., & Al-Busaidi, K. A. (2019). Impacts of Salalah Tourism Festival on Oman tourism. *Journal of Tourism and Cultural Change*, *17*(1), 1–17. doi:10.1080/14766825.2017.1382264

Ali, F., Park, E., Kwon, J., & Chae, B. (2019). 30 Years of contemporary hospitality management: Uncovering the bibliometrics and topical trends. *International Journal of Contemporary Hospitality Management*, *31*(7), 2641–2665. doi:10.1108/IJCHM-10-2018-0832

Alkaya, A., & Şahin, F. (2018). Bilgi paylaşımının teknoloji kabul modeli üzerinden incelenmesi: Bir sosyal ağ uygulaması. *Uşak Üniversitesi Sosyal Bilimler Dergisi*, *11*(1), 11–21.

Alkin, K., Bulu, M., & Kaya, H. (2007). İller arası rekabet endeksi: Türkiye'deki illerin rekabetçilik seviyelerinin göreceli olarak ölçülebilmesi için bir yaklaşım. *İstanbul Ticaret Üniversitesi. Sosyal Bilimler Dergisi*, *11*, 221–235.

Allen, J., O'Toole, W., & McDonnell, I. (2005). *Festival and special event management* (3rd ed.). John Wiley and Sons Inc.

Allen, J., O'Toole, W., McDonnell, I., & Harris, R. (2008). *Organização e Gestão de Eventos* (3rd ed.). Elsevier Editora.

Allspaw, J. (2008). *The Art of Capacity Planning*. O'Reilly Media, Inc.

Al-Maamari, A., & Al-Hinai, A. (2021). The role of tourism in promoting the Sultanate of Oman: An analytical study. *International Journal of Management, Technology, and Social Sciences*, *6*(2), 57–67.

Almeyda-Ibáñez, M., & George, B. P. (2017). The evolution of destination branding: A review of branding literature in tourism. *Journal of Tourism* [JTHSM]. *Heritage & Services Marketing*, *3*(1), 9–17.

Alok, C. (2016, February 20). A toast to SulaFest. *Business Standard*. https://www.business-standard.com/article/beyond-business/a-toast-to-sulafest-116021901007_1.html

Al-Rawahi, N. A., Al-Badi, A. H., & Al-Shihi, H. (2018). The impact of the Muscat Festival on Oman tourism. *Journal of Tourism and Cultural Change*, *16*(4), 435–450.

Althuwaini, S. (2022). The Effect of Social Media Activities on Brand Loyalty for Banks: The Role of Brand Trust. *Administrative Sciences*, *12*(4), 148. doi:10.3390/admsci12040148

Amara, D. (2017). Events tourism: a potential to build a tourist destination branding: the case of cultural events in Egypt. *International Journal of Heritage. Tourism and Hospitality*, *11*(2), 1–12.

Ambavale, R., & Surtı, N. (2015). Is brand everythıng? Issues and challenges of brandıng in global market. *International Journal of Application or Innovation in Engineering & Management*, *4*(1), 14–18.

Amrhein, S., Nirappil, F., Leone, J., & Dupree, J. (2021, September 6). In Florida, a summer of death and resistance as the coronavirus rampaged. *The Washington Post*. www.washingtonpost.com/health/2021/09/05/florida-coronavirus-delta-surge

Andersson, T. D., & Getz, D. (2008, September). Stakeholder management strategies of festivals. []. Taylor & Francis Group.]. *Journal of Convention & Event Tourism*, *9*(3), 199–220. doi:10.1080/15470140802323801

Anderton, C. (2022). Festivals. In *The Bloomsbury Handbook of Popular Music, Space and Place*. Bloomsbury Publishing. doi:10.5040/9781501336317.ch-028

Anim, O., & Titus, E. (2012). The relevance of tourism on the economic development of Cross River state, Nigeria. *Journal of Geography and Regional Planning*, *5*(1), 14–20. doi:10.5897/JGRP11.122

ANP. (2022, May 16). More than 1.1 million people visited the Keukenhof in 2022. *NL Times*. www.nltimes.nl/2022/05/16/11-million-people-visited-keukenhof-2022

Antón, C., Camarero, C., & Garrido, M. J. (2019). What to Do After Visiting a Museum? From Post-consumption Evaluation to Intensification and Online Content Generation. *Journal of Travel Research*, *58*(6), 1052–1063. doi:10.1177/0047287518793040

Arcodia, C., & Whitford, M. (2006). Festival attendance and the development of social capital. *Journal of Convention & Event Tourism*, *8*(2), 1–18. doi:10.1300/J452v08n02_01

Aria, M., & Cuccurullo, C. (2017). Bibliometrix: An R-tool for comprehensive science mapping analysis. *Journal of Informetrics*, *11*(4), 959–975. doi:10.1016/j.joi.2017.08.007

Arkoudis, S. (2022). How technology amplifies music experiences in festivals. *Tuned Global*. https://blog.tunedglobal.com/how-technology-music-experiences-festivals

Artun, E. (2007). Türk Halk Kültüründe Hıdırellez. *Ç.Ü. Türkoloji, 1575*, 1-9. Retrieved from https://turkoloji.cu.edu.tr/HALKBILIM/erman_artun_turk_halk_kulturunde_hidrellez.pdf

Associated Press. (2022, June 29). Essence fest is back in New Orleans after two-year hiatus. *U.S. News & World Report*. www.usnews.com/news/entertainment/articles/2022-06-29/essence-fest-is-back-in-new-orleans-after-two-year-hiatus

Atay, L., & Dülgaroğlu, O. (2017). Markalaşma açısından Cunda Adası'nın destinasyon kişiliğinin tespitine yönelik bir çalışma. *Balıkesir Üniversitesi Sosyal Bilimler Enstitüsü Dergisi*, *20*(38), 383–396. doi:10.31795/baunsobed.645146

Australian Bureau of Statistics. (n.d.) *Statistical language - Quantitative and qualitative data*. Australian Bureau of Statistics. www.abs.gov.au/ websitedbs/D3310114.nsf/Home/Statistical+Language+-+quantitative+and+qualitative+data

Avcıkurt, C. (2009). *Turizm sosyolojisi genel ve yapısal yaklaşım*. Detay Yayıncılık.

Avcılar, M. Y., & Kara, E. (2015). Şehir markası kavramı ve marka şehir yaratma stratejilerine yönelik literatür incelemesi. *Sosyal ve Beşeri Bilimler Araştırmaları Dergisi*, *34*, 76–94.

Axelsen, M., & Swan, T. (2010). Designing festival experiences to influence visitor perceptions: The case of a wine and food festival. *Journal of Travel Research*, *49*(4), 436–450. doi:10.1177/0047287509346796

Ay, A. (2010). Kirkpinar Oil Wrestling: Oil Wrestling, Edirne, Turkey. Ceren Kitapevi.

Aydemir, B., Saylan, U., & Aydoğmuş, F. (2014). Turizmde rekabet: seyahat ve turizm (T&T) rekabet edebilirlik raporunda Türkiye'nin ve Avrupa'nın değerlendirilmesi . *Çukurova Üniversitesi İktisadi ve İdari Bilimler Fakültesi Dergisi, 18* (1), 1-15.

Aydın, G. (2020). Social media engagement and organic post effectiveness: A roadmap for increasing the effectiveness of social media use in hospitality industry. *Journal of Hospitality Marketing & Management*, *29*(1), 1–21. doi:10.1080/19368623.2019.1588824

Ayeni, O. (2023, January 5). World Cup a boon for tourism in Gulf countries: CSR Gulf report. *The Peninsula*. https://thepeninsulaqatar.com/article/05/01/2023/world-cup-a-boon-for-tourism-in-gulf-countries-csrgulf-report

Baade, R. A., & Matheson, V. A. (2016). Going for the Gold: The Economics of the Olympics. *The Journal of Economic Perspectives*, *30*(2), 201–218. doi:10.1257/jep.30.2.201

Babb, N. (2021, May 5). "Baby won't you please come home": Studying ethnoracial segregation trends in New Orleans pre and Post Hurricane Katrina. *Journal of Public and International Affairs*. jpia.princeton.edu/news/baby-wont-you-please-come-home-studying-ethnoracial-segregation-trends-new-orleans-pre-and-post

Babin, B. J., & Griffin, M. (1998). The nature of satisfaction: An updated examination and analysis. *Journal of Business Research*, *41*(2), 127–136. doi:10.1016/S0148-2963(97)00001-5

Bagaeen, S. (2007). Brand Dubai: The Instant City; or The Instantly Recognizable City. *International Planning Studies*, *12*(2), 173–197. doi:10.1080/13563470701486372

Bairwa, M., & Patawari, S. (2021). Festivals for Sustainable Tourism Development: A Case Study of Marwar Region of Rajasthan, India. In LOCAL FOOD AND COMMUNITY EMPOWERMENT THROUGH TOURISM (pp.82-96) Eureka Publications.

Bakan, Ö., & Canöz, K. (2017). Kültürlerarası İletişim Bağlamında Türk Üniversite Öğrencilerinin Japonlara Yönelik Kalıp Yargıları. *Selçuk Üniversitesi SBE Dergisi*, (37), 91-107. Retrieved from https://dergipark.org.tr/en/download/article-file/1724897

Baker, W. J. (1986). Brand Familiarity and Advertising: Effects on the Evoked Set and Brand Preferences. *NA: Advances in Consumer Research*, *13*, 637–642.

Balaji (2022) *Personal Communication* [WhatsApp]. Whatsapp.

Balakrishnan, M. S., & Michael, S. (2020). Branding Major Sporting Events as Tourism Products: The Case of FIFA World Cup. *Journal of Destination Marketing & Management, 15*, 100383. doi:10.1016/j.jdmm.2019.100383

Ball, D., & Tasaki, L. H. (1992). The Role and Measurement of Attachment in Consumer Behavior. *Journal of Consumer Psychology, 1*(2), 155–172. doi:10.120715327663jcp0102_04

Baloğlu, Ş., & Mangaloğlu, M. (2001). Tourism destination images of Turkey, Egypt, Greece and Italy As Perceived by US-based tour operators and travel agents. *Tourism Management, 22*(1), 2. doi:10.1016/S0261-5177(00)00030-3

Baloglu, S., & McCleary, K. W. (1999). A model of destination image formation. *Annals of Tourism Research, 26*(4), 868–897. doi:10.1016/S0160-7383(99)00030-4

Barnes, S. J., Mattsson, J., & Sørensen, F. (2014). Destination brand experience and visitor behavior: Testing a scale in the tourism context. *Annals of Tourism Research, 48*, 121–139. doi:10.1016/j.annals.2014.06.002

Bashir, A., Singh, R., & Ps, S. (2022). Mapping the foundation and trends in tourism research: A bibliometric review of literature in tourism: An international interdisciplinary journal. *Tourism: An International Interdisciplinary Journal, 70*(3), 480–492. doi:10.37741/t.70.3.10

Basit, T. N. (2003). Manual or Electronic? The Role of Coding in Qualitative Data Analysis. *Educational Research, 45*(2), 143–154. doi:10.1080/0013188032000133548

Baysal, K., Özgürel, G., & Çeken, H. (2016). Aydın yöresindeki yerel etkinliklerin kırsal turizm açısından değerlendirilmesi. *International Journal of Social and Economic Sciences, 6*(1), 100–108.

Beer, S., & Rosini, N. J. (2022, January 14). *Licenses and rights in the post-COVID festival landscape*. Documentary.org. www.documentary. org/column/licenses-and-rights-post-covid-festival-landscape

Beerli, A., & Martin, D. (2004). Tourist Characteristics ans The Perceived İmage of Destinations: A Quantitative Analysis- a case study of Lanzorate, Spain. *Tourism Management, 25*(5), 623–636. doi:10.1016/j.tourman.2003.06.004

Benckendorff, P., & Zehrer, A. (2013). A network analysis of tourism research. *Annals of Tourism Research, 43*, 121–149. doi:10.1016/j.annals.2013.04.005

Berbel-Pineda, J. M., Palacios-Florencio, B., Ramírez-Hurtado, J. M., & Santos-Roldán, L. (2019). Gastronomic Experience as a Factor of Motivation in the Tourist Movements. *International Journal of Gastronomy and Food Science, 18*, 1–10. doi:10.1016/j.ijgfs.2019.100171

Berg, B. L., & Lune, H. (2017). *Qualitative Research Methods for the Social Sciences*. Pearson.

Beyaz Kömürcü, G. (2013). *Etkinlik turizmi çeşidi olarak festivaller: Bozcaada Yerel Tatlar Festivali örneği, Çanakkale Onsekiz Mart Üniversitesi, Turizm İşletmeciliği anabilim dalı*. Basılmamış Yüksek Lisans Tezi.

Bezborodkina, N. (2019). IoT for Event Management: What's in for Event Planners. *KeepGo*. https://business.keepgo.com/blogs/articles/iot-for-event-management-what-s-in-for-event-planners

Bialkowski, C. (2015). Economic impact of mega sports events on tourism in developing countries. *Procedia Economics and Finance, 24*, 343–348. doi:10.10162212-6571(15)00600-8

Bibicioiu, S., & Cretu, R. (2013). Enotourism: A niche tendency within the tourism market. *Scientific Papers. Series Management, Economic, Engineering in Agriculture and Rural Development, 13*(2), 31–40.

Bieger, T., Beritelli, P., & Laesser, C. (2009). Size matters! Increasing DMO effectiveness and extending tourist destination boundaries. *Tourism: An International Interdisciplinary Journal, 57*(3), 309–327.

Bigné, J. E., Sánchez, M. I., & Sánchez, J. (2001). Tourism image, evaluation variables and after purchase behaviour: Inter-relationship. *Tourism Management*, *22*(6), 607–616. doi:10.1016/S0261-5177(01)00035-8

Bilar, E. (2022). Kırkpınar Yağlı Güreşleri Tarihi. *Edirne Kent ve Kültür Bilinci*. Retrieved from https://edirnekultur.com/detay/tarihi-863

Binns, T., & Nel, E. (2002). Supporting local economic development in post-apartheid South Africa. *Local Economy*, *17*(1), 8–24. doi:10.1080/02690940110073800

Bjeljac, Ţ. (2010). *Turističke manifestacije u Srbiji* [*Tourist events in Serbia*]. Serbian Academy of Sciences and Arts.

Björk, P., & Kauppinen-Räisänen, H. (2014). Culinary-gastronomic tourism–a search for local food experiences. *Nutrition & Food Science*, *44*(4), 294–309. doi:10.1108/NFS-12-2013-0142

Björk, P., & Kauppinen-Räisänen, H. (2016). Local food: A source for destination attraction. *International Journal of Contemporary Hospitality Management*, *28*(1), 177–194. doi:10.1108/IJCHM-05-2014-0214

Blain, C., Levy, S. E., & Ritchie, J. B. (2005). Destination branding: Insights and practices from destination management organizations. *Journal of Travel Research*, *43*(4), 328–338. doi:10.1177/0047287505274646

Blichfeldt, B. S., & Halkier, H. (2013). Mussels, tourism and community development: A case study of place branding through food festivals in rural North Jutland, Denmark. *European Planning Studies*, *22*(8), 1587–1603. doi:10.1080/09654313.2013.784594

Bonardi, L., Ludovici, A. A., & Furlani, R. (2006). *Alps and tourism: Finding the balance point*. Academic Press.

Boniello, K., & Byrne, K. J. (2022, April 24). Beloved Gilroy Garlic Festival shutting down after four decades. *New York Post*. www.nypost.com/ 2022/04/23/gilroy-garlic-festival-shutting-down-indefinitely/

Boniface, P. (2003). *Tasting Tourism: Traveling for Food and Drink*. Ashgate Publishing.

Boo, S., Busser, J., & Baloglu, S. (2009). A model of customer-based brand equity and its application to multiple destinations. *Tourism Management*, *30*(2), 219–231. doi:10.1016/j.tourman.2008.06.003

Boring, N. (2020, March 18). *France: Government imposes confinement measures to stem spread of coronavirus*. The Library of Congress. www.loc.gov/item/global-legal-monitor/2020-03-18/france-government-imposes-confinement-measures-to-stem-spread-of-coronavirus/

Bornhorst, T., Ritchie, J. B., & Sheehan, L. (2010). Determinants of tourism success for DMOs & destinations: An empirical examination of stakeholders' perspectives. *Tourism Management*, *31*(5), 572–589. doi:10.1016/j.tourman.2009.06.008

Botti, A., Monda, A., & Vesci, M. (2018). Organizing Festivals, Events and Activities for Destination Marketing. In Tourism Planning and Destination Marketing. Emerald Publishing Limited. doi:10.1108/978-1-78756-291-220181010

Bowdin, G. A. J., Allen, J., O'Toole, W., Harris, R., & McDonnell, I. (2012). *Events Management* (2nd ed.). Butterworth-Heinemann. doi:10.4324/9780080964317

Bowman, C., & Ambrosini, V. (2000). Value creation versus value capture: Towards a coherent definition of value in strategy. *British Journal of Management*, *11*(1), 1–15. doi:10.1111/1467-8551.00147

Boyd, C. (2018, October 31). *Tomorrowland festival goers affected by Data Breach: Malwarebytes labs*. Malwarebytes. www.malwarebytes.com/ blog/news/2018/10/tomorrowland-data-breach

Bozyer, Ü. (2012). Turizm Sosyolojisi. In S. G. S. T. Tartışmaları (Ed.), *Muammer Tuna*. Barkod.

Bracey, D. (2020, August 23). *From pandemic to protests: How food businesses are responding*. Bon Appétit. www.bonappetit.com/story/food-businesses-covid-19

Brakus, J. J., Schmitt, B. H., & Zarantonello, L. (2009). Brand experience: What is it? How is it measured? Does it affect loyalty? *Journal of Marketing*, *73*(3), 52–68. doi:10.1509/jmkg.73.3.052

Breiby, M. A., & Slatten, T. (2018). The role of aesthetic experiential qualities for tourist satisfaction and loyalty. *International Journal of Culture, Tourism and Hospitality Research*, *12*(1), 1–14. doi:10.1108/IJCTHR-07-2017-0082

Brida, J. G., Disegna, M., & Osti, L. (2013). The effect of authenticity on visitors' expenditure at cultural events. *Current Issues in Tourism*, *16*(3), 266–285. doi:10.1080/13683500.2012.674105

Brida, J. G., & Scuderi, R. (2013). Determinants of tourist expenditure: A review of microeconometric models. *Tourism Management Perspectives*, *6*, 28–40. doi:10.1016/j.tmp.2012.10.006

Brooks, R. (2020). *The Future of Events: Glastonbury 2050*. UKF. https://ukf.com/words/the-future-of-events-glastonbury-2050/27005

Brouder, P., Teixeira, R., Ioannides, D., & Ioannides, M. (2019). Complementary destinations and visitor experiences: The case of Oman and Qatar. *Journal of Destination Marketing & Management*, *12*, 13–22.

Brown, G., & Chalip, L. Jago. L., & T. Mules, (2004). Developing Brand Australia: Examining the Role of Events. In N. Morgan, A. Pritchard, and R. Pride (Ed.), Destination Branding: Creating the Unique Destination Proposition (Amsterdam: Elsevier), (pp.279–305). A. Oxford, UK: Butterworth-Heinemann.

Brown, G., Chalip, L., Jago, L., & Mules, T. (2004). Developing brand Australia: Examining the role of events. In N. Morgan, A. Pritchard, & R. Pride (Vol. Eds.), Destination branding: Creating the unique destination proposition. Vol. 2. Destination branding: Creating the unique destination proposition (pp. 279–305). Oxford: Elsevier Butterworth-Heinemann.

Bruton, M. (2021, February 9). X Games' Pandemic Pivot, Embrace Of New Social Media Platforms Pay Off With Tv And Digital Viewership Spike. *Forbes*. www.forbes.com/sites/michellebruton/2021/02/09/x-games-pandemic-pivot-embrace-of-new-social-media-platforms-pay-off-with-tv-and-digital-viewership-spike/

Bruwer, J. (2002). Wine and Food Events: A Golden Opportunity to Learn More about Wine Consumers. *NA: The Australian and New Zealand Wine Industry Journal*, *17*(3), 92–99.

Brzakoviš, A., Brzakoviš, T., & Brzakoviš, P. (2019). Creating brand market value. *The Annals of the Faculty of Economics in Subotica*, *55*(42), 17–33.

Buabeng-Andoh, C. (2012). Factors influencing teachers' adoption and integration of information and communication technology into teaching: A review of the literature. *International Journal of Education and Development using ICT, 8*(1).

Budeanu, A. (2007). Sustainable tourist behaviour–a discussion of opportunities for change. *International Journal of Consumer Studies*, *31*(5), 499–508. doi:10.1111/j.1470-6431.2007.00606.x

Buhalis, D. (2000). Marketing the competitive destination of the future. *Tourism Management*, *21*(1), 97–116. doi:10.1016/S0261-5177(99)00095-3

Buhalis, D. (2001). Tourism in Greece: Strategic analysis and challenges. *Current Issues in Tourism*, *4*(5), 440–480. doi:10.1080/13683500108667898

Buhalis, D., & Law, R. (2008). Progress in information technology and tourism management: 20 years on and 10 years after the Internet—The state of eTourism research. *Tourism Management, 29*(4), 609–623. doi:10.1016/j.tourman.2008.01.005

Burmann, C., Hegner, S., & Riley, N. (2009). Towards an identity-based branding. *Marketing Theory, 9*(1), 109–114. doi:10.1177/1470593108100065

Çakmakcı, E., & Öztürk, Ş. (2019). Toplumsal Değişimin Bir Faktörü Olarak Turizm. *Süleyman Demirel Üniversitesi, Fen-Edebiyat Fakültesi Sosyal Bilimler Dergisi, 47*(2), 1-17. Retrieved from https://dergipark.org.tr/tr/download/article-file/841854 doi:10.35237/sufesosbil.544714

Camarero, C., Garrido, M., & Vicente, E. (2010). Components of art exhibition equity for internal and external visitors. *Tourism Management, 31*(4), 495–504. doi:10.1016/j.tourman.2009.05.011

Camilleri, M. (2018). Travel marketing, tourism economics and the airline product: An introduction to theory and practice. Springer Nature. doi:10.1007/978-3-319-49849-2

Can, A. S., Ekinci, Y., & Pino, G. (2021). Joint brand advertising for emerging heritage sites. *Annals of Tourism Research, 91*, 103294. doi:10.1016/j.annals.2021.103294

Capital, A. (2022, August). *GCC Hospitality Industry Report*. Retrieved from https://alpencapital.com/research/2022/alpen-capital-gcc-hospitality-industry-report-aug-2022.php

Capriello, A., & Rotherham, I. D. (2011). Building a preliminary model of event management for rural communities. *Journal of Hospitality Marketing & Management, 20*(3-4), 246–264. doi:10.1080/19368623.2011.562411

Cardoso, L., Dias, F., de Araújo, A. F., & Marques, M. I. A. (2019). A destination imagery processing model: Structural differences between dream and favourite destinations. *Annals of Tourism Research, 74*, 81–94. doi:10.1016/j.annals.2018.11.001

Carlsson, B. (2001). Depicting Experiences. *Scandinavian Journal of Educational Research, 45*(2), 125–143. doi:10.1080/00313830120052723

Čavić, S., & Mandarić, M. (2021). Authenticity of gastronomic events as a function of branding a destination – *Hotel and Tourism Management,9*(1), 89-101

Celebi, D., Pirnar, I., & Eris, E. D. (2020). Bibliometric analysis of social entrepreneurship in gastronomy tourism. Tourism. *An International Interdisciplinary Journal, 68*(1), 58–67. doi:10.37741/t.68.1.5

Çelik, H., Başer Baykal, N., & Kılıç Memur, H. N. (2020). Qualitative Data Analysis and Fundamental Principles. *Journal of Qualitative Research in Education, 8*(1), 379–406. doi:10.14689/issn.2148-2624.1.8c.1s.16m

Celik, S., & Cetinkaya, M. Y. (2013). Festivals In Event Tourism: The Case of International Izmir Art Festival. *International Journal of Contemporary Economics & Administrative Sciences, 3*(1), 1–21.

Centers for Disease Control and Prevention. (2022, August 16). *CDC Museum Covid-19 Timeline*. Centers for Disease Control and Prevention. www.cdc.gov/museum/timeline/covid19.html

Chalip, L. (2004). Beyond impact: a general model for sport event leverage. In B. W. Ritchie & D. Adair (Eds.), *Sport tourism: interrelationships, impacts and issues* (pp. 226–252). Channel View Publications. doi:10.21832/9781873150672-014

Chalip, L., & Costa, C. (2005). Sport event tourism and the destination brand: Towards a general theory. *Sport in Society, 8*(2), 218–237. doi:10.1080/17430430500108579

Chalip, L., Green, B. C., & Hill, B. (2003). Effects of sport event media on destination image and intention to visit. *Journal of Sport Management, 17*(3), 214–234. doi:10.1123/jsm.17.3.214

Chametzky, B. (2016). Coding in Classic Grounded Theory: I've Done an Interview; Now What? *Sociology Mind, 6*(04), 163–172. doi:10.4236m.2016.64014

Chang, R. C., Kivela, J., & Mak, H. A. (2011). Attributes that Influence the Evaluation of Travel Dining Experience: When East Meets West. *Tourism Management, 32*(2), 307–316. doi:10.1016/j.tourman.2010.02.009

Chang, W., & Yuan, J. (2011). A taste of tourism: Visitors' motivations to attend a food festival. *Event Management, 15*(1), 13–23. doi:10.3727/152599511X12990855575024

Chan-Olmsted, S. M., Cho, M., & Lee, S. (2013). User perceptions of social media: A comparative study of perceived characteristics and user profiles by social media. *Online Journal of Communication and Media Technologies, 3*(4), 149–178. doi:10.29333/ojcmt/2451

Charters, S. K. (2002). Who is the Wine Tourist? *NA: Tourism Management, 23*, 311–319.

Chaturvedi, D. (2010). Tourism in India: Ensuring buoyancy and sustainability. *Yojana, 54*, 5–7.

Chaudhary, M. (2010). Training Men to Manage Tourism. *Yojana, 54*, 8–10.

Chaudhuri, A., & Holbrook, M. B. (2001). The chain of effects from brand trust and brand affect to brand performance: The role of brand loyalty. *Journal of Marketing, 65*, 81–93.

Chaulagain, S. W., Wiitala, J., & Fu, X. (2019). The impact of country image and destination image on US tourists ' travel intention. *NA: Journal of Destination Marketing and Management, 12*, 1–11. doi:10.1016/j.jdmm.2019.01.005

Chavez, R. (2022, February 16). *For New Orleans, the return of Mardi Gras is critical 'for our pocketbooks and our souls'*. PBS. www.pbs.org/newshour/nation/for-new-orleans-the-return-of-mardi-gras-is-critical-for-our-pocketbooks-and-our-souls

Chebli, L., & Gharbi, A. (2014). The Impact of the Effectiveness of Sponsorship on Image and Memorizing: Role of Congruence and Relational Proximity. *Procedia: Social and Behavioral Sciences, 109*, 913–924. doi:10.1016/j.sbspro.2013.12.564

Chekalina, T., Fuchs, M., & Lexhagen, M. (2018). Destination brand promise: The core of customer-based brand equity modeling. *Tourism Analysis, 23*(1), 93–107. doi:10.3727/108354218X15143857349800

Chen, H., Jiang, W., Yang, Y., Yang, Y., & Man, X. (2017). State of the art on food waste research: A bibliometrics study from 1997 to 2014. *Journal of Cleaner Production, 140*, 840–846. doi:10.1016/j.jclepro.2015.11.085

Chen, X., You, S. E., Lee, T. J., & Li, X. (2021). The influence of historical nostalgia on a heritage destination's brand authenticity, brand attachment, and brand equity. *International Journal of Tourism Research, 23*(6), 1176–1190. doi:10.1002/jtr.2477

Chen, Y. C., King, B., & Lee, H. (2018). Experiencing the destination brand: Behavioral intentions of arts festival tourists. *Journal of Destination Marketing & Management, 10*, 61–67. doi:10.1016/j.jdmm.2018.06.004

Chen, Z. (2018). A Pilot Study of the Co-creation Experience in Traditional Cantonese Teahouses in Hong Kong. *Journal of Heritage Tourism, 13*(6), 506–527. doi:10.1080/1743873X.2018.1444045

Chhabra, D., Erin, S., & Frederick, W. (2003). The significance of festivals to rural economies: Estimating the economic effects of Scottish Highland Games in North Carolina. *Journal of Travel Research, 41*(4), 421–427. doi:10.1177/0047287503041004012

Chhabra, D., Sills, E., & Rea, P. (2002). Tourist expenditures at heritage estivals. *Event Management*, *7*(4), 221–230. doi:10.3727/152599502108751613

Chi, H. K., Huang, K. C., & Nguyen, H. M. (2020). Elements of destination brand equity and destination familiarity regarding travel intention. *Journal of Retailing and Consumer Services*, *52*, 101728. doi:10.1016/j.jretconser.2018.12.012

Chiu, C. M., Huang, H. Y., & Yen, C. H. (2010). Antecedents of online trust in online auctions. *Electronic Commerce Research and Applications*, *9*(2), 148–159. doi:10.1016/j.elerap.2009.04.003

Choi, K., Kang, H. J., & Kim, C. (2021). Evaluating the efficiency of Korean festival tourism and its determinants on efficiency change: Parametric and non-parametric approaches. *Tourism Management*, *86*, 104348. doi:10.1016/j.tourman.2021.104348

Choo, H., & Park, D. B. (2020). Comparison between local and non-local visitors for local food festivals. *Asia Pacific Journal of Tourism Research*, *25*(6), 692–705. doi:10.1080/10941665.2020.1760905

Choudhury, T. (2021). Experience the Magical Cherry Blossom Festival 2021 In Shillong. [Online News]. *East Mojo*. https://www.eastmojo.com/meghalaya/2021/11/26/experience-the-magical-cherry-blossom-festival-2021-in-shillong/

Choudhury, N. (2014). World wide web and its journey from web 1.0 to web 4.0. *International Journal of Computer Science and Information Technologies*, *5*(6), 8096–8100.

Chow, A. R. (2022). Coachella Gave Fans a Wild AR Experience. It Could Be the Future of Concerts. *Time*. https://time.com/6168688/coachella-enters-the-metaverse/

Çiçek, E., & Pala, U. (2017). Destinasyon markalaşması üzerine bir araştırma: Tarsus örneği. *Çağ Üniversitesi Sosyal Bilimler Dergisi, 14*(2), 25-47.

Cimbaljević, M., Stankov, U., Demirović, D., & Pavluković, V. (2021). Nice and smart: Creating a smarter festival – the study of EXIT (Novi Sad, Serbia). *Asia Pacific Journal of Tourism Research*, *26*(4), 415–427. doi:10.1080/10941665.2019.1596139

Ciurkot, A. (2016). How Beacons Helped the Reeperbahn Festival Boost In-App Music Streaming by 86%. *Kontakt*. https://kontakt.io/blog/beacons-at-music-festivals/

Civelek, A. (2014). Turizmin Sosyal Yapıya ve Sosyal Değişmeye Etkileri. *Selçuk Üniversitesi Sosyal Bilimler Meslek Yüksekokulu Dergisi, 13*(1-2), 331-350. Retrieved from https://dergipark.org.tr/en/pub/selcuksbmyd/issue/11301/135146

Clark, C. R., Doraszelski, U., & Draganska, M. (2009). The effect of advertising on brand awareness and perceived quality: An empirical investigation using panel data. *Quantitative Marketing and Economics*, *7*(2), 207–236. doi:10.100711129-009-9066-z

Çoban, M., & Süer, S. (2018). Destinasyon markalaşmasında festival turizminin rolü: Alaçatı Ot Festivali üzerine bir araştırma. *International Journal of Contemporary Tourism Research*, *2*(1), 58–67.

Cobanoglu, C., Doğan, S., & Güngör, M. Y. (2021). Emerging Technologies at the Events. In *Impact of ICTs on Event Management and Marketing* (pp. 53–68). IGI Global. doi:10.4018/978-1-7998-4954-4.ch004

Cohen, E. (1979a). A phenomenology of tourist experiences. *Sociology*, *13*(2), 179–201. doi:10.1177/003803857901300203

Cohen, E. (1979b). Rethinking the sociology of tourism. *Annals of Tourism Research*, *6*(1), 18–35. doi:10.1016/0160-7383(79)90092-6

Cohen, E. (1988). Authenticity and commoditization in tourism. *Annals of Tourism Research, 15*(3), 371–386. doi:10.1016/0160-7383(88)90028-X

Cole, S. T., & Scott, D. (2004). Examining the mediating role of experience quality in a model of tourist experiences. *Journal of Travel & Tourism Marketing, 16*(1), 79–90. doi:10.1300/J073v16n01_08

Coletto, D. (2020). As the pandemic goes on, the threat to live music grows. *Abacus*. abacusdata.ca/live-music-threat-pandemic-music-canada/

Colliers International. (2020). *Oman: A complementary destination to Qatar*. Retrieved from https://www2.colliers.com/en-OM/Research/Oman-A-Complementary-Destination-to-Qatar

Connell, J., & Meyer, D. (2015). Visitor attractions and events: Responding to seasonality. *Tourism Management, 46*, 283–298. doi:10.1016/j.tourman.2014.06.013

Connor, S. M. (2006). Food-related advertising on preschool television: Building brand recognition in young viewers. *Pediatrics, 118*(4), 1478–1485. doi:10.1542/peds.2005-2837 PMID:17015538

Cooper, M. (2009). In B. Prideaux & M. Cooper (Eds.), *River tourism in the south Asian subcontinent. River Tourism* (pp. 23–40).

Corder, J. (2023, January 5). *Oman hotel industry outlook for 2023: A fresh start*. Hotelier Middle East. Retrieved from https://www.hoteliermiddleeast.com/business/oman-hotel-industry-outlook-for-2023

Cordova, R. (2014, May 28). *Maya Angelou's 2011 'Arizona Republic' interview*. AZCentral.com. www.azcentral.com/story/entertainment/books/2014/05/28/maya-angelou-arizona-republic-interview/9682587/

Correia, A., Barros, C. P., & Silvestre, A. L. (2007). Golf tourism repeat choice behaviour in the Algarve: A mixed logit approach. *Tourism Economics, 13*(1), 111–127. doi:10.5367/000000007779784489

Cottrill, C., Gault, P., Yeboah, G., Nelson, J. D., Anable, J., & Budd, T. (2017). Tweeting Transit: An examination of social media strategies for transport information management during a large event. *Transportation Research Part C, Emerging Technologies, 77*, 421–432. doi:10.1016/j.trc.2017.02.008

Coursaris, C. K., Osch, W. V., & Balogh, B. A. (2016 January). Do Facebook likes lead to shares or sales? Exploring the empirical links between social media content, brand equity, purchase intention, and engagement. *Proceedings of the 49th Hawaii International Conference on System Sciences (HICSS)*,(pp. 3546-3555). IEEE. 10.1109/HICSS.2016.444

Crespi-Vallbona, M., & Richards, G. (2007). The meaning of cultural festivals: Stakeholder perspectives in Catalunya. *International Journal of Cultural Policy, 13*(1), 103–122. doi:10.1080/10286630701201830

Creswell, J. W. (2014). *A concise introduction to mixed methods research*. SAGE publications.

Creswell, J. W. (2017). *Nitel Araştırmacılar İçin 30 Temel Beceri* (H. Özcan, Trans.). Anı Yayıncılık.

Crockett, S. R., & Wood, L. (1999). Brand western australia: A totally ıntegrated approach to destination branding. *Journal of Vacation Marketing, 5*(2), 276–289. doi:10.1177/135676679900500307

Crofton, S. O., & Parker, R. D. (2012). Do Twitter and Facebook matter? Examining the economic impact of social media marketing in tourism websites of Atlantic Canada. *Journal of Tourism Research & Hospitality, 1*(4). doi:10.4172/2324-8807.1000107

Crompton, J. L. (1979). Motivations for Pleasure Vacation. *Annals of Tourism Research*, *6*(4), 408–424. doi:10.1016/0160-7383(79)90004-5

Crompton, J. L., & McKay, S. L. (1997). Motives of visitors attending festival events. *Annals of Tourism Research*, *24*(2), 425–439. doi:10.1016/S0160-7383(97)80010-2

Crouch, G. I. (2007). *Modelling destination competitiveness. A survey and analysis of the impact of competitiveness attributes*. Academic Press.

Crouch, G. I., & Ritchie, J. B. (1999). Tourism, competitiveness, and societal prosperity. *Journal of Business Research*, *44*(3), 137–152. doi:10.1016/S0148-2963(97)00196-3

Cruz, F. G. S., Tito, J. C., Perez-Galvez, J. C., & Medina-Viruel, M. J. (2019). Gastronomic Experiences of Foreign Tourists in Developing Countries. The Case in the City of Oruro (Bolivia). *Heliyon*, *5*, 1–8. PMID:31312731

Cudny, W. (2013). Festival Tourism – The Concept, Key Functions and Dysfunctions in the Context of Tourism Geography Studies. *Geographical Journal*, *65*(2), 105-118.

Cudny, W. (2013). Festival tourism – the concept, key functions and dysfunctions in the context of tourism geography studies. *The Geographical Journal*, *65*(2), 105–118.

Cudny, W. (2013). Festival Tourism - The Concept, Key Functions and Dysfunctions in the context of Tourism Geography studies. *The Geographical Journal*, *65*(2), 105–118. https://www.sav.sk/journals/uploads/03101218Cudny.pdf

Cudny, W. (2013). Festival tourism–the concept, key functions and dysfunctions in the context of tourism geography studies. *The Geographical Journal*, *65*(2), 105–118.

Cudny, W. (2014). The Phenomenon of Festivals: Their Origins, Evolution, and Classifications, Anthropos 2014. www.jstor.org/stable/pdf/43861801.pdf

Cudny, W. (2016). The Concept, Origins and Types of Festivals. In W. Cudny (Ed.), *Festivalisation of Urban Spaces: Factors, Processes and Effects* (pp. 11–42). Springer International Publishing., doi:10.1007/978-3-319-31997-1_2

Culha, O. (2020). The effect of food festival quality on place attachment and destination recommendation intention through festival experience and festival satisfaction: The case of the Didim International Olive Festival. *Journal of Convention & Event Tourism*, *21*(5), 387–416. doi:10.1080/15470148.2020.1775743

Cvent. (2022). *Cvent Event Diagramming®: Better events by design*. Cvent. https://www.Cvent.com/en/supplier-venue/event-diagramming-software

Dada, Z. A., Najar, A. H., & Gupta, S. K. (2022). Pro-Poor Tourism as an Antecedent of Poverty Alleviation: An Assessment of the Local Community Perception. *International Journal of Hospitality and Tourism Systems*, *15*(1).

Dalgın, T., & Oruç, M. C. (2015). *Turistik tüketicilerin tercihlerinde sosyal medyanın etkisi: Amasya ilindeki 4 yıldızlı bir konaklama işletmesinde uygulama. Proceedings of the 16th National Tourism Congress*, Çanakkale.

Dann, G. (1981). Tourism Motivations: An Appraisal. *Annals of Tourism Research*, *8*(2), 189–219. doi:10.1016/0160-7383(81)90082-7

Darian, S. G. (1978). The Ganges in myth and history. University of Hawaii Press. .

Das, B., & Acharjee, S. (2013). Problems and prospects of cultural tourism: A case study of Assam, India. *International Journal of Physical and Social Sciences*, *3*, 455-462.

Das, G., & Mukherjee, S. (2016). A measure of medical tourism destination brand equity. *International Journal of Pharmaceutical and Healthcare Marketing*, *10*(1), 104–128. doi:10.1108/IJPHM-04-2015-0015

Davidson, R. (2011). Web 2.0 as a marketing tool for conference centres. *International Journal of Event and Festival Management*, *2*(2), 117–138. doi:10.1108/17582951111136559

Davis, W. (2020). Magdalena: River of dreams. Random House.

Davras, Ö., & Özperçin, İ. (2021). The relationships of motivation, service quality, and behavioral intentions for gastronomy festival: The mediating role of destination image. *Journal of Policy Research in Tourism, Leisure & Events*, 1–14. doi:10.1080/19407963.2021.1968881

De Chernatony, L., & McDonald, M. (1992). *Creating powerful brands*. Butterworth-Heinemann.

De Ly, F., & Fontaine, M. (2006). *Drafting international Contracts: An Analysis of Contract Clauses*. Academic Press.

de Souza, J., Mendes-Filho, L., & Buhalis, D. (2020). Evaluating the effectiveness of tourist advertising to improve the competitiveness of destinations. *Tourism Economics*, *26*(6), 1001–1020. doi:10.1177/1354816619846748

Deanna, J. R., & Robert, A. F. (1998). Overview and perspectives. *The ecology of industry: sectors and linkages*, 1.

Del Barrio, M. J., Devesa, M., & Herrero, L. C. (2012). Evaluating intangible cultural heritage: The case of cultural festivals. City. *Cultura e Scuola*, *3*(4), 235–244. doi:10.1016/j.ccs.2012.09.002

Del Chiappa, G., & Atzeni, M. (2016). The role of emotions in a sport event experience. *Tourismos*, *11*(5).

Del Chiappa, G., Melis, G., & Atzeni, M. (2017). Driving first-time and repeat spectators to cultural events. In A. Correia, M. Kozak, J. Gnoth, & A. Fyall (Eds.), *Co-creation and wellbeing in tourism* (pp. 47–59). Springer. doi:10.1007/978-3-319-44108-5_4

Del Chiappa, G., Napolitano, E., & Atzeni, M. (2019). Perceived Authenticity, Satisfaction and Behavioural Intentions at Wineries. *Micro and Macro Marketing*, *1*, 119–138.

Delgado-Ballester, E., Manuera-Aleman, J. L., & Yague-Guillen, M. J. (2003). Development and validation of a brand trust scale. *International Journal of Market Research*, *45*(1), 35–54.

Demetriou, D. (February 12, 2019). Everything you need to know about seeing cherry blossoms in Japan. [online newspaper]. *Telegraph*. https://www.telegraph.co.uk/travel/destinations/asia/japan/articles/japan-cherry-blossom-sakura-guide

Demirci Orel, F., & Arık, A. (2020). Sosyal medya pazarlamasının tüketici satın alma niyetine etkisinin teknoloji kabul modeli aracılığıyla incelenmesi. *Erciyes Üniversitesi İktisadi ve İdari Bilimler Fakültesi Dergisi*, *57*(57), 205–232. doi:10.18070/erciyesiibd.701115

Demirören News Agency. (2022). *Edirne'de 1.5 ton bedava ciğer dağıtıldı, vatandaş birbiriyle yarıştı*. Retrieved from https://www.dha.com.tr/yerel-haberler/edirne/edirnede-1-5-ton-bedava-ciger-dagitildi-vatan-2069957

Deng, Q., & Li, M. (2014). A model of event–destination image transfer. *Journal of Travel Research*, *53*(1), 69–82. doi:10.1177/0047287513491331

Deniz, E. (2013). *Markalaşma ve reklam*. İstanbul: Kumsaati Yayın Dağıtım LTD. ŞTİ.

Department of Tourism Annual Report. (2018/2019). https://www.gov.za/sites/default/files/gcis_document/202002/department-tourism-annual-report-2018-2019.pdf

Derrett, R. (2003). Festivals & regional destinations: How festivals demonstrate a sense of community & place. *Rural Society, 13*(1), 35–53. doi:10.5172/rsj.351.13.1.35

Dickson, S., & Hall, T. E. (2006). An Examination of Whitewater Boaters' Expectations: Are Pretrip and Post-Trip Measures Consistent? *Leisure Sciences: An Interdisciplinary Journal, 28*(1), 1–16. doi:10.1080/01490400590962452

Dimanche, F. (2002). The contribution of special events to destination brand equity. In K. W. Wöber (Ed.) *City Tourism 2002: Proceedings of European Cities Tourism's International Conference* (pp. 73-80). Springer. 10.1007/978-3-7091-6125-8_8

Dimitrova, M., & Yoveva, I. (2014). *Food and wine events in Europe - a stakeholder approach.* Routledge.

Director of Economics and Statistics. (2022). Economic Survey of Assam, 2021-22. Assam.

Disegna, M., & Osti, L. (2016). Tourists' expenditure behaviour: The influence of satisfaction and the dependence of spending categories. *Tourism Economics, 22*(1), 5–30. doi:10.5367/te.2014.0410

Dixon, S. (2022, February 8). *Topic: Social media use during coronavirus (COVID-19) worldwide.* Statista. www.statista.com/topics/7863/social-media-use-during-coronavirus-covid-19-worldwide/#topicHeader__wrapper

Do Thi, H., & Dombroski, K. (2022). Diverse more-than-human approaches to climate change adaptation in Thai Binh, Vietnam. *Asia Pacific Viewpoint, 63*(1), 25–39. doi:10.1111/apv.12329

Doğan, M., Aksu, M., Çelik, Özge., & Kaymak, H. (2019). Cittaslow markasının turizme etkisi: Seferihisar'daki konaklama işletmeleri yöneticileri üzerine bir araştırma. *International Journal of Social and Economic Sciences, 4*(2), 06–11.

Döğüş, S. (2015). Anadolu'da Hızır-İlyas Kültür ve Hıdrellez Geleneği. *Türk Kültürü ve Hacı Bektaş Velî Araştırma Dergisi, 74*, 77-100. Retrieved from https://hbvdergisi.hacibayram.edu.tr/index.php/TKHBVD/article/view/1721

Dolnicar, S., & Zare, S. (2020). COVID19 and Airbnb – disrupting the disruptor. *Annals of Tourism Research, 83*, 102961. doi:10.1016/j.annals.2020.102961 PMID:32834218

Domínguez-Quintero, M., González-Rodríguez, M. R., & Paddison, B. (2020). The mediating role of experience quality on authenticity and satisfaction in the context of cultural-heritage tourism. *Current Issues in Tourism, 23*(2), 248–260. doi:10.1080/13683500.2018.1502261

Dore, L., & Crouch, G. I. (2003). Promoting destinations: An exploratory study of publicity programmes used by national tourism organisations. *Journal of Vacation Marketing, 9*(2), 137–151. doi:10.1177/135676670300900203

Dowd, T. J., Liddle, K., & Nelson, J. (2004). Music festivals as scenes: Examples from serious music, women's music, and skatepunk. Music scenes: Local, translocal and virtual, pp. 149–67.

Dowling, R. K. (2006). *Cruise ship tourism.* Cabi. doi:10.1079/9781845930486.0000

Dragani, R. (2019). *Augmented reality will change concerts forever.* Verizon. https://www.verizon.com/about/our-company/fourth-industrial-revolution/augmented-reality-will-change-concerts-forever

Drury, G. (2008). Social media: Should marketers engage and how can it be done effectively? *Journal of Direct, Data and Digital Marketing Practice, 9*(3), 274–277. doi:10.1057/palgrave.dddmp.4350096

Du Rand, G. E., & Heath, E. (2006). Towards a Framework for Food Tourism as an Element of Destination Marketing. *Current Issues in Tourism, 9*(3), 206–234. doi:10.2167/cit/226.0

Du Rand, G. E., Heath, E., & Alberts, N. (2003). The role of local and regional food in destination marketing: A South African situation analysis. *Journal of Travel & Tourism Marketing, 14*(3/4), 97–112. doi:10.1300/J073v14n03_06

Duran, E. (2013). A swot analysis on sustainability of festivals: The case of International Troia Festival. *Journal of International Social Research, 6*(28).

Durmaz, Y. (2017). The impact of psychological factors on consumer buying behavior and an empirical application in Turkey. *Asian Social Science, 10*(6), 194–204.

Echtner, C., & Ritchie, J. R. B. (1991). The meaning and measurement of destination image. *Journal of Tourism Studies, 2*(2), 2–12.

Edensor, T. (2016). *National Identity, Popular Culture and Everyday Life*. Bloomsbury Publishing.

Edinburgh International Festival. (2022). *About*. EIF. https://www.eif.co.uk/about/media .

Edirne Belediyesi. 2022). *10. Uluslararası Edirne bando ve ciğer festivali başlıyor*. Retrieved from https://www.edirne.bel.tr/icerik/10uluslararasi-edirne-bando-ve-ciger-festivali-basliyor

Edirne Municipality. (2022). *Kırkpınar Yağlı Güreşleri Festivali Tarihi*. Retrieved from https://edirnekirkpinar.com/tr/yazi/6-genel-bilgiler/36-tarihi

Ehsani, M., Sadeqi Sogdel, H., Heidary, A., & Jamshidi, A. (2012). The impact of hosting sporting events on sport tourism development: A case study. *International Journal of Research in Business and Social Science, 1*(1). Retrieved from http://www.ssbfnet.com

Ekinci, Y., Dawes, P. L., & Massey, G. R. (2008). An extended model of the antecedents and consequences of consumer satisfaction for hospitality services. *European Journal of Marketing, 42*(1), 35–68. doi:10.1108/03090560810840907

Ekinci, Y., & Hosany, S. (2006). Destination personality: An application of brand personality to tourism destinations. *Journal of Travel Research, 45*(2), 127–139. doi:10.1177/0047287506291603

Ekinci, Y., Sirakaya-Turk, E., & Preciado, S. (2013). Symbolic consumption of tourism destination brands. *Journal of Business Research, 66*(6), 711–718. doi:10.1016/j.jbusres.2011.09.008

Ellis, A., Park, E., Kim, S., & Yeoman, I. (2018). What is Food Tourism? *Tourism Management, 68*, 250–263. doi:10.1016/j.tourman.2018.03.025

Emekli, G. (2008). *Sürdürülebilir turizmde çeşitlendirmenin önemi: Çeşme-İzmir*. V. Ulusal Coğrafya Sempozyumu Bildiriler Kitabı.

Emir, O., & Toker, A. (2020). Turizmin Toplumsal Etkileri I. *Açıköğretim Fakültesi Yayını, 2875*, 51-66. Retrieved from https://ets.anadolu.edu.tr/storage/nfs/SOS109U/ebook/SOS109U-20V1S1-8-0-0-SV1-ebook.pdf

Enterprise Event Marketing. (2016). *Technology - Key experiential marketing stats for 2016*. EEM. https://www.slideshare.net/BEcauseExperientialMarketing/technology-key-experiential-marketing-stats-for-2016

Erbay, R., Demirkol, C., & Yıldırım Özmutlu, S. (2016). *Trakya Bölgesinde Festival, Fuar ve Panayır Organizasyonlarının Bölgenin Sosyo-Ekonomik Yapısına Etkileri*. Namik Kemal University Scientific Research Project. NKUBAP.00.13.AR.12.03. Retrieved from http://acikerisim.nku.edu.tr:8080/xmlui/bitstream/handle/20.500.11776/867/NKUBAP.00.13.AR.12.03.pdf?sequence=1

Ersun, N., & Arslan, K. (2011). Turizmde destinasyon seçimini etkileyen temel unsurlar ve pazarlama stratejileri. *Marmara Üniversitesi İktisadi ve İdari Bilimler Dergisi*, *31*(2), 229–248.

Escalas, J. E. (2004). Narrative processing: Building consumer connections to brands. *Journal of Consumer Psychology*, *14*(1-2), 168–180. doi:10.120715327663jcp1401&2_19

Escalas, J. E., & Bettman, J. R. (2003). You Are What They Eat: The Influence of Reference Groups on Consumer Connections to Brands. *Journal of Consumer Psychology*, *13*(3), 339–348. doi:10.1207/S15327663JCP1303_14

Eşitti, Ş., & Işık, M. (2015). Sosyal medyanın yabancı turistlerin Türkiye'yi tatil destinasyonu olarak tercih etmelerine etkisi. *Karadeniz Uluslararası Bilimsel Dergi*, *1*(27), 11–33.

Eventbrite. (2015). *Millennials and the new era of food, wine and beer festival*. Eventbrite. www.eventbrite.com/academy/millennials-and-the-new-era-of-food-wine-beer-festivals/

Eventbrite. (2016). *The Buzz on Beacons: Introducing the Latest Festival Technology*. Eventbrite. https://www.Eventbrite.com/blog/the-buzz-on-beacons-introducing-the-latest-festival-technology-ds00/

Eventbrite. (2019). *Event Life Cycle: Before, During & After an Event*. Eventbrite. Eventbrite. https://www.Eventbrite.com/blog/event-life-cycle-ds00/

Eventbrite. (2022a). *5 Questions to Ask When Choosing an RFID Vendor for Your Event*. Eventbrite. https://www.Eventbrite.com/blog/choosing-rfid-vendor-ds0c/

Eventbrite. (2022b). *Exciting Event Technology You Can Use to Enrich Your Attraction*. Eventbrite. https://www.Eventbrite.com/blog/exciting-event-technology-you-can-use-to-enrich-your-attraction-ds00/

Everett, S., & Aitchison, C. (2008). The role of food tourism in sustaining regional identity: A case study of Cornwall, Southwest England. *Journal of Sustainable Tourism*, *16*(2), 150–167. doi:10.2167/jost696.0

Executive Office of the Mayor. (2023, January 9). *Mayor Bowser unveils DC's comeback plan*. Government of the District of Columbia Muriel Bowser, Mayor. mayor.dc.gov/release/mayor-bowser-unveils-dcs-comeback-plan

Fahimnia, B., Sarkis, J., & Davarzani, H. (2015). Green supply chain management: A review and bibliometric analysis. *International Journal of Production Economics*, *162*, 101–114. doi:10.1016/j.ijpe.2015.01.003

Fathabadi, H., Nejad, M. R. O., & Alizadeh, H. (2017). An investigation of the affecting tourism destination brand equity. *Asian Journal of Social Sciences & Humanities*, *6*(1), 101–113.

Fathabadi, H., Nejad, M. R. O., & Alizadeh, H. (2017). An investigation of the factors affecting tourism destination brand equity. *Asian Journal of Social Sciences & Humanities*, *6*(1), 101–113.

Fedorikhin, A., Park, C. W., & Thomson, M. (2008). Beyond fit and attitude: The effect of emotional attachment on consumer responses to brand extensions. *Journal of Consumer Psychology*, *18*, 281– 291.

Felsenstein, D., & Fleischer, A. (2003). Local festivals and tourism promotion: The role of public assistance and visitor expenditure. *Journal of Travel Research*, *41*(4), 385–392. doi:10.1177/0047287503041004007

Fennel, D. (2008). *Ecotourism* (3rd ed.). Routledge.

Fernández, J. A. S., Azevedo, P. S., Martin, J. M. M., & Martin, J. A. R. (2020). Determinants of tourism destination competitiveness in the countries most visited by international tourists: Proposal of a synthetic index. *Tourism Management Perspectives*, *33*, 100582. doi:10.1016/j.tmp.2019.100582

Ferns, B. H., & Walls, A. (2012). Enduring travel involvement, destination brand equity, and travelers' visit intentions: A structural model analysis. *Journal of Destination Marketing & Management, 1*(1–2), 27–35. doi:10.1016/j.jdmm.2012.07.002

Fields, K. (2002). Demand for the Gastronomy Tourism Product: Motivational Factors. In A-M. Hjalager ve G. Richards (Eds.). Tourism and Gastronomy, 36–50. London & New York: Routledge.

Fırat, A., & Kömürcüoğlu, F. (2015). Muğla şehir markası ve imaj algısı: Muğla Sıtkı Koçman Üniversitesi öğrencileri üzerine bir alan araştırması. *Yönetim Bilimleri Dergisi, 13*(26), 285–304.

Flood, G. (2003). The Saiva Traditions. *The Blackwell Companion to Hinduism*, 200–228.

Flowers, A. (2021) Lalbagh Botanical Garden Bangalore, History and Attraction. *Arena Flowers*. https://www.arenaflowers.co.in/blogs/news/lalbagh-botanical-garden-bangalore

Folgado-Fernández, J., Hernández-Mogollón, J., & Duarte, P. (2016). Destination image and loyalty development: The impact of tourists' food experiences at gastronomic events. *Scandinavian Journal of Hospitality and Tourism*, 1–19.

Fotiadis, A., & Vassiliadis, C. (Eds.). (2020). *Principles and Practices of Small-scale Sport Event Management*. IGI Global. doi:10.4018/978-1-7998-4757-1

Fotis, J., Buhalis, D., & Rossides, N. (2012). Social media use and impact during the holiday travel planning process. Springer-Verlag, 13-24.

Freire, J. R. (2012). Destination branding in the Middle East: Politics, religion, and commerce. *Journal of Destination Marketing & Management, 1*(1-2), 74–82.

Frey, B. S. (1994). The economics of music festivals. *Journal of Cultural Economics, 18*(1), 29–39. doi:10.1007/BF01207151

Fujisan Marketing. (2018). Geofencing: What Is It, And How Is It Used? *Medium*. https://medium.com/@fujisanmarketing/geofencing-what-is-it-and-how-is-it-used-1b2754499316

Fyall, A., Garrod, B., & Wang, Y. (2012). Destination collaboration: A critical review of theoretical approaches to a multi-dimensional phenomenon. *Journal of Destination Marketing & Management, 1*(1-2), 10–26. doi:10.1016/j.jdmm.2012.10.002

Garcia, S. E. (2020, March 21). The hottest parties in town are now online. *The New York Times*. www.nytimes.com/2020/03/21/arts/d-nice-instagram.html

Garcia, J. A., Gomez, M., & Molina, A. (2012). A destination branding model: An empirical analysis based on stakeholders. *Tourism Management, 3*(3), 646–666. doi:10.1016/j.tourman.2011.07.006

Gartner, W. C. (1996). Tourism Development: Principles processed and policies. Van Nostrand Reinhold.

Getz, D. (1998). Event tourism and the authenticity dilemma, Ed. William F. Theobald Global Tourism, Butterworth-Heinemann; 2nd edition.

Getz, D. (2010). The Nature and Scope of Festival Studies. *International Journal of Event ManagementResearch, 5*, 1-47. doi:10.1108/17852951011029298

Getz, D. (1991). *Festivals, special events, and tourism*. Van Nostrand Reinhold.

Getz, D. (1997). *Event management and Event Tourism*. Cognizant Communication Corporation.

Getz, D. (1997). *Event management and event tourism*. Van Nostrand Reinhold.

Getz, D. (2000). *Explore wine tourism: management, development & destinations*. Cognizant Communication Corporation.

Getz, D. (2008). Event tourism: Definition, evolution, and research. *Tourism Management*, *29*(3), 403–428. doi:10.1016/j.tourman.2007.07.017

Getz, D. (2010). The Nature and Scope of Festival Studies. *International Journal of Event Management Research*, *5*(1), 1–47.

Getz, D., & Brown, G. (2006). Critical Success Factors for Wine Tourism Regions: A Demand Analysis. *NA: Tourism Management*, *27*(1), 146–158. doi:10.1016/j.tourman.2004.08.002

Getz, D., & Page, S. J. (2016). *Event tourism: Definition, evolution, and research*. Routledge.

Getz, D., & Page, S. J. (2016). Progress and prospects for event tourism research. *Tourism Management*, *52*, 593–631. doi:10.1016/j.tourman.2015.03.007

Ghana Tourism Authority. (2020, December 21). *Afrochella Block Party 2020*. Visit Ghana. www.visitghana.com/events/afrochella-block-party/

Ghimire, K., & Schewe, C. D. (2017). Mega-events, tourism development and perceived community well-being. *Current Issues in Tourism*, *20*(4), 401–417. doi:10.1080/13683500.2015.1126275

Ghosh, M. M., Ghosh, A., & Ghosh, M. (2016). Impact and influence of culture on brands in Indian market. *IOSR Journal of Business and Management*, *18*(7), 54–60.

Giango, M. K., Hintapan, R., Suson, M., Batican, I., Quiño, L., Capuyan, L., Anoos, J. M., Batoon, J., Aro, J. L., Maturan, F., Yamagishi, K., Gonzales, G., Burdeos, A., & Ocampo, L. (2022). Local support on sports tourism development: An integration of emotional solidarity and social exchange theory. *Sustainability*, *14*(19), 12898. doi:10.3390u141912898

Giannopoulos, A., Piha, L., & Skourtis, G. (2021). Destination branding and co-creation: A service ecosystem perspective. *Journal of Product and Brand Management*, *30*(1), 148–166. doi:10.1108/JPBM-08-2019-2504

Gibson, C., Connell, J., Waitt, G., & Walmsley, J. (2011). The Extent and Significance of Rural Festivals. In *Festival Places: Revitalising Rural Australia (Tourism and Cultural Change)* (pp. 3–24). Channel View Publications. doi:10.21832/9781845411688-004

Gilmore, F. (2002). A country—can it be repositioned? Spain—the success story of country branding. *Journal of Brand Management*, *9*(4/5), 281–293. doi:10.1057/palgrave.bm.2540078

Givens, D. (2020, December 30). *Afrochella co-founder talks the future of live events amid covid-19*. Black Enterprise. www.blackenterprise.com/afrochella-co-founder-abdul-karim-abdulla-talks-the-future-of-live-events-amid-covid-19/

Glasner, E., & Weaver, J. (2020, September 19). *What worked and what wowed the mostly virtual TIFF 2020 film fest*. CBCnews. www.cbc.ca/news/entertainment/tiff-2020-wrap-1.5730261

Gligorijević, J. (2022). Practices and ideological effects of music festival branding: The case study of Serbia's two major music festivals, Exit and Gu č a. In *Musik & Marken* (pp. 117–144). Springer Fachmedien Wiesbaden. doi:10.1007/978-3-658-36472-4_6

Globalpayments. (2020). *What is the Difference Between NFC and RFID?* Global Payments Integrated. https://www.Globalpaymentsintegrated.com/en-us/blog/2020/04/21/what-is-the-difference-between-nfc-and-rfid

Gnoth, J. (1997). Tourism Motivation and Expectation Formation. *Annals of Tourism Research*, *24*(2), 283–304. doi:10.1016/S0160-7383(97)80002-3

Goeldner, C. R., & Ritchie, J. R. B. (2006). *Tourism: Principles, Practices and Philosophies*. John Wiley and Sons, Inc.

Goffi, G., Cucculelli, M., & Masiero, L. (2019). Fostering tourism destination competitiveness in developing countries: The role of sustainability. *Journal of Cleaner Production*, *209*, 101–115. doi:10.1016/j.jclepro.2018.10.208

Goh, K. Y., Heng, C. S., & Lin, Z. (2013). Social media brand community and consumer behavior: quantifying the relative ımpact of user- and marketer- generated content, In *2016 49th Hawaii international conference on system sciences (HICSS)*, IEEE.

Goicochea, E. (2017). *3 Music Festivals that use Location-based Marketing to create amazing mobile experiences*. MOCA. https://www.mocaplatform.com/blog/3-music-festivals-use-proximity-marketing-create-amazing-experiences

Goldblatt, J. (2002) *Special Events: Twenty First Century Global Event Management*. John Wiley & Sons, Inc.

Gómez, M., Lopez, C., & Molina, A. (2015). A model of tourism destination brand equity: The case of wine tourism destinations in Spain. *Tourism Management*, *51*, 210–222. doi:10.1016/j.tourman.2015.05.019

Gonzalez, C. (2022, April 16). *Coachella 2022: Festival's return leaving local businesses dancing*. FOX 11 Los Angeles. www.foxla.com/news/coachella-2022-festivals-return-leaving-local-businesses-dancing

González-Mansilla, Ó., Berenguer-Contri, G., & Serra-Cantallops, A. (2019). The impact of value co-creation on hotel brand equity and customer satisfaction. *Tourism Management*, *75*, 51–65. doi:10.1016/j.tourman.2019.04.024

Goossens, C. (2000). Tourism Information and Pleasure Motivation. *Annals of Tourism Research*, *27*(2), 301–321. doi:10.1016/S0160-7383(99)00067-5

Gordin, V., & Trabskaya, J. (2013). The role of gastronomic brands in tourist destination promotion: The case of St. Petersburg. *Place Branding and Public Diplomacy*, *9*(3), 189–201. doi:10.1057/pb.2013.23

Goulding, C. (2000). The commodification of the past, postmodern pastiche, and the search for authentic experiences at contemporary heritage attractions. *European Journal of Marketing*, *34*(7), 835–853. doi:10.1108/03090560010331298

Govindarajan, N. (2014). The Tigalas of Bendakaluru. *Deccan Herald*. https://www.deccanherald.com/content/378918/tigalas-bendakaluru.html

Gözgeç Mutlu, H., Avcıkurt, C., & Filiz, A. (2020). Festival katılımcılarının motivasyonları ve turistik deneyimleri. *International Journal of Contemporary Tourism Research*, *4*(2), 219–233.

Grappi, S., & Montanari, F. (2011). The role of social identification and hedonism in affecting tourist re-patronizing behaviours: The case of an Italian festival. *Tourism Management*, *32*(5), 1128–1140. doi:10.1016/j.tourman.2010.10.001

Grayson, K., & Martinec, R. (2004). Consumer perceptions of iconicity and indexicality and their influence on assessments of authentic market offerings. *The Journal of Consumer Research*, *31*(2), 296–312. doi:10.1086/422109

Griffiths, M., & Korstanje, M. E. (2021). Issues of authenticity in religious and spiritual tourism. In D. H. Holsen & D. J. Timothy (Eds.), *The Routledge Handbook of Religious and Spiritual Tourism* (pp. 276–287). Routledge. doi:10.4324/9780429201011-23

Guinness World Records. (2021). *Oldest Competition, Wrestling*. Retrieved from https://www.guinnessworldrecords.com/world-records/73717-oldest-competition-wrestling

Gulf News. (2021). *Oman sees potential in complementing Qatar's World Cup offerings*. Retrieved from https://gulfnews.com/oman/oman-sees-potential-in-complementing-qatars-world-cup-offerings-1.1622710583378

Gündüz Alptürker, İ., & Alptürker, H. (2021). Gelenekten Geleceğe Yöresel Kültürün Taşınmasında Festivallerin Rolü. *MANAS Sosyal Araştırmalar Dergisi, 10*(2), 1409-1422. Retrieved from https://dergipark.org.tr/tr/download/article-file/1299408

Güngördü, E. (2007). Türkiye'nin turizm coğrafyası. Ankara. Ankara: Asil Yayın Dağıtım.

Gunsoy, E., & Hannam, K. (2013). Festivals, community development and sustainable tourism in the Karpaz region of Northern Cyprus. *Journal of Policy Research in Tourism, Leisure & Events*, *5*(1), 81–94. doi:10.1080/19407963.2013.774201

Gupta, K., & Gulla, A. (2010). Internet deployment in the spiritual tourism industry: The case of Vaishno Devi Shrine. *Worldwide Hospitality and Tourism Themes*, *2*(5), 507–519. doi:10.1108/17554211011090120

Güripek, E. (2013). *Turizm destinasyonlarının rekabet gücünün artırılmasında stratejik destinasyon yönetimi: Çeşme Alaçatı destinasyonu üzerine bir uygulama*. Doktora Tezi, Dokuz Eylül Üniversitesi, Sosyal Bilimler Enstitüsü.

Gursoy, D., Nunkoo, R., & Yolal, M. (Eds.). (2020). *Festival and Event Tourism Impacts* (1st ed.). Routledge. doi:10.4324/9780429274398

Guttentag, D. A. (2015). Airbnb: Disruptive innovation and the rise of an informal tourism accommodation sector. *Current Issues in Tourism*, *18*(12), 1192–1217. doi:10.1080/13683500.2013.827159

Gwinner, K., & Bennett, G. (2008). The impact of brand cohesiveness and sport identification on brand fit in a sponsorship context. *Journal of Sport Management*, *22,* 410–426.

Hair, J., Hult, G., Ringle, C., & Sarstedt, M. (2016). *A primer on partial least squares structural equation modeling (PLS-SEM).* Sage publications.

Hall, C. M., & Sharples, L. (2003). The Consumption of Experiences or the Experiences of Comsumption? An Introduction to the Tourism of Taste. In M. C. Hall, L. Sharples, R. Mitchell, N. Macionis ve B. Cambourne (Eds.). Food Tourism Around the World: Development, Management and Markets. Oxford: Butterworth Heinemann, 1–24.

Hall, C. M. (1989). Hallmark events and the planning process. In G. J. Syme, B. J. Shaw, D. M. Fenton, & W. S. Mueller (Eds.), *The Planning and Evaluation of Hallmark Events*. Avebury.

Hall, C. M. (1992). *Hallmark tourist events: impacts, management & planning*. Belhaven Press.

Hall, C. M., & Mitchell, R. (2005). Gastronomic tourism: Comparing food and wine tourism experiences. In M. Novelli (Ed.), *Niche tourism: Contemporary issues, trends and cases* (pp. 73–88). Butterworth-Heinemann. doi:10.1016/B978-0-7506-6133-1.50014-7

Hall, D. (1999). Destination branding, niche marketing and national image projection in Central and Eastern Europe. *Journal of Vacation Marketing*, *5*(3), 227–237. doi:10.1177/135676679900500303

Handayani, B., & E Korstanje, M. (2017). Place Brand Authenticity in Social Media Interaction: A Postmodern Perspective. *Études caribéennes*, (37-38).

Hangal, R. (2022). *Personal Communication.* [Discussion].

Hankinson, G. (2005). Destination brand images: A business tourism perspective. *Journal of Services Marketing*, *19*(1), 24–32. doi:10.1108/08876040510579361

Hankinson, G. (2007). The management of destination brands: Five guiding principles based on recent developments in corporate branding theory. *Journal of Brand Management, 14*(3), 240–254. doi:10.1057/palgrave.bm.2550065

Hanna, S., & Rowley, J. (2015). Towards a model of the place brand web. *Tourism Management, 48*, 100–112. doi:10.1016/j.tourman.2014.10.012

Harb, A. A., Fowler, D., Chang, H. J. J., Blum, S. C., & Alakaleek, W. (2019). Social media as a marketing tool for events. *Journal of Hospitality and Tourism Technology, 10*(1), 28–44. doi:10.1108/JHTT-03-2017-0027

Haviland, W. A. (1999). *Cultural Anthropology* (9th ed.). Harcourt Barace College Publishers.

Heithaus, H. H. (2021, November 13). Artis-naples protest decries COVID-19 mandates, but for small audience. *Naples Daily News*. www.naplesnews.com/story/news/local/2021/11/13/rally-artis-naples-protests-covid-policy-and-evokes-nazi-germany/85950560
02/

Hemmonsbey, J., & Tichaawa, T. M. (2018). The effects of major sport event leveraging for tourism and destination branding: The case of South Africa as an emerging destination. *Geo Journal of Tourism and Geosites, 11*(1), 213–225.

Hemmonsbey, J., & Tichaawa, T. M. (2020). Brand messages that influence the sport tourism experience: The case of South Africa. *Journal of Sport & Tourism, 24*(3), 177–194. doi:10.1080/14775085.2020.1822200

Henderson, J. C. (2009). Food tourism reviewed. *British Food Journal, 111*(4), 317–326. doi:10.1108/00070700910951470

Hermann, U. P., Boshoff, L., & Ncala, T. T. (2019). Understanding beer festival attendee motivations to a craft beer festival in South Africa. *Journal for New Generation Sciences, 17*(2), 18–30.

Hernández-Mogollón, J. M., Duarte, P. A., & Folgado-Fernández, J. A. (2018). The contribution of cultural events to the formation of the cognitive and affective images of a tourist destination. *Journal of Destination Marketing & Management, 8*, 170–178. doi:10.1016/j.jdmm.2017.03.004

Hernández-Mogollón, J., Folgado-Fernández, J., & Duarte, P. (2014). Event tourism analysis and state of the art. *European Journal of Tourism. Hospitality and Recreation, 5*(2), 83–102.

Hinch, T., & Ito, E. (2018). Sustainable sport tourism in Japan. *Tourism Planning & Development, 15*(1), 96–101. doi:10.1080/21568316.2017.1313773

Hiscock, J. (2001). Most Trusted Brands. Marketing, 32–33.

History.com Editors. (2010, January 25). *Mardi Gras 2022*. www.history.com/topics/holidays/mardi-gras

Hjalager, A. M. (2002). A Typology of Gastronomy Tourism. In A. M. Hjalager ve G. Richards (Eds.). Tourism and Gastronomy. Londra ve New York: Routledge, 21–35.

Hjalager, A. M., & Corigliano, M. A. (2000). Food for tourists-Determinants of an image. *International Journal of Tourism Research, 2*(4), 281–293. doi:10.1002/1522-1970(200007/08)2:4<281::AID-JTR228>3.0.CO;2-Y

Hodur, N. M., & Leistritz, F. L. (2006). Estimating the economic impact of event tourism: A review of issues and methods. *Journal of Convention & Event Tourism, 8*(4), 63–79. doi:10.1300/J452v08n04_05

Holcomb, B. (1999). Marketing cities for tourism. In D. Judd & S. Fainstein (Eds.), *The Tourist City* (pp. 54–70).

Holmes, J. G. (1991). trust and the appraisal process in close relationships. In W. H. Jones, & D. Perlman (Eds.), Advances in personal relationships 2 (pp. 57–104). London: Jessica Kingsley.

Homburg, C., Klarmann, M., & Schmitt, C. (2010). Brand Awareness in Business Markets: When is It Related to Firm Performance? *International Journal of Research in Marketing*, *27*(3), 201–212. doi:10.1016/j.ijresmar.2010.03.004

Honey, M. (1999). *Ecotourism and sustainable development: Who owns paradise?* Island Press.

Hong, I. B., & Cho, H. (2011). The impact of consumer trust on attitudinal loyalty and purchase intentions in B2C e-marketplaces: Intermediary trust vs. seller trust. *International Journal of Information Management*, *31*(5), 469–479. doi:10.1016/j.ijinfomgt.2011.02.001

Hookings, M. (2019). Guest Blog: How technology is enhancing festivals. Event Industry News. https://www.eventindustrynews.com/news/guest-blog-how-technology-is-enhancing-festivals

Horne, J., & Manzenreiter, W. (2006). An introduction to the sociology of sports mega-events. *The Sociological Review*, *54*(2, suppl), 1–24. doi:10.1111/j.1467-954X.2006.00650.x

Horner, S., & John, S. (2005). Leasure marketing a global perspective, İngiltere: Elsevier.

Horng, J. S., Liu, C. H., Chiu, H. Y., & Tsai, C. Y. (2012a). The role of international tourist perceptions of brand equity and travel intention in culinary tourism. *Service Industries Journal*, *32*(16), 2607–2621. doi:10.1080/02642069.2011.594879

Horng, J. S., Liu, C. H., Chou, H. Y., & Tsai, C. Y. (2012b). Understanding the impact of culinary brand equity and destination familiarity on travel intentions. *Tourism Management*, *33*(4), 815–824. doi:10.1016/j.tourman.2011.09.004

Horng, J. S., & Tsai, C. T. (2012). Constructing indicators of culinary tourism strategy: An application of resource-based theory. *Journal of Travel & Tourism Marketing*, *29*(8), 796–816. doi:10.1080/10548408.2012.730945

Horng, J. S., & Tsai, C. T. S. (2010). Government websites for promoting East Asian culinary tourism: A cross-national analysis. *Tourism Management*, *31*(1), 74–85. doi:10.1016/j.tourman.2009.01.009

Howat, G., & Murray, D. (2002). The role of critical incidents to complement service quality information for a sports and leisure centre. *European Sport Management Quarterly*, *2*(1), 23–46. doi:10.1080/16184740208721910

Hsu, T.-K., Tsai, Y.-F., & Wu, H.-H. (2009). The Preference Analysis for Tourist Choice of Destination: A Case Study of Taiwan. *Tourism Management*, *30*(2), 288–297. doi:10.1016/j.tourman.2008.07.011

Hsu, Y. L. (2012). Facebook as international eMarketing strategy of Taiwan hotels. *International Journal of Hospitality Management*, *31*(3), 972–980. doi:10.1016/j.ijhm.2011.11.005

Hu, M. R., & Lee, A. D. (2020). Airbnb, COVID-19 risk and lockdowns: Global evidence. SSRN *Electronic Journal, 2*. www.doi.org/10.2139/ssrn.3589141 doi:10.2139/ssrn.3589141

Huang, J. Z., Li, M., & Cai, L. A. (2010). A model of community-based festival image. *International Journal of Hospitality Management*, *29*(2), 254–260. doi:10.1016/j.ijhm.2009.10.010

Huang, S., & Hsu, C. H. (2009). Effects of travel motivation, past experience, perceived constraint, and attitude on revisit intention. *Journal of Travel Research*, *48*(1), 29–44. doi:10.1177/0047287508328793

Hu, B. (2007). *Informal Institutions and Rural Development in China*. Routledge. doi:10.4324/9780203947449

Hudson, S., Roth, M. S., Madden, T. J., & Hudson, R. (2015). The effects of social media on emotions, brand relationship quality, and word of mouth: An empirical study of music festival attendees. *Tourism Management*, *47*, 68–76. doi:10.1016/j.tourman.2014.09.001

Huh, J. (2002). *Tourist satisfaction with cultural/heritage sites: The Virginia Historic Triangle* [Thesis, Virginia Polytechnic Institute and State University, Blacksburg, VA].

Hui, T. K., Wan, D., & Ho, A. (2007). Tourists' satisfaction, recommendation and revisiting Singapore. *Tourism Management, 28*(4), 965–975. doi:10.1016/j.tourman.2006.08.008

Hutabarat, P. M. (2022). Music tourism potentials in Indonesia: Music festivals and their roles in city branding. *Journal of Indonesian Tourism and Policy Studies, 7*(1).

Ian, Y., Robertson, M., Ali-Knight, J., Drummond, S., & McMahon-Beattie, U. (2004). *Festival and Events Management*. Elsevier.

İlban, M. O., & Kömür, T. (2019). Destinasyon markalaşmasında festival turizminin rolü: Ayvalık Uluslararası Zeytin Hasat Günleri Örneği. *Seyahat ve Otel İşletmeciliği Dergisi, 16*(2), 274–295.

İlgüner, M. (2006). *Türkiye 'de marka yaratma ve yaratmanın altın kuralları*. Rota Yayınları.

Im, H. H., Kim, S. S., Elliot, S., & Han, H. (2012). Conceptualizing destination brand equity dimensions from a consumer-based brand equity perspective. *Journal of Travel & Tourism Marketing, 29*(4), 385–403. doi:10.1080/10548408.2012.674884

İmirgi, A. (2005). Festival kavramıüzerinedüşünceler. *Milli Folklor, 17*(65), 29–36.

Indian Express 4th Dec. 2022. Rajasthan govt to organise folk art festivals from panchayat to state level to boost tourism. Retreived from https://indianexpress.com/article/lifestyle/destination-of-the-week/rajasthan-govt-to-organise-folk-art-festivals-from-panchayat-to-state-level-to-boost-tourism- 8305324/

Jackson, L. (2008). Residents' perceptions of the impacts of special event tourism. *Journal of Place Management and Development, 1*(3), 240–255. doi:10.1108/17538330810911244

Jacobsen, J., Kr, S., & Munar, A. M. (2012). Tourist information search and destination choice in a digital age. *Tourism Management Perspectives, 1*, 39–47. doi:10.1016/j.tmp.2011.12.005

Jago, K. L. (1997). Special events and tourism behaviour: A conceptualisation and an empirical analysis from a values perspective [Unpublished doctoral dissertation]. Victoria University.

Jago, K. L., & Shaw, R. N. (2000). Consumer Perceptions of Special Events: A Multi-Stimulus Validation. *Journal of Travel & Tourism Marketing, 8*(4), 1–24. doi:10.1300/J073v08n04_01

Jago, L., Chalip, L., Brown, G., Mules, T., & Ali, S. (2003). Building Events Into Destination Branding: Insights From Experts. *Event Management, 8*(1), 3–14. doi:10.3727/152599503108751658

Jain, V. (2016). Candy of Cannabis sativa L. - A traditional intoxicant in Udaipur, Rajasthan. *J Trad Folk Pract, 4*, 129–131.

Jain, V. (2017b). Ethno-biology of exhibits in a mega handicraft fair of Rajasthan. *Indian Journal of Traditional Knowledge, 16*(3), 524–532.

Janeczko, B., Trevor M., & Brent, R. (2002). *Estimating the economc impacts of festivals and events: A Research Guide*. National Library of Australia Cataloguing in Publication Data, Research Report.

Jang, J. (2013). The Effects of Quality and Satisfaction on Awareness and Behavioral Intentions:Exploring the Role of a Wine Festival. *Kentucky: Journal Of Travel Research.*

Jang, S. S., Bai, B., Hong, G. S., & O'Leary, J. T. (2004). Understanding travel expenditure patterns: A study of Japanese pleasure travelers to the United States by income level. *Tourism Management, 25*(3), 331–341. doi:10.1016/S0261-5177(03)00141-9

Jang, S. S., & Ham, S. (2009). A double-hurdle analysis of travel expenditure: Baby boomer seniors versus older seniors. *Tourism Management*, *30*(3), 372–380. doi:10.1016/j.tourman.2008.08.005

Janiszewska, K., & Insch, A. (2012). Strategic importance of brand positioning in the place brand concept: Elements, structure and application capabilities. *Journal of International Students*, *5*(1), 9–19. doi:10.14254/2071-8330.2012/5-1/2

Japutra, A., Ekinci, Y., & Simkin, L. (2014). Exploring brand attachment, its determinants, and outcomes. *Journal of Strategic Marketing*, *22*(7), 1–15. doi:10.1080/0965254X.2014.914062

Jayaram, S. (2021). *Bangalore's Lalbagh: A Chronicle of the Garden and the City*. Visual Arts Collective.

Jayaswal, T. (2009). *Events-induced tourism: A protocol analysis*. [Master's Thesis, Auckland University of Technology].

Jayswal, T. (2008). Events Tourism: Potential to build a brand destination. *Conference on Tourism in India – Challenges Ahead*, (pp. 251–262). D-Space. http://dspace.iimk.ac.in/xmlui/handle/2259/178

Jex, S. M., & Britt, T. W. (2014). *Organizational psychology: A scientist-practitioner approach*. John Wiley & Sons.

Jiang, W. H., Li, Y. Q., Liu, C. H., & Chang, Y. P. (2017). Validating a multidimensional perspective of brand equity on motivation, expectation, and behavioural intention: A practical examination of culinary tourism. *Asia Pacific Journal of Tourism Research*, *22*(5), 524–539. doi:10.1080/10941665.2017.1287106

Ji, Z., Xu, H., & Cui, Q. (2022). Tourism and poverty alleviation in Tibet, China: The role of government in enhancing local linkages. *Asia Pacific Journal of Tourism Research*, *27*(2), 173–191. doi:10.1080/10941665.2021.2020312

Johnson, C. K. (2023, March 10). Pandemic 3 years later: Has the COVID-19 virus won? *The Washington Post*. www.washingtonpost.com/health/2023/03/10/covid-pandemic-anniversary/f3252752-bf55-11ed-9350-7c5fccd598ad_story.html

Jude, O. C., Uchenna, O., & Ngozi, E. (2018). Impact of religious tourism in host communities: The case of Awhum Monastery. *American Journal of Social Sciences*, *6*(3), 39–47.

Juhasz, A. (2021, February 12). *What Mardi Gras looks like during the COVID-19 pandemic*. NPR. www.npr.org/2021/02/12/967458788/what-mardi-gras-looks-like-during-the-covid-19-pandemic

Kádár, B., & Klaniczay, J. (2022). Branding built heritage through cultural urban festivals: An instagram analysis related to sustainable co-creation, in Budapest. *Sustainability*, *14*(9), 5020. doi:10.3390u14095020

Kakaei, H., Nourmoradi, H., Bakhtiyari, S., Jalilian, M., & Mirzaei, A. (2022). Effect of covid-19 on food security, hunger, and food crisis. *COVID-19 and the Sustainable Development Goals*, 3–29. doi:10.1016/B978-0-323-91307-2.00005-5

Kalkstein-Silkes, C. A. (2007). *Food and food related festivals in rural destination branding*. [Unpublished Ph. D. Thesis, Purdue University, Indiana.]

Kalkstein-Silkes, C. A. (2007). *Food and food-related festivals in rural destination branding*. [Unpublished doctoral dissertation, Purdue University, Indiana].

Kamath, S. U. (1983). *Karnataka State Gazetteer, Part II*. Government of Karnataka Publication.

Kaplan, A. M., & Haenlein, M. (2010). Users of the world, unite! The challenges and opportunities of Social Media. *Business Horizons*, *53*(1), 59–68. doi:10.1016/j.bushor.2009.09.003

Karagöz, D. (2011). Etkinlik turizmi ve etkinlik turizmi bağlamında yabancı ziyaretçi harcamalarının ekonomiye etkisi: Formula 1 2005 Türkiye Grand Prix örneği [Doctoral dissertation, Anadolu University, Turkey].

Karar, A. (2010). Impact of pilgrim tourism at Haridwar. *The Anthropologist, 12*(2), 99–105. doi:10.1080/09720073.2010.11891138

Kasapi, I., & Cela, A. (2017). Destination Branding: A Review of the City Branding Literature. *Mediterranean Journal of Social Sciences, 8*(4), 129–142. Advance online publication. doi:10.1515/mjss-2017-0012

Kasimati, E., & Dawson, P. (2002). Assessing the impact of the summer Olympics on Greece's tourism industry. *Journal of Travel Research, 41*(4), 385–390.

Kastenholz, E., & Lima, J. (2011). The Integral Rural Tourism Experience from the Tourist's Point of View – A Qualitative Analysis of Its Nature and Meaning. *Tourism & Management Studies, 7*, 62–74.

Kates, R. W. (2011). What kind of a science is sustainability science? *Proceedings of the National Academy of Sciences of the United States of America, 108*(49), 19449–19450. doi:10.1073/pnas.1116097108 PMID:22114189

Keleş, H. (2021). *Turizmin Yerel Halkın Yaşam Kalitesi, Memnuniyeti ve Turizme Desteği Üzerindeki Etkileri: Manavgat Örneği* [Doctoral Thesis]. Necmettin Erbakan Üniversitesi Sosyal Bilimler Enstitüsü Turizm İşletmeciliği Anabilim Dalı, Turizm İşletmeciliği Bilim Dalı, Konya. Tez no: 704246. Retrieved from https://tez.yok.gov.tr/UlusalTezMerkezi/tezSorguSonucYeni.jsp

Keller, K. (1998). *Strategic brand management: Building. Measuring and Managing Brand Equity*. Prentice Hall Inc.

Keller, K. L. (1993). Conceptualizing, measuring, and managing customer-based brand equity. *Journal of Marketing, 57*(1), 1–22. doi:10.1177/002224299305700101

Keller, K. L. (2003). Understanding brands, branding and brand equity. *Interactive Marketing, 5*(1), 7–20. doi:10.1057/palgrave.im.4340213

Kendir, H. (2020). Gastronomik Lezzetlere Yönelik Çevrimiçi Turist Deneyimlerinin İncelenmesi: Edirne Tava Ciğeri Örneği. *Safran Kültür ve Turizm Araştırmaları Dergisi, 3*(3), 408-423. Retrieved from https://dergipark.org.tr/tr/pub/saktad/issue/59328/845512

Keskin, H. D., & Yıldız, S. (2010). Tüketicilerin marka tercihlerinde etkili olan faktörler ile marka imajının marka değeri üzerindeki etkileri: Trabzon örneği. *Dumlupınar Üniversitesi Sosyal Bilimler Dergisi, 27*, 1–13.

Keskin, S. N. (2018). *Katılımcılarının Motivasyonlarına Göre Pazar Bölümlendirmesi: Alaçatı Ot Festivali Örneği. (Yüksek Lisans Tezi)*. Mersin Üniversitesi Sosyal Bilimler Enstitüsü.

Keukenhof. (2020, May 12). *Keukenhof closes successful virtual season*. Keukenhof. www.keukenhof.nl/en/news/keukenhof-sluit-succesvol-virtueel-seizoen/

Khan, S. & M., F. M. (2022). Indian Tourism and Hospitality, the FTAs, Domestic Tourism, GDP and Tourism Promotion. In Chowdhary, N., (Ed.), *Indian Tourism, Diaspora Perspectives* (pp. 3-19). Emerald Publishing Limited. doi:10.1108/978-1-80262-937-820221001

Kikuchi, T., & Koo, W. W. (2017). Mega sporting events and tourism: Brazil's image and the 2014 FIFA World Cup. *Journal of Travel Research, 56*(3), 292–303.

Kılıçhan, R., & Köşker, H. (2020). Participation Motivation of Visitors of Local Food-Themed Gastronomy Festival. *Journal of Tourism and Gastronomy Studies, 8*(1), 78–91. doi:10.21325/jotags.2020.537

Kim, B., Kim, S., & King, B. (2020). Religious tourism studies: Evolution, progress, and future prospects. *Tourism Recreation Research, 45*(2), 185–203. doi:10.1080/02508281.2019.1664084

Kim, H., Borges, M. C., & Chon, J. (2006). Impacts of environmental values on tourism motivation: The case of FICA, Brazil. *Tourism Management, 27*(5), 957–967. doi:10.1016/j.tourman.2005.09.007 PMID:32287716

Kim, K., Uysal, M., & Chen, J. S. (2001). Festival visitor motivation from the organizers' points of view. *Event Management, 7*(2), 127–134. doi:10.3727/152599501108751533

Kim, M., Chung, N., & Lee, C. (2011). The effect of perceived trust on electronic commerce: Shopping online for tourism products and services in South Korea. *Tourism Management, 32*(2), 256–265. doi:10.1016/j.tourman.2010.01.011

Kim, S. S., Choe, J. Y. J., & Petrick, J. F. (2018). The effect of celebrity on brand awareness, perceived quality, brand image, brand loyalty, and destination attachment to a literary festival. *Journal of Destination Marketing & Management, 9*, 320–329. doi:10.1016/j.jdmm.2018.03.006

Kim, S. S., & Morrsion, A. M. (2005). Change of images of South Korea among foreigntourists after the 2002 FIFA World Cup. *Tourism Management, 26*(2), 233–247. doi:10.1016/j.tourman.2003.11.003

Kim, S., Moon, J., & Choe, J. (2016). Comparison of destination brand equity models of competitive convention cities in East Asia. *Journal of Convention & Event Tourism, 17*(4), 318–342. doi:10.1080/15470148.2016.1151394

Kim, S., Savinovic, A., & Brown, S. (2013). Visitors' Motivations in Attending an Ethnic Minority Cultural Festival: A Case Study of the Feŝta Croatian Food and Wine Festival, South Australia. *Event Management, 17*(4), 349–359. doi:10.3727/152599513X13769392444828

Kim, Y. G., Eves, A., & Scarles, C. (2009). Building a Model of Local Food Consumption on Trips and Holidays: A Grounded Theory Approach. *International Journal of Hospitality Management, 28*(3), 423–431. doi:10.1016/j.ijhm.2008.11.005

Kim, Y. G., Eves, A., & Scarles, C. (2013). Empirical Verification of a Conceptual Model of Local Consumption at a Tourist Destination. *International Journal of Hospitality Management, 33*, 484–489. doi:10.1016/j.ijhm.2012.06.005

Kim, Y. H., Goh, B. K., & Yuan, J. (2010). Development of a Multi-Dimensional Scale for Measuring Food Tourist Motivations. *Journal of Quality Assurance in Hospitality & Tourism, 11*(1), 56–71. doi:10.1080/15280080903520568

Kim, Y., Suh, B., & Eves, A. (2010). The relationships between food-related personality traits, satisfaction, and loyalty among visitors attending food events and festivals. *International Journal of Hospitality Management, 29*(2), 216–226. doi:10.1016/j.ijhm.2009.10.015

Kitterlin, M., & Yoo, M. (2014). Festival motivation and loyalty factors. *Tourism & Management Studies, 10*(1), 119–126.

Kivela, J., & Crotts, J. C. (2005). Gastronomy Tourism: A Meaningful Travel Market Segment. *Journal of Culinary Science & Technology, 4*(2/3), 39–55. doi:10.1300/J385v04n02_03

KIZILIRMAK, İ. (2006). Türkiye'de düzenlenen yerel etkinliklerin turistik çekicilik olarak kullanilmasina yönelik bir inceleme. [An investigation on the use of local events organized in Turkey as a tourist attraction.]. *Manas Üniversitesi Sosyal Bilimler Dergisi, 8*(15), 181–196.

Knott, B., Fyall, A., & Jones, I. (2015). The nation branding opportunities provided by a sport mega-event: South Africa and the 2010 FIFA World Cup. *Journal of Destination Marketing & Management, 4*(1), 46–56. doi:10.1016/j.jdmm.2014.09.001

Knott, B., Fyall, A., & Jones, I. (2017). Sport mega-events and nation branding: Unique characteristics of the 2010 FIFA World Cup, South Africa. *International Journal of Contemporary Hospitality Management, 29*(3), 900–923. doi:10.1108/IJCHM-09-2015-0523

Knott, B., & Hemmonsbey, J. (2015). The strategic value of sport for an African city brand. *African Journal for Physical, Health Education, Recreation, and Dance, 21*, 191–204.

Kocaman, S. (2012). *Destinasyon yönetimi kapsamında marka kimliğine etki eden faktörlerin marka imajına etkisi: Alanya Örneği.* Yayınlanmamış Doktora Tezi, Akdeniz Üniversitesi.

Koç, E. (2012). *Tüketici Davranışı ve Pazarlama Stratejileri Global ve Yerel Yaklaşım.* Seçkin Yayıncılık.

Kockel, U., Craith, M. N., Clopot, C., & Tjarve, B. (2019). Heritages, identities and Europe: Exploring cultural forms and expressions. In *Heritage and Festivals in Europe*. Taylor & Francis. doi:10.4324/9780429202964-1

Kodaş, B. (2021). Web Revolution and Events: Development and Progress, Birdir, K, Birdir, S, Dalgıç, A, and Toksöz, D. (Eds), In Impact of ICTs on Event Management and Marketing (pp. 18-32). IGI Global

Kodaş, D. (2018). *The relationship between gastronomy experience, gastronomy motivation, destination satisfaction and destination brand equity: case of Gaziantep.* [Doctoral dissertation, Anadolu University, Institute of Social Sciences, Eskisehir.]

Kodaş, D. (2021). Kodaş, D. (2021). Understanding Food Consumption Motivations: An Interpretivist Approach Using User-Generated Content. In Cihan Cobanoglu, Ebru Gunlu Kucukaltan, Muharrem Tuna, Alaattin Basoda & Seden Dogan (Eds.), Advances in Managing Tourism Across Continents (pp. 623–636). USA: USF M3 Publishing.

Kodaş, D. (2021a). Festival Turizmi ve Festival Turizmi Motivasyonları. Tören, E. (Ed) Kültür Turizmi Motivasyonları. (pp. 119-147). Gazi Kitabevi: Ankara.

Kodaş, B. (2021). Web Revolution and Events: Development and Progress. In K. Birdir, S. Birdir, A. Dalgıç, & D. Toksöz (Eds.), *Impact of ICTs on Event Management and Marketing* (pp. 18–32). IGI Global. doi:10.4018/978-1-7998-4954-4.ch002

Kodaş, B., & Aksoy, Z. (2022). Understanding of tourists' memorable local food experiences: A netnography study. *Journal of Economy Culture and Society, 66*, 379–396.

Kodaş, B., & Sekizkardeş, M. (2022). Mardin'e Gelen Yerli Ziyaretçilerin Gastronomi Deneyimlerinin Demografik. Özelliklere Göre İncelenmesi. *Journal of Tourism and Gastronomy Studies, 10*(4), 2862–2880.

Kodaş, D., & Özel, Ç. H. (2022). (Article in Press). Antecedents of Gastronomy Destination Brand Equity: An Examination of Gastronomy Experience, Motivation, and Destination Satisfaction. [AHTR]. *Advances in Hospitality and Tourism Research.*

Koetsier, J. (2021, June 30). Virtual events up 1000% since COVID-19, with 52,000 on just one platform. *Forbes.* www.forbes.com/sites/johnkoetsier/2020/05/27/virtual-events-up-1000-with-52000-on-just-one-platform/?sh=29792fa7a233

Köksal, Y., & Özdemir, Ş. (2013). Bir İletişim Aracı Olarak Sosyal Medya'nın Tutundurma Karması İçerisindeki Yeri Üzerine Bir İnceleme. *Süleyman Demirel Üniversitesi İktisadi ve İdari Bilimler Fakültesi Dergisi, 18*(1), 323–337.

Kolar, T., & Zabkar, V. (2010). A consumer-based model of authenticity: An oxymoron or the foundation of cultural heritage marketing? *Tourism Management, 31*(5), 652–664. doi:10.1016/j.tourman.2009.07.010

Konecnik, M., & Gartner, W. C. (2007). Customer-based brand equity for a destination. *Annals of Tourism Research, 34*(2), 400–421. doi:10.1016/j.annals.2006.10.005

Koo, J., & Lee, Y. (2018). Sponsor-event congruence effects: The moderating role of sport involvement and mediating role of sponsor attitudes. *Sport Management Review.*

Korkmaz, M. (2001). Orman kaynaklarına doğa turizmi etkinliklerinin ekonomik çözümlemeleri. *Turkish Journal of Forestry*, *2*(1), 111–134.

Korstanje, M. E. (2009). Reconsidering the roots of event management: Leisure in ancient Rome. *Event Management*, *13*(3), 197–203. doi:10.3727/152599509790029783

Ko, S., Kang, S., Kang, H., & Lee, M. J. (2018). An Exploration of Foreign Tourists' Perceptions of Korean Food Tour: A Factor-Cluster Segmentation Approach. *Asia Pacific Journal of Tourism Research*, *23*(8), 833–846. doi:10.1080/10941665.2018.1494613

Kotler, P., & Bowen, J. T. ve Makens, J.C. (2010). Marketing for Hospitality and Tourism (5th edition). Pearson.

Kotler, P., & Gertner, D. (2002). Country as brand, product, and beyond: A place marketing and brand management perspective. *Journal of Brand Management*, *9*(4), 249–261. doi:10.1057/palgrave.bm.2540076

Ko, Y. J., Kim, K., Claussen, C. L., & Kim, T. H. (2008). The effects of sport involvement: Sponsor awareness and corporate image on intention to purchase sponsors' products. *International Journal of Sports Marketing & Sponsorship*, *9*(2), 79–94. doi:10.1108/IJSMS-09-02-2008-B004

Kozak, M. (2001). Repeaters' behavior at two distinct destinations. *Annals of Tourism Research*, *28*(3), 784–807. doi:10.1016/S0160-7383(00)00078-5

Kozak, M., & Rimmington, M. (2000). Tourist satisfaction with Mallorca, Spain, as an off-season holiday destination. *Journal of Travel Research*, *38*(3), 260–269. doi:10.1177/004728750003800308

Krajíčková, A., & Šauer, M. (2018). Differences in motivation of food festivals visitors: A view from the Czech Republic. *Geographica Pannonica*, *22*(3), 189–200. doi:10.5937/gp22-17050

Krakhmal, K. (2023). To The Study Of The Origins Of The Origin Of Local Civilizations In The Length Of The Syrdarya River. *Galaxy International Interdisciplinary Research Journal*, *11*(2), 208–218.

Kruger, M., & Viljoen, A. (2021). Terroir wine festival visitors: Uncorking the origin of behavioural intentions. *Current Issues in Tourism*, *24*(5), 1–21. doi:10.1080/13683500.2019.1667310

KTVG. (2014) *Karnataka Tourism Vision Group Recommendations to the Government of Karnataka: January 2014.* Kreto. https://www.kreto.in/images/Vision_Group.pdf

Kubickova, M., & Martin, D. (2020). Exploring the relationship between government and destination competitiveness: The TALC model perspective. *Tourism Management*, *78*, 104040. doi:10.1016/j.tourman.2019.104040

Kulkarni, S., & Bhopatkar, A. (2015). *The Impact of Festivals in promoting cultural tourism: A case study of Ganesh Utsav in Pune City, Maharashtra.* RTD. http://rtd7.org/wp- content/uploads/2013/09/Alka-Bhopatkar-paper.pdf

Kumar, S. (2022). Personal Communication [Discussion]

Kumar, S., & Sadarangani, H. P. (2018). A Narrative Inquiry of Consumption. Shailendra Pratap Jain and Akshaya Vijayalakshmi, (eds) ACR Asia-Pacific Advances in Consumer Research 12, 79-79. Duluth, MN: Association for Consumer Research.

Kumar, D. A. (2021). *Socio-Economic Benefits from Indian Winery.* International Journal of Engineering and Management Research. doi:10.31033/ijemr.11.4.3

Kumar, P., Mishra, J. M., & Kumar, G. S., & R, R. (2020). Destination marketing & management organizations in India. *Solid State Technology*, *63*(6).

Kumar, S. (2017). A Conceptual Paper on Dark Social–Social Cognitive and Social Capital Theory Perspective. *International Journal of Advanced Studies in Computer Science and Engineering*, *6*(9), 1–11.

Kumar, S., Baishya, K., Sadarangani, P. H., & Samalia, H. V. (2020). Cultural Influence on e-Government Development. *Electronic Journal of Information Systems Evaluation*, *23*(1), 17–33. doi:10.34190/EJISE.20.23.1.002

Kumar, S., Baishya, K., Sreen, N., Sadarangani, P. H., & Samalia, H. V. (2021). Impact of national culture on E-government development: A longitudinal study. [JGIM]. *Journal of Global Information Management*, *29*(2), 1–22. doi:10.4018/JGIM.2021030101

Kumar, S., & Dhir, A. (2020). Associations between travel and tourism competitiveness and culture. *Journal of Destination Marketing & Management*, *18*, 100501. doi:10.1016/j.jdmm.2020.100501

Kumar, S., Giridhar, V., & Sadarangani, P. (2019). A cross-national study of environmental performance and culture: Implications of the findings and strategies. *Global Business Review*, *20*(4), 1051–1068. doi:10.1177/0972150919845260

Kumar, S., Giridhar, V., Sadarangani, P. H., & Gogoi, B. J. (2018). A Cross-Cultural Study on Tourism: Strategies and Implications of Findings. In *Global Conference on Business and Economics (GLOBE 2018)* (p. 5).

Kumar, S., & Sadarangani, P. (2021). An empirical study on shopping motivation among generation Y Indian. *Global Business Review*, *22*(2), 500–516. doi:10.1177/0972150918807085

Kumar, S., & Sadarangani, P. H. (2018). Study of shopping motivation and buying behaviour among generation Y in India. *ISSN*, *7832*(2362), 341.

Kundra, S., Kumar, S., Sreen, N., & Dwivedi, R. (2020). A qualitative inquiry on determinants of employee satisfaction: Evidence from it firms. [IJM]. *International Journal of Management*, *11*(8), 1207–1225.

Lagos, K., & Wang, Y. (2022). International tourism and poverty alleviation: Cross-country evidence using panel quantile fixed effects approach. *Journal of Travel Research*, 00472875221119978. doi:10.1177/00472875221119978

Lang, B. (2022, August 30). Sundance will return in-person in 2023 with later Digital Component, unveils Covid Safety Measures. *Variety*. www.variety.com/2022/film/news/sundance-dates-2023-digital-covid-safety-measures-1235354472/

Lattanzio, R. (2020, March 14). *Cannes won't decide whether to cancel the festival until April 15 as France shutters theaters*. IndieWire. www.indiewire.com/2020/03/cannes-coronavirus-cancellation-decision-1202217815/

Laurell, C., & Björner, E. (2018). Digital festival engagement: On the interplay between festivals, place brands, and social media. *Event Management*, *22*(4), 527–540. doi:10.3727/152599518X15300559276930

Law, R., Buhalis, D., & Cobanoglu, C. (2014). Progress on information and communication technologies in hospitality and tourism. *International Journal of Contemporary Hospitality Management*, *26*(5), 727–750. doi:10.1108/IJCHM-08-2013-0367

Lee, I., & Lee, T. J. (2009). The success factors of using events in destination branding: A case study of Woodford Folk Festival. Paper presented at the *8th Asian Pacific forum for graduate students research in tourism,* Seoul, Korea.

Lee, J. W. (2015). Unpacking brand management superiority: Examining the Interplay among brand orientation and market-linking and controlling mechanisms. [Master Thesis, Tasmanian School of Business and Economics].

Lee, C. K., Lee, Y. K., & Wicks, B. E. (2004). Segmentation of festival motivation by nationality and satisfaction. *Tourism Management*, *25*(1), 61–70. doi:10.1016/S0261-5177(03)00060-8

Lee, C. S., Martin, D., Hsieh, P. F., & Yu, W. C. (2020). Principles of value creation in event tourism: Enhancing the competitiveness of regional clusters. *Journal of Global Scholars of Marketing Science, 30*(4), 437–453. doi:10.1080/21639159.2020.1784771

Lee, C.-C., Zeng, M., & Luo, K. (2022). The Impact of Green Development on Modernization in China: Evidence from 108 Cities in the Yangtze River Economic Belt. *Emerging Markets Finance & Trade, 58*(13), 3664–3688. doi:10.1080/1540496X.2022.2067475

Lee, C.-K., Lee, Y.-K., & Lee, B. (2005). Korea's destination image formed by the 2002 World Cup. *Annals of Tourism Research, 32*(4), 839–858. doi:10.1016/j.annals.2004.11.006

Lee, I., & Arcodia, C. (2011). The role of regional food festivals for destination branding. *International Journal of Tourism Research, 13*(4), 355–367. doi:10.1002/jtr.852

Leenders, M. A. A. M. (2010). The relative importance of the brand of music festivals: A customer equity perspective. *Journal of Strategic Marketing, 18*(4), 291–301. doi:10.1080/09652541003768061

Lee, T. H., & Jan, F. H. (2022). How does tourism image affect visitor perceptions of a festival's identity? Influence analysis of three aboriginal festivals in Taiwan. *Journal of Destination Marketing & Management, 24*, 100704. doi:10.1016/j.jdmm.2022.100704

Lee, W., Sung, H., Suh, E., & Zhao, J. (2017). The effects of festival attendees' experiential values and satisfaction on re-visit intention to the destination: The case of a food and wine festival. *International Journal of Contemporary Hospitality Management, 29*(3), 1005–1027. doi:10.1108/IJCHM-10-2015-0559

Lee, W., Xiong, L., & Hu, C. (2012). The effect of Facebook users' arousal and valence on intention to go to the festival: Applying an extension of the technology acceptance model. *International Journal of Hospitality Management, 31*(3), 819–827. doi:10.1016/j.ijhm.2011.09.018

Lee, Y. K., Lee, C. K., Lee, S. K., & Babin, B. J. (2008). Festivalscapes and patrons' emotions, satisfaction, and loyalty. *Journal of Business Research, 61*(1), 56–64. doi:10.1016/j.jbusres.2006.05.009

Lee, Y.-J. (2015). Creating Memorable Experiences in a Reuse Heritage Site. *Annals of Tourism Research, 55*, 155–170. doi:10.1016/j.annals.2015.09.009

Leiesen, B. (2000). Image sementation: The case of a tourism destination. *Journal of Services Marketing, 15*(9), 51.

Lepak, D. P., Smith, K. G., & Taylor, M. S. (2007). Value creation and value capture: A multilevel perspective. *Academy of Management Review, 32*(1), 180–194. doi:.2007.23464011 doi:10.5465/amr

Leung, D., Law, R., Van Hoof, H., & Buhalis, D. (2013). Social media in tourism and hospitality: A literature review. *Journal of Travel & Tourism Marketing, 30*(1-2), 3–22. doi:10.1080/10548408.2013.750919

Levickaite, R. (2010). Generations X, Y, Z: How social networks form the concept of the world without borders (the case of Lithuania). *LIMES: Cultural Regionalistics, 3*(2), 170–183.

Li, J., Dai, G., Tang, J., & Chen, Y. (2020). Conceptualizing Festival Attractiveness and Its Impact on Festival Hosting Destination Loyalty: A Mixed Method Approach. *Sustainability, 12*(8), 3082. doi:10.3390/su12083082

Li, M., Huang, Z., & Cai, L. A. (2009). Benefit segmentation of visitors to a rural community-based festival. *Journal of Travel & Tourism Marketing, 26*(5-6), 585–598. doi:10.1080/10548400903163152

Linderová, I., Scholz, P., & Almeida, N. (2021). Attitudes of Local Population Towards the Impacts of Tourism Development: Evidence From Czechia. *Frontiers in Psychology*, *12*, 1-11. Retrieved from https://www.frontiersin.org/articles/10.3389/fpsyg.2021.684773/full doi:10.3389/fpsyg.2021.684773

Litvin, S. W., & Fetter, E. (2006). Can a festival be too successful? A review of Spoleto, USA. *International Journal of Contemporary Hospitality Management*, *18*(1), 41–49. doi:10.1108/09596110610641966

Liu, C. H. (2020). Integration of different perspectives of culinary brand equity. *Journal of Hospitality and Tourism Management*, *45*, 152–161. doi:10.1016/j.jhtm.2020.08.007

Li, X. R., & Vogelsong, H. (2006). Comparing methods of measuring image change: A case study of a small-scale community festival. *Tourism Analysis*, *10*(4), 349–360. doi:10.3727/108354206776162769

Li, X., & Petrick, J. F. (2005). A review of festival and event motivation studies. *Event Management*, *9*(4), 239–245. doi:10.3727/152599506776771526

Li, X., Wang, D., & Li, Y. (2019). Exploring the effects of destination branding on tourist revisit intention: The role of emotional attachment. *Journal of Destination Marketing & Management*, *12*, 29–39. doi:10.1016/j.jdmm.2019.01.002

Lohmann, M. (2004). New demand factors in tourism. *European Tourism Forum Budapest/Hungary*, 10.

Longkumer, A. (2016). *Visualising national life, the Hornbill festival as culture and politics.* doi:10.23912/978-1-910158-55-5-3012

Lopez, C., & Leenders, M. A. A. M. (2018). Building a local identity through sellout crowds: The impact of brand popularity, brand similarity, and brand diversity of music festivals. *Journal of Strategic Marketing*, 1–16.

López-Guzmán, T., Lotero, C. P. U., Gálvez, J. C. P., & Rivera, I. R. (2017). Gastronomic festivals: Attitude, motivation and satisfaction of the tourist. *British Food Journal*, *119*(2), 267–283. doi:10.1108/BFJ-06-2016-0246

Lorenzini, E., Calzati, V., & Giudici, P. (2011). Territorial brands for tourism development. *Annals of Tourism Research*, *38*(2), 540–560. doi:10.1016/j.annals.2010.10.008

Lu, J., & Nepal, S. K. (2009). Sustainable tourism research: An analysis of papers published in the Journal of Sustainable Tourism. *Journal of Sustainable Tourism*, *17*(1), 5–16. doi:10.1080/09669580802582480

Lu, L., Chi, C. G., & Liu, Y. (2015). Authenticity, involvement, and image: Evaluating tourist experiences at historic districts. *Tourism Management*, *50*, 85–96. doi:10.1016/j.tourman.2015.01.026

MacCannell, D. (1973). Staged authenticity: Arrangements of social space in tourist settings. *American Journal of Sociology*, *79*(3), 589–603. doi:10.1086/225585

MacCannell, D. (1976). *The tourist: A new theory of the leisure class.* Universidade da California Press.

MacCannell, D. (1977). *The Tourist.* Schockon.

Madhav, B., & Thakare, M. L. (2015). *A perspective study of wine tourism in Nashik district.* Journal of contemporary research.

Mai, N. (2021). Increase tourism value through value chain concept. *Destination: Business and Management Review.* https://destination-review.com/en/increase-tourism-value-through-value-chain-concept/

Mair, J., & Weber, K. (2019). Event and festival research: A review and research directions. *International Journal of Event and Festival Management, 10*(3), 209–216. doi:10.1108/IJEFM-10-2019-080

Ma, L., & Lew, A. A. (2012). Historical and geographical context in festival tourism development. *Journal of Heritage Tourism, 7*(1), 13–31. doi:10.1080/1743873X.2011.611595

Malär, L., Krohmer, H., Hoyer, W. D., & Nyffenegger, B. (2011). Emotional brand attachment and brand personality: The relative importance of the actual and ideal selves. *Journal of Marketing, 75*(4), 35–52. doi:10.1509/jmkg.75.4.35

Mandariš, M. (2016). Strategijski brend menadžment [Strategic Brand Management]. Vrnjaţka Banja: University of Kragujevac, Faculty of Hotel Management and Tourism in Vrnjaţka Banja.

Mandariš, M., Miliševiš, S., & Sekuliš, D. (2017). Traditional values in the function of promotion of Šumadija and Pomoravlje as rural tourism destinations. *Economics of Agriculture, 64*(2), 787–803.

Mangold, W. G., & Faulds, D. J. (2009). Social media: The new hybrid element of the promotion mix. *Business Horizons, 52*(4), 357–365. doi:10.1016/j.bushor.2009.03.002

Manhas, P. S., Manrai, L. A., & Manrai, A. K. (2016). Role of tourist destination development in building its brand image: A conceptual model. *Journal of Economics, Finance and Administrative Science, 21*(40), 25–29. doi:10.1016/j.jefas.2016.01.001

Marcussen, C. H. (2011). Determinants of tourist spending in cross-sectional studies and at Danish destinations. *Tourism Economics, 17*(4), 833–855. doi:10.5367/te.2011.0068

Mardi Gras for families: New Orleans. (n.d.). www.neworleans.com/events/holidays-seasonal/mardi-gras/mardi-gras-for-families/

Marrocu, E., Paci, R., & Zara, A. (2015). Micro-economic determinants of tourist expenditure: A quantile regression approach. *Tourism Management, 50*, 13–30. doi:10.1016/j.tourman.2015.01.006

Marzano, G., & Scott, N. R. (2006). Consistency in Destination Branding: The Impact of Events. *Global Events Congress and Event Educators Forum*, Brisbane, Australia.

Masiello, B., Bonetti, E., & Izzo, F. (2020). Multiple identities of a festival: Intended, communicated and perceived brand personality in the social media environment. *International Journal of Contemporary Hospitality Management, 32*(2), 749–768. doi:10.1108/IJCHM-11-2018-0937

Mason, M. C., & Paggiaro, A. (2012). Investigating the role of festivalscape in culinary tourism: The case of food and wine events. *Tourism Management, 33*(6), 1329–1336. doi:10.1016/j.tourman.2011.12.016

Mastan (2022). Personal Communication [Discussion].

Mathur, A., Moschis, G. P., & Lee, E. (2003). Life events and brand preference changes. *Journal of Consumer Behaviour, 3*(2), 2. doi:10.1002/cb.128

Matzler, K. E., Fuller, J., & Mooradian, T. A. (2011). Personality, Person-Brand Fit, and Brand Community: An Investigation of Individuals, Brands, and Brand Communities. *Journal of Marketing Management, 27*(9-10), 874–890. doi:10.1080/0267257X.2010.543634

Mazanek, J. (2010). Managing the heterogeneity of city tourists. In J. Manzanek & K. Wöber (Eds.), *Analysing International City Tourism*. Springler Verlag. doi:10.1007/978-3-211-09416-7_6

Mbaiwa, J. E. (2003). The socio-economic and environmental impacts of tourism development on the Okavango Delta, north-western Botswana. *Journal of Arid Environments, 54*(2), 447–467. doi:10.1006/jare.2002.1101

Mccartney, G., & Osti, L. (2007). From Cultural Events to Sport Events: A Case Study of Cultural Authenticity in the Dragon Boat Races. *Journal of Sport & Tourism, 12*(1), 25–40. doi:10.1080/14775080701496750

McClure, J., Wu, T., & Scarr, S. (2020, July 1). New normal: How far is safe enough? *Reuters*. www.graphics.reuters.com/HEALTH-CORONAVIRUS/SOCIALDISTANCING /qzjvqenmyvx/

Mcintosh, S. (2019). How Event Diagramming Is Shaping The Event World. *Purple Pass*. https://www.purplepass.com/blog/how-event-diagramming-is-shaping-the-event-world/

McKercher, B., & So-Ming, B. C. (2001). Cultural Distance and Participation in Cultural Tourism. *Pacific Tourism Review, 5*(1), 23–32.

Medlik, S. (2003). Dictionary of travel. Tourism and Hospitality (3. Baskı). Oxford: Butterworth-Heinemann Publications.

Melnick, K. (2019). *Coachella Debuts Its First Interactive AR Stage*. VR Scout. https://vrscout.com/news/coachella-ar-interactive-stage/

Mende, M., Bolton, R. N., & Bitner, M. J. (2013). Decoding customer-firm relationships: How attachment styles help explain customers' preferences for closeness, repurchase intentions, and changes in relationship breadth. *JMR, Journal of Marketing Research, 50*(1), 125–142. doi:10.1509/jmr.10.0072

Merriam, S. B. (2009). *Qualitative Research: A Guide to Design and Implementation*. Jossey-Bass.

Miles, M. B., & Huberman, A. M. (2016). *Genişletilmiş Bir Kaynak Kitap: Nitel Veri Analizi* (S. Akbaba Altun & A. Ersoy, Trans.). Pegem Akademi.

Miller, G. A. (2003). Consumerism in sustainable tourism: A survey of UK consumers. *Journal of Sustainable Tourism, 11*(1), 17–39. doi:10.1080/09669580308667191

Ministère de l'Europe et des Affaires étrangères. (2021, November). *Tourism*. France Diplomacy - Ministry for Europe and Foreign Affairs. www.diplomatie.gouv.fr/en/french-foreign-policy/tourism/

Ministry of Heritage and Tourism. (2022, October 25). *The World Cup is Coming to Oman with the Football Fan Festival*. Retrieved from https://mht.gov.om/ar/news/82/THE-WORLD-CUP-IS-COMING-TO-OMAN-WITH-THE-FOOTBALL-FAN-FESTIVAL

Mishra, A., & Kukreti, M. (2021). *Challenges and prospects for Oman in the making of luxury tourism destination*. Routledge.

Mitchell, R., & Hall, C. M. (2004). Consuming tourists: Food tourism consumer behaviour. In *Food tourism around the world* (pp. 72–92). Routledge.

Mkono, M., Markwell, K., & Wilson, E. (2013). Applying Quan and Wang's structural model of the tourist experience: A Zimbabwean netnography of food tourism. *Tourism Management Perspectives, 5*, 68–74. doi:10.1016/j.tmp.2012.10.007

Mohammadi, A., & Jazaei, M. (2016). Reviewing the sport tourism, push and pull factors in Zanjar province, Iran. *International Journal of Research in Social Sciences and Humanities, 6*(3), 2249–4671.

Mohanty, P. (2022). *Managing Events and Festivals Post-COVID-19: Maintaining the Status Quo or Resetting for the Sustainability Agenda*. CABI International. doi:10.1079/9781789248685.0002

Moise, D., & Cruceru, A. F. (2014). An empirical study of promoting different kinds of events through various social media networks websites. *Procedia: Social and Behavioral Sciences, 109*, 98–102. doi:10.1016/j.sbspro.2013.12.426

Moore, S. (2020). *Glastonbury's Shangri-La team launch new VR festival Lost Horizon*. NME. https://www.nme.com/news/music/glastonbury-shangri-la-team-launch-new-vr-festival-lost-horizon-2683835

Morgan, N. J., Pritchard, A., & Pride, R. (2010). Destination branding. 2ndrev. (Ed.), Oxford: Elsevier.

Morgan, M. (2008). What Makes a Good Festival? Understanding the Event Experience. *Event Management, 12*(2), 81–93. doi:10.3727/152599509787992562

Morgan, M., Elbe, J., & Curiel, E. (2009). Has the experience economy arrived? The views of destination managers in three visitor-dependent areas. *International Journal of Tourism Research, 11*(2), 201–216. doi:10.1002/jtr.719

Morgan, N., Pritchard, A., & Pride, R. (2004). *Destination branding: Creating the unique destination proposition*. Elsevier Butterworth-Heinemann.

Morgan, R., & Hunt, S. (1994). The Commitment-Trust Theory of Relationship Marketing. *Journal of Marketing, 58*(3), 20–38. doi:10.1177/002224299405800302

Morrison, A. M., & Anderson, D. J. (2002). Destination branding. Paper presented at the *Missouri Association of Convention & Visitor Bureaus Annual Meeting*. MAC.

Morrison, A. (1998). Small firm statistics: A hotel sector focus. *Service Industries Journal, 18*(1), 132–142. doi:10.1080/02642069800000008

Morrison, A. (2013). *Marketing and managing tourism destinations*. Routledge. doi:10.4324/9780203081976

Mossberg, L., & Getz, D. (2006). Stakeholder influences on the ownership and management of festival brands. *Scandinavian Journal of Hospitality and Tourism, 6*(4), 308–326. doi:10.1080/15022250601003273

Moyo, D. (2009). *Dead aid: Why aid is not working and how there is a better way for Africa*. Macmillan.

Ms. Hiral Sheth, M. A. (2014). Wine Festival and Wine Consumerism - A Perspective Study on Influence and awareness in the Growth and Acceptance of Bangalore. *International Journal of Engineering Technology, Management and Applied Sciences, 2*.

Myagmarsuren, O., & Chen, C. F. (2011). Exploring relationships between destination brand equity, satisfaction, and destination loyalty: A case study of Mongolia. *Journal of Tourism. Hospitality & Culinary Arts, 3*(2), 81–94.

Nair, V. (2022, December 28). The aviation sector in the Sultanate of Oman benefited from the FIFA Football World Cup. *Oman Observer*. https://www.omanobserver.om/article/1130231/oman/oman-benefits-from-fifa-world-cup-travel-iata

Nam, J., Ekincy, Y., & Whyatt, G. (2011). Brand equity, brand loyalty, and customer satisfaction. *Annals of Tourism Research, 38*(3), 1009–1030. doi:10.1016/j.annals.2011.01.015

Naramski, M., Szromek, A. R., Herman, K., & Polok, G. (2022). Assessment of the Activities of European Cultural Heritage Tourism Sites during the COVID-19 Pandemic. *Journal of Open Innovation, 8*(1), 55. doi:10.3390/joitmc8010055

Narmadha, V., & Anuradha, A. (2022). Paradigm change in tourist preferences towards evolving new normal tourism and travel trends. In Handbook of Research on Changing Dynamics in Responsible and Sustainable Business in the Post-COVID-19 Era. IGI Global. doi:10.4018/978-1-6684-2523-7.ch019

National Centre for Statistics and Information. (2020). *Tourism in Figures 2019*. Retrieved from https://data.gov.om/OMN-2020/tourism-in-figures-2019

National Centre for Statistics and Information. (2021). *Tourism in Oman*. Retrieved from https://www.ncsi.gov.om/Elibrary/Tourism%20in%20Oman%202020%20.pdf

National Park Service. (2021). *History of Cherry trees*. NPS. https://www.nps.gov/subjects/cherryblossom/history-of-the-cherry-trees.htm

Nergusa, A. L., & Coros, M. M. (Eds.). (2016). Destination management organizations (DMO's) roles and performance – literature review. In *Proceedings of the 2nd International Scientific Conference SAMRO 2016 – News, Challenges and Trends in Management of Knowledge-based Organizations*. Editura Economica.

New Orleans & Company. (n.d.). *Late night itinerary: New Orleans*. www.neworleans.com/plan/itineraries/late-night/

News Nine. (2022). Bengaluru Karaga Festival: procession to visit Mastan Saab Dargah amid objections by Hindutva outfits. *News Nine*. https://www.news9live.com/state/karnataka/bengaluru-karaga-festival-procession-to-visit-mastan-saab-dargah-amid-objection-by-hindutva-outfits-163427?infinitescroll=1

Nezakati, H., Amidi, A., Jusoh, Y. Y., Moghadas, S., Aziz, Y. A., & Sohrabinezhadtalemi, R. (2015). Review of social media potential on knowledge sharing and collaboration in tourism industry. *Procedia: Social and Behavioral Sciences*, *172*, 120–125. doi:10.1016/j.sbspro.2015.01.344

NFCdirect. (2019). *What are the uses of NFC at festivals?* NRF Direct. https://www.NFCdirect.co.uk/blog/what-are-the-uses-of-nfc-at-festivals.htm

Ngernyuang, K., & Wu, P. Y. (2020). Using social media as a tool for promoting festival tourism. *AIRCC's. International Journal of Computer Science and Information Technologies*, *12*(3), 17–32. doi:10.5121/ijcsit.2020.12302

Nguyen, D. T., & Vu, H. T. (2022). Measuring Attitudes Toward Sponsor And Purchase Intention. *Cogent Business & Management*, *9*(1), 2049961. doi:10.1080/23311975.2022.2049961

Nicholson, R., & Pearce, D. G. (2001). Why do people attend events: A comparative analysis of visitor motivations at four South Island events. *Journal of Travel Research*, *39*(4), 449–460. doi:10.1177/004728750103900412

Nikolaeva, J., Bogoliubova, N., Fokin, V., Baryshnikov, V., Klevtsov, P., & Eltc, E. (2017). Cultural mega-events as an international, cultural and political tool. *Codrul Cosminului*, *23*(2), 291–304.

Obermiller, C. (1985). Varieties of Mere Exposure: The Effects of Processing Style on Affective Response. *NA: Journal of Consumer Research*, *12*(June), 17–30. doi:10.1086/209032

Observer, O. (2022, December 15). *Oman are participating at the Camel and World Cup 2022 activities*. Retrieved from https://www.omanobserver.om/article/1129800/sports/oman-taking-part-at-camel-and-world-cup-2022-event

Oestergaard, F. (2019, July 19). How Data Rocks at Smukfest 2019. *IBM Nordic Blog*. Www.Ibm.Com. https://www.ibm.com/blogs/nordic-msp/smukfest-2019/

Oğan, O., & Yasak, Ü. (2020). Küreselleşme bağlamında mekansal kent kimliği ve markalaşmanın kent turizmine etkisi. *Türk Coğrafya Dergisi*, (74), 97–105.

Ohajionu, U. C., & Mathews, S. (2016). Social media and advertising: An exploratory study on the lifestyle of generation Y consumers in Malaysia. *TIJ's Research Journal of Social Science & Management*, *6*(1), 31–39.

Oklobdžija, S. (2015). The role of events in tourism development. *Bizinfo (Blace), 6*(2), 83–97. doi:10.5937/BIZINFO1502083O

Olimpia, B. (2007). Variables of the image of tourist destination. *Annals of the University of Oradea. Economic Science Series, 7*(2), 559–564.

Oliver, R. L. (1977). Effect of Expectation and Disconfirmation on Postexposure Product Evaluations: An Alternative Interpretation. *The Journal of Applied Psychology, 62*(4), 480–486. doi:10.1037/0021-9010.62.4.480

Oliver, R. L. (1997). *Satisfaction: A behavioral perspective on the consumer.* McGrawHill.

Oman Ministry of Tourism. (2020). *Tourism Strategy 2040.* Retrieved from https://www.omantourism.gov.om/wps/portal/mot/tourism/home/about-us/our-strategy/tourism-strategy-2040/

Oman News Agency. (2022, September 7). *Oman unveils details of joint programme for FIFA World Cup Qatar 2022.* Retrieved from https://www.omannews.gov.om/NewsDescription/ArtMID/392/ArticleID/74803/Oman-unveils-details-of-joint-programme-for-FIFA-World-Cup-Qatar-2022

Ooi, C. S., & Pedersen, J. (2010). City branding and film festivals: Re-evaluating stakeholder's relations. *Place Branding and Public Diplomacy, 6*(4), 316–332. doi:10.1057/pb.2010.30

Organ, K., Koenig-Lewis, N., Palmer, A., & Probert, J. (2015). Festivals as agents for behaviour change: A study of food festival engagement and subsequent food choices. *Tourism Management, 48*, 84–99. doi:10.1016/j.tourman.2014.10.021

Osmankovic, J., Kenjic, V., & Zrnic, R. (2010). Destinantion managment: Consensus for competitiveness. *Proceedings of Tourism and Hospitality Management Conference*, 513-525.

Osman, S. (2022). Technology Application in Tourism Fairs, Festivals and Events in Malaysia. In A. Hassan (Ed.), *Technology Application in Tourism Fairs, Festivals and Events in Asia* (pp. 301–311). Springer Singapore., doi:10.1007/978-981-16-8070-0_18

Ostroff, C., & Judge, T. A. (2007). *Perspectives on organizational fit.* Psychology Press. doi:10.4324/9780203810026

Otto, J. E., & Ritchie, J. R. B. (1996). The Service Experience in Tourism. *Tourism Management, 17*(3), 165–174. doi:10.1016/0261-5177(96)00003-9

Özaltaş Serçek, G., & Serçek, S. (2015). Destinasyon markalaşmasında gastronomi turizmin yeri ve önemi. *Journal Of Tourism Theory And Research, 1*(1), 22–40. doi:10.24288/jttr.202819

Özdemir, G. (2014). Destinasyon yönetimi ve pazarlaması (1.Baskı). Ankara:Detay Yayıncılık.

Özdemir, G., & Çulha, O. (2009). Satisfaction and loyalty of festival visitors. *Anatolia, 20*(2), 359–373. doi:10.1080/13032917.2009.10518914

Özilhan Özbey, D., & Başer, H. H. (2015). Kent marka imajı oluşumunda etkili unsurların önem ve yeterlilik derecelerinin Konya açısından incelenmesi. *Uluslararası Sosyal Araştırmalar Dergisi, 8*(39), 923–932.

Pala, N. (2001). *Turizmin Sosyo-Kültürel Sonuçları Açısından Bir İnceleme (Ballıca Mağarası Örneği)* [Master Thesis]. Sakarya Üniversitesi Sosyal Bilimler Enstitüsü, Sosyoloji Ana Bilim Dalı, Kurumlar Sosyolojisi Bilim Dalı, Sakarya. Tez no:111986. Retrieved from https://tez.yok.gov.tr/UlusalTezMerkezi/tezSorguSonucYeni.jsp

Pamir, H. Y. (1997). *Turizmin Sosyal Yapıya ve Sosyal Değişmeyi Etkileri* [Master Thesis]. İstanbul Üniversitesi Sosyal Bilimler Enstitüsü, Turizm Ana Bilim Dalı, İstanbul. Tez no 64564. Retrieved from https://tez.yok.gov.tr/UlusalTezMerkezi/tezSorguSonucYeni.jsp

Panda S.N., Panda P. & Majhi R. (2017): An empirical analysis of socio-economic growth and development by festival tourism: open-air theatre dhanuyatra, *International Journal of Physical and Social Science, 7*(5).

Paraschak, V. (1991). Sport festivals and race relations in the Northwest Territories of Canada. In G. Jarvie (Ed.), *Sport, racism, and ethnicity*. Falmer Press.

Paris, C. M., Lee, W., & Seery, P. (2010, February 10-12). The role of social media in promoting special events: acceptance of Facebook 'events'. [Conference presentation]. ENTER2010 eTourism Conference, Lugano, Switzerland.

Park, C. W., Eisingerich, A. B., & Park, J. W. (2013). Attachment–Aversion (A.A.) Model of Customer–Brand Relationships. *Journal of Consumer Psychology, 23*(2), 229–248. doi:10.1016/j.jcps.2013.01.002

Park, C. W., MacInnis, D. J., & Priester, J. (2006). Beyond attitudes: Attachment and consumer behavior. *Seoul Journal of Business, 12*, 3–35.

Park, C. W., MacInnis, D. J., Priester, J., Eisingerich, A. B., & Iacobucci, D. (2010). Brand attachment and brand attitude strength: Conceptual and empirical differentiation of two critical brand equity drivers. *Journal of Marketing, 74*(6), 1–17. doi:10.1509/jmkg.74.6.1

Park, E., Choi, B. K., & Lee, T. J. (2019a). The role and dimensions of authenticity in heritage tourism. *Tourism Management, 74*, 99–109. doi:10.1016/j.tourman.2019.03.001

Parker, R. D., Sears, D., & Smith, R. K. (2012). International design concepts in Internet tourism marketing: Comparing web-design practices in Atlantic Canada and New England. *Academy of Marketing Studies Journal, 16*(2), 65.

Park, K. S., Reisinger, Y., & Kang, H. J. (2008). Visitors' motivation for attending the South Beach wine and food festival, Miami Beach, Florida. *Journal of Travel & Tourism Marketing, 25*(2), 161–181. doi:10.1080/10548400802402883

Park, K., Park, J. Y., & Back, R. M. (2019b). Determinants of Marathoners' event expenditures: An in-depth exploration of past experience. *Journal of Hospitality and Tourism Insights, 2*(2), 110–120. doi:10.1108/JHTI-09-2018-0062

Park, S., Woo, M., & Nicolau, J. L. (2020). Determinant factors of tourist expenses. *Journal of Travel Research, 59*(2), 267–280. doi:10.1177/0047287519829257

Paspati, A. G. (1870). *Études sur les Tchingianés ou Bohémiens de l'Empire Ottoman*. Constantinopol.

Patowary, D. K. N. (n.d.). *The Kalika Purana and the Yogini Tantra: A Comparative Study in Reference to Great Indian Rivers*. 4.

Patton, M. Q. (2001). *Qualitative Research & Evaluation Method*. Sage Publications.

Patton, M. Q. (2002). Two decades of developments in qualitative inquiry: A personal, experiential perspective. *Qualitative Social Work: Research and Practice, 1*(3), 261–283. doi:10.1177/1473325002001003636

Paul, K. I., Polglase, P. J., Nyakuengama, J. G., & Khanna, P. K. (2002). Change in soil carbon following afforestation. *Forest Ecology and Management, 168*(1-3), 241–257. doi:10.1016/S0378-1127(01)00740-X

Pearce, D. G. (1992). *Tourist organizations*. Longman Group UK Ltd.

Pearce, P. L., & Moscardo, G. M. (1986). The concept of authenticity in tourist experiences. *The Australian and New Zealand Journal of Sociology*, *22*(1), 121–132. doi:10.1177/144078338602200107

Pentina, I., Zhang, L., & Basmanova, O. (2013). Antecedents and consequences of trust in a social media brand: A cross-cultural study of Twitter. *Computers in Human Behavior*, *29*(4), 1546–1555. doi:10.1016/j.chb.2013.01.045

Periš, G., & Mandariš, M. (2020). Challenges in tourist destination branding in Serbia: The case of Prolom Banja. *Ekonomika Preduzeća*, *68*(5-6), 442–456. doi:10.5937/EKOPRE2006442P

Perry, B., Ager, L., & Sitas, R. (2020). Cultural heritage entanglements: Festivals as integrative sites for sustainable urban development. *International Journal of Heritage Studies*, *26*(6), 603–618. doi:10.1080/13527258.2019.1578987

Persing, S. (2022, January 14). *Hospitals in southwest Florida complying with vaccine mandate after Supreme Court decision.* Wink News. www.winknews.com/2022/01/14/hospitals-in-southwest-florida-complying-with-vaccine-mandate-after-supreme-court-decision/

Peter, S., Anandkumar, V., & Peter, S. (2013). Role of shopping festivals in destination branding: A tale of two shopping festivals in the United Arab Emirates. *Anatolia*, *24*(2), 264–267. doi:10.1080/13032917.2013.799077

Pike, S. (2004). *Destination marketing organisations.* Routledge. doi:10.4324/9780080494463

Pike, S. (2005). Tourism Destination Branding Complexity. *Journal of Product and Brand Management*, *14*(4), 258–259. doi:10.1108/10610420510609267

Pike, S. (2009). Destination brand positions of a competitive set of near-home destinations. *Journal of Travel Research*, *48*(2), 139–151.

Pike, S. (2010). Destination brand positioning - A comparative analysis of the positioning of competitive sets. *Journal of Hospitality Marketing & Management*, *19*(2), 112–132.

Pike, S. (2010). Destination branding case study: Tracking brand equity for an emerging destination between 2003 and 2007. *Journal of Hospitality & Tourism Research (Washington, D.C.)*, *34*(1), 124–139. doi:10.1177/1096348009349820

Pike, S., & Bianchi, C. (2016). Destination branding equity for Australia: Testing a model of CBBE in short – hauland long –haul markets. *Journal of Hospitality & Tourism Research (Washington, D.C.)*, *40*(1), 114–134. doi:10.1177/1096348013491604

Pike, S., & Page, S. J. (2014). Destination marketing organizations and destination marketing: A narrative analysis of the literature. *Tourism Management*, *41*, 202–227. doi:10.1016/j.tourman.2013.09.009

Pine, B. J., & Gilmore, J. H. (1999). *The experience economy.* Harvard Business School Press.

Pine, B. J., & Gilmore, J. H. (1999). *The experience economy: Work is theatre and every business a stage.* Harvard Business Press.

Pine, B. J., & Gilmore, J. H. (1999). *The Experience Economy: Work is Theatre and Every Business a Stage.* Harvard Business School Press.

Pırnar, I., Kurtural, S., & Tütüncüoğlu, M. (2019). Festivals and destination marketing: An application from Izmir City. *Journal of Tourism. Heritage & Services Marketing*, *5*(1), 9–14. doi:10.5281/zenodo.2640987

Piva E., Cerutti S., Prats L., & Raj R. (2017). *Enhancing Brand Image through Events and Cultural Festivals: The Perspective of the Stresa Festival's Visitors.*

Piva, E., Cerutti, S., Prats, L., & Raj, R. (2017). Enhancing Brand Image through Events and Cultural Festivals: the Perspective of the Stresa Festival's Visitors. *Almatourism - Journal of Tourism, Culture and Territorial Development, 8*(15 SE-Articles and reports), 99–116. doi:10.6092/issn.2036-5195/6888

Piva, L., Cerutti, S., Prats, L., & Raj, R. (2017). Enhancing brand image through events and cultural festivals: The perspective of the Stresa Festival's visitors. *Journal of Tourism. Culture and Territorial Development*, *8*(15), 99–116.

Pizam, A., & Ellis, T. (1999). Customer satisfaction and its measurement in hospitality enterprises. *International Journal of Contemporary Hospitality Management, 11*(7), 326–339. doi:10.1108/09596119910293231

PlotProjects. (2022a). *10 Ways to Use Geofencing and Beacons for Events, Conferences and Festivals to Increase Attendee Engagement*. PlotProjects. https://www.plotprojects.com/blog/10-ways-use-geofencing-beacons-events-conferences-festivals/

PlotProjects. (2022b). *Beacon Technology – What it is & Why You Need it | 2022*. PlotProjects. https://www.plotprojects.com/blog/beacon-technology-why-the-market-is-booming/

PlotProjects. (2022c). *Geofencing Software – How to Set up Geofencing in 2022 | Free Trial*. PlotProjects. https://www.plotprojects.com/blog/how-to-set-up-geofencing/

Powers, T., Advincula, D., Austin, M. S., Graiko, S., & Snyder, J. (2012). Digital and social media in the purchase decision process: A special report from the Advertising Research Foundation. *Journal of Advertising Research*, *52*(4), 479–489. doi:10.2501/JAR-52-4-479-489

Prasad, A. (2014). Will Nurseries become just a memory. *The New Indian Express, Bangalore Edition.*

Prentice, R., & Andersen, V. (2003). Festival as a creative destination. *Annals of Tourism Research*, *30*(1), 7–30. doi:10.1016/S0160-7383(02)00034-8

Presenza, A., Sheehan, L., & Ritchie, J. B. (2005). Towards a model of the roles and activities of destination management organizations. *Journal of Hospitality, Tourism, and Leisure Science*, *3*(1), 1–16.

Preuss, H., & Solberg, H. A. (2017). The impact of the 2010 FIFA World Cup in South Africa on neighboring countries. *Journal of African Business*, *18*(4), 483–501. doi:10.1080/15228916.2017.1365717

Prideaux, B., Timothy, D. J., & Cooper, M. (2009). Introducing river tourism: Physical, ecological and human aspects. *River Tourism, 1.*

Prossack, A. (2021, November 6). The power of asking the right questions. *Forbes*. www.forbes.com/sites/ashiraprossack1/2021/11/05/the-power-of-asking-the-right-questions/?sh=73d5e308217e

Qatar News Agency. (2022, September 7). *Oman Announces Accompanying Programs for FIFA World Cup Qatar 2022*. Qatar News Agency. https://www.qna.org.qa/en/News-Area/News/2022-09/07/0073-oman-announces-accompanying-programs-for-fifa-world-cup-qatar-2
022R

Quan, S., & Wang, N. (2004). Towards a structural model of the tourist experience: An illustration from food experiences in tourism. *Tourism Management*, *25*(3), 297–305. doi:10.1016/S0261-5177(03)00130-4

Quinn, B. (2003). Symbols, practices and myth-making: Cultural perspectives on the Wexford Festival Opera. *Tourism Geographies*, *5*(3), 329–349. doi:10.1080/14616680309710

Quinn, B. (2006). Problematising “festival tourism”: Arts festivals and sustainable development in Ireland. *Journal of Sustainable Tourism, 14*(3), 288–306. doi:10.1080/09669580608669060

Quinn, B. (2009). Festivals, events and tourism. Festivals, events and tourism. In T. Jamal & M. Robinson (Eds.), *The SAGE Handbook of Tourism Studies* (pp. 483–503). Sage. doi:10.4135/9780857021076.n27

Quinn, B. (2010). Arts festivals, urban tourism and cultural policy. *Journal of Policy Research in Tourism, Leisure & Events, 2*(3), 264–279. doi:10.1080/19407963.2010.512207

Rahman, M. K., Masud, M. M., Akhtar, R., & Hossain, M. M. (2021). Impact of community participation on sustainable development of marine protected areas: Assessment of ecotour -ism development. *International Journal of Tourism Research.*

Rahman, M. K., Rana, M. S., Akter, S., & Hassan, A. (2022). Technology innovation as a destination branding tool for festivals and events tourism. In *Technology Application in Tourism Fairs, Festivals and Events in Asia* (pp. 135–147). Springer Singapore. doi:10.1007/978-981-16-8070-0_8

Raina, A. K., & Agarwal, S.K. (2004). The essence of tourism development. India: Sarup & Sons.

Rakshitha, R. (2019). This year, Karaga may follow a different route. *Deccan Herald.* https://www.deccanherald.com/city/top-bengaluru-stories/this-year-karaga-may-follow-a-different-route-728440.html

Ramshaw, G. (2016). Food, Heritage and Nationalism. In D. J. Timothy (Ed.), *Heritage Cuisines Traditions, Identities and Tourism* (pp. 53–64). Routledge.

Ratten, V. (2017). Mega-events and entrepreneurship: A research agenda. *Journal of Entrepreneurship in Emerging Economies, 9*(2), 246–259. doi:10.1108/JEEE-07-2016-0031

Rein, I., & Shields, B. (2007). Place branding sports: Strategies for differentiating emerging, transitional, negatively viewed and newly industrialised nations. *Journal of Place Branding and Public Diplomacy, 3*(1), 73–85. doi:10.1057/palgrave.pb.6000049

Reisinger, Y. (2009). *International Tourism: Cultures and Behavior* (1st ed.). Elsevier. Retrieved from https://www.researchgate.net/publication/330919283_Sociocultural_Impacts_of_Tourism_on_Residents_of_World_Cultural_Heritage_Sites_in_China

Reiss, J., Abbott, C., & Gold, S. F. (2001). SACRED WATERS: A Pilgrimage Up the Ganges River to the Source of Hindu Culture (Book Review). *Publishers Weekly, 248*(42), 68–68.

Revilla Hernández, M., Santana Talavera, A., & Parra López, E. (2016). Effects of co-creation in a tourism destination brand image through Twitter. *Journal of Tourism. Heritage & Services Marketing, 2*(2), 3–10. doi:10.5281/zenodo.376341

Richards, G. (2002). Gastronomy: An Essential Ingredient in Tourism Production and Consumption? In A. M. Hjalager ve G. Richards (Eds.). Tourism and Gastronomy. London: Routledge, 3–20.

Richards, G., & King, B. (2022). The experience of cultural festivals: Evidence from Hong Kong. *Journal of Policy Research in Tourism, Leisure & Events, 14*(3), 296–309. doi:10.1080/19407963.2022.2033249

Richards, G., & Leal Londoño, M. D. P. (2022). Festival cities and tourism: Challenges and prospects. *Journal of Policy Research in Tourism, Leisure & Events, 14*(3), 219–228. doi:10.1080/19407963.2022.2087664

Richards, G., & Wilson, J. (2004). The impact of cultural events on city image: Rotterdam, cultural capital of Europe 2001. *Urban Studies (Edinburgh, Scotland)*, *41*(10), 1931–1951. doi:10.1080/0042098042000256323

Richards, G., & Wilson, J. (2006). Developing creativity in tourist experiences: A solution to the serial reproduction of culture? *Tourism Management*, *27*(6), 1209–1223. doi:10.1016/j.tourman.2005.06.002

Ringle, C., Wende, S. & Becker, J.-M. (2015). SmartPLS 3.

Ritchie, B. J. R., & Ritchie, J. R. R. (1998). The branding of tourism destinations. *Annual Congress Of The International Association Of Scientific Experts in Tourism*, 1- 31.

Ritchie, J. R. B., & Crouch, G. I. (2003). *The competitive destination: A sustainable tourism perspective*. CABI. doi:10.1079/9780851996646.0000

Roberts, J. H., & Lattin, J. M. (1991). Development and testing of a model of consideration set composition. *JMR, Journal of Marketing Research*, *28*(4), 429–440. doi:10.1177/002224379102800405

Robertson, M., Chambers, D., & Frew, E. (2007). Events and festivals: Current trends and issues. *Managing Leisure*, *12*(2–3), 99–101. doi:10.1080/13606710701406485

Robinson, R. N., & Clifford, C. (2012). Authenticity and festival foodservice experiences. *Annals of Tourism Research*, *39*(2), 571–600. doi:10.1016/j.annals.2011.06.007

Ruiz-Real, J. L., Uribe-Toril, J., & Gázquez-Abad, J. C. (2020). Destination branding: Opportunities and new challenges. *Journal of Destination Marketing & Management*, *17*, 100453. doi:10.1016/j.jdmm.2020.100453

Saayman, M. (2011). Motives for attending the Cultivaria arts festival. *S.A. Journal for Research in Sport Physical Education and Recreation*, *33*(1), 109–120. doi:10.4314ajrs.v33i1.65492

Saçlı, Ç. (2020). Festival Memnuniyeti ile Kültürel Etkileşim Arasındaki İlişki Üzerine Bir Araştırma: Silifke Uluslararası Müzik ve Folklor Festivali Örneği. *Türk Turizm Araştırmaları Dergisi*, *4*(1), 505-521. Retrieved from https://www.tutad.org/index.php/tutad/article/view/189/188

Saldaña, J. (2013). *The Coding Manual for Qualitative Researchers*. Sage.

Santich, B. (2004). The study of gastronomy and its relevance to hospitality education and training. *International Journal of Hospitality Management*, *23*(1), 15–24. doi:10.1016/S0278-4319(03)00069-0

Sazak, Ş., & Eyüboğlu Erşen, A. (2009). Bir Kültür Endüstrisi Olarak "Kırkpınar Güreş Festivali". ["Kırkpınar Wrestling Festival" as a Cultural Industry.]

Schmitt, B. H. (1999). *Experiential marketing: How to get customers to sense, feel, think, act and relate to your company and brands*. The Free Press.

See-To, E. W. K., & Ho, K. K. W. (2014). *Value co-creation and purchase intention in social network sites: The role of electronic Word-of-Mouth and trust: A theoretical analysis*. Computers in Human Behavior. doi:10.1016/j.chb.2013.10.013

Sekaran, U. (2003). Research Methods for Business: A Skill Bulding Approach (4th ed.). John Wiley & Sons Inc.

Selby, M. (2004). *Understanding Urban Tourism: Image, Culture and Experience*. IB Tauris.

Seminatore, C. (2021). *Geofencing for Events*. Get GeoFencing. https://www.getgeofencing.com/geofencing-for-events-for-targeted-audiences/

Sen, C. T. (2022). *Ashoka and the Maurya Dynasty: The History and Legacy of Ancient India's Greatest Empire*. Reaktion Books.

Şengül, S., & Genç, K. (2016). Festival Turizmi Kapsaminda Yöresel Mutfak Kültürünün Destekleyici Ürün Olarak Kullanilmasi: Mudurnu İpekyolu Kültür Sanat ve Turizm Festivali Örneği. *Pamukkale Üniversitesi Sosyal Bilimler Enstitüsü Dergisi*, (23), 79–89.

Sethi, I., & Chawla, A. (2014). Influence of cultural, social and marketing factors on the buying behavior of telecom users: A comparative study of rural, semi-urban and urban areas in and around Chandigarh. *Journal of Marketing Management*, *2*(1), 97–110.

Sever, N. S., & Ataizi, M. (2017). Yeni medya ortam ve araçlarının bireylerin yeni yerler öğrenmeleri ve deneyimlemelerine katkıları üzerine öneriler. *Kurgu*, *25*(1), 51–64.

Shah, H. (2020). *Powering Smukfest with IBM Data and AI*. Linkdin. https://www.linkedin.com/pulse/powering-smukfest-ibm-data-ai-hetu-shah/

Shahzalal, M. (2016). Positive and Negative Impacts of Tourism on Culture: A Critical Review of Examples from the Contemporary Literature. *Journal of Tourism, Hospitality and Sports*, *20*, 30-34. Retrieved from https://core.ac.uk/download/pdf/234696865.pdf

Shaikh, N. (2021). How to use beacons to enhance in-event experience? *Peer Bits*. https://www.peerbits.com/blog/how-to-beacons-to-enhance-in-event-experience.html

Sharma, P. M. (2017). *Fairs & Festivals: A Pull Factor for Domestic Tourist*. Utilizing the Social Media Tool.

Sharma, S. (2022). Smart Tourism-A Solution for Tourism Challenges in Himachal. *Journal of Smart Tourism*, *2*(2), 21–32. doi:10.52255/SMARTTOURISM.2022.2.2.4

Sharoff, R. (2021, April 2). *How covid-19 and trade tensions have affected the U.S. and China*. University of Chicago News. news.uchicago.edu/story/how-covid-19-and-trade-tensions-have-affected-us-and-china

Sharples, L. (2008). Apples, cider and celebration. In C. M. Hall & L. Sharples (Eds.), *Food and Wine Festivals and Events Around the World* (pp. 134–145). Butterworth-Heinemann. doi:10.1016/B978-0-7506-8380-7.00009-9

Sharpley, R. (2006). Tourism - A vehicle for development? In R. Sharpley & D. J. Telfer (Eds.), *Tourism and development - Concepts and issues* (p. 214). Viva Books Private Limited. doi:10.21832/9781845414740-003

Sharpley, R., & Sundaram, P. (2005). Tourism: A sacred journey? The case of ashram tourism, India. *International Journal of Tourism Research*, *7*(3), 161–171. doi:10.1002/jtr.522

Shee, P. (2020): Impact of fairs and festivals on local economy: A case study on tarakeswar town in Hugli district of West Bengal. *International Journal of Scientific and Technology Research, 9*(3),

Sheehan, L. R., Ritchie, J. R. B., & Hudson, S. (2007). The destination promotion triad: Understanding the asymmetric stakeholder interdependencies between the city, the hotels and the DMO. *Journal of Travel Research*, *46*(1), 64–74. doi:10.1177/0047287507302383

Sheehan, L., Vargas-Sánchez, A., Presenza, A., & Abbate, T. (2016). The use of intelligence in tourism destination management: An emerging role for DMOs. *International Journal of Tourism Research*, *18*(6), 549–557. doi:10.1002/jtr.2072

Sheeraz, M., Qadeer, F., Masood, M., & Hameed, I. (2018). Self-Congruence Facets and Emotional Brand Attachment: The Role of Product Involvement and Product Type. *Pakistan Journal of Commerce and Social Sciences*, *12*(2), 598–616.

Sheinin, D. A., Varki, S., & Ashley, C. (2011). The Differential Effect of Ad Novelty and Message Usefulness on Brand Judgments. *Journal of Advertising*, *40*(3), 5–17. doi:10.2753/JOA0091-3367400301

Shenoy, S. S. (2005). Food Tourism and the Culinary Tourist. [Unpublished Doctoral Thesis, Graduate School of Clemson University, ABD.]

Shetty, K., & Fitzsimmons, J. R. (2022). The effect of brand personality congruence, brand attachment, and brand love on loyalty among HENRYs in the luxury branding sector. *Journal of Fashion Marketing and Management*, *26*(1), 21–35. doi:10.1108/JFMM-09-2020-0208

Shi, H., Liu, Y., Kumail, T., & Pan, L. (2022). Tourism destination brand equity, brand authenticity and revisit intention: The mediating role of tourist satisfaction and the moderating role of destination familiarity. *Tourism Review*, *77*(3), 751–779. doi:10.1108/TR-08-2021-0371

Shin, H., Lee, H., & Perdue, R. R. (2018). The congruity effects of commercial brand sponsorship in a regional event. *Tourism Management*, *67*, 168–179. doi:10.1016/j.tourman.2018.01.016

Shipway, R., & Fyall, A. (2012). *International sport events: Impacts, experience and Identities*. Routledge.

Shivagupta, K. (2022). Personal Communication [WhatsApp].

Shone, A., & Parry, B. (2004). *Successful Event Management. A Practical Handbook* (2nd ed.). Thompson.

Shone, A., & Parry, B. (2004). *Successful event management: A practical handbook*. Thomson Learning.

Shoval, N., & Birenboim, A. (2019). Customization and augmentation of experiences through mobile technologies: A paradigm shift in the analysis of destination competitiveness. *Tourism Economics*, *25*(5), 661–669. doi:10.1177/1354816618806428

Silvers, J. R. (2004). *Professional event coordination*. John Wiley and Sons.

Simeng, Z., Yizhou, Z., & Raj, V. (2020). Explore the Improvement of the Management of China's International Film Festivals Based on Artificial Intelligence. *AICSconf '20: Proceedings of the 2020 Artificial Intelligence and Complex Systems Conference*, (pp. 63–67). ACM. 10.1145/3407703.3407716

Simms, A. (2012). Online user-generated content for travel planning - different for different kinds of trips?. *e-Review of Tourism Research*, 10(3), 76-85.

Simpson, M. C. (2008). Community benefit tourism initiatives—A conceptual oxymoron? *Tourism Management*, *29*(1), 1–18. doi:10.1016/j.tourman.2007.06.005

Singh, S. (2001). Indian tourism: Policy, performance and pitfalls. *Tourism and the Less Developed World: Issues and Case Studies*, 137–149.

Singh. S. S. (2017). No carpet of colours at Shillong festival, as cherry blossoms fail to appear [online newspaper]. *The Hindu*. https://www.thehindu.com/news/national/other-states/meghalaya-no-carpet-of-colours-at-shillong-international-festival-as-cherry-blossoms/

Singjai, K., Winata, L., & Kummer, T. F. (2018). Green initiatives and their competitive advantage for the hotel industry in developing countries. *International Journal of Hospitality Management*, *75*, 131–143. doi:10.1016/j.ijhm.2018.03.007

Sinha, R., & Murdia, M. (2018): Heritage Festivals: An important factor to promote cultural tourism (A study of Murshidabad, West Bengal). AACH, 3(2).

Sinha, M. R., & Murdia, M. (2018). Local Residents notion on the contribution of cultural tourism in Murshidabad, West Bengal. *IJRAR–International Journal of Research and Analytical Reviews*, *5*, 1430–1438.

Skinner, T. (2020). *BBC announce pop-up Glastonbury channel as part of 50th anniversary coverage plans*. NME. https://www.nme.com/news/music/bbc-announce-glastonbury-50th-anniversary-celebration-plans-2676162

Smith, J. (2022). *Oman Leverages FIFA 2022 Football Cup for Destination Branding*. Oman Tourism News. Retrieved from https://www.omantourismnews.com/oman-leverages-fifa-2022-football-cup-for-destination-branding/

Smith, S., & Costello, C. (2009). Culinary Tourism: Satisfaction With a Culinary Event Utilizing Importance-performance Grid Analysis. *Journal of Vacation Marketing*, *15*(2), 99–110. doi:10.1177/1356766708100818

Smith, S., & Costello, C. (2009). Segmenting Visitors to a Culinary Event: Motivations, Travel Behavior, and Expenditures. *Journal of Hospitality Marketing & Management*, *18*(1), 44–67. doi:10.1080/19368620801989022

Smith, S., Costello, C., & Muenchen, R. (2010). Influence of push and pull motivations on satisfaction and behavioral intentions within a culinary tourism event. *Journal of Quality Assurance in Hospitality & Tourism*, *11*(1), 17–35. doi:10.1080/15280080903520584

Sohn, E., & Yuan, J. (2013). Who are the Culinary Tourists? An Observation at a Food and Wine Festival. *International Journal of Culture, Tourism and Hospitality Research*, *7*(2), 118–131. doi:10.1108/IJCTHR-04-2013-0019

Song, H., & Cheung, C. (2010). Factors affecting tourist satisfaction with theatrical performances: A case study of the romance of the song dynasty in Hangzhou, China. *Journal of Travel & Tourism Marketing*, *27*(7), 708–722. doi:10.1080/10548408.2010.519675

Song, H., Wang, J., & Han, H. (2019). Effect of image, satisfaction, trust, love, and respect on loyalty formation for name-brand coffee shops. *International Journal of Hospitality Management*, *79*, 50–59. doi:10.1016/j.ijhm.2018.12.011

Soteriades, M. D., & Dimou, I. (2011). Special events: A framework for efficient management. *Journal of Hospitality Marketing & Management*, *20*(3-4), 329–346. doi:10.1080/19368623.2011.562418

Sotiriadis, M. D. (2015). Culinary tourism assets and events: Suggesting a strategic planning tool. *International Journal of Contemporary Hospitality Management*, *27*(6), 1214–1232. doi:10.1108/IJCHM-11-2013-0519

Spiller, K. (2012). It tastes better because consumer understandings of UK farmers market food. *Appetite*, *59*(1), 100–107. doi:10.1016/j.appet.2012.04.007 PMID:22521516

Splash. (2022). Event Data and Technology. *Splash*. https://Splashthat.com/resources/event-marketing-guide/event-data-and-technology

Spoolman, S. (2022). Wisconsin Waters: The Ancient History of Lakes, Rivers, and Waterfalls. Wisconsin Historical Society.

Sreen, N., Mukherjee, S., Jebarajakirthy, C., Kumar, S., & Sharma, H. (2022). Role of personal values on revisiting green hotels: A hierarchical value map approach. *Tourism Recreation Research*, •••, 1–21. doi:10.1080/02508281.2022.2076190

Srikanth, P., & Ram, M. R. (2013). Economic impact of festivals: Evidence from Diwali effect on Indian stock market. *Researchers World: Journal of Arts, Science and Commerce*, *IV*(2), 27–37.

Srinivas (2022). Personal Communication [Discussion].

Srivastava, S. (2016). Socio economic impact of religious tourism-a case study of mahakumbh mela 2013 at Allahabad. *International Journal of Management and Social Sciences*, *4*(1), 205–212.

Stainton, H. (2018). *TEFL tourism in South East Asia: A sustainable solution?* Tourism and Sustainable Development.

Stanley, J., & Stanley, L. (2015). *Food Tourism a Practical Marketing Guide*. CABI. doi:10.1079/9781780645018.0000

Stavska, Y. (2021). The Features of Festival Tourism as a Component of Event Tourism. *The Scientific Heritage*, (66), 23–32. doi:10.24412/9215-0365-2021-66-3-23-32

Steinbach, J. (1995). River related tourism in Europe—An overview. *GeoJournal*, *35*(4), 443–458. doi:10.1007/BF00824354

Stejskal, P. (2015). *Music Festival Sponsorship: In Investigation of the Role of Brand-Event Congruence.* [Master Thesis, Graduate School of Communication. Universiteit van Amsterdam].

Stone, A., Winsor, M., Margolin, J., & Doom, J. (2019, July 29). *Gilroy Garlic Festival shooting: Suspect Santino William Legan, 19, identified as gunman who allegedly killed 3 people*. ABC News. www.abcnews.go.com/US/active-shooter-incident-garlic-festival-california/story?id=64624542

Stone, M. J., Soulard, J., Migacz, S., & Wolf, E. (2018). Elements of memorable food, drink, and culinary tourism experiences. *Journal of Travel Research*, *57*(8), 1121–1132. doi:10.1177/0047287517729758

Subedi, Y. (2015). Prospects and Problems of Religious Tourism in Nepal A Case Study of RamnadiDham (Ramdi), MalungaTunibotVDC, Syangja District, Nepal [PhD Thesis, Central Department of Rural Development Tribhuvan University, Kathmandu].

Swart, K., Cazorla Milla, A., & Mataruna-Dos-Santos, L. J. (2021). From the desert to the city: How is Dubai branding itself through sustainable sport events? *Worldwide Hospitality and Tourism Themes*, *13*(1), 34–48. doi:10.1108/WHATT-08-2020-0101

Tanses, G. (1999). *Türkçe-İngilizce, İngilizce-Türkçe dizinli reklam terimleri ve kavramları sözlüğü*. Adam Yayınları.

Tasci, A. D. A., & Kozak, M. (2006). Destination brands vs. destination image – do we know what we mean? *Journal of Vacation Marketing*, *12*(4), 299–317. doi:10.1177/1356766706067603

Tavakoli, R., & Wijesinghe, S. N. (2019). The evolution of the web and netnography in tourism: A systematic review. *Tourism Management Perspectives*, *29*, 48–55. doi:10.1016/j.tmp.2018.10.008

Taylor, B. (2014, July 23). Wisdom of Warren Buffett: On innovators, imitators, and idiots. *Harvard Business Review*. www.hbr.org/2008/10/wisdom-of-warren-buffet-on-imi

Terziš, S., & Đališ, I. (2019). Analysis of brand valuation based on a combination of valuation models. *The Annals of the Faculty of Economics in Subotica*, *55*(42), 33–48.

The 2600-year-old history of Purmandal Temple . (n.d.). U4U-Voice. http://u4uvoice.com/the-2600-year-old-history-of-purmandal-temple/

The Mayor's Office of Cultural Economy. (2017, May). *2016 New Orleans Cultural Economy Snapshot*. nola.gov/cultural-economy/documents/2016-ce-snapshot-electronic/

The New Orleans Jazz & Heritage Festival and Foundation, Inc. (2021, October 14). *Our history*. The New Orleans Jazz & Heritage Foundation. www.jazzandheritage.org/about-us/our-history/

The United States Government. (2022, April 5). *Fact sheet: The Biden administration accelerates whole-of-government effort to prevent, detect, and treat long covid*. The White House. www.whitehouse.gov/briefing-room/statements-releases/2022/04/05/fact-sheet-the-biden-administration-accelerates-whole-of-government-effort-to-prevent-detect-and-treat-long-covid/

Thidell, Å., & Leire, C. (2004). *Consumer perceptions, understanding and use of product related environmental information-A literature review of the Nordic knowledge base*. Academic Press.

Thomas, A. (Ed.). (2020). *Participation and role of women in events and festivals: A study on the contributions of women in revitalizing and sustaining the folk potential of fairs and festivals in India. In Event Tourism in Asian Countries: Challenges and Prospects*. Apple Academic Press Incorporated.

Thompson, N. (2016). *How technology is giving more life to festivals*. UK Tech. https://www.uktech.news/news/technology-trends-give-festivals-life-beyond-physical-now-20160727

Thompson, K. J., & Schofield, P. (2009). Segmenting and Profiling Visitors to the Ulaanbaatar Naadam Festival by Motivation. *Event Management*, *13*(1), 1–15. doi:10.3727/152599509789130601

Thomson, M., MacInnis, D. J., & Park, C. W. (2005). The ties that bind: Measuring the strength of consumers' emotional attachments to brands. *Journal of Consumer Psychology*, *15*(1), 77–91. doi:10.120715327663jcp1501_10

Tichaawa, T., Bob, U., & Swart, K. (2018). Africa and sports tourism. In J. Nauright & M. Amara (Eds.), *Sport in the African world* (pp. 185–204). Routledge. doi:10.4324/9781351212755-11

Tikkanen, I. (2007). Maslow's Hierarchy and Food Tourism in Finland: Five Cases. *British Food Journal*, *109*(9), 721–734. doi:10.1108/00070700710780698

Times of Oman. (2021). *Oman sees 37% jump in tourist arrivals in January 2021*. Retrieved from https://timesofoman.com/article/oman-sees-37-jump-in-tourist-arrivals-in-january-2021

Times, E. (2017, March 15). Wine Tourism. *Economic Times*, p. 1.

Timothy, D. J. (2009). River-based Tourism in the USA: Tourism and Recreation on the Colorado and Mississippi Rivers. *River Tourism*, 41–54.

Tinsley, R., & Lynch, P. (2001). Small tourism business networks and destination development. *International Journal of Hospitality Management*, *20*(4), 367–378. doi:10.1016/S0278-4319(01)00024-X

Tiwari, P., & Chowdhary, N. (2022). Technology and Crowd Management at Events: a Case Study of Kumbh Festival in India. In CABI Books. CABI Books. doi:10.1079/9781789248685.0015

Toker, A. (2022). Kent kimliği inşaasında festivallerin rolü. *Bingöl Üniversitesi Sosyal Bilimler Enstitüsü Dergisi*, (23), 121–140.

Tokol, T. (1994). *Pazarlama yönetimi*. 5. Basım Bursa: Uludağ Üniversitesi Yayınları.

Tom Dieck, M. C., Jung, T. H., & Rauschnabel, P. A. (2018). Determining visitor engagement through augmented reality at science festivals: An experience economy perspective. *Computers in Human Behavior*, *82*, 44–53. doi:10.1016/j.chb.2017.12.043

Tomorrowland. (2016). *Tomorrowland 2016, UNITE Trailer*. [Video]. Youtube. https://youtu.be/Mdit0COM0ZM

Töre, Ö. (2022, March 30). *World's biggest flower park in Netherlands opens*. Focus on Travel News. www.ftnnews.com/tours/43688-world-s-biggest-flower-park-in-netherlands-opens

Tosun, C., & Timothy, D. J. (2003). Arguments for community participation in the tourism development process. *Journal of Tourism Studies*, *14*(2), 2–14.

Tour My India. (n.d.). Purmandal Temple Jammu: Its History, Mythology & Purmandal Fair. *Tour My India*. https://www.tourmyindia.com/states/jammu-kashmir/purmandal.html

Tourism, K. (n.d.) *Bengaluru Karaga*. Karnataka Tourism. https://www.karnatakatourism.org/destinations/bengaluru-karaga/

Travkina, E., & Sacco, P. (2020, September 7). *Culture shock: Covid-19 and the cultural and creative sectors*. www.oecd.org/coronavirus/policy-responses/culture-shock-covid-19-and-the-cultural-and-creative-sectors-08da9e0e/

Tse, P., & Crotts, J. C. (2005). Antecedents of Novelty Seeking: International Visitor's Propensity to Experiment Across Hong Kong's Culinary Traditions. *Tourism Management*, *26*(6), 965–968. doi:10.1016/j.tourman.2004.07.002

Türkay, O. (2014). *Destination Management: Approaches and Tools from A Management Science Perspective*. Detail Broadcast.

Turkish Ministry of Culture and Tourism General Directorate of Research and Development. (2022). *Kırkpınar Yağlı Güreş Festivali*. Retrieved from https://aregem.ktb.gov.tr/TR-268224/kirkpinar-yagli-gures-festivali.html

Turkish Ministry of Culture and Tourism. (2022). *EDEN 2008-Turizm ve Somut Olmayan Kültürel Miras*. Retrieved from https://eden.ktb.gov.tr/TR-246372/eden-2008---turizm-ve-somut-olmayan-kulturel-miras.html

Turkish Patent Institute. (2022). *Coğrafi İşaret Tescil Belgesi*. Retrieved from https://ci.turkpatent.gov.tr/Files/GeographicalSigns/128.pdf

Türkmendağ, T., & Türkmendağ, Z. (2022). Technology Use and Innovation Strategies in Event Tourism. In C. M. Q. Ramos, S. Quinteiro, & A. R. Gonçalves (Eds.), *ICT as Innovator Between Tourism and Culture* (pp. 104–121). IGI Global. doi:10.4018/978-1-7998-8165-0.ch007

Türkmenoğlu, G. Ö. (2017). *IMC Strategies of Festivals in Destination.* NA: Strategic Innovative Marketing, Springer Proceedings.

Turnbull, D., & Uysal, M. (1995). An Exploratory Study of German Visitors to the Caribbean: Push and Pull Motivations. *Journal of Travel & Tourism Marketing*, *4*(2), 85–92. doi:10.1300/J073v04n02_07

Turner, V. (1982). Introduction. In V. Turner (Ed.), *Celebration: Studies in Festivity and Ritual* (pp. 11–29). Smithsonian Institution Press.

Türten, B. (2021). Film festivalleri ve turizm ilişkisi: Belgesel film yönetmenleri üzerine bir çalışma. *Erciyes İletişim Dergisi*, *8*(1), 205–230. doi:10.17680/erciyesiletisim.810902

Tylor, E. B. (1871). *Primitive Culture: Researches into the Development of Mythology, Philosophy, Religion, Art and Custom.* John Murray. Retrieved from https://books.google.com.tr/books?hl=tr&lr=&id=RUMBAAAAQAAJ&oi=fnd&pg=PA1&dq=tylor+1871+culture+pdf&ots=X_u4fqQSiZ&sig=8LynCxMkaXp5CkcTzkJ9-cH7D-8&redir_esc=y#v=onepage&q=culture&f=false

Uçar, N. (2010). *Turizmin Kırsal Alana Etkisi: Gökçeada Örneği* [Master Thesis]. İstanbul Üniversitesi Sosyal Bilimler Enstitüsü Antropoloji Anabilim Dalı, İstanbul. Tez no: 262553. Retrieved from https://tez.yok.gov.tr/UlusalTezMerkezi/tezSorguSonucYeni.jsp

Uğur, İ., Gökkaya, S., & Acar, A. (2018). Yerli turistlerin destinasyon imajına ilişkin tekrar ziyaret etme niyetleri: Safranbolu üzerine bir çalışma. *Safran Kültür ve Turizm Araştırmaları Dergisi*, *1*(1), 29–40.

Unal, N. (2001). Rekabet düzeyi en yüksek sektör turizm. *Standard. Yıl*, *40*(470), 27.

UNESCO Turkish National Commission. (2022). *UNESCO İnsanlığın Somut Olmayan Kültürel Mirası Temsilî Listesi.* Retrieved from https://www.unesco.org.tr/Pages/126/123/UNESCOİnsanlığın-Somut-Olmayan-Kültürel-Mirası-Temsilî-Listesi

United Nations Department of Economic and Social Affairs. (n.d.). *Least developed countries (ldcs) | Department of Economic and Social Affairs.* United Nations. www.un.org/development/desa/dpad/least-developed-country-category.html

Unurlu, C. (2021). The Integration of social media Into Event Tourism. In *Impact of ICTs on Event Management and Marketing* (pp. 69–85). IGI Global. doi:10.4018/978-1-7998-4954-4.ch005

UNWTO-United Nations World Tourism Organization. (2012). UNWTO Tourism Highlights. UNWTO.

Ürkmez, D., & Akbulut, D. (2020). Yerel Yönetimlerde Bir Halkla İlişkiler Aracı Olarak Festivallerin Kullanımı. *Karadeniz İletişim Araştırmaları Dergisi*, *10*(2), 1-27. Retrieved from https://dergipark.org.tr/tr/download/article-file/1299100

Urry, J. (1990). *The tourist gaze: Leisure and travel in contemporary societies.* Sage Publications.

Urry, J. (2002). *Tourist gaze* (2nd ed.). Sage Publications.

Usakli, A., & Baloglu, S. (2011). Brand personality of tourist destinations: An application of self congruity theory. *Tourism Management*, *32*(1), 114–127. doi:10.1016/j.tourman.2010.06.006

Ussmartcards. (2019). *How can NFC be used at festivals?* US Smart Cards. https://www.usmartcards.com/2019/05/29/how-can-nfc-be-used-at-festivals/

Valle, P. O., Correia, A., & Rebelo, E. (2008). Determinants of tourism return behaviour. *Tourism and Hospitality Research*, *8*(3), 205–219. doi:10.1057/thr.2008.19

Valle, P. O., Silva, J. A., Mendes, J., & Guerreiro, M. (2006). Tourist satisfaction and destination loyalty intention: A structural and categorical analysis. [IJBSAM]. *International Journal of Business Science and Applied Management*, *1*(1), 25–44.

Valls, J. F., Ouro, A., Freund, D., & Andrade Suárez, M. J. (2012). Analysis of social media platforms and their potential value for the tourism industry. 2nd Advances in Hospitality & Tourism Marketing & Management Conference (AHTMM) 2012, Corfu (Greece).

Van Harssel, J. (2005). *Glossary-Destination Management Organization. In Fundamentals of Destination Management and Marketing*. Educational Institute American Hotel & Lodging Association.

Van Niekerk, M. (2017). Contemporary issues in events, festivals and destination management. *International Journal of Contemporary Hospitality Management*, *29*(3), 842–847. doi:10.1108/IJCHM-01-2017-0031

van Vliet, H. (2021). Why do we go to festivals? Festivals and Experience. Amsterdam University of Applied Science. doi:, 2(19799.80805). doi:10.13140/RG

Varghese, B. (2016). A strategic evaluation on competency of Karnataka destinations through Destination Management Organizations. *American Journal of Industrial and Business Management*, *6*(02), 102–108. doi:10.4236/ajibm.2016.62010

Varkaris, E., & Neuhofer, B. (2017). The influence of social media on the consumers' hotel decision journey. *Journal of Hospitality and Tourism Technology*, *8*(1), 101–118. doi:10.1108/JHTT-09-2016-0058

Vecco, M. (2010). A definition of cultural heritage: From the tangible to the intangible. *Journal of Cultural Heritage*, *11*(3), 321–324. doi:10.1016/j.culher.2010.01.006

Venue, S. (2021). *Karaga festival and its rituals*. Select Venue. https://selectvenue.in/blog/karaga-festival-and-its-rituals/

Verma, R., Stock, D., & McCarthy, L. (2012). Customer preferences for online, social media, and mobile innovations in the hospitality industry [Electronic version]. *Cornell Hospitality Quarterly*, *53*(3), 183–186. doi:10.1177/1938965512445161

Visser, G. (2005). Let's be festive: Exploratory notes on festival tourism in South Africa. *Urban Forum, 16*, 155–175.

Volgger, M., & Pechlaner, H. (2014). Requirements for destination management organizations in destination governance: Understanding DMO success. *Tourism Management*, *41*, 64–75. doi:10.1016/j.tourman.2013.09.001

Vukic, M., Kuzmanovic, M., Martic, M., Mihajlovic, D., & Kostic-Stankovic, M. (2015). Preferences of consumers towards food festivals: Empirical study. Paper presented at *34th International Conference on Organizational Science Development*, Portoroz, Slovenija.

Vukic, M., Kuzmanovic, M., & Vukic, K. (2016). Gastronomic festivals as A form of branding Serbia as a tourist destination. In *Proceedings of the Xv International Symposium Symorg* (pp. 913-919).

Vukonić, B. (1998). Religious tourism: Economic value or an empty box? *Zagreb International Review of Economics & Business*, *1*(1), 83–94.

Walimbe, O. (2019, January 10). How events like SulaFest are helping boost Nashik's growth. *Just News*. https://justnashik.com/2019/01/10/how-events-like-sulafest-are-helping-boost-nashiks-growth/

Walker, P. (2017). Sustainability: A discipline and a political agenda? *Environmental Hazards*, *16*(2), 93–98. doi:10.1080/17477891.2017.1296810

Wallace, E., Buil, I., & de Chernatony, L. (2017). Consumers' self-congruence with a "Liked" brand: Cognitive network influence and brand outcomes. *European Journal of Marketing*, *51*(2), 367–390. doi:10.1108/EJM-07-2015-0442

Wang, C. H. (2003). Predicting tourism demand using fuzzy time series and hybrid grey theory. *Tourism Management*, *25*(3), 369.

Wang, H., & He, G. (2022). Rivers: Linking nature, life, and civilization. *River*, *1*(1), 25–36. doi:10.1002/rvr2.7

Wang, N. (1999). Rethinking authenticity in tourism experience. *Annals of Tourism Research*, *26*(2), 349–370. doi:10.1016/S0160-7383(98)00103-0

Wang, Y. (2007). Customized authenticity begins at home. *Annals of Tourism Research*, *34*(3), 789–804. doi:10.1016/j.annals.2007.03.008

Wang, Y. D., & Emurian, H. H. (2005). An overview of online trust: Concepts, elements, and implications. *Computers in Human Behavior*, *21*(1), 105–125. doi:10.1016/j.chb.2003.11.008

Wang, Y., & Davidson, M. C. (2010). A review of micro-analyses of tourist expenditure. *Current Issues in Tourism*, *13*(6), 507–524. doi:10.1080/13683500903406359

Wan, Y. K. P., & Chan, S. H. J. (2013). Factors that affect the levels of tourists' satisfaction and loyalty towards food festivals: A case study of Macau. *International Journal of Tourism Research, 15*(3), 226–240. doi:10.1002/jtr.1863

Ward, B. (2021, December 10). *2021 Sundance Film Festival Announces Additional Films*. Sundance.org. www.sundance.org/blogs/news/2021-sundance-film-festival-announces-additional-films/

Warnick, R. B., Bojanic, D. C., & Xu, F. (2015). Using a trade market analysis technique to refine measurements for economic impact analysis of special events. *Journal of Travel Research, 54*(1), 52–65. doi:10.1177/0047287513513160

Wells, W. D. (2014). *Measuring advertising effectiveness*. Psychology Press. doi:10.4324/9781315806099

WHO. (2017, October 10). WHO. *Times of India*. https://timesofindia.indiatimes.com/india/Average-Indian-male-consumes33-litres-of-alcohol-/year-WHO/articleshow/3513936
0.cms

Wight, P. (1998). *Tools for sustainability analysis in planning and managing tourism and recreation in the destination*. Academic Press.

Wijaya, S. (2014). *Encounters with Local Food: The Culinary Experiences of International Visitors in Indonesia*. [Unpublished Doctoral Thesis, College of Business Victoria University, Melbourne, Australia].

Wijaya, S., King, B., Nguyen, T. H., & Morrison, A. (2013). International visitor dining experiences: A conceptual framework. *Journal of Hospitality and Tourism Management, 20*, 34–42. doi:10.1016/j.jhtm.2013.07.001

Williams, H. A., Williams, R. L. Jr, & Omar, M. (2014). Gastro-tourism as destination branding in emerging markets. *International Journal of Leisure and Tourism Marketing, 4*(1), 1–18. doi:10.1504/IJLTM.2014.059257

Wilson, O. (2021, April 5). *Driven by young talent, X Games Aspen 2021 TV viewership, social and digital content experience significant growth*. ESPN Press Room U.S. www.espnpressroom.com/us/press-releases/2021/02/driven-by-young-talent-x-games-aspen-2021-tv-viewership-social-and-digit
al-content-experience-significant-growth/

Wong, I. A., Wan, Y. K. P., & Qi, S. (2015). Green events, value perceptions, and the role of consumer involvement in festival design and performance. *Journal of Sustainable Tourism, 23*(2), 294–315. doi:10.1080/09669582.2014.953542

Wong, J., Wu, H. C., & Cheng, C. C. (2015). An empirical analysis of synthesizing the effects of festival quality, emotion, festival image and festival satisfaction on festival loyalty: A case study of Macau Food Festival. *International Journal of Tourism Research, 17*(6), 521–536. doi:10.1002/jtr.2011

Wong, P. P. W. (2018). Role of components of destination competitiveness in the relationship between customer-based brand equity and destination loyalty. *Current Issues in Tourism, 21*(5), 504–528. doi:10.1080/13683500.2015.1092949

Wood, E. H., & Masterman, G. (2008). *Event marketing: Measuring an experience*. In *7th International Marketing Trends Congress*, Venice.

Wood, L. (2000). Brands and brand equity: Definition and management. *Management Decision, 38*(9), 662–669. doi:10.1108/00251740010379100

World Food Travel Association. (2019). *What Is Food Tourism?* WFTA. https://www.worldfoodtravel.org/cpages/what-is-food-tourism

World Health Organization. (2022, May). *Global excess deaths associated with covid-19, January 2020 - December 2021.* World Health Organization. www.who.int/data/stories/global-excess-deaths-associated-with-covid-19-january-2020-december-2021

World Population Review. (n.d.). *New Orleans, Louisiana population 2022.* New Orleans, Louisiana Population 2022 (Demographics, Maps, Graphs). www.worldpopulationreview.com/us-cities/new-orleans-la-population

World Tourism and Travel Council. (2022). *Economic Impact Reports.* Author.

World Tourism Organization. (2019). *International Tourism Highlights* (2019 Edition). UNWTO. doi:10.18111/9789284421152

Wray, M., Dredge, D., Cox, C., Buultjens, J., Hollick, M., Lee, D., ... Lacroix, C. (2010). *Sustainable regional tourism destinations: Best practice for management, development and marketing.* CRC for Sustainable Tourism Pty Ltd.

Wu, H. C., Wong, J. W. C., & Cheng, C. C. (2014). An empirical study of behavioral intentions in the food festival: The case of Macau. *Asia Pacific Journal of Tourism Research, 19*(11), 1278–1305. doi:10.1080/10941665.2013.844182

WUSA 9. (2023, January 9). *Mayor Bowser reveals DC comeback plan.* www.wusa9.com/article/news/local/dc/mayor-muriel-bowser-reveals-dc-comeback-plan/65-05ecee64-8758-47d2-ac21-de8b9e00135d

Xia, H., Vu, H. Q., Law, R., & Li, G. (2020). Evaluation of hotel brand competitiveness based on hotel features ratings. *International Journal of Hospitality Management, 86*, 102366. doi:10.1016/j.ijhm.2019.102366

Xing, X., & Chalip, L. (2006). Effects of Hosting a Sports Event on Destination Brand: A Test of Co-branding and Match-up Models. *Sport Management Review, 9*(1), 49–78. doi:10.1016/S1441-3523(06)70019-5

Yağbasan, M., & Demirbağ, İ. (2017). Kültürlerarası İletişime Stratejik Bir Yaklaşım: Kültürlerarası Empati (Bir Kültürlerarası İletişim Projesinin Sonuçları). *Gümüşhane Üniversitesi İletişim Fakültesi Elektronik Dergisi, 5*(2), 619-640. Retrieved from https://dergipark.org.tr/tr/download/article-file/357717 doi:10.19145/e-gifder.320646

Yalım, F. (2018). Yavaş şehir (citaslow) hareket ekseninde kent markalaşması ve kent iletşimi: Kırklareli "Vize" Yavaş Şehir örneği. *Trakya Üniversitesi İktsadi ve İdari Bilimler Fakültesi Dergisi*, 1- 28

Yamini, C. S. (2022). Karaga festival of Bengaluru: All you need to know. *Hindustan Times.* https://www.hindustantimes.com/cities/bengaluru-news/karaga-festival-of-bengaluru-all-you-need-to-know-101648034899002.html

Yang, F. X., Wong, I. K. A., Tan, X. S., & Wu, D. C. W. (2020). The role of food festivals in branding culinary destinations. *Tourism Management Perspectives, 34*, 100671. doi:10.1016/j.tmp.2020.100671

Yavuz, G., & Enes, K. (2021). Social Media and Events: Before, During, and After. In Impact of ICTs on Event Management and Marketing (pp. 139-155). IGI Global.

Yayla, Ş. (2017). *Yöresel yemek festivaline katılım motivasyonu ile festivallerin destinasyon imajına ve davranışsal niyetlere etkisi* [Master's thesis, ESOGÜ, Sosyal Bilimler Enstitüsü].

Yenipinar, U., & Yildiz, E. (2016). Festivals as cultural heritage: the Mesir festival of Manisa. R. Efe, Global Issues and Trends in Tourism, 323.

Yeoman, I., Robertson, M., Ali-Knight, J., Drummond, S., & McMahon-Beattie, U. (2017). *Festival and Events Management: An International Arts and Culture Perspective*. Routledge.

Yılmaz, Y., & Çizel, B. (2000). *Türk turizminde günü kurtarmak değil marka yaratmak. Türkiye Turizm Araştırma Enstitüsü, 1*. Ulusal Türkiye Sempozyumu Tebliğ Kitapçığı.

Yoon, Y. S., Lee, J. S., & Lee, C. K. (2010). Measuring festival quality and value affecting visitors' satisfaction and loyalty using a structural approach. *International Journal of Hospitality Management, 29*(2), 335–342. doi:10.1016/j.ijhm.2009.10.002

Yoon, Y., & Uysal, M. (2005). An examination of the effects of motivation and satisfaction on destination loyalty: A structural model. *Tourism Management, 26*(1), 45–56. doi:10.1016/j.tourman.2003.08.016

Yuan, J. M., Morrison, A. M., Cai, L. A., & Linton, S. (2008). *A model of wine tourist behaviour: A festival approach*. NA. *International Journal of Tourism Research, 10*(3), 10. doi:10.1002/jtr.651

Yükselen, C., & Güler, G. E. (2009). *Antakya marka kent Görüş Ve öneriler*. Detay Yayıncılık.

Yu, S.-Y., Li, W.-J., Zhou, L., Yu, X., Zhang, Q., & Shen, Z. (2023). Human disturbances dominated the unprecedentedly high frequency of Yellow River flood over the last millennium. *Science Advances, 9*(8), eadf8576. doi:10.1126ciadv.adf8576 PMID:36812330

Zaczkiewicz, A. (2017, April 26). *Researchers: Use coachella, burning man to test new concepts*. WWD. www.wwd.com/business-news/marketing-promotion/coachella-trendalytics-report-10875558

Zaltman, G., LeMasters, K., & Heffring, M. (1982). *Theory construction in marketing: Some thoughts on thinking*. John Wiley & Sons.

Zamparini, L. (2018). The Role of Gastronomy and Typical Foods in the Tourism Experience. In A. Isoni, M. Troisi ve M. Pierri (Eds.). Food Diversity Between Rights, Duties and Autonomies, 191–202. sSpringer. doi:10.1007/978-3-319-75196-2_11

Zeithaml, V. A., Berry, L., & Parasuraman, A. (1996). The behavioural consequences of service quality. *Journal of Marketing, 60*(2), 31–46. doi:10.1177/002224299606000203

Zeithaml, V. A., Bitner, M. J., & Gremler, D. D. (2017). *Services Marketing: Integrating Customer Focus Across the Firm*. McGraw-Hill Education.

Zeng, B., & Gerritsen, R. (2014). What do we know about social media in tourism? *A review. Tourism Management, 10*, 27–36.

Zhang, C. X., Fong, L. H. N., Li, S., & Ly, T. P. (2019). National Identity and Cultural Festivals in Postcolonial Destinations. *Tourism Management, 73*, 94–104. doi:10.1016/j.tourman.2019.01.013

Zhang, S. N., Li, Y. Q., Liu, C. H., & Ruan, W. Q. (2021). Does live performance play a critical role in building destination brand equity – a mixed-method study of 'impression Dahongpao'. *Journal of Retailing and Consumer Services, 59*, 102392. doi:10.1016/j.jretconser.2020.102392

Zhang, Y., & Murphy, P. (2009). Supply-chain considerations in marketing underdeveloped regional destinations: A case study of Chinese tourism to the Goldfields region of Victoria. *Tourism Management, 30*(2), 278–287. doi:10.1016/j.tourman.2008.07.004

Zhuang, X., Yao, Y., & Li, J. (2019). Sociocultural Impacts of Tourism on Residents of World Cultural Heritage Sites in China. *Sustainability*, *11*(840), 1–19. doi:10.3390u11030840

About the Contributors

Sharad Kumar Kulshreshtha (b. 1978) is an Assistant Professor at the Department of Tourism and Hotel Management, North-Eastern Hill University, Shillong-Meghalaya (India). He has acclaimed publications research papers in various books, journals of national and international repute and participated in national and international Seminars and Conferences. including International Conference on Wine Market and Cultures of Consumption in SHTM the Hong Kong Polytechnic University, Hong Kong. His area of research interest is Event Management, Health & Wellness, Sustainable Tourism. He has also been invited as a resource person in various universities and business associations of India to deliver talks as subject experts. He has co-edited books on 'Emerging Dynamics of India Tourism & Hospitality: Transformation and Innovation' and Global Developments in Healthcare and Medical Tourism Advance in Hospitality, Tourism, and the Service Industry (AHTSI) Book Series, IGI Global, U.S.A., Virus Outbreaks, and Tourism Mobility: Strategy to Encounter Global Health Hazards, Publisher: Emerald Publishing Limited under Book Series: Tourism Security Safety and Post Conflict Destinations Series.

* * *

Shruti Arora, Ph.D., is currently working as a guest faculty in the Department of Commerce and Management, University of Kota, Kota, Rajasthan, India and had an experience of 12 years in education industry. Her core subjects are marketing, general management, international business management, and customer relationship management. She has attended various International Conferences and her publication includes various chapters in edited book like Routledge, Emerald, Springer and research papers in national and international UGC approved and peer-reviewed journals. She has authored a book on "Event Management and Marketing: Theory, Practical Approaches and Planning" with ISBN: 978-93-86608-61-1 in 2018 and two edited under Apple Academic Press, CRC Press and one edited book with NOVA Science Publishers, USA. She is working as a Co-PI with Dr. Anukrati Sharma on a Major Research Project under Mahatma Gandhi National Council of Rural Education, Ministry of Human Resource Development Government of India. She is appointed as the BRICS Community of Practice (CoP) Researcher for promoting cross-BRICS nations' collaborations in co-authorship, co-supervision, co-examining, co-publishing in DHET and SCOPUS Accredited Journals, co-conferencing participation and hosting rotating BRICS workshops, seminars and guest lectures in the academic years 2023 and 2024. The term "Food Walk Tourism" is included in Encyclopedia of Tourism Management and Marketing, editor in Chief, Professor Dimitrios Buhalis: Bournemouth University, Edward Elgar Publishing Limited, ISBN: 9781800377479, e-ISBN: 9781800377486 in 2022.

Aydemir Ay was born in Edirne. He received a diploma from Edirne lhami Ertem High School. From 1997 to 2001, he studied Tourism and Hotel Management at Eastern Mediterranean University. After graduating, he began teaching part-time at Trakya University in Turkey. In 2002, he was a consultant to the Governor of Edirne and a full-time academician at Trakya University. In 2014, he completed his MBA at Istanbul Beykent University. He began working as the Tourist Consultant for Edirne Municipality in 2015 and continues to this day. He is the author of the Amazon Books-sold book "Kirkpinar Oil Wrestling." The Gourmand Cookbook Prize recognized Edirne Jewish Food, also written by Aydemir Ay - Yilmaz Seim, as the best cookbook in the world. This work has been translated into Hebrew, Ladino, English, and Turkish. The book explores the 1457 II. Mehmed Edirne Festival and the 1675 IV. Mehmed Edirne Festival. Aydemir Ay is still a PhD candidate at Trakya University, Edirne, in the Tourism Management Department. He has served as a consultant for numerous hotels, tourism organizations, and food and beverage businesses. In addition, he creates English promotional and tourism publications, does academic research, and composes articles about Edirne. Tourism, tourists and local public perception, cross-border tourism, the history of Edirne, the cuisine of Edirne, minority cultures, the history of Kirkpinar Oil Wrestling, lessening poverty, education for and economic strengthening of low-income communities are his particular areas of interest.

Rachel Chanel Clarke is an American entrepreneur and human rights activist. As the CEO of Trigo Networks, Rachel uses her position to help and motivate other women to aspire to careers at the executive level. She launched Trigo Networks, which provides services for industry leaders, charities, and governments to support philanthropy, well-being, and culture through positive content and events, media dissemination, tourism advisory, and consulting. Trigo Networks serves clients in over 20 countries, including Brazil, Colombia, Germany, Italy, and South Africa. She speaks around the world promoting gender equality and advocating for young people including at the Moses Mabhida Stadium in Durban, South Africa, Yale University in CT, USA, and Georgetown University in Washington, DC, USA. She served as the Senior Advisor for the Asociación Sol y Luna in the Sacred Valley of the Incas, Peru, which assists people in rural Peru with holistic services, including serving as the only school in the province to accept children with disabilities and created the opportunity for children and young people with disabilities to be allowed to participate in a 176-year-old city-wide celebration in which they were previously excluded. Since, her philanthropic outreach has expanded; she established the RCC Globalization Program, which teaches children and young people in at-risk communities and developing nations entrepreneurial skills with a focus on sustainability, global consciousness and responsibility, diversity, equity, and inclusion. The program serves multiple areas, including Los Angeles County, Haiti, Peru, Nigeria, and South Africa. It has gained tremendous support from major companies, NGOs, global business leaders, and other individuals. Rachel travels the world promoting peace, the love of Christ, and cultivating ways to connect local governments and multinational business authorities to under-served populations of children, young people, and women.

Aarti Dangwal is a Ph.D. holder in management with over 15 years of academic experience. She is working as an Assistant Professor at Chitkara University, Rajpura, Punjab, India. She is actively involved in research and academic co-curricular and educational development activities. Her research areas are tourism, finance, and business.

Debendra Nath Dash is the Assistant Director (Research and Networking) at the Mahatma Gandhi National Council of Rural Education (formerly NCRI), Ministry of Education, Government of India. He has 22 years of Research experience. Till date he has more than 16 publications to his credit in various national and international journals and Co editor for 9 Action Research Books. Also he has one books to his credit. He has participated more than 40 national and international workshops/seminars/conferences as a resource person and presented papers. He is having multiple degrees with B Ed, M A (Education, History and Sociology) M Phil & Ph D in Education. He has Post Graduate Diplomas in Rural Development, Post Graduate Diploma in Distance Education and P G Diploma Human Resource Management. He was awarded University Medal for his MA & MPhil degree.

Giacomo Del Chiappa (PhD) is Associate Professor of Marketing at the Department of Economics and Business, University of Sassari (Italy), where he teaches "Tourism Marketing and Management", "Destination Management and Digital Marketing Lab" and "Service and Retail Marketing". He is also Senior Research Fellow at the School of Tourism & Hospitality of University of Johannesburg (South Africa). Giacomo has covered the role of scientific coordinator of the Competence Center "Advanced Robotics and enabling digital Technology and Systems – ARTES 4.0", University of Sassari. He has also acted as the co-chair of the Tourism, Culture and Arts Marketing SIG of the Italian Marketing Association, where he currently sits in the Board of Directors. Yet, he is Associate Researcher, National Research Council (CNR) - Institute for Studies on the Mediterranean. His research is related to destination governance and branding, tourism marketing, hotel marketing, tourism sustainability, consumer/tourist behavior, services marketing, retail and digital marketing. He is Co-editor-in-chief of the European Journal of Tourism Research and member of numerous editorial boards of national and international scientific journals. He is author and co-author of around 200 national and international publications. Among others, he has published in: Annals of Tourism Research, Journal of Service Management, Journal of Services Marketing, Journal of Business Research, Tourism Management, Journal of Travel Research, International Journal of Contemporary and Hospitality Management, International Journal of Hospitality Management. He regularly teaches in PhD programs, Masters and EMBA in Italy and abroad in the areas of consumer behavior, tourist behavior, destination management and hospitality marketing and management.

Jeet Dogra acquired both Master's (Gold Medalist) and Ph.D. in Tourism Management from the University of Jammu, India. He has also qualified for the UGC NET & Junior Research Fellowship (JRF). Moreover, he holds Post Graduations in Management; Marketing Management, and Computer Applications. As an alumnus of IIM Lucknow - NOIDA Campus in Marketing Research, he has international research publications in Tourism Destination Branding and Marketing. Along with he has successfully completed research projects and consultancy works in Tourism for the Marketing Research Division, Ministry of Tourism, Jabalpur Smart City Limited, etc.

Vishal Jain is an Assistant Professor in the University of Technology and Applied Sciences, Sultanate of Oman. He holds PhD, MBA, and PGDCA. He has 25 years of rich experience in Industry and Academia. He has taught several courses like Management, Marketing, Entrepreneurship, Operations, Research, and Quantitative etc. His academic contributions include a number of articles, chapters and conference papers. His main research areas include marketing, retailing, and social sciences.

Emrah Keskin, PhD, is an Associate Professor at the Department of Gastronomy and Culinary Arts at Nevsehir HBV University. His research interests include gastronomy, food identity, food and service quality, tourist behaviour, destination marketing and the interaction between residents and tourists in destinations.

Shaheed Khan is academically qualified in the areas of and armed with Post Graduate Degrees in History, Tourism, Business Administration and Law and a Research Degree in Tourism Management; Shaheeds' focus area has been Tourism, Ecotourism, Mergers and Acquisitions and Human Resources. He also has a Micro-specialization in Strategy Management from Indian Institute of Management Kozhikode and a Corporate Program in Leadership Management from Harvard School of Publishing. He brings to the table multiple facets of functioning that supports in any area of academics and practice in the corporate world. Having worked in the Government, Private and Social Sectors, Shaheed has got to broad understanding of Tourism, Travel and Hospitality (TTH)., Shaheed currently is the Head, Research, Training and Advocacy with Dharthi NGO and Dharthi Learning Centre, Pondicherry. Dharthi has its Headquarters in Bangalore and works with Private and Government organizations in the domain of Environment, Forests, Communities and Tourism. Shaheed had worked in the states of Karnataka, Tamil Nadu, Pondicherry, Maharashtra, and Madhya Pradesh for Dharthi. He is also an Adjunct Faculty in the domain of Tourism, General Management, M&A in Educational Institutions (Central and State University, Deemed to be and Private Universities), and engages Class session for Corporates in Leading IT and manufacturing Companies of India. He has authored over 15 articles in Journals of repute and contributed to Chapters in Edited Books as well. He has worked with other stakeholders to ensure for the Ecotourism segment and has been supporting the cause of Sustainable Tourism.

Betül Kodaş is assistant of professor department of tourism management in the Faculty of Tourism, Mardin Artuklu University, Turkey. She hold Bachelor's degree from School of Tourism and Hotel Management in Adnan Menderes University (Aydın/Turkey), Master's degree from Dokuz Eylül University and Ph.D. from Eskişehir Osmangazi University, Turkey. Her research interests are tourism management, organizational behaviour, destination managment and marketing, travel management.

Davut Kodaş is an assistant of professor at the Faculty of Tourism of Mardin Artuklu University, Turkey. He holds a Bachelor's degree from the School of Tourism and Hotel Management, Mersin University (Turkey), and a Master's and PhD from Anadolu University, Turkey. His research interests include gastronomy tourism, destination marketing, travel behaviour and sociology of tourism.

Mohit Kukreti is a Ph.D. holder in human resource development (HRD) with more than 25 years of international experience in higher education administration, academic quality assurance, and higher education institution (HEI) strategic planning. He has worked as the Program Director of the IBA Program with the Directorate General of 6 CAS colleges at the Ministry of Higher Education, Oman, and is currently associated with the University of Technology and Applied Sciences (College of Economics and Business Administration CAS - Ibri). He has produced several policy documents, funded projects and refereed journal articles. His research areas are HRD, HEI policies and academic quality, and tourism, etc. ().

Ruchika Kulshrestha is working with Amity School of Hospitality, Amity University, Gurugram (Haryana) as Assistant Professor and Programme Coordinator (Tourism). She has done her Ph.D. (Tourism) from Jiwaji University, MBA (Tourism) from IIPS Indore and qualified UGC NET (Tourism Management). She has also got the opportunity to be as a session chair at the places like IIM Kashipur, DSMS Durgapur. Her research interest pertains to areas of Heritage Tourism, Destination management, Tourism Marketing, Tourist Behaviour. She has published her research work in reputed international journals and also presented papers in various International conferences. Her articles are published in Economic times, travel daily news and other good magazines. She has been also awarded for her contribution in Tourism Education. In order to enhance knowledge and to learn new transitions, she is actively involved in the Faculty's development programmes.

Sushant Kumar is an Assistant Professor at Indian Institute of Management Lucknow, India. He has rich experience in publishing research papers and case studies. His research has appeared in the Journal of destination Marketing and Management, Journal of Consumer Marketing, Journal of Retailing and Consumer Research, Journal of Destination Marketing and Management, Journal of Business and Industrial Marketing, Journal of Cleaner Production, Food Quality and Preference, Journal of Global Information Management, Benchmarking: an International Journal among other outlets. He has also published cases with Ivey Publishing.

Antonio Loriga has a PhD in Marketing and Business. Research topic is "Tourism and Destination Marketing". Nowadays, subject expert at the department of business and economics of the University of Sassari in "business management and tourism marketing."

Amitabh Mishra has over 15 years of academic and administrative expertise in international higher education. He is an assistant professor in the Department of Business Administration at the Sultanate of Oman's Government University of Technology and Applied Sciences (College of Economics and Business Administration CAS - Ibri). He has vast experience in tourism, hospitality and business administration, having published over 20 research papers/articles in a variety of peer-reviewed and Scopus-indexed publications. He has also authored a couple of books. Tourism, hospitality and business management are among his areas of interest.

Ester Napolitano (PhD) is a marketing consultant working with the University of Sassari. She worked as a junior expert in territorial and tourism marketing for the MedArtSal project with the CUEIM Foundation (University Consortium for Industrial and Managerial Economics) (Italy). Her research interests include tourism marketing, consumer behaviour, food and wine tourism.

Pedro Espírito Santo holds a PhD in Management-Science Applied to Decision-making and recognized as an expert in marketing and advertising. He is currently a professor at the University of Aveiro (ESTGA department) in Management and Marketing. He regularly collaborates as a consultant in companies from different sectors of activity, in strategic business development, sales and marketing and operational and logistics optimization. He is also a trainer of middle and upper management in companies and author of several articles and scientific studios published in national and international journals and events.

Nevres Sezen, PhD candidate, is a student at the Department of Gastronomy and Culinary Arts at Nevsehir HBV Universty. Her research interests include gastronomy branding, destination marketing, tourist behavior, tourist experiences.

Anukrati Sharma is the Head & Associate Professor of the Department of Commerce and Management at the University of Kota, Rajasthan, India. She is the Director of the Skill Development Centre of the same University. She is also Dean (Honorary) of two faculties, Tourism and Hospitality and Aviation and Aerospace at Rajasthan Skill University, (A Govt. State University) in Jaipur. In 2015, she received a Research Award from the University Grants Commission(UGC), New Delhi, for her project "Analysis of the Status of Tourism in Hadoti and Shekhawati Region/Circuit (Rajasthan): Opportunities, Challenges and Future Prospects." Her doctorate from the University of Rajasthan is in Tourism Marketing, and she completed her dissertation research on Tourism in Rajasthan--Progress &Prospects. She has two postgraduate degree specialties --one in International Business (Master of International Business) and the other in Business Administration (Master of Commerce). Her special interest areas are Tourism, Tourism Marketing, Strategic Management, and International Business Management. Dr. Sharma is a featured author of Routledge publisher. Dr. Sharma is the Book Series Editor of "Building the Future of Tourism" under Emerald Publishing. U.K. She is also the book series Editor of Insights in Tourism Book Series, Routledge. She has many edited books under IGI Global (Scopus Index), Apple Academic Press (CRC Press a Taylor and Francis Group), CABI, Springer Nature Singapore, Emerald Publishing U.K., Routledge, Nova Science Publishers, USA. She is also working on a Major Research Project under Mahatma Gandhi National Council of Rural Education, Ministry of Human Resource Development Government of India. A member of 10 professional bodies, she has attended a number of national and international conferences and presented 45 papers. She has been invited as Keynote, Speaker, Panel member by different countries such as Sri Lanka, Uzbekistan, Nepal, and Turkey. She has been invited as Visiting Professor at Kazakhstan and Uzbekistan.

Sahil Sharma is currently a PhD Candidate at the Himachal Pradesh University, Shimla, India. His research interests lie in Data Science and Tourism Research. He is also passionate on topics such as Technology, Digitalisation, Smart Tourism, Marketing, and Destination Management. He has been awarded the Junior Research Fellowship (JRF) by the University Grants Commission (UGC) of India. Previously he has done an MBA in Tourism Management from the Indian Institute of Tourism and Travel Management (IITTM), Gwalior and a BA degree in Tourism Administration from the Government Degree College (GDC), Kullu.

Surabhi Srivastava is a scrupulous and analytical researcher with 21 years of teaching experience in heritage tourism marketing and management. She is an Associate Professor in Om Kothari Institute of research and Management, Kota, Rajasthan. She did MBA(tourism), Ph.D. (Tourism), from University of Lucknow and PDF (Tourism) from University of Kota(UGC Sponsored). She engrossed herself through organizing workshops, seminars and presentations, educating students, peers and superiors on exploring new trends of tourism development. She contributed as author of 4 books (National publication) and editor of 2 books (International publication) on Heritage tourism, dedicated towards research findings, gaps and limitations of heritage tourism academia. She had been awarded first prize as researcher (2015), young researcher (2017), Global women inspiration award (2022) for exploring tribal tourism, nature tourism and agri tourism for community development.

Sibel Sü Eröz is Associate Professor at Kırklareli University, Faculty of Tourism in Kırklareli, Turkey. She was born in Balıkesir, Turkey. After completion of her elementary, middle and high school education in İstanbul, she obtained her Bachelor's degree in Tourism and Hotel Management from Balıkesir University in 1996. She received her Master's degree in Tourism Management from Balıkesir University, Institute of Social Sciences in 1999. She received her PhD from Uludağ University, department of Business/Management and Organization in Institute of Social Sciences in 2011. She published research papers in various academic journals in English and in Turkish on her primary research interests which are including organizational behaviour, tourism management and organization, public relation, social media. She teaches courses in organizational behaviour, tourism management and organization, public relation, tourist product diversification and, tourism policy and planning.

Freeda Maria Swarna M. has over 15 years of experience in the Sustainability, Conservation Science, Environment, Communities and Tourism domains. With Post Graduate Degrees in Tourism and Ecology & Environment, and a Research Degree in Tourism Management; her strength lies in combining Applied Sciences and the integration of sound research with a robust implementation. As Director, Dharthi NGO was nominated to the Karnataka Ecotourism Development Board as a member of the General Body, where during her three-year tenure, she was instrumental in conducting Cumulative Impact assessment (CIA) studies, having consultations with stakeholders for a robust Ecotourism Policy for the State of Karnataka. Freeda is in the process of authoring two books, one on Ecotourism, and has published articles and papers to her credit. With a penchant for writing, she is in the process of putting up her blogsite called , which will contain impactful stories of her travels, people and places.

Veena Tewari is currently working as an Asst. Professor at the College of Business and Economics, University of Technology and Applied Sciences, Ibri, Sultanate of Oman. She is a Senior Fellow of the Higher Education Academy, UK. She has a mixed experience of 20 years in teaching, research, and consultancy. She has been on teaching assignments in East Europe, Middle East, Southeast Asia, and Northeast Africa. She has been an extensive traveler and has been to various countries by virtue of consultancy and research projects. She has publications both at the international and national level in various conferences, journals, book chapters and seminars. Presently she is serving as the editor in chief, board and advisory member for various journals, Marketing organizations and conferences. Her areas of expertise are consumer behavior in services, corporate social responsibility, entrepreneurship, and use of IT in management. ORCID iD 0000-0003-4025-9909.

Iadonlang Tynsong is an assistant professor at Martin Luther Christian University in Shillong who teaches travel and tourism. His work focuses particularly on event tourism, festival management, entrepreneurship development, and their impacts on the culture and sustainable livelihood of the local community.

Kaplan Uğurlu is Associate Professor at Kırklareli University, Faculty of Tourism in Kırklareli, Turkey. After his bachelor's degree at Uludağ University, Turkey (BSc in Tourism and Hotel Management), he completed his master's degree at the University of Surrey, England (MSc in International Hotel Management). He received his PhD at Marmara University, Turkey (PhD in Production Management and Marketing). He received his associate professor degree in 2019 from the Interuniversity Board Presidency of Turkey, Ankara. He has been working in Turkey for twelve years as an academician. He has worked as a senior manager for 28 years in the tourism sector. He is specialized in marketing, finance, accounting, cost controlling and hotel openings. He has more than 65 papers presented and published in national and international congresses, journals, and books. His academic research interests include tourism and hotel management, tourism and hotel marketing, accounting, and finance. Besides, he has theoretical and practical knowledge, experience, and qualifications in the fields of hotel openings, smart building project systems, smart tourism and marketing applications, sustainable tourism, and cultural studies.

Index

A

Adi-Shakti 171

B

Bibliometric analysis 110, 112, 115-118, 124-126
Brand 1-2, 4-11, 13-31, 35, 37-42, 54, 56, 58, 73, 89-90, 93, 96, 98-109, 112-113, 153, 173, 176, 180-182, 196-202, 204-206, 208, 211-213, 215, 224-226, 229, 234, 237, 241, 285-286, 288, 291-292, 295, 298-299, 301-302
Brand Awareness 5, 10, 16, 18-19, 25, 95-96, 98, 100, 197-198, 200-201
Brand Connection 26
Brand Equity 17-19, 23, 25, 31, 39, 41, 95-96, 98-109, 181
Brand Image 5, 10, 19, 21-22, 27, 30, 93, 95-96, 98-101, 107, 176, 180, 200-201, 226, 237, 288, 291, 298
Brand Loyalty 19, 29-31, 38-39, 41, 95-96, 98-101
Brand Quality 95
Branding 1-25, 31, 39, 42, 44, 58-59, 72-73, 76-78, 82, 88, 90, 93, 95-108, 111, 113, 147, 162, 169, 180, 196-197, 199-202, 210-216, 221, 224-225, 229, 232, 238, 245, 256, 285-286, 288-289, 292-297, 300-302

C

Case Study 14-15, 18, 20-21, 23, 43-44, 46-47, 54, 94, 104-108, 142, 154-155, 195-197, 215, 219, 240-241, 243, 286, 300
Cattle Fair Festivals 43-44, 50, 53
Community 15, 41, 45, 58, 60, 64, 69-70, 78, 96-97, 106, 110-113, 145, 147-149, 151-154, 156-158, 161-165, 169, 171, 182, 188, 191-193, 202, 204, 207, 212-214, 217, 221, 223, 225-226, 231-233, 235-237, 242-243, 248, 254, 256, 265, 269-270, 288, 297, 300
Community Engagement 256
Culinary 11, 18, 23-25, 74, 96-99, 101-102, 105-109, 129, 142-144, 177
Culinary Festival 95-96, 99, 101-102, 109
Cultural Balance 263-266, 270, 272, 274, 276-277, 279-280
Cultural Exchange 96, 218, 263-265, 270, 272, 274-275, 279-280, 291
Cultural Exploration 26, 29-30, 35
Cultural Interaction 263-266, 269-270, 273-274, 276, 278-281
cultural Tourism 77, 112, 126, 143, 146-147, 153, 155, 228, 233, 241
Culture 1-2, 5, 8, 10-11, 15, 17, 22-23, 25, 29, 55, 58-60, 64-65, 77-78, 91, 93-94, 96-97, 99, 106-107, 111, 113-114, 129, 132, 134-135, 137-139, 144-146, 148-149, 151-153, 156, 160, 162, 164-169, 174, 186, 190, 192, 194, 201-202, 205, 210-212, 215, 217, 219, 224, 233-234, 236, 239, 241, 246, 253, 262, 264-265, 267-268, 279-280, 283, 288, 290-294, 296-297, 302
Culture and Arts Festival 25

D

Dekho Apna Desh 167, 171
Destination 1-25, 31, 39, 41, 44, 46, 55-56, 58-62, 64-65, 69-72, 74-78, 93-94, 96-109, 111, 113, 127-129, 132, 134, 136, 139-142, 145, 147, 149, 153, 162-163, 166-169, 173-174, 177-180, 183, 186, 189, 191-193, 196-202, 205, 207-208, 210-216, 218, 221, 224-233, 235-246, 248-249, 253, 257, 266-267, 270-271, 279, 281, 285-289, 291-293, 295-296, 298-302
Destination Brand Equity 17, 19, 23, 25, 31, 95-96, 98-101, 103-107, 109
Destination Branding 1, 3, 6-7, 9-23, 25, 58-59, 72, 77, 93, 95-108, 111, 180, 201-202, 210-213, 215-216, 224-225, 229, 232, 285-286, 288-289, 300-302

Destination Image 5, 8-9, 15-17, 60, 95-104, 106, 145, 174, 178-179, 201, 208, 226
Destination Management Organisations (DMOs) 214, 227-228
Destination Marketing and Branding 77, 196, 199, 214
Destination Promotion 76, 104, 180, 199, 242
Destinations 1-15, 17, 21, 23, 25, 27, 42, 44-46, 54, 58-60, 62, 70-71, 74, 78, 95-104, 108, 110-113, 127-129, 138, 140, 147, 151, 153, 162, 166-170, 172-175, 177-180, 187, 192, 197, 199-201, 211, 213, 215, 218, 225, 227-230, 232, 234-237, 240-243, 246, 267, 284, 286, 288-289, 291, 298, 300-301
Developing Nations 244
Development 2, 7-8, 11, 13, 15, 17-19, 21-24, 27, 29-30, 39, 44-45, 55, 58, 77-78, 93, 96-97, 99, 104-105, 107, 110-111, 113-114, 124-125, 138, 142, 145, 151-152, 154-155, 167, 174-175, 180-181, 183-184, 186-187, 189-195, 197-198, 201-204, 207-208, 211, 213, 215-216, 226-231, 234-243, 247-248, 253, 260, 262-267, 282-283, 286-289, 291-292, 294, 296-297, 300
Digital Marketing 82, 172, 178, 182
Digital Marketing tools 172

E

Edirne 263-264, 266-271, 274, 276, 279-282
Emerging Nations 244, 249
Eno-Tourism 196
Entertainment 4, 9-10, 12, 29, 45, 58, 86, 98, 128, 132, 134, 139, 148, 153, 174, 205-207, 228, 244-245, 247-248, 252, 258-259, 264, 291, 294, 297
Entertainment Business 244
Event Promotion 70, 172-173
Event Tourism 15, 17, 19, 71, 73, 76-78, 92, 94, 103-104, 124, 147, 154, 182, 184, 211, 221, 224-226, 235, 241, 243, 286, 288, 300
Events 1-12, 14-25, 27-30, 32, 37-39, 45-48, 53, 58-62, 64, 67, 69-72, 74-79, 81-85, 88-99, 101, 103-108, 110-113, 123, 125-126, 129, 145-149, 153-154, 162-163, 165-166, 169, 172-174, 176-185, 190, 194, 197-202, 205, 208-213, 215-217, 221, 223-229, 231, 233-238, 243-246, 248-251, 257, 259-260, 263-264, 267-270, 279-280, 285-289, 293-295, 297-302
External Socialization 26, 30

F

Failures 43, 253
Fairs 22, 83, 93, 113, 145-153, 155, 185, 189, 227-229, 232-233, 236-238, 240, 243
Fairs and Festivals 145-153, 155, 189, 227-229, 232-233, 236-238, 240, 243
Festival 1-3, 6-14, 16-22, 24-31, 35, 37-55, 57, 71, 73-74, 76-78, 80-84, 86-113, 115, 123-129, 132-140, 143-149, 152-160, 162-164, 166, 168-171, 174, 177-180, 182-184, 186, 188, 190, 192-193, 196-197, 199-202, 205-218, 220-226, 228-229, 234, 236, 238-239, 241, 244-264, 267-270, 272, 274, 276, 279-283, 286-288, 293-294, 297-299, 301-302
Festival Tourism 7, 24, 54, 92, 96, 110, 112, 147, 154-155, 179, 184, 186, 190, 192-193, 211-212, 225-226, 263-264, 267, 269-270, 274, 276, 280-281
Festivals 1-3, 5, 7-32, 35, 37-39, 42-55, 76-87, 89-95, 97-116, 118-119, 123-129, 132-135, 138-140, 145-155, 159, 162, 165-166, 168-169, 172-174, 177-180, 184-186, 189-190, 192, 194, 196-197, 199-202, 204, 206-208, 210-216, 220-229, 232-234, 236-238, 240, 243-253, 256-257, 263-268, 270-272, 276, 279-282, 284
FIFA World Cup 2022 285, 289, 291-295, 297-299

G

Gastronomic Festival 127-129, 132, 134-136, 138-140
Gastronomic Tours 127, 129, 138
Gastronomy Festival 25, 103, 105
Gastronomy Tourism 96-97, 125, 127-129, 131-132, 134, 138-140, 142, 269
Gupt Ganga 186, 188-189

H

Habba 164, 168, 171
Hakki 168, 171
Heritage 3, 8, 14, 19, 27, 39, 44, 54, 56, 60, 72-74, 93, 109-116, 118-119, 123-126, 129, 141, 143, 145, 148-149, 151-156, 162, 166-169, 210, 224-226, 233, 238, 240, 261, 264, 267-268, 279-280, 284, 288, 292-294, 296-298, 301

I

India 8, 43-44, 46, 53, 55, 76, 85, 93-94, 110, 118, 145-147, 151, 154-156, 158, 162, 166-169, 171, 186-189, 192, 194-199, 202-206, 209-210, 214, 219, 223, 227-228, 230-234, 236-238, 240-241, 243, 285
Institutionalized Social Change 246

J

Jammu and Kashmir 186-187, 189
Jatre 156-157, 160-166, 171

K

Karaga 156-160, 162-170
Katti 157, 171
Kere 161, 171
Kote 161, 171

L

Live Events 244, 248, 259

M

Mahime 164, 171
Marketing and Promotion 227, 238, 248
Mega Events 285-287, 289, 299
Meghalaya 210-212, 214-219, 221-225
Music festivals 9, 11, 17-21, 26-28, 30-31, 35, 38-39, 83, 91, 93, 148, 249

N

Naadu 165, 171
Naari-Shakti 159, 171
Nashik 196-197, 203-205, 207-209
North East India 223, 228, 232-233, 240

O

Oman 285-302

P

Perceived Authenticity 57-60, 62-64, 68-72
Pete (pet) 171
Pilgrimage 29, 167, 186, 188, 191, 194, 237
Pooja 157, 165, 171
Punya Bhoomi 164, 171

R

Regional Development 2, 186, 190, 197-198
Religious Festival Tourism 186, 192
Rural Areas 57, 59, 61-62, 70-71, 129

S

Sardinia 57, 62, 65, 70
Small-Scale Event 57, 70, 79
Smart Festival 76, 78
Smart Technology 76, 78, 85, 91
Smart Tourism 76, 78, 94
Social Interaction 174-176, 179, 245, 251, 264, 280
Social Media 20, 30, 38, 41, 73, 79, 82-83, 88-90, 104, 106, 136, 155, 169, 172-173, 175-185, 201, 206-207, 213, 221, 223, 251, 253, 259, 294-295, 298
Social Media Networks 172-173, 175, 177-180, 183
Socio-Demographics 57-59, 61-63, 66, 70-71
Sonepur 43, 46, 49-53
Sponsor-Festival Congruence 26
Sports Festival 25
Sustainable Tourism 25, 104, 125, 145-147, 149, 154-155, 242-243, 292, 297
Swadesh Darshan 167, 171, 238

T

Temples 156, 158-159, 161-162, 168, 189-190
Thigalas 156-158, 163-165, 169, 171
Tota 161, 171
Tourism 1-2, 4-7, 9-25, 27, 29, 31, 39-42, 44-46, 49, 51-52, 54-62, 64-65, 70-75, 77-78, 82, 92-100, 102-113, 115, 121, 124-126, 128-129, 131-132, 134, 138-145, 147, 149, 151-156, 162-163, 165-217, 220-229, 231-248, 252-253, 259, 261-267, 269-270, 274, 276, 279-302
Tourism Festivals 210, 214, 216, 220-223, 244, 247
Tourist Expectations 127
Tourist Motivations 127-128, 131, 142
Traditions 9, 44, 58, 62, 64-65, 78, 98, 111-113, 129, 143-146, 148-149, 163-165, 169, 194, 210-211, 219, 224, 233-234, 236, 238, 265, 268, 293, 295
Turkey 1, 14, 16, 19, 95, 97, 99, 125, 127-129, 136-138, 172, 179, 263, 267-269, 279, 281, 287

V

Value Creation 199, 227, 235, 240-241
Veerakumaras 157, 160, 171
VOS viewer 110, 112

W

Wine Tourism 17, 98, 104, 108, 128, 196-208

www.ingramcontent.com/pod-product-compliance
Lightning Source LLC
LaVergne TN
LVHW060408300926
849162LV00006B/1373
9781668463567